THE GEORGE FISHER BAKER
NON-RESIDENT LECTURESHIP
IN CHEMISTRY AT
CORNELL UNIVERSITY

CHEMISTRY IN
TWO DIMENSIONS
SURFACES

CHEMISTRY IN TWO DIMENSIONS
SURFACES

by GABOR A. SOMORJAI

Professor of Chemistry
University of California at Berkeley

Cornell University Press | ITHACA AND LONDON

First published 1981 by Cornell University Press.
Published in the United Kingdom by Cornell University Press, Ltd., Ely House, 37 Dover Street, London W1X 4HQ.

International Standard Book Number 0-8014-1179-3
Library of Congress Catalog Card Number 80-21443
Printed in the United States of America
Librarians: Library of Congress cataloging information appears on the last page of the book.

To my wife, Judy

Preface

The last time the properties of surfaces were discussed in the George Fisher Baker Non-Resident Lectures was in 1955, when Professor Paul Emmett delivered the lectures. Surface chemistry by that time had already had a major impact on our lives. The application of surface-active agents—soaps and detergents, adhesives and lubricants, for example— had become commonplace. The study and better understanding of colloid systems led to important advances in food processing and soil chemistry and in the paper, paint, and rubber industries. Surface chemistry played key roles in electrochemistry and in the development of inhibitors of chemical corrosion. Heterogeneous catalysts were employed in most chemical technologies—ranging from ammonia synthesis to the production of gasoline and the making of polymers—to facilitate achieving chemical equilibrium through the use of surface reactions.

The importance of surfaces was recognized from the very beginning of the development of the chemical sciences. Determinations of the surface tension of liquids, the amounts of gases adsorbed in porous solids, and the amounts of solids vaporized were already possible a century ago, and these experimental quantities could be related to surface thermodynamic parameters. Gibbs developed much of the framework of surface thermodynamics that we employ today.[1] Adhesion and friction as well as lubrication were already important concerns during the latter part of the nineteenth century. Progress in surface science was rapid during the first four decades of this century. The synthesis of ammonia over iron that was "promoted," that is, improved, by additives such as potassium and calcium and the oxidation of ammonia over platinum focused attention on transition metals as catalysts.[2] Hydrogenation of

[1] J. W. Gibbs, *Collected Works*, vol. 1, Longmans-Green, New York, 1928, pp. 184–328.
[2] A Mittasch, Adv. Catal. **2**, 81 (1950).

carbon monoxide over ruthenium, nickel, iron, and thorium oxide catalyst surfaces became one of the most important sources of gasoline, methane, and other chemicals in Germany and in much of Europe before and during World War II.[3] Adsorption and gas–surface interactions became better understood in connection with the development of the light bulb[4] and later the gas mask.[5] The properties of surface space charge and electrical double layers at surfaces were uncovered and explored in connection with electrochemical processes[6] and during the studies of colloids.[7] The various surface-characterization techniques that were developed during this period have provided much macroscopic information about surfaces (surface areas, average heats of adsorption, rates and activation energies for surface reactions).[8]

During the next stage of development surface chemistry could not maintain its preeminent role at the frontiers of chemical research. Much research turned toward the investigation of molecular properties, utilizing rapidly developing spectroscopic techniques and X-ray diffraction. Then information on molecular structure was related to the dynamics of chemical reactions, which, with the advent of relaxation spectroscopy and molecular beam scattering techniques, could also be scrutinized on the molecular scale. Surface chemistry could not participate in this development for a number of reasons. The volume density of a solid, such as ice, is about $\rho = 3 \times 10^{22}$ molecules/cm^3; the surface concentration, (A), is about (A) $= \rho^{2/3} \simeq 10^{15}$ molecules/cm^2. Defining the surface to be studied as the topmost layer of atoms, one must obtain detectable signals from 10^{15} atoms or molecules in the background of 10^{22} atoms or molecules to obtain surface information. Because of the low scattering cross section of a monolayer of molecules, the molecules are not amenable to study by most of the experimental techniques that successfully use electromagnetic radiation to study molecular properties in the gas or in the solid state. Usually, large surface areas, in the range 1 to 10^2 m^2, were necessary to obtain detectable signals from adsorbed atoms or molecules during studies by infrared spectroscopy, nuclear magnetic resonance,

[3]H. H. Storch, N. Golumbic, and R. B. Anderson, *The Fischer–Tropsch and Related Synthesis,* Wiley, New York, 1951.

[4]I. Langmuir, *Collected Works,* Pergamon Press, New York, 1962, vol. 2, pp. 161–399.

[5]S. Brunauer, *The Adsorption of Gases and Vapors,* Princeton University Press, Princeton, N.J., 1940; J. H. de Boer, *The Dynamical Character of Adsorption,* Clarendon Press, Oxford, 1953.

[6]A Fromkin, Z. Phys. Chem. **103,** 55 (1923); G. Gony, J. Phys. **9,** 457 (1910); P. Debye and E. Huckel, Phys. Z. **24,** 185 (1923).

[7]H. R. Kruyt, *Colloid Science,* Elsevier, Amsterdam, 1952.

[8]N. K. Adam, *The Physics and Chemistry of Surfaces,* Oxford University Press, London, 1930.

and conventional X-ray diffraction (although these large areas are not necessary when the recently available high-intensity synchrotron is used as a source of radiation). The effort needed for the preparation of clean or reproducible surfaces of such high areas was formidable. It was often necessary to send the same sample of large surface areas from laboratory to laboratory to compare or reproduce different measurements.

During the 1950s, marked changes began to take place in surface chemistry. These were connected largely with the development of the electronics and computer industries and with the rise of aerospace technology. Less expensive and faster electronic devices could be fabricated by miniaturization, which necessitated the use of an ever-increasing surface, A, to volume, V, ratio, A/V. In fact, there are recent reports of light-emitting solid-state devices that are nearly one atomic layer thick. Thus surface characterizations and the study of the physical-chemical properties of the surface layer by and large controlled the rate of development in semiconductor-device technology. Space exploration necessitated the development of ultrahigh-vacuum technology (pressure of less than 10^{-8} torr), which permitted the preparation of clean surfaces and more reproducible surface studies. An explosive development of new techniques yielded atomic-scale information on the atomic and electronic structure, composition, and oxidation states of all types of surfaces. Surface chemical analysis that had eluded the chemist for so long can now be carried out with the sensitivity of less than 1 percent of a monolayer (less than 10^{13} atoms/cm^2) over an area of much less than 1 mm^2 (10^6 to 10^8 Å^2). It is no longer necessary to study large-surface-area samples (often more than 10^2 m^2/g) to obtain detectable surface signals. A 1-cm^2 surface is sufficient for most fundamental surface chemical studies.

Because of the importance of surface studies in so many areas in addition to the chemical sciences and technologies, and because of the interdisciplinary nature of these studies, we refer to the field today more frequently as surface science than as surface chemistry or surface physics. Indeed, modern surface chemistry, which includes atomic-scale scrutiny of surface chemical phenomena, has its roots in solid-state physics and vacuum sciences.

In the past 10 years there has been an accelerated development of our understanding of surfaces on the atomic scale. Modern surface science has emerged, and its impact on various technologies is beginning to be felt. In the period we are now entering, various physical-chemical phenomena in which surfaces play a key role are being studied on the atomic scale with the techniques of modern surface science. These range from corrosion to heterogeneous catalysis and involve efforts to learn how to passivate the surface against external attack or to activate it to carry out chemical reactions selectively.

This development cannot come fast enough, since in most of our schemes of producing, storing, and transporting energy, surfaces play important roles. Nuclear-energy conversion, thermonuclear fusion, fossil-fuel generation and conversion, and solar and geothermal energy conversions all involve the use of surfaces and interfaces either as active components or for passivation. Surfaces also aid in controlling pollution, which is often the by-product of energy-conversion processes. Since energy research and development will be the major societal concern for the foreseeable future, intensive research in the field of surface chemistry will have particular importance and significance.

This book serves as a depository of information obtained by surface science research in the last two decades. The field is developing so rapidly that much important information and many new techniques have become available between the completion and publication of this book. It is my hope that this book will be useful as a textbook for an advanced course on surfaces and as a reference for professionals in the field.

In Chapter 1 we review the reasons why surfaces are unique chemical systems so that *Chemistry in Two Dimensions* is of special significance. This chapter presents a brief summary of the unique physical-chemical surface properties based on what we know about the dynamics of surface atoms, the thermodynamics of surfaces, and the electrical properties of surfaces. Detailed discussions of these fundamental surface properties are outside the scope of this book. Several textbooks cover these subjects in depth. Also, my introductory textbook, *Introduction to Surface Chemistry,* treats these and related topics in detail.

In Chapter 2 the various experimental techniques used most frequently in surface studies are discussed. We then begin our review of the composition and atomic structures of clean solid surfaces (Chapters 3 and 4). Following this, the structure and chemical bonding of adsorbed monolayers of atoms and molecules are discussed (Chapters 5 and 6). After treating these static properties of surfaces, we review the dynamics of gas–surface interactions, including the kinetics of elementary surface reactions (Chapter 7). Then we turn to heterogeneous catalysis (Chapter 8) and to the application of modern surface science to three problems of heterogeneous catalysis: hydrocarbon conversion over platinum (Chapter 9), the hydrogenation of carbon monoxide (Chapter 10), and photochemical, thermodynamically uphill surface reactions (Chapter 11).

The Baker Lectures provide the opportunity to discuss a major field of chemistry from the perspective of one of its practitioners. I have used the lectures to present my view of modern surface chemistry, emphasizing those areas in which much of my research is concentrated. I am grateful for this opportunity. Science is created by individuals who put their personal stamp on it through their hard and careful work, their perspective, their emotions, and their dreams. Research in physical science is a creative and emotional experience. Our simplistic vision of what is to be is constantly shattered by nature's telling us what is through "unexpected" experimental results. This erratic course of following our erroneous extrapolations, then modifying our views and experimental direction every so often, is the unavoidable path to new discoveries. It is exciting and challenging but also trying, as we pay dearly for becoming emotionally attached to a pet idea or thought process. Every researcher dreams of doing "the last experiment first," but it never happens that way. We must try and fail often before grasping the key questions to be asked and the definitive experiments to be performed. Surface science presents unique challenges and the added thrill that comes with having a direct impact on technology and on the life around us. This field of chemical sciences is so much a part of our everyday life that the path from discovery to useful application can generally be rather short.

This book also reflects 15 years of my research at the University of California at Berkeley and elsewhere, and during this period modern surface science has undergone an explosive development. Most of my research is carried out by hard-working and outstanding graduate students and postdoctoral fellows, and their insight is part of every page of this book. I could not have accomplished much of this research without the steady funding I have received through the Materials and Molecular Research Division of the Lawrence Berkeley Laboratory. The support for my research has come mostly from the Division of Basic Energy Sciences of the Department of Energy and from its predecessor, the Research Office of the Atomic Energy Commission. I have enjoyed the friendship and the support of my colleagues in the College of Chemistry in Berkeley. The stimulating and intensive atmosphere of the Berkeley campus provides an ideal place for me to carry out research and teach.

Many of the illustrations and tables were taken from the over 200 papers that were published by my collaborators and me. I am indebted to the American Chemical Society, the American Institute of Physics, Academic Press, Springer-Verlag, Inc., Marcel Dekker, Inc., and North-Holland Publishing Co. for permission to reproduce this material.

Preface

I am grateful for the warm hospitality of my faculty colleagues in the Department of Chemistry at Cornell University, who made my stay in Ithaca during the fall of 1977 an enjoyable and memorable experience.

G. A. SOMORJAI

Berkeley, California

Contents

Preface   7

1. Surfaces—Favorite Media of Evolution   21

Principles of Operation   *21*
Some of the Unique Physical-Chemical Properties of Surfaces   *25*
   The Surface of a Solid Is Heterogeneous on the
      Atomic Scale   *25*
   The Heterogeneous Surface Is Covered with a
      Near-Monolayer of Adsorbate under Most
      Experimental Conditions   *27*
   Two-Dimensional Phase Approximation   *29*
   The Surface Free Energy Is Always Positive   *30*
   Surface Dipole   *34*
   References   *35*

2. Tools of the Surface Scientist's Trade   37

Sample Preparation   *37*
   Generation of Ultrahigh Vacuum (UHV) and Controlled
      Pressures for Surface Studies   *39*
Principles of Surface Analysis by Electrons, Atoms, and Ions   *40*
Methods Sensitive to Atomic Geometry at Surfaces   *45*
   Low-Energy Electron Diffraction (LEED)   *45*
   Reflection High-Energy Electron Diffraction and
      Medium-Energy Electron Diffraction (RHEED and
      MEED)   *54*
   Electron Microscopy   *56*
   Field-Ion Microscopy (FIM)   *58*
   Low-, Medium-, and High-Energy Ion Scattering (LEIS,
      MEIS, and HEIS)   *60*
   Atomic Scattering and Diffraction   *61*
   Surface-Sensitive Extended X-ray Absorption Fine
      Structure (SEXAFS)   *65*

Contents

X-ray and Ultraviolet Photoelectron Spectroscopies of
Surfaces (XPS and UPS)   65

Ion-Neutralization Spectroscopy (INS)   70

Surface Penning Ionization (SPI)   70

Techniques Sensitive to Electron Distribution at Surfaces:
Work-Function Measurements   71

Methods Sensitive to Chemical Composition at Surfaces   72

Auger Electron Spectroscopy (AES)   72

Thermal Desorption Spectroscopy (TDS)   77

Ellipsometry   79

Secondary-Ion Mass Spectroscopy (SIMS)   81

Ion-Scattering Spectroscopy (ISS)   82

Methods Sensitive to Vibrational Structure of Surfaces   84

Infrared Spectroscopies (IR) and Raman Spectroscopy   84

High-Resolution Electron Energy Loss Spectroscopy
(HREELS)   86

Electron Tunneling Spectroscopy   87

Modeling of Surfaces   89

Model Calculations   89

Clusters   90

Techniques for Studies of the Dynamics of Surface Processes   91

References   95

**3. Composition of Surfaces: Thermodynamic Guidelines
and Experimental Results**   **100**

Role of Surface Forces in the Control of Surface Composition   100

Surface Composition of Alloys from Model Calculations   107

Empirical Correlation   111

How the Surface Composition Is Measured   112

Experimental Data and Correlation to the Various Models
Predicting Surface Enrichment   116

Surface Composition of Systems with Complex Phase
Diagrams: The Au-Sn System   116

δ Phase   118

ζ Phase   119

α Phase   119

Surface-Phase Diagram   120

Bimetallic Clusters   123

Effect of the Ambient Condition on the Surface Composition   123

References   124

4. Structure of Clean Surfaces   **126**

Results of Surface Crystallography Studies   126

Conversion of the Diffraction Pattern to a Surface Structure;

Notation of the Surface Structures on Low-Miller-Index
  Crystal Planes — *127*
Surface Structures of Clean Low-Miller-Index Metal Surfaces — *135*
Surface Structure of Alloys — *147*
Structure of Low-Miller-Index Semiconductor Surfaces — *148*
  Elemental Semiconductors — *148*
  Compound Semiconductors — *151*
Structure of Ionic Crystal Surfaces — *152*
Structure of Oxide Surfaces — *153*
Structure of Molecular Crystal Surfaces — *155*
Structure of Inert-Gas Crystal Surfaces — *156*
Stepped, High-Miller-Index Surfaces — *156*
Nomenclature of High-Miller-Index Surfaces — *162*
Structure and Stability of High-Miller-Index Surfaces — *162*
Studies of Antiferromagnetic Phase Transformations by
  LEED — *168*
Surface-Atom Vibrations: The Debye-Waller Factor — *169*
References — *171*

5. Structure of Adsorbed Monolayers on Solid Surfaces — **176**

Introduction — *176*
Principles of Monolayer Adsorption — *177*
Principles of Ordering of Adsorbed Monolayers — *181*
  Causes of Ordering — *181*
  Degree of Ordering — *184*
  Effect of Temperature on the Ordering of Adsorbed
    Monolayers — *187*
  Effect of Surface Irregularities on Ordering — *189*
Unit Cells of Adsorbed Monolayers — *192*
Review of Ordered Adsorbate Structures — *194*
  Introduction — *194*
  Metals on Metals — *195*
  Surface Structures of Small Molecules on Low- and
    High-Miller-Index Surfaces — *209*
  Ordered Organic Monolayers — *245*
Surface Crystallography of Ordered Monolayers of Atoms — *252*
  Effect of the Adsorbate on the Clean Substrate Surface
    Structure — *252*
  Adsorption Geometry of Atoms — *267*
Surface Crystallography of Ordered Multiatomic and
  Molecular Monolayers — *274*
  Coadsorption of Atoms — *275*
  Dissociative Adsorption of Carbon Monoxide — *276*
  Adsorption Geometry of Molecules — *276*
References — *281*

Contents

6. The Surface Chemical Bond — 283

Introduction — 283
Heats of Adsorption of Small Molecules on Solid Surfaces — 284
Sequential Filling of Surface Binding Sites by Adsorbates — 295
Temperature-Dependent Changes in the Character of the Surface Chemical Bond — 299
Pressure Dependence of Chemical Bond Breaking on Surfaces — 302
Effects of Surface Irregularities on Bonding — 303
Effects of Other Adsorbates on the Bonding of Adsorbed Gases — 305
Surface Compounds — 306
Cluster Model of the Surface Chemical Bond — 308
References — 311
Appendix Tables — 314

7. Energy Transfer in Gas–Surface Interactions — 331

Introduction — 331
T–V_s Energy Transfer — 336
Effect of Diatomic Adsorbates on the T–V_s Energy Transfer — 338
R–V_s Energy Transfer — 338
V–V_s Energy Transfer — 342
Gas–Surface Interaction Potential — 342
 Alkali–Halide and Graphite Surfaces — 343
 Metal Surfaces — 344
 Rainbow Peaks — 345
Elementary Surface Processes: Introduction — 346
Modes of Signal Detection — 348
Analysis of the Molecular Beam Surface Scattering and Thermal Desorption Data — 349
Adsorption — 353
Surface Diffusion — 359
Desorption — 360
Surface Reactions — 361
Corrosion Reactions — 366
Catalyzed Surface Reactions — 367
 H_2–D_2 Exchange — 368
 Dissociation Reactions — 374
 Exothermic Small-Molecule and Atom-Recombination Reactions — 376
References — 377

8. Catalyzed Surface Reactions: Principles   **381**

Catalytic Action   *381*
Selective Catalysis   *382*
Kinetic Expressions   *383*
Tabulated Kinetic Parameters for Catalytic Reactions   *386*
Large Surface Area   *387*
Surface Intermediates   *389*
Active Sites   *392*
Acid or Base Character of a Catalyst   *395*
Catalyst Additives   *396*
Deactivation and Regeneration   *399*
Some Frequently Used Concepts of Catalysis   *400*
The Business of Catalysis   *402*
Synthetic Approach to Catalytic Chemistry   *405*
Techniques to Characterize and Study the Reactivity of
    Small-Area Catalyst Surfaces   *408*
  High-Pressure Reactors   *408*
  Low-Pressure Reactors   *409*
References   *412*

9. Hydrocarbon Conversion on Platinum   **479**

Introduction   *479*
Structures of Clean, Single-Crystal Platinum Surfaces with
    Distinguishable Catalytic Activities   *484*
Structure-Sensitive Deactivation of Clean Platinum Surfaces
    during Hydrocarbon Reactions   *486*
Structure Sensitivity of the H—H, C—H, and C—C Bond
    Breaking of Clean Platinum Crystal Surfaces   *488*
Identical Reaction Rates for Single-Crystal and Dispersed
    Platinum Catalysts for Structure-Insensitive Reactions at
    High Pressures: The Ring Opening of Cyclopropane   *493*
Effects of Additives That Modify the Catalytic Behavior of
    Platinum   *494*
  The Carbonaceous Deposit   *494*
  Oxygen Additive to Platinum Surfaces   *499*
Reconstruction of the Platinum Catalyst Surface Structure
    under the Reaction Conditions   *507*
Building of Platinum Catalysts   *510*
References   *514*

Contents

10. Catalytic Hydrogenation of Carbon Monoxide   516

Brief History   *516*
Thermodynamic Considerations   *519*
Different Reaction Paths and Selectivity during the
     Hydrogenation and Insertion of Carbon Monoxide   *525*
Methanation   *526*
Methanol Formation   *530*
Polymerization Reactions   *531*
Formation of Oxygenated Organic Molecules   *539*
Insertion of CO Molecules (Carbonylation) to Form C_2 or C_3
     Products   *541*
Research Directions in the Near Future   *542*
References   *543*

11. Photochemical Surface Reactions   545

Introduction, Photosynthesis, and the Photoelectrochemical
     Dissociation of Water   *545*
Photocatalytic Dissociation of Water Using $SrTiO_3$ Crystals   *551*
Studies of Hydrogen Photoproduction on $SrTiO_3$ Crystals   *554*
Studies of the Location of Hydrogen Production on
     Platinized and Metal-Free $SrTiO_3$ Crystals   *557*
Hydroxide Concentration Dependence of Hydrogen
     Photoproduction   *558*
Electron Spectroscopy Studies of the Active Surface Species
     for the Photodissociation of Water on $SrTiO_3$ Crystal
     Surfaces   *560*
Interaction of Ti^{3+} Surface Sites with Oxygen, Hydrogen,
     and Water in Dark and in Light   *561*
Identification of Hydroxyl Groups on the $SrTiO_3$ Surface   *563*
Mechanistic Considerations   *563*
References   *565*

Index   567

CHEMISTRY IN
TWO DIMENSIONS
SURFACES

1. Surfaces—
Favorite Media of Evolution

Principles of Operation

In our attempts to use surfaces for energy conversion, charge transport, or chemical reactions, we are actually following nature, which is filled with surface-active devices. Systems with a high surface-to-volume ratio (A/V) have played important roles in evolution and in our lives. The human brain (Figure 1.1) has a large area—almost tenfold larger than the surface area of that of an ape although its volume is larger by only sevenfold.[1] Our bone structure, stomach lining, and skin are all large A/V systems. The coral reef (Figure 1.2) and the leaf (Figure 1.3), where much of photosynthesis takes place, are large A/V systems. Their extensive surface area provides optimum rates for the absorption of solar radiation and for the adsorption of the reactants CO_2 and H_2O and for the desorption of O_2. It appears that an increase in the A/V ratio of various systems on our planet and in human evolution leads to optimum reaction rates and increased chemical selectivity.

Shown in Figure 1.4 is a man-made large-A/V-ratio device, an integrated circuit used in a small computer built on a silicon single-crystal surface. This device has a very large number of active circuit elements. The use of this and other large-A/V-ratio devices (catalysts, ion exchangers, semipermeable membranes) has revolutionized technology in many areas that have a significant impact on our lives.

Other large A/V systems frequently consist of tiny particles deposited on large-surface-area porous solids along internal walls or suspended in liquids. Industrial catalysts are often made of metal particles 10 to 150 Å in diameter deposited in porous solids. The soil, our blood, and most paints are examples of stable colloid systems in which small solid particles of $\sim 10^3$ Å diameter are suspended in solution. For these highly dispersed systems a parameter other than the A/V ratio is used to obtain a figure of merit as to the amount of surface available. This is called

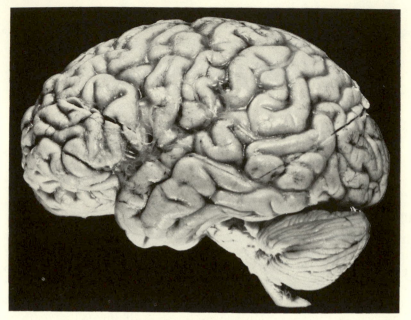

Figure 1.1. Human brain.

"dispersion," the ratio of the number of surface atoms to the total number of atoms of the particle. If the dispersion approaches unity, all of the atoms or molecules are located on the surface; small dispersion values indicate the presence of large particles in which only a small fraction of the total number of atoms are surface atoms. Van Hardeveld and Hartog[2] have calculated the variation of the dispersion with particle size for close-packed cubic packing of spherical particles. This is shown in Figure 1.5. The dispersion drops rapidly for particles greater than 100 Å in radius.

We have given numerous examples of large-surface-area systems. Some of these function at the *solid–gas interface* (most heterogeneous catalysts, integrated circuitry), others at the *solid–liquid interface* (colloids, the brain). The mechanical properties of solids are very much dependent on the chemical properties of interfaces between crystalline grains (grain boundaries). In this circumstance we are concerned with *solid–solid interfaces.* It is customary to divide surface chemistry into areas of study involving solid–gas, solid–liquid, and solid–solid interfaces. Although many of the physical-chemical phenomena to be studied are identical at the different interfaces, the experimental techniques that are

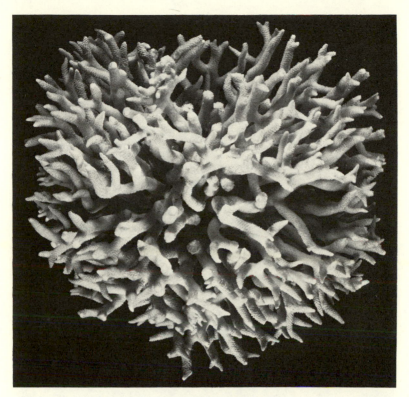

Figure 1.2. Coral.

Figure 1.3. Closeup view of a leaf.

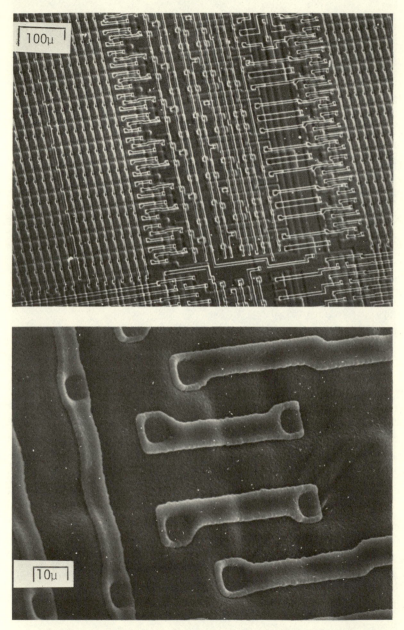

Figure 1.4. Computer logic integrated circuit fabricated on a silicon single crystal.

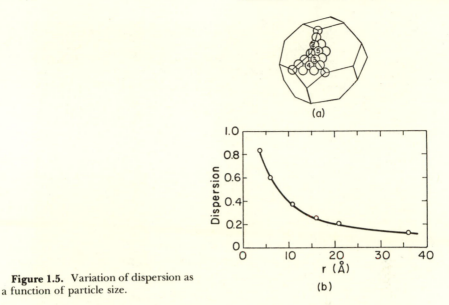

Figure 1.5. Variation of dispersion as a function of particle size.

used to study them are quite different. This is the primary reason for classifying them in this way.

In our discussion of surfaces we concentrate on the properties of clean surfaces and of the solid–gas interface because most of the techniques that provide us with molecular-scale information are useful only under these circumstances. This is unfortunate, because the properties of solid–liquid and solid–solid interfaces are just as important in our everyday life. Biological and electrochemical processes take place predominantly at solid–liquid interfaces, whereas the mechanical properties of solids (embrittlement, crack propagation, corrosion) are controlled by the chemistry at solid–solid interfaces. It is hoped that these other areas of surface science will be opening up to atomic-scale scrutiny in the near future. But, clearly, they will also benefit from our understanding of the properties of the solid–vacuum and solid–gas interfaces.

Some of the Unique Physical-Chemical Properties of Surfaces

The Surface of a Solid Is Heterogeneous on the Atomic Scale

A model of the atomic structure of the surface has emerged from many studies over the past three decades. Figure 1.6 depicts schematically the various surface sites that are identified by experiments.[3] On the

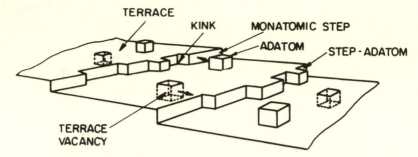

Figure 1.6. Model of a heterogeneous solid surface, depicting different surface sites. These sites are distinguishable by their number of nearest neighbors.

heterogeneous solid surface, atoms in terraces are surrounded by the largest number of nearest neighbors. Atoms in steps have fewer nearest neighbors, and atoms at kink sites have even fewer. Kink, step, and terrace atoms have large equilibrium concentrations on any real surface. Point defects such as adatoms and vacancies are also present and are important participants in the atomic transport along the surface, although their equilibrium concentrations are much less than 1 percent of a monolayer, even at the melting point.

A great deal of experimental evidence from studies of transition metal and oxide surfaces indicates that *different types of surface sites have different chemistries.*[4] This is exhibited in the large differences in the heats of adsorption of molecules at the various sites and in their differing ability to break large binding-energy chemical bonds (H—H, C—H, N—O, N≡N and C=O bonds). Their effects have been explained theoretically in terms of large variations in the localized charge density distribution of atoms at the various sites as a result of the structural differences (variations in crystal-field splitting) and the appearance of large surface dipoles due to the redistribution of the charge density of the electron gas at these various sites in metals.[5] Herein lies one of the important reasons for the diversity of surface chemistry. The overall rate and product distribution in a surface reaction is the result of the rates at each surface site and the different products that form there. From measurements of these macroscopic parameters for reactions over the heterogeneous surface, it is very difficult to identify the elementary chemical processes associated with each site. Since the preparation of the surface establishes the relative concentration of each site, the chemical properties depend intimately on surface preparation.

26

The Heterogeneous Surface Is Covered with a
Near-Monolayer of Adsorbate under Most
Experimental Conditions

In our earth environment, surfaces are never clean or void of
adsorbates. They are covered with a layer of atoms or molecules that
come mostly from the ambient atmosphere. On approaching a surface,
each atom or molecule encounters a net attractive potential. The process
that involves trapping the atom or molecule that is incident on the surface is called "adsorption." It is always an exothermic process (although
for historical reasons the heat of adsorption, ΔH_{ads}, is always denoted as
having a positive sign, unlike ΔH for an exothermic process, which is
negative by the usual thermodynamic convention).

It is often important to determine the time necessary to cover a surface
with a monolayer of adsorbate at any given ambient pressure. From the
kinetic theory of gases we obtain a relationship between the flux of
incident molecules striking the surface of unit area per second and the
ambient pressure as

$$F(\text{molecules/cm}^2/\text{sec}) = 3.52 \times 10^{22} P(\text{torr})(MT)^{-1/2} \qquad (1)$$

where M is the average molecular weight of the gaseous species and T
the temperature.[6] Substituting $P = 3 \times 10^{-6}$ torr and using the values
$M = 28$ g/mol and $T = 300$ K, we obtain $F \sim 10^{15}$ molecules/cm^2/sec.
Thus at this low pressure 10^{15} molecules are incident on the surface each
second or may adsorb if each and every one is trapped by the attractive
surface potential. Since the surface concentration of atoms is about 10^{15}
cm^{-2} at this pressure, the surface may be covered with a monolayer of
gas within seconds. For this reason the unit of gas exposure is 10^{-6}
torr-sec, which is also called the Langmuir. The surface may be covered
with a monolayer of adsorbate during 10^{-6} torr-sec $\equiv$ 1 Langmuir $\equiv$ 1 L
exposure if all the incident atoms or molecules that collide with the
surface will "stick" to it. At much higher pressures the surface may be
covered in a fraction of a second since the exposures are much greater
than 1 L. At pressures less than 10^{-8} torr there may be 10^2 to 10^3 sec
before the surface becomes completely covered. Thus this low-pressure
regime is required to maintain a clean surface condition. Ultrahigh-
vacuum conditions that are required to obtain and maintain clean surfaces correspond to pressures of less than 10^{-8} torr.

The adsorbate may impart to the surface unique chemical properties
by blocking sites or changing the oxidation states of surface atoms. The
presence of adsorbates changes the nature of bonding of incoming reac-

tants, reaction intermediates, and product molecules as well. Such an adsorbed overlayer is schematically represented in Figure 1.7. Not only the chemical but also the mechanical properties of the surface (friction, adhesion, resistance to mechanical or chemical attack) are affected by the presence of the adsorbate. Manipulating the adsorbed layer by depositing chemically active additives permits a great deal of control of important surface properties such as catalysis and corrosion inhibition. Carbon, hydrocarbons, oxygen, sulfur, and water are the most common adsorbates on surfaces exposed to the ambient conditions of our planet.

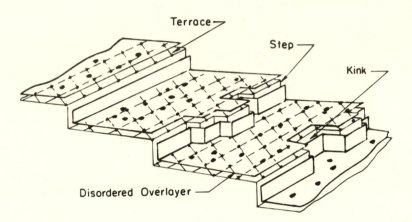

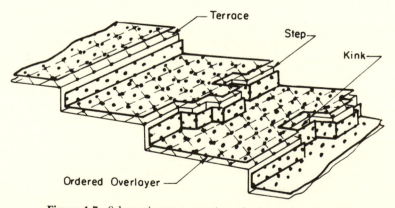

Figure 1.7. Schematic representation of a heterogeneous solid surface covered with a monolayer of disordered or ordered adsorbate.

Two-Dimensional Phase Approximation

A great deal of exchange takes place among atoms and molecules that are adsorbed at the different surface sites. The reason for this lies in the low activation energies for transport along the surface as compared to the high values for desorption into the gas phase or for diffusion into the bulk. Activation energies for surface diffusion of atoms from one step site along the terrace to another are frequently one-half or less of the activation energies for bulk diffusion or heats for desorption into the gas phase. As a result, a great deal of movement of molecules occurs along the surface from one site to another by surface diffusion during their resident time there. Therefore, we may assume equilibrium among molecules in the various surface sites in most circumstances. The long residence times, τ, are forced on the system as a result of the large desorption energies, since

$$\tau = \tau_0 \exp\left(\frac{\Delta H_{des}}{RT}\right) \qquad (2)$$

where τ_0, which is associated with surface-atom vibrations, is frequently of the order of 10^{-12} sec; ΔH_{des} is the desorption energy; T is the temperature; and R is the gas constant.[7] τ can be 1 sec or longer at 300 K for $\Delta H > 15$ kcal. The surface concentration of adsorbed molecules [(A); molecules/cm^2] is given by the product of the residence time τ and the incident gas flux F.

$$(A) = \tau F \qquad (3)$$

Because of the long residence time of various gases that are common in the earth's atmosphere (oxygen, water vapor, hydrocarbons), the surface may be viewed as a two-dimensional phase that is well protected from exchange with the gas or the bulk of the condensed phase by large potential-energy barriers, while transport and chemical exchange along the surface is facile. This is shown schematically in Figure 1.8.

Theories that assume equilibrium among atoms at the various surface sites and among different adsorbates have been successful in explaining the nature of evaporation,[8] crystal growth,[9] and adsorption [10] in many systems. There are systems and experimental conditions, of course, for which the two-dimensional phase approximation is not appropriate. Surface reactions at high temperatures or exothermic reactions, where much of the chemical energy may be retained by the desorbing products, would belong to this category.

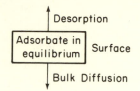

Exceptions:

Low pressure–high surface temperature studies (molecular beam–surface scattering)

Exothermic surface reactions?

Figure 1.8. Adsorbates on the surface which exist as a two-dimensional phase protected by large potential barriers from desorption and from bulk diffusion.

The Surface Free Energy Is Always Positive

The energy necessary to create a unit area of the surface is always positive.[11] Thus a solid or liquid would have a lower total energy without a surface if this were possible. The magnitude of the specific surface free energy (or surface energy),[12] γ (ergs/cm^2), depends on the chemical bonding of the solid or liquid. For metals the surface energies are in the range 10^3 ergs/cm^2. This is about 14 kcal/g-atom for surfaces with atom concentrations of 10^{15} cm^{-2}. For most ionic solids and oxides the surface free energies have a range of a few hundred of ergs/cm^2: for water, ~82 ergs/cm^2 at 300 K; and for hydrocarbons, considerably less. The surfaces of fluorinated hydrocarbons are among those with the lowest surface free energy. A list of specific surface free energies of selected solids and liquids is given in Table 1.1. The total surface energy of any system is the product of the specific surface energy and the available surface area, $\gamma \times A$.

Some very important consequences result from this positive surface free energy. It provides the thermodynamic driving force for changes of shape and the chemical composition of the surface. Condensed systems, in equilibrium, minimize their total surface free energy by assuming shapes with the smallest possible surface area.[12] Also, the surface is covered at all times with a substance that minimizes the surface free energy. This is one reason why surfaces are so readily covered with a layer of organic material. A layer of fluorocarbon polymer, for example, adheres to metal or oxide surfaces very well and provides a nonsticking coating for many applications. In fact, upon scratching this surface layer with a knife and exposing some of the metal or oxide surface, a self-healing process takes place as the organic molecules again completely cover the

Table 1.1. Surface tension of selected solids and liquids

Material	γ (ergs/cm^2)	T (°C)
W (solid)	2900	1727
Nb (solid)	2100	2250
Au (solid)	1410	1027
Ag (solid)	1140	907
Ag (liquid)	879	1100
Fe (solid)	2150	1400
Fe (liquid)	1880	1535
Pt (solid)	2340	1311
Cu (solid)	1670	1047
Cu (liquid)	1300	1535
Ni (solid)	1850	1250
Hg (liquid)	487	16.5
LiF (solid)	340	−195
NaCl (solid)	227	25
KCl (solid)	110	25
MgO (solid)	1200	25
CaF$_2$ (solid)	450	−195
BaF$_2$ (solid)	280	−195
He (liquid)	0.308	−270.5
Na (liquid)	9.71	−195
Xenon (liquid)	18.6	−110
Ethanol (liquid)	22.75	20
Water (liquid)	72.75	20
Benzene (liquid)	28.88	20
n-Octane (liquid)	21.80	20
Carbon tetrachloride (liquid)	26.95	20
Bromine (liquid)	41.5	20
Acetic acid (liquid)	27.8	20
Benzaldehyde (liquid)	15.5	20
Nitrobenzene (liquid)	25.2	20
Perfluoropentane (liquid)	18.6	−110

bare surface by surface diffusion to minimize the total surface free energy.

For multicomponent systems, the constituent that has the lowest surface free energy segregates to the surface.[13] As a result, alloys, for example, have different compositions at the surface than in the bulk. This phenomenon will be discussed in some detail in Chapter 3. Wetting or lack of adhesion is determined by whether the spreading of one type of molecule on the surface reduces or increases the total surface free energy.[14] When a liquid droplet is placed on a solid surface, the relative magnitudes of the surface energies at the solid–liquid, solid–gas, and liquid–gas interfaces determine whether the liquid will spread to cover the solid surface completely (wetting) or whether it will remain a droplet of nearly spherical shape. If the surface energy at the solid–gas interface is much greater than at the solid–liquid interface, complete wetting occurs. When the solid–liquid interfacial energy is very large, the system will want to make as little solid–liquid interface as possible. The liquid forms a nearly spherical droplet and there is no wetting. By measuring the contact angle between the liquid and the solid surfaces, one obtains a figure of merit for the wetting or lubricating ability of the liquid on a solid surface.[15]

Adhesion is best between two substances with the highest surface free energy. By joining them, the free energy of the system is lowered considerably by eliminating the high positive free energy surfaces. Clean metal surfaces readily stick to each other as a result of their high values of σ. Lubrication of surfaces involves the deposition of low-surface-energy materials such as larger organic molecules. When these lubricated surfaces are brought into contact, very little energy is gained; thus they are readily separated.

Another important phenomenon that owes its existence to positive surface free energy is nucleation.[16] In the absence of a condensed phase, it is very difficult to nucleate one from vapor atoms because the small particles that would form have a very high surface area and dispersion and, as a result, a very large surface free energy. The total energy of a small spherical particle has two major components: its positive surface free energy, which is proportional to $\pi r^2 \gamma$, where r is the radius of the particle, and its negative free energy of formation of the particle with volume V. The volumetric energy term is proportional to $-r^3 \ln (P/P_{eq})$, where P is the pressure over the system and P_{eq} is the equilibrium vapor pressures.

$$\Delta G(\text{total}) = - \left(\frac{4\pi r^3}{3V_m} \right) RT \ln \left(\frac{P}{P_{eq}} \right) + 4\pi r^2 \gamma \tag{4}$$

where V_m is the molar volume of the forming particle. Initially, the atomic aggregate is very small and the surface-free-energy term is the larger of the two terms. In this circumstance a condensate particle cannot form from the vapor even at relatively high saturation: $P > P_{eq}$. Similarly, a liquid may be cooled below its freezing point without solidification occurring.

Above a critical size of the spherical particles the volumetric term becomes larger and dominates since it decreases as $\sim r^3$, while the surface-free-energy term increases only as $\sim r^2$. Therefore, a particle that is larger than this critical size grows spontaneously at $P > P_{eq}$. This is shown in Figure 1.9. Because of the difficulty of obtaining this critical size, which involves as many as 30 to 100 atoms or molecules, *homogeneous* nucleation is very difficult indeed.[17] To avoid this problem one adds to the system particles of larger than critical size that "seed" the condensation or solidification. The use of small particles to precipitate water vapor in clouds to start rain and the use of small crystallites as seeds in crystal growth are two examples of the application of *heterogeneous* nucleation.[18]

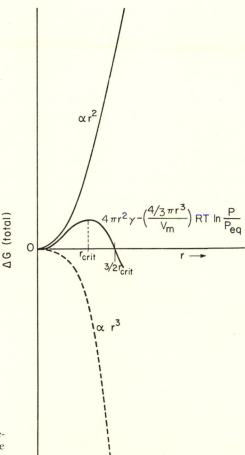

Figure 1.9. Free energy of homogeneous nucleation as a function of particle size.

Surface Dipole

The surface atoms are in an anisotropic environment. Atoms of the same type are on one side of the surface and on the other side are atoms of different charge density or a vacuum. This asymmetry gives rise to a redistribution of charge density at the surface that, for a metal, may be explained as follows. In the bulk of a metal, for example, each electron lowers its energy by pushing the other electrons aside to form an "exchange correlation hole." This attractive interaction, V_{exc}, is lost when the electron leaves the solid, so there is a sharp potential barrier at the surface.[19] Quantum mechanically, the electrons are not totally trapped at the surface and there is a small probability for them to spread out into the vacuum. This is shown in Figure 1.10. This charge redistribution induces a surface dipole, V_{dip}, that modifies the barrier potential.[20] The work function, ϕ, which is the minimum energy necessary to remove an electron at the Fermi energy, E_f, from the metal into vacuum, is given by $\phi = V_{exc} + V_{dip} - E_f$.

The magnitude of this induced surface dipole is different at various sites on the heterogeneous metal surface.[21] For example, a step site on a tungsten surface has a dipole of 0.37 Debye (D) per edge atom in magnitude as measured by work-function-change studies. At a tungsten adatom on the surface there is a dipole moment of 1 D, three times as large as for an atom at a step. Similar variations of the surface dipole from surface site to surface site were found for platinum as well.

There is a net, static charge separation at the surface of ionic solids due to the large polarization caused by the localized surface ions in the asymmetric surface environment.[22] Separation of negative and positive charges gives rise to the formation of a space-charge region at semiconductor and insulator surfaces of all types. The anisotropy of surface-charge density results in the formation of localized electronic surface states.[23] At solid–liquid interfaces the surface-charge separation induces

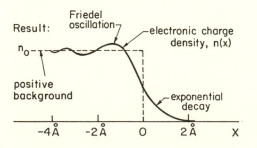

Figure 1.10. Charge density oscillation and redistribution at a metal–vacuum interface.

an electrical double layer at the liquid side of the interface as well. At colloid surfaces this electrical double layer is responsible for keeping the colloid system stable. Breakdowns in the space-charge layer by agitation or by ion exchange lead to precipitation and coagulation of colloids. This phenomenon is of major consequence in soil, in the food industries, and in human biology.

The surface space charge influences the transport of neutral atoms as well as of charged particles (electrons and ions) along the surface from site to site. An adsorbed atom experiences different dipole fields as it diffuses along the surface. The surface dipole also provides an important mechanism for bonding of incident atoms or molecules. By polarizing the adsorbed species in the strong electric field of the surface dipoles, or by partial charge transfer, strong chemical bonds may be formed at the surface, and in some cases surface ionization occurs. The application of externally applied fields to enhance the strength of the surface dipole leads to field emission of electrons from the solid or field ionization of incoming atoms or molecules. These phenomena will be discussed in Chapter 2.

References

1. T.S. Bok, *Histonomy of the Cerebral Cortex,* Elsevier, Amsterdam, 1959, pp. 237–239.
2. R. Van Hardeveld and F. Hartog, Surf. Sci. **15,** 189 (1969).
3. G.A. Somorjai, *Principles of Surface Chemistry,* Prentice-Hall, Englewood Cliffs, N.J., 1972, p. 31.
4. G.A. Somorjai, Adv. Catal. **26,** 1 (1977).
5. (a) L.L. Kesmodel and L.M. Falicov, Solid State Commun. **16,** 1201 (1975); (b), Y.W. Tsang and L.M. Falicov, J. Phys. **C9,** 51 (1976).
6. J.H. deBoer, *The Dynamical Character of Adsorption,* Clarendon Press, Oxford, 1953.
7. F.C. Tompkins, *Chemisorption of Gases on Metals,* Academic Press, London, 1978.
8. J.P. Hirth and A.M. Pound, *Condensation and Evaporation,* Pergamon Press, Oxford, 1963.
9. W.K. Burton, N. Cabrera, and F.C. Frank, Phil. Trans. Roy. Soc. (Lond.) **243A,** 299 (1951).
10. M.W. Roberts and C.S. McKee, *Chemistry of the Metal–Gas Interface,* Oxford University Press, New York, 1978.
11. J.W. Gibbs, *Collected Works,* vol. 1, Longmans-Green, New York, 1928.
12. C. Herring, in *Structure and Properties of Solid Surfaces,* eds. R. Gomer and C.S. Smith, University of Chicago Press, Chicago, 1953.
13. R. Defay, I. Prigogine, A. Bellemans, and D.H. Everett, *Surface Tension and Adsorption,* Wiley, New York, 1966.

14. (a) J.J. Bickerman, *The Science of Adhesive Joints,* Academic Press, New York, 1968; (b) F.P. Bowden and D. Tabor, *Friction and Lubrication of Solids,* Clarendon Press, Oxford, 1964.

15. (a) A.W. Adamson, *Physical Chemistry of Surfaces,* Wiley, New York, 1976; (b) D. Tabor, in *Surface Physics of Materials,* ed. J.M. Blakely, Academic Press, New York, 1975.

16. A.C. Zettlemoyer, ed., *Nucleation,* Marcel Dekker, New York, 1969.

17. R.S. Bradley, *Nucleation and Phase Change,* Quart. Rev. (Lond.) **5,** 315 (1951).

18. F.C. Frank, Disc. Faraday Soc. **5,** 48 (1949).

19. J.M. Blakely, *Introduction to the Properties of Crystal Surfaces,* Pergamon Press, Elmsford, N.Y., 1973.

20. (a) S.R. Morrison, *The Chemical Physics of Surfaces,* Plenum Press, New York, 1977; (b) N.D. Lang and W. Kohn, Phys. Rev. **B3,** 1215 (1971).

21. H. Wagner, *Physical and Chemical Properties of Stepped Surfaces,* Springer Tracts of Modern Physics, Springer-Verlag, New York, 1978.

22. Y.W. Tsang and L.M. Falicov, Phys. Rev. **B12,** 2441 (1975).

23. F. Koch, Surf. Sci. **80,** 110 (1979).

2. Tools of the Surface Scientist's Trade

In this chapter we review the various experimental techniques that are used to study the arrangement of atoms and molecules at surfaces. First we discuss some of the important aspects of sample preparation, since this is a particularly critical step in surface studies. Then we discuss the principles involved in the measurements of surface-specific physical quantities. Since each of the many techniques of surface analysis is sensitive to only a few of the surface properties (such as relative atomic position, electronic levels, chemical composition, binding energies, or vibration frequencies), we classify these techniques according to the surface characteristics to which they are most sensitive.

Sample Preparation

Studies of surfaces all begin with sample preparation. Ideally, we would like to use surfaces whose surface composition and atomic structure are uniform. For studies of thermodynamically stable surfaces it is often necessary to heat the specimen to be used for surface studies to an elevated temperature in a controlled, chemically inert ambient (or in vacuum) to achieve equilibrium of the surface and bulk compositions and to remove excess defects that were introduced during previous specimen preparation. Frequently, we desire to study ordered single-crystal surfaces. In this circumstance single-crystal rods or boules that were grown by zone refining, vapor transport, or strain annealing are oriented by the back-reflection Laué technique and then cut along the desired crystal face. Diamond blade or spark erosion cutting techniques are used most frequently for this purpose. The cutting treatment often severely damages the near-surface regions of the specimen (especially if the material is soft), rendering it amorphous. This damaged layer should be removed by chemical or electrochemical dissolution (etching), which should not affect the surface orientation. The preparation of uniform surfaces also includes repeated polishing by fine mesh alumina or carbide particles followed by repeated etching.[1] The sample so prepared is placed into the reaction chamber using suitable holders that permit heating or cooling and accurate positioning. Usually, a thermocouple is also attached, for accurate temperature determination. A typical single-crystal sample placed in the middle of an ultrahigh vacuum chamber prepared for surface studies is shown in Figure 2.1.

Frequently, the preparation of a specimen requires unique experimental conditions and must be performed inside the experimental chamber. For example, for studies of argon or xenon single-crystal surfaces, or for preparation of surfaces of solid benzene, methane, or other high-vapor-pressure materials, a low-temperature environment is needed.[2] For the preparation of a certain phase of solids that undergo phase transformations (iron, cobalt, uranium, etc.), controlled temperature ranges are again required during preparation. In these circumstances the specimen may be prepared by vaporization (or vapor transport) onto a well-ordered substrate (generally, an ordered single-crystal

Figure 2.1. Small-surface-area sample mounted in a UHV chamber prepared for surface studies.

surface) that is held at the desired temperature. The specimen is grown epitaxially until a multilayer (20 to 2000 Å) deposit is produced inside the experimental chamber.[3]

The surface of the specimen that is prepared outside the reaction chamber is usually covered with a thick layer of carbonaceous deposit by the time it is placed in the chamber. Alternatively, impurities from the bulk of the sample may be diffused to the surface upon heating and segregate there. The most frequently detectable impurities that segregate to the surface in surface studies are carbon, sulfur, silicon, oxygen, and aluminum. Some of these impurities may be removed by chemical treatment, by heating the sample in flowing oxygen or hydrogen, and so on, to a pressure and temperature where there is a significant rate of solid–gas reactions that remove the impurity without changing the chemical composition of the sample. For the removal of other impurities, the sample surface is usually bombarded with ions of inert gas (argon, xenon, or krypton) that are generated inside the experimental chamber using gas pressures in the range 10^{-5} torr.[4] Ion fluxes with energies of 100 to 5000 V and sufficient intensities are used to remove 1 to 100 layers/sec at the near-surface region and to remove impurities. The structural damage introduced by the high-energy ion impact can be annealed by heating the specimen, thus allowing the surface atoms to move back into their equilibrium position by surface diffusion. In field-ion microscopy, to be discussed later, small-diameter ($\sim 10^{-4}$-cm) crystalline tips are utilized.[5] X-ray absorption fine-structure studies permit the use of high-surface-area porous samples as well as crystalline surfaces.[6] Since heterogeneous metal catalysts are frequently deposited on large-surface-area oxides, this technique can be employed for the studies of supported catalysts as well.

Generation of Ultrahigh Vacuum (UHV) and Controlled Pressures for Surface Studies

There are two main reasons why a high vacuum (10^{-9} to 10^{-4} torr) must be maintained around the samples during some phase of the surface chemical experiment. First, it is often desirable to start our investigation with initially clean surfaces, and ultrahigh vacuum (less than 10^{-8} torr) is needed to achieve a surface that is free from adsorbed gases. Second, many of the surface characterization techniques use electrons or ions as probing particles to reveal the surface structure, composition, and oxidation state. These particles need a long mean free path (greater than 10 cm) to be able to strike or exit the sample and then reach the detector without colliding with gas-phase molecules. For this reason, pressures lower than 10^{-3} torr must be used. High vacuum may be generated with many different pumping devices[7] (oil diffusion pump, vacuum ionization pump, sublimation pump, turbomolecular pump, cryo-pump) and may be maintained indefinitely in a leak-free chamber usually built out of nonmagnetic stainless steel. Vacuum technology has reached a level of sophistication where obtaining high vacuum in a short time (less than 1 hr) has become a simple and reliable procedure.[8]

Often, we need to place a specimen in a high-pressure environment ($\geqslant$ one atm.) after surface characterization in high vacuum to carry out surface studies. The sample may be enclosed by a small high-pressure cell operated by hydraulic pressure or a threaded drive mechanism.[9] The same apparatus can also be used for studies of reactions and processes at the solid–liquid interface.[10] The liquid can readily be introduced and then pumped out after the study, and the surface can be studied by the various surface diagnostic techniques before and after the experiments. Using the same principle, isolation cells that are capable of containing high pressures around the sample inside the UHV chamber can be constructed in a variety of geometries.

Principles of Surface Analysis by Electrons, Atoms, and Ions

In surface studies one is confronted with the difficulty of detecting a small number of surface atoms in the presence of a large number of bulk atoms; a typical solid surface has 10^{15} atoms/cm^2 as compared with 10^{23} atoms/cm^3 in the bulk. To probe the properties of solid surfaces with more conventional methods using electromagnetic radiation, one needs to use powders with very high surface-to-volume ratios, so that surface effects become dominant. However, this technique suffers from the distinct disadvantage of an entirely uncontrolled surface structure and composition, which are known to play an important role in surface chemical reactions. It is thus desirable to use specimens with well-defined surfaces, which generally means a small surface area of the order of 1 cm^2, and to examine them with tools that are surface-sensitive.

It turns out that electrons with energies in the range 10 to 500 eV are ideally suited for this purpose. Figure 2.2 shows a plot of the mean free path for inelastic scattering in solids as a function of the electron energy.[11] The curve that is often called the "universal curve" exhibits a broad minimum in the energy range between 10 and 500 eV, with the corresponding mean free path on the order of 4 to 20 Å. Electron emission from solids with energy in this range must therefore originate from the

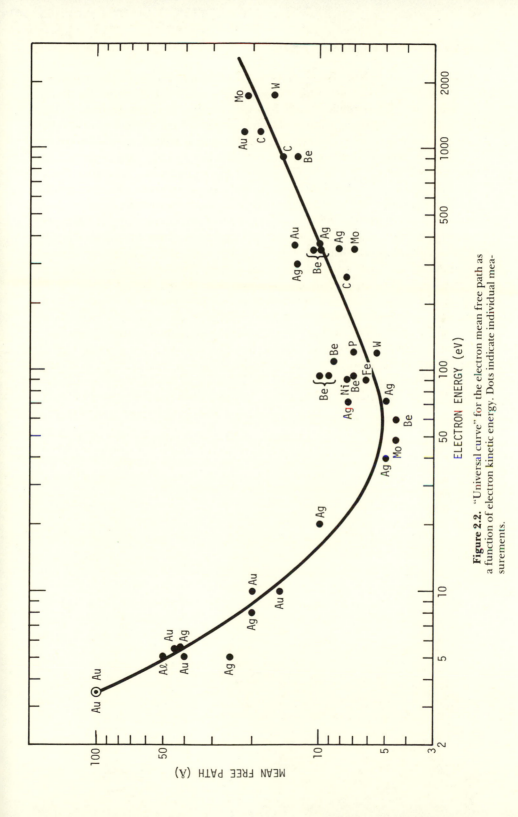

Figure 2.2. "Universal curve" for the electron mean free path as a function of electron kinetic energy. Dots indicate individual measurements.

top few atomic layers. By extension, all experimental techniques involving the incidence onto and/or emergence from surfaces of electrons having energy between 10 and 500 eV are thus surface-sensitive. Suppose that we collimate a monoenergetic beam of electrons of energy E_p and allow it to strike a solid surface. A typical plot of the number of scattered electrons, $N(E)$, as a function of their kinetic energy, E, is shown in Figure 2.3. The $N(E)$ versus E curve is usually dominated by a strong peak at low energies due to secondary electrons created as a result of inelastic collisions between the incident electrons and electrons bound to the solid. The rest of the spectrum consists of two major features which turn out to be extremely useful in surface characterization: (1) the elastic peak that is utilized in low energy electron diffraction, and (2) the inelastic peaks that are used to obtain information about the vibrational structure and electronic structure of surface atoms.

In addition to electrons of low energy (5 to 500 eV), atoms and ions also provide information about surface atoms almost exclusively when scattered from solids or liquids at low energies. Atoms or molecules with thermal energy (~0.01 eV) or higher energies (up to a few electron volts) can be directed at the surface as well-collimated beams from effusion cells or from nozzle-beam sources.[12] (These are discussed in more detail in Chapter 7.) Their penetration depth is even lower than that of elec-

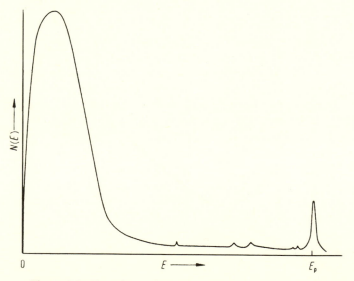

Figure 2.3. Experimental number of scattered electrons [$N(E)$] of energy E versus electron energy, E, curve.

trons, because of their much larger mass and lower kinetic energies. For all practical purposes the atoms scatter from the topmost layer of atoms at the surface and thus only contain information about surface atoms. A small fraction of them scatter elastically, and since their de Broglie wavelength is smaller than the interatomic distance, they diffract and are utilized to obtain information about the atomic surface structure (see the next section). Most of the atoms or molecules lose energy in the scattering process. Therefore, measurements of their angular distribution and velocity after scattering reveal the nature of energy transfer between the translational energy of the incident atoms and the vibrational modes of the surface atoms. In the case of incident molecules, energy transfer takes place not only between translational and vibrational modes of the incident and surface atoms, respectively, but also between the internal modes of the molecules (rotation, vibration) and the surface atom vibration.[13] These energy-transfer processes are of great importance in studies of gas–surface interactions, both reactive and nonreactive.

Similarly, the scattering of low-energy ions in the range 1 to 10^3 eV provides a great deal of information about the surface atoms. Upon scattering, ions at low energy capture electrons from surface atoms. This process is the basis of ion-neutralization spectroscopy (INS), a technique that is used to study the electronic structure of surfaces.[14] At higher energies the energy loss and change of direction of incident ions is primarily determined by the mass of the surface atoms. Thus measurements of the angle and energy of the scattered ion beams yield the surface composition (ISS).[15] At even higher ion energies, atoms and ions are sputtered from the surface with high efficiency. Sputtering of surface atoms by ions is used to remove or deposit thin layers of solids (thin films), and the ionized fraction of the sputtered atoms is used for analysis of the surface composition (secondary-ion mass spectroscopy, SIMS).[15]

The scattering of electrons, atoms, and ions, both elastic and inelastic, serves as a probe of many important properties of surface atoms, their atomic structure, electronic structure, composition, and oxidation states.

Many techniques using electrons, atoms, and ions have been developed and are being used for surface analysis. Low-energy electron diffraction (LEED), in which electrons are elastically scattered off a surface, has been the most successful technique among those for surface crystallography.[16] Inelastically scattered electrons also provide surface-structural information in a method called high-resolution electron energy loss spectroscopy (HREELS).[17,18] Secondary electrons ejected from the surface by incident electrons can also be used, especially for chemical composition analysis in Auger electron spectroscopy (AES), an important and essential tool in many surface investigations.[19] Impinging light, ions, or atoms can cause electrons to leave the surface in ways characteristic of the surface structure; thus photoemission, ion neutralization spectroscopy, surface penning ionization (SPI), and many other techniques have been applied to surface analysis. Low-energy atomic and ionic scattering off surfaces are uniquely surface-sensitive processes in which the scattering particles only come into contact with the outermost surface atoms, since no penetration through atoms occurs. Light reflection can also be turned into a surface-sensitive tool, as in infrared spectroscopy.

It must be emphasized that all the surface analysis techniques developed so far have particular, often stringent, limitations. Some are more sensitive to chemical composition, for example AES; others to relative atomic positions, for example LEED, angle-resolved photoemission, and SPI; others to vibration modes, for example HREELS and IR; and still others to electronic levels, for example, angle-integrated photoemission and INS. To extricate the various properties of any given surface it has become necessary to use several complementary and/or supporting techniques in parallel. AES is a basic method used in nearly all studies; in addition, LEED, together with thermal desorption spectroscopy (TDS), and photoemission or LEED, together with HREELS and work-function measurements, might be used. The scheme of an apparatus equipped with several surface probes is shown in Figure 2.4.

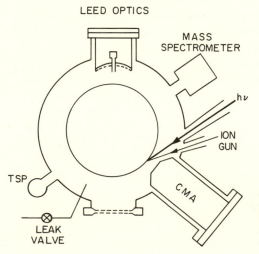

Figure 2.4. Scheme of apparatus equipped with several surface probes, which include low-energy electron diffraction (LEED) and a cylindrical mirror analyzer (CMA) for electron energy analysis.

We now briefly describe the mechanisms, capabilities, and limitations of the principal techniques used in surface analysis, classifying them by the nature of the information obtained by using them.

Methods Sensitive to Atomic Geometry at Surfaces

Low-Energy Electron Diffraction (LEED)

LEED has yielded a large amount of structural information and will therefore be treated relatively extensively here. In LEED, electrons of well-defined (but variable) energy and direction of propagation diffract off a crystal surface. Usually, only the elastically diffracted electrons are considered, and we shall do so here as well. The electrons are scattered mainly by the individual atom cores of the surface and produce, because of the quantum-mechanical wave nature of electrons, wave interferences that depend strongly on the relative atomic positions of the surface under examination.

The de Broglie wavelength of electrons, λ, is given by the formula λ (in Å) $= \sqrt{150/E}$, where E is measured in eV. In the energy range 10 to 500 eV the wavelength then varies from 3.9 to 0.64 Å, smaller or equal to the interatomic distances in most circumstances. Thus the elastically scattered electrons can diffract to provide information about the periodic surface structure. The LEED experiment is carried out as follows: a monoenergetic beam of electrons (energy resolution approximately 0.2 eV) in the range 10 to 500 eV is incident on one face of a single crystal. Roughly 1 to 5 percent of the incoming electrons are elastically scattered, and this fraction is allowed to impinge on a fluorescent screen. The scheme of the LEED experiment is displayed in Figure 2.5. If the crystal surface is well ordered, a diffraction pattern consisting of bright, well-defined spots will be displayed on the screen. The sharpness and overall intensity of the spots are related to the degree of order on the surface. When the surface is less ordered, the diffraction beams broaden and become less intense, while some diffuse brightness appears between the beams. A typical set of diffraction patterns from a well-ordered surface is shown in Figure 2.6.

The electron beam source commonly used has a coherence width of about 100 Å. This means that sharp diffraction features are obtained only if the regions of well-ordered atoms ("domains") are of (100 Å)^2 or larger. Diffraction from smaller domains gives rise to beam broadening and finally to the disappearance of detectable diffraction from a disordered (liquidlike) surface.[20]

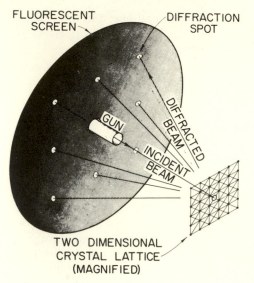

FLUORESCENT SCREEN

DIFFRACTION SPOT

DIFFRACTED BEAM

GUN

INCIDENT BEAM

TWO DIMENSIONAL CRYSTAL LATTICE (MAGNIFIED)

Figure 2.5. Scheme of the low-energy electron diffraction experiment.

One may distinguish between "two-dimensional" LEED and "three-dimensional" LEED. In two-dimensional LEED one observes only the shape of the diffraction pattern (as seen and easily photographed on a fluorescent screen). The bright spots appearing in this pattern correspond to the points of the two-dimensional reciprocal lattice belonging to the repetitive crystalline surface structure [i.e., they comprise a (reciprocal) map of the surface periodicities]. They therefore inform us about the size and orientation of the surface unit cell; this is important information, since the presence of, for example, reconstruction-induced and overlayer-induced new periodicities, the so-called superlattices, is made immediately visible. This information also includes the presence or absence of periodic steps in the surface.[21] The background in the diffraction pattern contains information about the nature of any disorder present on the surface.[22] As in the analogous case of X-ray crystallography, the two-dimensional LEED pattern in itself does not allow one to predict the internal geometry of the unit cell (although good guesses can sometimes be made): that requires an analysis of the *intensities* of diffraction.[20] Nevertheless, two-dimensional LEED already can give a very good idea of essential features of the surface geometry, in addition to those mentioned before. Thus one may follow the variation of the diffraction pattern as a function of exposure to foreign atoms. It is often possible to obtain semiquantitative values for

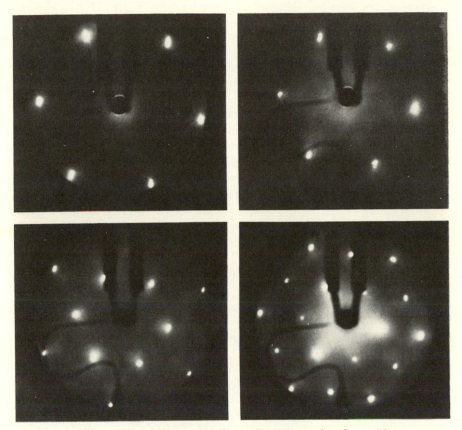

Figure 2.6. LEED patterns from a Pt(111) crystal surface at 51-, 63.5-, 160-, and 181-eV incident electron energies.

the coverage, for the attractive and/or repulsive interactions between adsorbates,[23] and for some details of island formation.[24] The variation of the diffraction pattern with changing surface temperature also provides information about these interactions (in particular at order–disorder transitions[24,25]), while the variation with electron energy is sensitive to quantities such as surface roughness perpendicular to the surface and step heights.

In three-dimensional LEED the information obtained from the two-dimensional pattern is supplemented by the intensities of the diffraction spots that are measured as a function of the changing electron energy and changing direction of incidence. From an analysis of these curves the precise location of atoms or molecules in the surface with respect to their surface neighbors and their neighbors in the second plane beneath the surface plane is determined.

Measurements of the diffracted electron beam intensities can be carried out by various techniques, which include photographing[26] the fluorescent screen or collecting the electrons at any given angle of emission.[27] The resultant intensity versus electron energy curves (usually called I–V curves) or I–θ or I–ϕ curves (for variation of the polar or azimuthal incidence angles, respectively) serve as the basis for surface structural analysis. A set of I–V curves from a Pt(111) crystal face is displayed in Figure 2.7. They exhibit pronounced peaks and valleys, indicative of constructive and destructive interference of the electron beam scattered from atomic planes parallel to the surface as the electron wavelength is varied. Often, Bragg peaks (due to simple interference between electrons backscattered from different atomic planes, as in X-ray diffraction) can be identified. However, in addition to these and also overlapping with these, there are usually extra peaks due to multiple scattering of electrons through the surface lattice. The strong multiple scattering of the diffracted electrons is due to their large collision cross sections, which are about 10^6 times larger than that of X-rays. The behaviors of diffracted low-energy electrons and X-rays are compared schematically in Figure 2.8. Because of the large scattering cross section, a large fraction of the electrons are backscattered; hence there is a great deal of surface sensitivity as compared to X-rays. However, because of the large backscattering intensity, there is a finite probability that the backscattered electron will scatter again before exiting the surface. This gives rise to the multiple scattering peaks that are readily observable in LEED intensity analysis.[20]

The presence of well-defined peaks and valleys in I–V curves indicates that LEED is indeed not a purely two-dimensional surface diffraction technique. There is a finite penetration, and diffraction takes place in the first three to five atomic layers. The depth of penetration affects peak widths markedly; the shallower the penetration, the broader is the diffraction peak. By simulating such I–V curves numerically with the help of a suitable theory, it is often possible to determine the relative positions of surface atoms (including, therefore, bond lengths and bond angles).[28] It may also be possible to verify the thermal vibration state of surface atoms.[28] However, a chemical identification of the surface atoms is not possible with LEED.

The analysis of LEED intensities requires a theory of the diffraction process. This is a nontrivial point because multiple scattering of the

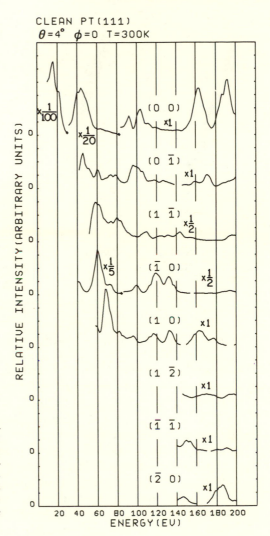

CLEAN PT(111)
$\theta = 4°$ $\phi = 0$ T=300K

RELATIVE INTENSITY (ARBITRARY UNITS)

$x\frac{1}{100}$

$x\frac{1}{20}$

(0 0) x1

(0 $\bar{1}$) x1

(1 $\bar{1}$) $x\frac{1}{2}$

$x\frac{1}{5}$ ($\bar{1}$ 0) $x\frac{1}{2}$

(1 0) x1

(1 $\bar{2}$) x1

($\bar{1}$ $\bar{1}$) x1

($\bar{2}$ 0) x1

20 40 60 80 100 120 140 160 180 200
ENERGY (EV)

Figure 2.7. Experimental intensity I versus electron energy eV curves for electron diffraction from a Pt(111) surface. Beams are identified by different labels (j,k), representing reciprocal lattice vectors parallel to the surface. Here the angle of incidence was 4° from the surface normal.

LEED electrons by the surface is always present and is not easy to represent in a theory. Even a simple single-scattering theory (such as is used in X-ray crystallography) must justify its validity with respect to the neglect of multiple scattering. Clearly, the computational effort increases with the complexity of the theory, and so the present situation has arisen, in which a variety of theories of different complexity coexist, each justifying its existence by a different compromise between

DIFFRACTION

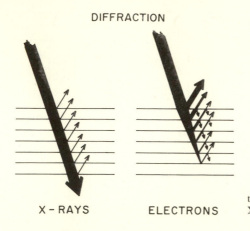

X – RAYS   ELECTRONS

Figure 2.8. Schematic comparison of the scattering behavior of diffracted X-rays and low-energy electrons.

computational effort, ease of use, amount of experimental data required, reliability, accuracy, type of information produced, and range of applicability.

In order of increasing computational complexity, the following primary theoretical methods are used today in LEED:

1. Simple kinematical theory (single s-wave, isotropic scattering, as in X-ray diffraction theory, with inner potential correction).
2. Method 1 with an anisotropic atomic form factor.
3. Method 1 or 2 with averaging of experimental data (to average away multiple-scattering effects).[29]
4. Methods 1, 2, or 3 with Fourier transformation from momentum space to coordinate space.[30]
5. Quasi-dynamical theory (this includes multiple scattering between, but not within, atomic layers parallel to the surface).[31]
6. Iterative dynamical theory (multiple scattering is iterated to convergence; examples are the renormalized forward scattering[28] and reverse scattering perturbation[32] methods).
7. Full dynamical theory[28] (multiple scattering is included in closed form; examples are Beeby's matrix inversion, the layer-doubling, and the Bloch-wave methods; the first two of these assume a crystal of finite thickness, which is increased until convergence of the results in the case of layer doubling).
8. Spin-polarized LEED theory[33] (relativistic spin-dependent effects are added to a dynamical theory).
9. LEED theory for disordered surfaces[34] (effects of disorder in the surface structure are added to a full dynamical theory).

Roughly speaking, the LEED theories in the order listed above give increasing accuracy and reliability of the structural results. Few of these methods can produce a complete surface structure with a single calculation; with most methods a trial-and-error search for the actual structure must be undertaken (giving rise to the desirability of independent hints about the structure from other surface analysis techniques). Most results of surface crystallography by LEED to date have been obtained with iterative or full dynamical theories. With these theories as they stand today, the limitations on the possibilities are the following:

1. Unit cell areas are limited to about 25 Å [the equivalent of a (2 × 2) superstructure on a low-index face of a simple metal].
2. The number of atoms in one unit cell per layer parallel to the surface is limited to about 4.
3. The accuracy in distances perpendicular to the surface is, depending on the case, about 0.1 Å or better (for comparison, atomic vibration amplitudes at room temperature are usually of the order of 0.1 Å).
4. The accuracy in distances parallel to the surface is of the order 0.2 Å, unless a well-defined symmetrical atomic position can be assumed, in which case one assumes no uncertainty.
5. The resulting accuracy in bond lengths varies from less than 0.05 Å (for bonds more or less parallel to the surface, assuming no uncertainty in the bonding site) to 0.2 Å; this translates to a relative uncertainty of less than 2 percent to 10 percent of the bond length.

In order to indicate the theoretical ideas involved in LEED crystallography, we now outline the main dynamical (i.e., multiple-scattering) methods used to compute I–V curves for comparison with the experiment (see also references 28a and 28b).

The crystal surface is imagined to consist of individual atomic layers parallel to the surface. Whenever convenient, the LEED electrons between these layers are represented by a set of plane waves (to each diffracted beam corresponds one plane wave), as the electron–solid interaction potential is assumed to be a constant between the layers. These plane waves are diffracted any number of times by these individual atomic layers, whose diffraction properties are discussed below and assumed known here. The multiple scattering between layers is treated usually in one of three ways:

1. In the Bloch-wave method the periodicity of the crystal underneath the surface region, which has a different structure, is exploited. The Bloch theorem applies and enables the electronic eigenfunctions (the

Bloch waves) to be determined. These eigenfunctions are then matched across the surface region to the conditions outside the surface (consisting of one incident beam and a set of reflected beams): this matching fixes the intensities of the reflected beams.

2. In the layer-doubling method the diffraction properties of pairs of layers are determined exactly from those of the individual layers. This is done by summing up the multiple scattering between the layers as in a geometrical series, but using matrix inversion rather than series expansion. By repeating this combination of layers, the crystal can be built up layer by layer until convergence of the surface reflectivities. (In the periodic bulk each step can double the thickness of the growing slab of layers.) This procedure converges because of the presence of electron "absorption"; most of the incoming electrons lose energy as they move through the crystal surface and are therefore removed from the flux of elastically scattered electrons that we are interested in. (This absorption is simulated by a mean free path or by an imaginary part of the electron energy.) The layer-doubling method is computationally more efficient than the Bloch-wave method, and it is more flexible in terms of allowing the variation of the surface structure, as is necessary in a structural search.

3. In the renormalized forward scattering (RFS) method, substantial computation time is saved by recognizing and exploiting the fact that many multiple-scattering processes are too weak to contribute significantly to the diffracted intensities. Namely, *back*scattering off any atomic layer is usually weak (forward scattering is not), and therefore scattering paths are ordered according to an increasing number of such backscatterings. This method is cast in a convenient iterative form, providing a most efficient computation scheme for interlayer multiple scattering. (Nonconvergence, however, occurs in cases of very strong scattering and small interlayer spacings.)

The individual layer diffraction properties needed as input to the methods described above are obtained as follows. The multiple scattering between the atoms of a given layer can be summed up exactly to produce a matrix inversion, in a way analogous to the treatment of interlayer multiple scattering in the layer-doubling method. Computationally, this is manageable only when the individual atoms of the crystal surface are assumed to be spherical and spherical waves may be used between the atoms. One therefore uses an electron–solid interaction potential consisting of nonoverlapping, spherically symmetrical regions with a constant interstitial value; such a potential is called a "muffin-tin potential." This approach is used in all current dynamical LEED computations for treating individual atomic layers. These layers may have more than one atom per unit cell, and these atoms need not be coplanar.

In fact, it is possible to consider the entire surface as a single thick layer composed of perhaps five individual layers (since only finite electron penetration occurs) that would be treated by this matrix inversion method in angular momentum space. However, this solution gives rise to matrix dimensions that rapidly exceed the possibilities of all existing computers. Also, the matrix inversion would have to be repeated for each different surface geometry, a waste that is largely overcome by using plane waves between atomic layers. Generally speaking, plane waves are used as often as possible because they offer clear computational advantages.

A perturbation expansion version of this matrix inversion method in angular momentum space has been introduced with the reverse scattering perturbation (RSP) method, in which the ideas of the RFS method are used. The matrix inversion is replaced by an iterative, convergent expansion that exploits the weakness of the electron backscattering by any atom and sums over significant multiple-scattering paths only. This method can be applied to individual atomic layers or to a thick layer representing the entire surface, but remains relatively time-consuming compared to the plane-wave methods except when the separation between individual layers becomes small ($\lesssim 0.5$ Å).

To obtain the above-mentioned *layer* diffraction properties, one needs as input the single-atom scattering properties. These, in the case of spherically symmetrical atoms, are given by a set of phase shifts (which are species- and energy-dependent). The phase shifts are obtained by a numerical integration of the Schrödinger equation with an atomic potential, which in turn has to be generated from first principles, including electrostatic and exchange-correlation effects in the electron–atom interaction, as well as the effects of neighboring atoms.

In practice, the atomic scattering properties are modified by including a Debye–Waller factor, which represents the effect of the thermal vibrations of the surface atoms; thus the temperature correction is applied at each scattering in each chain of scatterings. Thermal atomic vibrations have in LEED an effect similar to that in X-ray diffraction: the intensity of the diffracted beams is decreased while the background reflection between beams increases. (Electron–phonon scattering can impart a change of momentum parallel to the surface that generates intensity in directions other than those of the beams.) Often, decrease in intensity of the beams also behaves as in X-ray diffraction in spite of multiple scattering: an exponential decrease is usually observed, which can be described by the Debye–Waller factor

$$\exp\left(\ \frac{-3|\Delta\mathbf{k}|^2 T}{mk_B\Theta_D{}^2}\ \right) \tag{1}$$

where $\Delta\mathbf{k}$ is the change in electron momentum, T the temperature, m the atomic mass, and k_B Boltzmann's constant. This factor involves a material-dependent constant, the Debye temperature, Θ_D, which quantifies the rigidity of the crystal lattice and thereby influences the amplitudes of the vibrations. Θ_D can be obtained experimentally for a surface through measurement of the Debye–Waller factor, and typical results are shown in Figure 2.9. The experimental Θ_D is observed to vary irregularly with the LEED electron energy, tending at high energies to a constant that lies in the neighborhood of the value of Θ_D for the bulk of the crystal.[35] The interpretation of this electron-energy-dependent Debye temperature is the following: At high energies the electrons penetrate deeply into the surface and sample the bulk properties of the crystal. As the energy is lowered, two effects are noticed. First, multiple scattering produces rapid variations in the experimental Θ_D with electron energy. Some decrease in Θ_D is also attributable to multiple scattering at low energies, but not enough. The second effect that causes a decrease in Θ_D at lower electron energies is due to larger thermal vibration amplitudes of surface atoms as compared with bulk atoms, which contribute more to the electron scattering with the shallower penetration of the low-energy electrons. One can estimate in this way that clean-surface atoms have vibrations enhanced by typically 50 percent in the direction perpendicular to the surface, and little enhanced in directions parallel to the surface, in good agreement with theoretical predictions. Because of the complication of multiple scattering, however, attempts to extract more precise information about surface vibrations have not met with success, not even in the simplest case, that of clean metal surfaces.

Reflection High-Energy Electron Diffraction and Medium-Energy Electron Diffraction (RHEED and MEED)

RHEED and MEED[14] differ from LEED in the range of energies used. While in LEED "low" energies of about 10 to 500 eV are used, in RHEED "high" energies of about 5 to 50 keV are used, with MEED bridging the intermediate-energy range. The surface sensitivity of LEED is guaranteed by the small mean free path ($\sim$5 to 10 Å) at the low

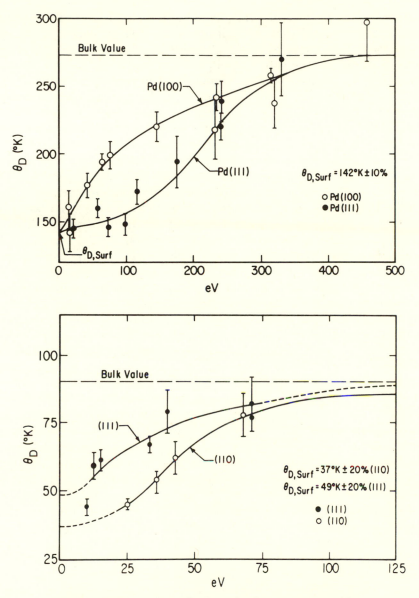

Figure 2.9. Experimental values of Debye temperatures (Θ_D) as a function of electron energy (eV) for surfaces of Pd (top) and Pb (bottom).

energies. At higher energies the mean free path increases (~20 to 100 Å for RHEED energies) and thus impairs the surface sensitivity on the atomic scale unless grazing angles of incidence and emergence are used. This is, therefore, the choice of experimental geometry in MEED and RHEED.[36a] Grazing angles of incidence put stronger requirements on the atomic scale planarity of the surface than the roughly perpendicular incidence directions used in LEED. Therefore, RHEED and MEED sample the near-surface layers. These techniques are particularly well suited to the study of the structure of thin films or thin surface coatings. The multiple scattering present in LEED is also present at the higher energies. Corresponding dynamical theories have been developed,[36b,c] but there is a lack of accurate experimental data to interpret.

In many studies the chemisorption and the surface reaction are just the first step in a series of solid-state reactions that take place as atoms from the surface move into the bulk. Corrosion, oxide, carbide, and other compound formations are generally initiated at the surface and then propagate into the bulk. There may be a concentration gradient of certain constituents at the surface in a multicomponent system that will influence the mechanical or chemical properties of that system. Hardening of materials and other forms of passivation treatment frequently involve introduction of certain substances only in the near-surface region. For the investigation of these problems, RHEED is a powerful technique.

Electron Microscopy

Electron-optical techniques are increasingly being applied under conditions of ultrahigh vacuum, allowing the study of surfaces under controlled circumstances.[37] The most significant developments for electron microscopy, however, have been in the imaging mode, where considerable enhancement of resolution has been obtained, leading to various new forms of operation.

Transmission electron microscopy (TEM) utilizes electrons that are accelerated in the 100-kV range form either a thermionic or field emission gun. These are used to illuminate a specimen that is typically 3 mm in diameter and $\leqslant$5000 Å thick. After transmission through the sample, the electrons are focused by an electromagnetic objective lens to form an image. Other lenses are also utilized in the optical column to image the diffraction pattern at the back focal plane of the objective and to magnify up to 1 million times the image produced by the objective. The final image is recorded photographically and can show clearly resolved detail in the 2- to 3-Å range.

Even though the TEM image is a two-dimensional projection of the specimen structure, it is highly sensitive to changes in thickness. Therefore, when enhanced by certain imaging techniques, surface steps and morphological irregularites can be examined visually. These imaging techniques employ either diffraction contrast for thick specimens ($\gtrsim 1500$ Å) or defocus contrast for thinner samples ($\gtrsim 500$ Å). Step structures having a height of ~ 10 Å may be revealed by either method, although the latter technique has successfully been used in the imaging of monolayer steps and of single, heavy element atoms on specially prepared substrates.

Scanning electron microscopy (SEM) utilizes a highly focused electron beam which is scanned over the surface of the specimen. Since penetration through the sample is not essential for this instrument, thicker samples (centimeter range) and lower accelerating potentials (low-kilovolt range) are commonly used. The most popular mode of operation is the emissive mode, which utilizes those electrons that have either been emitted by the specimen as secondaries or have been backscattered. Owing to the strong dependence of the number of collected electrons on incident illumination angle, surface topography is dramatically revealed by this technique. Resolution is primarily determined by the spot size of the focused electron beam and can be less than 100 Å. With thin enough specimens a detector may be placed in such a way that it collects the transmitted electron signal. This is the principle behind scanning transmission electron microscopy (STEM), which has allowed resolution down to 2 to 3 Å as a result of very small electron probe sizes. The electron–atom cross sections here also limit observability to heavy atoms on light substrates (typically carbon films). Moreover, the thin substrate films do not have single-crystal surfaces, but are amorphous.

Electron microscopy, in general, uses intense electron beams that can severely damage the surface, although at very high energies the electron–atom cross sections become smaller. However, in the scanning mode, each portion of the sample is subjected to the electron beam for a short time, reducing the destruction of the surface. Advances in electron microscope design utilizing cryogenically cooled electron optics are approaching the limits of resolution set by the astigmatism of the optics, and further improvements in optic design will improve these limits.

Since the first electron-microscopical observation of a heavy atom on a surface,[37a] different studies have looked at effects related to individual atomic adsorbates. These include diffusion along the surface (atoms can be tracked in real time), giving results in agreement with equivalent field-ion microscopy observations, and pair-spacing distributions, showing, for example, a peak in the distribution near 4 to 5 Å for uranium atoms on a carbon surface.[37] Clustering can be studied in some cases as well.

It would be interesting to extend such studies to other light atom substrates, such as the metals beryllium and aluminum, and to investigate step effects. Heavier-atom surfaces can also be analyzed in the form of thin films of monoatomic thickness on a lighter substrate, as has recently been done.[37d]

Field-Ion Microscopy (FIM)

In field-ion microscopy,[38] a hemispherical sample tip of radius $\sim 10^{-4}$ cm is imaged by allowing a gas (usually helium, but also hydrogen, neon, and others) to ionize at the surface of the tip under the influence of a strong applied electric field. The surface is positively charged in order to repel the ions formed. As the free atom approaches the surface, field ionization occurs at a critical distance, X_c, at the surface (about 4 to 8 Å) defined by

$$X_c \approx \frac{V_{ion} - \phi}{eE} \tag{2}$$

where V_{ion} is the ionization potential, ϕ is the work function, and eE is the applied electric field. The positive ions are repelled and accelerated onto a fluorescent screen, where a greatly magnified image of the tip surface is displayed. The schematic representation of the experiment is shown in Figure 2.10. The ionization probability depends strongly on the local field variations induced by the atomic structure of the surface. Protruding atoms generate appreciably stronger ionization than atoms embedded in close-packed atomic planes and so produce individual bright spots on the screen. The imaging by the ions from the sample tip to the screen occurs with very little motion tangential to the tip surface, especially at low temperatures ($T \sim 21$ K is often used for that reason), allowing a resolution of 2 to 3 Å. The use of small-radius tips is needed to produce the large field required for ionization, but also permits the immense magnification of this microscope. The tip surface is directly imaged with magnification of about 10^7.

Only a limited class of materials withstand the strong electric field at the sample tip without desorption or evaporation of the surface. Thus mainly refractory metals (W, Ta, Ir, Re, for example) are used. Adsorbates are equally affected by field desorption, greatly restricting the range of usable substrate–adsorbate systems that can be studied by FIM.

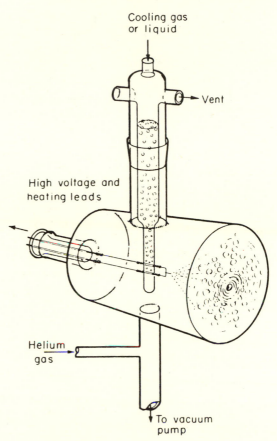

Figure 2.10. Scheme of the field-ion-microscopy technique.

Field desorption or field evaporation, however, is also used to obtain information about surface composition. The field-vaporized atoms and molecules are detected by a mass spectrometer. The combination of field-ion microscopy and field desorption permits the site-by-site determination of the surface composition. Although the properties of surfaces under high electric-field conditions may differ from those of the field-free state, FIM has been very helpful, nevertheless, in understanding many properties of metal surfaces.

These properties include the structure of the clean metal and of adsorbates, metal-on-metal surfaces, including defect structures, thermal disordering, atom–atom interactions, two-dimensional cluster formation and growth, and atomic surface diffusion (since real-time observation of the change of location of individual atoms is possible). As an example, we

59

may cite the study by FIM of the self-diffusion and correlated motions of Rh atoms on Rh substrates.[38f] The self-diffusion shows strong anisotropies on crystal faces that are channeled, such as fcc(110), with a large mobility along the troughs and a low mobility across the channels.

Concerning the analysis of the detailed geometrical surface structure, FIM unfortunately does not provide the depth information required to investigate the coordination of surface atoms to underlying atoms (layer registries cannot be determined) or to measure bond lengths.

Low-, Medium-, and High-Energy Ion Scattering (LEIS, MEIS, and HEIS)

Ion scattering at low energies (≤ 2 keV), medium energies (~ 50 to 500 keV), and high energies ($\gtrsim 500$ keV) has been used to study surface structures.[39] In LEIS[39a] high surface specificity is obtained because the very large cross sections for ion scattering ensure scattering off the outermost atomic layer only. Mutual shadowing of surface atoms is exploited to investigate their atomic positions. However, the physics of the scattering process at these energies is not well understood, leading to uncertainties in position determinations of the order of 0.5 Å. This does allow gross (but important) observations, such as whether adsorbed atoms lie tucked away between substrate atoms [for example, in the channels of fcc(110) surfaces] or instead are situated in more exposed positions [for example, on the ridges of fcc(110) surfaces]. At high energies,[39c] attained by the use of ion accelerators, the cross sections become very small, allowing deep penetration into the surface. But surface sensitivity is maintained by using channeling (penetration along open channels in the bulk crystalline structure) and looking for the blocking of this channeling by surface atoms whose positions deviate from the bulk positions. This amounts to a kind of "triangulation," in which the directions of the lines connecting pairs of surface atoms are identified by looking for shadowing of one atom in each pair by another atom; these directions are then sufficient information to determine the relative positions of surface atoms. At high energies the well-understood Rutherford scattering is the dominant mechanism, simplifying the interpretation. At medium energies ion scattering gives additional information about surface composition, since backscattered ions lose an amount of energy that depends on the mass of the surface atom that is struck by the incident ion.[39b] A general problem with ion scattering is that the thermal vibrations of the surface atoms complicate the interpretation. Sometimes computer simulations of the scattering are made to sort out the structural from the thermal effects. In principle, accuracies of better than 0.1 Å in the determination of atomic positions are possible.

As an example of such work, the clean Ni(110) surface has been studied with MEIS,[39d] confirming the contraction of the topmost interlayer spacing observed by LEED. Adsorption of half a monolayer of sulfur was found to cancel that contraction and actually expand the topmost interlayer spacing beyond its bulk value. The position of the adsorbed sulfur in the deepest hollows of the substrate surface could also be determined, the result being in agreement with that of a previous LEED study.

Atomic Scattering and Diffraction

The de Broglie wavelength associated with helium atoms is given by

$$\lambda(\text{Å}) = \frac{h}{(2ME)^{1/2}} = \frac{0.14}{E(\text{eV})^{1/2}} \tag{3}$$

Thus atoms with thermal energy of about 0.02 eV have $\lambda = 1$ Å and can readily diffract from surfaces. A beam of atoms is chopped with a variable-frequency chopper before striking the surface. This way, an alternating intensity beam signal is generated at the mass spectrometer detector that is readily separated from the "noise" due to helium atoms in the background.

Three processes observed during the scattering of atomic beams[40] of helium are displayed in Figure 2.11. There is specular reflection of the helium atoms from the surface (the helium atoms scattering at an angle that is equal to the angle of incidence) (curve a). There is rainbow scattering (curve b) that results in the appearance of multiple peaks. This type of scattering may be viewed as the classical limit of diffraction; it is due to scattering of atoms by the varying surface potential. The third type of process (curve c) is diffraction, and it appears to be detectable from suitable surfaces for values of $d > \lambda > 0.15d$, where d is the interatomic distance in the surface, and when using fairly monochromatic incident atomic beams.

All three of these processes have been observed under various conditions of the beam scattering experiment. A typical diffraction pattern of helium from a LiF(100) surface[41] is shown in Figure 2.12. The first-order beams, although broad, are clearly discernible. Diffraction from LiF has also been observed using neon, hydrogen, and deuterium.[42] Diffraction of helium from other surfaces—tungsten (112), silver (111),

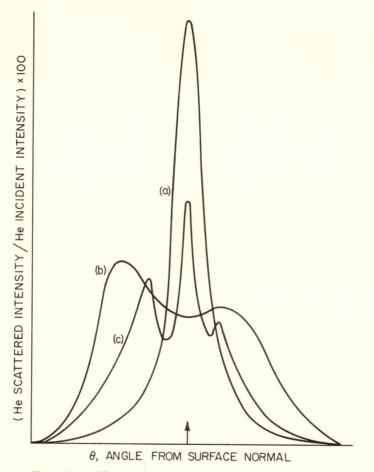

Figure 2.11. Three processes observed during the scattering of He atomic beams: (a) specular reflection; (b) rainbow scattering; (c) diffraction.

and tungsten carbide and silicon (111)—has also been detected. Another phenomenon, the presence of bound states for the incident helium atom, has also been detected during the helium–lithium–fluoride diffraction experiments. A fraction of the helium atoms appear to be trapped at the surface in weakly bound (2 to 12 cal) states but can readily translate along the surface before reemission without energy loss.[43] Such a phenomenon was predicted by Lennard-Jones and Devonshire in 1938.[44]

Rainbow scattering has been detected from high-Miller-index stepped

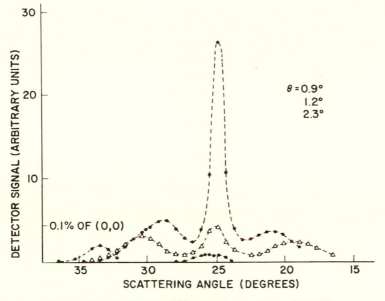

Figure 2.12. Typical He diffraction pattern from a LiF(100) surface.

platinum surfaces.[45] Typical rainbow scattering patterns are shown in Figure 2.13. The increase in intensity of the surface rainbows, as displayed by this figure, for an increase in the angle of incidence qualitatively follows the trend predicted from calculations by McClure for classical scattering for helium atoms. At grazing angles of incidence, the surface appears less spatially rough and thus allows a more intense elastic scattering contribution. At more normal angles of incidence, the elastic scattering distribution is less intense due to an increase in spatial roughening of the surface as seen by the helium atom. Rainbow scattering has been observed from a stepped platinum surface, whereas the smooth Pt(111) surface exhibits only specular scattering.

Helium, in general, gives strong specular scattering from ordered surfaces. As the surface temperature is increased, the specular beam intensity drops. This effect is similar to that observed for low-energy electrons and for X-rays and is due to surface atom vibrations that give rise to the Debye–Waller factor. However, the form of the Debye–Waller factor is different for atom beam diffraction as compared to the scattering of these other two surface probes. The slow, low-energy helium atoms encounter the attractive surface potential that has a well depth similar to the thermal energy of approaching atoms, and the tempera-

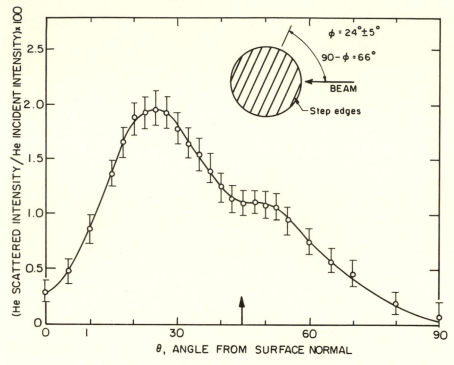

Figure 2.13. Typical rainbow scattering pattern from a stepped Pt crystal surface.

ture dependence of the scattering is influenced by the well depth of this potential. Beeby[46] has derived a formula for the temperature dependence of the helium beam intensities:

$$I = I_0 \exp\left(-\frac{25 M_g T D}{M_s \Theta_s^2}\right) \tag{4}$$

Here D is the depth of the atomic potential as sensed by the incident atom; M_g and M_s are the masses of the scattering atoms and surface atoms, respectively; and Θ_s is the surface Debye temperature. For the diffracting X-rays and electrons that have much higher energy as compared to helium atoms, the surface potential can well be neglected as having an unimportant effect on scattering. It appears, therefore, that the temperature dependence of atom scattering can yield information on the attractive potential that is operative during the solid–gas interaction.

Another important property of the specularly scattered fraction of atoms is their great sensitivity to surface disorder.[47] On scattering from a well-ordered surface, nearly 15 percent of the scattered helium atoms appear in the specular helium beam. This fraction decreases to 1 to 5 percent when the surface is disordered. Thus measurements of the fraction of specularly scattered helium can provide information on the degree of atomic disorder in the solid surface.

Surface-Sensitive Extended X-ray Absorption Fine Structure (SEXAFS)

In SEXAFS, incident X-rays of variable energy eject, for example, low-energy core Auger electrons[48] from substrate or adsorbate atoms, which by their small mean free path guarantee surface sensitivity on the atomic scale. It is possible to focus attention on adsorbates, for example, by considering only adsorbate Auger lines; it is sufficient that the adsorbate be different from substrate atoms and that suitable Auger lines are present. The yield of the ejected electrons is modulated as a function of incident energy due to interference between outgoing electrons and electrons backscattered from the neighboring atoms. By Fourier analysis of this modulation, the interatomic distances can be extracted with an accuracy that may exceed that of LEED. It is also possible to find the number of neighbors at the individual distances, thereby fixing the adsorption site through the coordination number.[48b]

One of the basic physical inputs to the SEXAFS analysis is a set of phase shifts for electron scattering off surface atoms. The uncertainty in these is one of the limiting factors in the accuracy of the method. However, in some cases comparison with experimental SEXAFS data from bulk material can help in circumventing the phase-shift uncertainty; essentially, the bulk and surface phase shifts are assumed equal, and these then divide out in the ratio of the surface to bulk data.

A drawback of SEXAFS is that synchrotron radiation is needed as a source of X-rays.

X-ray and Ultraviolet Photoelectron Spectroscopies of Surfaces (XPS and UPS)

We may divide this field arbitrarily into two classes—one studies the properties of valence electrons that are in the outermost shell, and the other investigates the properties of electrons in the inside shells of atoms. The principle of photoelectron spectroscopy is the excitation of electrons in an atom or molecule by means of X-rays into vacuum (XPS). This is shown schematically in Figure 2.14. The ejected "photoelectrons" have a kinetic energy E_{kin}, equal to

$$E_{kin} = h\nu - E_B \tag{5}$$

where $h\nu$ is the energy of the incident X-rays and E_B the binding energy of the ejected electron. These electrons are analyzed with high-resolution energy analyzers of various types (such as hemispherical, cylindrical mirror analyzers).[49] The X-ray source consists of an anode of a suitable material which is bombarded by energetic electrons that are emitted from the cathode. The emitted X-ray radiation can be made monochromatic either by diffraction or by taking advantage of the characteristic emission lines of the anode material, usually Al or Mg. The energies of these lines are 1253.6 eV for Mg with a full width at a half maximum (FWHM) of 0.7 eV, and 1486.6 eV for Al with a FWHM of 0.85 eV.

Another frequently used photon source is synchrotron radiation. When high-energy electrons are accelerated to speeds of 1 to 6 GeV,

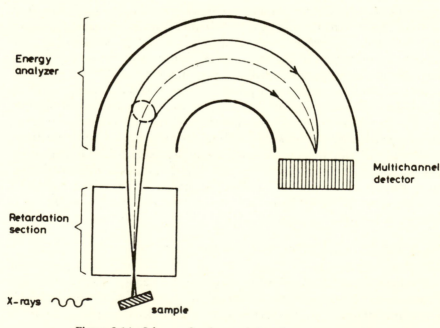

Figure 2.14. Scheme of a photoemission electron spectrometer.

electromagnetic radiation is emitted in the energy range 10 to 10^4 eV. Continuous radiation in this energy range, which has intensities more than two orders of magnitude higher than a conventional X-ray tube, provides a powerful probe of the electronic structure of atoms and molecules.

Low-energy radiation for studies of the valence-electron structure is also provided by the so-called vacuum-UV light source (UPS).[50] Its photon emission spectra is the result of deexcitation of excited atoms or ions. The emission lines are intense and have an FWHM of a few millivolts. The most commonly used low-energy photon sources are He II (40.8 eV), He I (21.22eV), Ne I (16.85 eV), Ne I (16.67 eV), Ar I (11.83 eV), and H (10.20 eV). The photon emission is produced by excitation of the bound atomic electrons by electric discharge in a continually flowing gas stream.

Eq. (5) gives a highly simplified relationship between the kinetic energy, E_{kin}, of the emitted photoelectrons and their binding energy. E_{kin} is modified by the work function of the energy analyzer and by several atomic parameters that are associated with the electron emission process. The ejection of one electron leaves behind an excited molecular ion. The electrons in the outermost and in other orbitals experience a change in the effective nuclear charge due to an alteration of screening by other electrons. This gives rise to satellite peaks near the main photoelectron peaks. Several other effects, including spin–orbit splitting, Jahn–Teller effect, and resonant absorption of the incident photon by the atom, influence the detected photoelectron spectra.

One of the most important applications of XPS is the determination of the oxidation state of elements at the surface. The electronic binding energies for inner-shell electrons shift as a result of changes in the chemical environment.[51] An example of these shifts can be seen in nitrogen, indicating the photoelectron energy for various chemical environments (Figure 2.15). These energy shifts are closely related to charge transfer in the outer electronic level. The charge redistribution of valence electrons induces changes in the binding energy of the core electrons, so that information on the valence state of the element is readily obtainable. A loss of negative charge (oxidation) is in general accompanied by an increase in the binding energy, E_B, of the core electrons.

The surface sensitivity of photoelectron spectroscopy is increased by collecting the emitted electrons that emerge at small angles to the surface plane. These are ejected mostly from atoms at the surface, as photoelectrons from deeper lying atomic planes would be strongly attenuated.[52]

Angularly resolved photoelectron spectroscopy[53] (ARUPS) not only

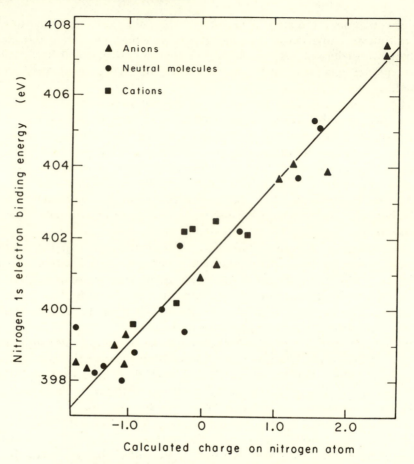

Figure 2.15. 1s electronic binding energy shifts in nitrogen, indicating the different photoelectron energies observed in various chemical environments.

provides enhanced surface sensitivity but also reveals directional effects due to the spatial distribution of electronic orbitals of atoms and molecules at the surface. By changing the angle of incidence and the angle of detection, the electronic orbitals from which the photoelectrons are ejected can be identified. ARUPS provides detailed information about the surface chemical bond that includes the direction of the bonding orbitals and the orientation of the molecular orbitals of adsorbed species on the surface.

UPS and XPS have been extensively used in surface analysis and have yielded much qualitative information about surface geometry.[54] Thus one can distinguish with these techniques between atoms that are in the adsorbed state, the same atoms that have penetrated the surface to form compounds, and those that have remained in intermediate stages (often interpreted as incorporation within the topmost surface layer), since such differences appear as shifts in initial-state levels. It is also possible to investigate the adsorption orientation and the structural modifications of molecules deposited on surfaces. Hence CO adsorbed on metals has been extensively studied with UPS. It is easy to distinguish between molecularly adsorbed and dissociatively adsorbed CO: the characteristic molecular 4σ, 1π, and 5σ levels are either present or absent, respectively, in the UPS spectrum (even if the 5σ and 1π levels often coincide in energy). In the case of molecular adsorption, it is usually clear that the 5σ orbital, which is located more toward the carbon end of the molecule, undergoes a larger energy shift upon adsorption than the 4σ orbital, which is located toward the oxygen end of the molecule. This strongly suggests that the molecule is bonded by its C end to the surface, with the O end sticking out away from the surface.

Although little theory is needed in the foregoing kind of analysis with UPS and XPS (unless a detailed understanding of relaxation energies and the like is sought[55]), the situation is quite different in ARUPS. It is now well established that final-state multiple-scattering effects are important there; a treatment of these processes along the lines of LEED theory (requiring well-ordered surfaces) is needed and is starting to produce encouraging results.[56] Also needed are an adequate treatment of the initial-state and the initial-to-final-state matrix elements and maybe of the refraction of the incident photons at the surface. It appears that ARUPS is sensitive to bonding symmetry more than to bond lengths[56e] and that an analysis of initial states (such as bonding orbitals and surface states) is indeed possible. ARUPS has been applied successfully to clean metal surfaces, where, for example, the d-band emission could be reasonably well reproduced. It has also been applied to atomic overlayers [such as S and Se on Ni(100)], confirming the binding site determined previously by LEED, as well as to molecular adsorption of CO on Ni(100): the CO is confirmed to adsorb by its C end to a single nickel atom, as predicted by LEED.

The computational effort is larger in ARUPS than in LEED, since more physical processes are involved, so that for the limited purpose of surface crystallography LEED seems more appropriate. Furthermore, ARUPS is best carried out using synchrotron radiation, which limits its availability.

Ion-Neutralization Spectroscopy (INS)

In INS,[57] slow, positively ionized noble-gas atoms (typically He^+ ions) are allowed to neutralize at a surface by attracting surface electrons. The energy liberated is transferred to other surface electrons (not belonging to the incoming ion), which can leave the surface and be detected. The probability of this two-electron process involves the self-convolution of the surface density of occupied states, the density of final states and matrix elements for electron tunneling to the ion and for ejection of the detected electrons (a final-state problem also encountered in LEED, SPI, and photoemission). The procedure is to extract by deconvolution from the measured emission probabilities the surface density of occupied states for energies between the Fermi level and about 10 eV below that, and to predict the relative atomic positions from information about the electronic structure of the surface; for example, adsorbate-induced peaks will occur that depend on the adsorbate and its position, as in UPS. This technique is primarily sensitive to the outermost atoms at the surface, in particular adsorbates, since the emitted electrons originate from those regions only. The difficulties in deconvoluting and interpreting the density-of-states information have limited the use of INS.

The technique has, however, been applied to the adsorption of O, S, and Se on Ni(100), Ni(110), and Ni(111). In the case of O on Ni(100), a substrate reconstruction was inferred, with penetration of the adsorbate into the topmost substrate layer, in disagreement with LEED results. A small distortion of the substrate was concluded for S on Ni(100), but the same adsorption site was found as with LEED.

Surface Penning Ionization (SPI)

A metastable helium-atom incident (with thermal kinetic energy) onto a surface allows a surface electron to tunnel to the unoccupied low helium energy level, enabling the excited helium electron to be emitted and detected.[58] Alternately, the helium deexcitation provides energy for the emission of a surface electron. This occurs with energy and angular distributions characteristic of the electronic structure and therefore geometry of the surface, in a way similar to photoemission. Whereas in photoemission the excitation occurs via a dipolar photon–electron interaction, in SPI it is an $|r_1 - r_2|^{-1}$ interaction that takes place between the deexcited and the emitted electrons, so that the selection rules are different. A further difference appears in the surface sensitivity. The photoemission excitation is sensitive to the entire region covered by the initial state of the electron to be excited, while the SPI excitation occurs where the empty incident-atom wave function starts to overlap with the surface wave functions, at the outer edge of the outermost atoms (final-state multiple scattering of the emitted electron is, however, similar in SPI and photoemission). SPI has similarities with INS, but is intrinsically easier to interpret, mainly because of the absence of a deconvolution of the surface density of occupied states; furthermore, SPI is even more surface sensitive than INS.

SPI is currently being applied to clean metal surfaces and to CO adsorbed on such surfaces.

Techniques Sensitive to Electron Distribution at Surfaces: Work-Function Measurements

The work function is the potential that an electron at the Fermi level must overcome to reach the level of zero kinetic energy in the vacuum. It is due to the interaction of an electron with the other charges of the surface through electrostatic, exchange, and correlation effects.[59] Any change of charge distribution at the surface will in general change the work function; indeed, charge redistributions perpendicular to the surface can produce sizable work-function changes (up to 2 to 3 eV in some cases), as is common when an adsorbed layer is deposited on a surface. Therefore, work-function measurements have become a sensitive technique for monitoring the charge state of the surface, usually as a function of coverage.

Experimentally, the work function itself can be measured with, among other methods, photoemission, since the work function appears as a clearly distinguishable threshold energy there. Changes in work function are often measured by the Kelvin method, which uses a vibrating probe as a variable capacitor.

Unfortunately, the work function is a rather complicated (and not fully understood) function of the surface composition and geometry. The work-function change is usually attributed to the formation of a dipole layer on the surface, such as occurs when charge flows from a substrate to an adsorbate, or vice versa. If ρ is the dipolar charge density (number of dipoles per unit area), d the dipole length (perpendicular to the surface), and e the electronic charge, one can write

$$\Delta\phi = 4\pi e\rho d \qquad (6)$$

This relation enables one to estimate the charge transfer upon adsorption from measured values of $\Delta\phi$ and values of d determined, for example, by LEED. However, the picture of dipoles consisting of two point charges a distance d apart is a drastic simplification of the actual continuous charge distribution at a surface. This use of work-function measurements is applied in the case of atomic adsorbates to determine the direction and magnitude of charge transfer between the substrate surface and the adsorbates.

The work-function change is not normally proportional to the concentration of adsorbates, (A), except at very low coverages ($\theta < 0.1$). The main reason is that dipoles mutually depolarize each other, the more so the higher their concentration. This effect is included in the following relation,[59c] based on considering the polarizability of the adsorbates,

$$\Delta\phi = \frac{4\pi e\rho_0\Theta}{1 + 9\alpha\Theta^{2/3}} \tag{7}$$

where ρ_0 is the dipolar charge density at $\theta = 1$, and α is proportional to the polarizability. This relation only rarely fits the experiment, however, because the charge distribution at surfaces is governed by more complicated processes than polarizability, such as charge transfer between substrate and adsorbate. Thus there are examples of sign reversal of the work-function change as coverage varies, while there is no detectable change in adsorption geometry (see Chapter 5).

Nevertheless, general systematic observations of $\Delta\phi$ are quite helpful. For example, the sign of the work-function change for atomic adsorption is mostly that implied by the magnitude of the ionization potential, electron affinity, or dipole moment of the adsorbates, as one would expect; furthermore, in atomic adsorption it appears that large atoms (such as alkali atoms) produce large work-function changes, mainly because of the size effect. Many organic molecules adsorbed on metal surfaces produce a decrease in work function, indicating the transfer of electrons from the molecules to the substrate; also, π-bonding is often implied for such molecules.

The most common usage of work-function changes is in the monitoring of the various stages of adsorption as a function of coverage. Often the work-function change will go through a minimum or maximum at particular coverages corresponding to the completion of an ordered atomic arrangement. The onset of adsorption in new adsorption sites may also be detected in this way.

Methods Sensitive to Chemical Composition at Surfaces

Auger Electron Spectroscopy (AES)

Auger electron spectroscopy (AES) is suitable for studying the composition of solid and liquid surfaces.[60] The sensitivity is about 1 percent of a monolayer, and it may be used with relative ease as compared with several other techniques of electron spectroscopy. At present, this is the most widespread technique for studies of surface composition. The Auger electron emission occurs in the following manner. When an energetic beam of electrons or X-rays (1000 to 5000 eV) strikes the atoms of a material, electrons, which have binding energies less than the incident beam energy, may be ejected from the inner atomic level. By this process a singly ionized, excited atom is created. The electron vacancy thus formed is filled by deexcitation of electrons from other electron energy states. The energy released in the resulting electronic transition can, by electrostatic interaction, be transferred to still another electron in the same atom or in a different atom. If this electron has a binding energy that is less than the energy transferred to it from the deexcitation of the previous process that involves the filling of the deep-lying electron vacancy, it will then be ejected into vacuum, leaving behind a doubly ionized atom. The electron that is ejected as a result of the deexcitation process is called an Auger electron, and its energy is primarily a function of the energy-level separations in the atom. These processes are schematically displayed in Figure 2.16.

There are essentially two types of experimental designs for AES differing in the detection system that is used. One is the retarding grid Auger analyzer, which is of the high-pass filter type. It has a rather low signal-to-noise ratio. The main advantage is that the electron optics of a low-energy electron diffraction system can also be used for obtaining the Auger spectrum. Thus LEED and Auger measurements can be performed using the same apparatus. The other popular design is called the

AUGER ELECTRON EMISSION

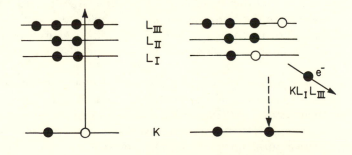

(a) EXCITATION   (b) ELECTRON EMISSION

Figure 2.16. Scheme of the Auger electron emission process.

cylindrical mirror analyzer, which is of the window filter type, which has a higher signal-to-noise ratio. One advantage of this type of detector is that it allows one to perform the Auger analysis in a short time (10^{-2} sec) compared to the 1 min or more required for the retarding grid analyzer. Both of these detectors with Auger electron spectroscopy systems are commercially available.

Most Auger spectroscopy studies of surfaces were carried out for surface chemical analysis. With the exception of hydrogen and helium, all other elements are detectable by Auger electron spectroscopy. Using this technique, surface segregation of various impurities has been found on a large number of systems, including stainless steel. It has been found that carbon, sulfur, and calcium are the most common impurities that segregate at the surface, and their removal is essential in many cases. In studies of alloy surfaces it has been found that the composition of the alloy at the surface is very different from the bulk composition. Typical Auger spectra from alloy surfaces[61] are shown in Figure 2.17. While the raw experimental data yield the electron intensity as a function of its energy (I versus eV), it is usually displayed as the second derivative of intensity, d^2I/dV^2, as a function of electron energy, eV. This way the Auger peaks are readily separated from the background, due to other electron loss processes that take place simultaneously. Many of these applications have opened up the field of surface science to definitive studies that have an impact on many areas, ranging from corrosion to heterogeneous catalysis. At present, almost every fundamental surface study includes Auger or photoelectron spectroscopy analysis of the surface composition as an integral part of the investigation.

By suitable analysis of the experimental data, as well as by the use of suitable reference surfaces, the Auger electron spectroscopy study can provide quantitative chemical analysis in addition to elemental compositional analysis of the surface. It is possible to separate the surface composition from the composition of layers below the surface by appropriate analysis of the Auger spectral intensities. In this way the surface composition as well as the composition in the near-surface region can be obtained.

The Auger peak intensity ratios can be calculated by properly taking into account the attenuation of the emitted Auger electrons by the atomic layers above. By comparing the calculated intensities with those obtained by experiments, quantitative chemical analysis can be performed. Since both the adsorbate and the substrate Auger peak intensities vary as the surface coverage changes, there is enough experimental information in most cases to detect when a monolayer coverage is reached and thus calibrate the amount adsorbed.

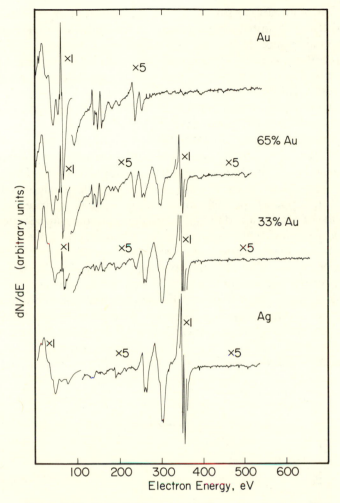

Figure 2.17. Typical Auger spectra from pure Au, two alloys, and pure Ag.

When chemical analysis is desired in the near-surface region, AES may be combined with ion sputtering to obtain a depth-profile analysis of the composition. Using high-energy ions, the surface is sputtered away layer by layer while, simultaneously, AES analysis detects the composition in depth. Sputtering rates of 100 Å/min are usually possible and the depth resolution of the composition is about 10 Å, which is mainly determined by the statistical nature of the sputtering process. In our example we show the Auger depth profile of Si thin films (~200 Å) deposited in vacuum onto a graphite substrate.[62] As is apparent from Figure 2.18, carbon diffused deeply into the silicon film and also silicon diffused into the carbon substrate during the deposition.

A different aspect of AES concerns shifts in the observed peak energies that are due to chemical shifts of atomic core levels (in a way analogous to XPS). In particular, studies of different oxidation states of oxygen at metal surfaces have shown chemical shifts that grow with the formation of higher oxidation states.[63]

An important concern in the application of electron scattering to probe the structure, composition, and oxidation states of surface atoms is the possibility of damage or chemical changes introduced by the incident electrons or photons. It appears that the energy density (energy/cm^3) that is deposited per unit time determines, to a large extent, the probability of "radiation damage." Incident photons appear to be less damaging than incident electrons of the same flux and energy because of their much lower scattering cross sections and, therefore, greater depth penetration. The radiation damage probability also depends on the ability of the excited surface atom or ion to transfer its excess energy to neighbor atoms rapidly before desorption or other chemical-bond-breaking processes are to occur. In fact, electron-stimulated desorption of adsorbed atoms and molecules is commonly observed and studied. Adsorbed monolayers on metal surfaces seem to undergo rapid deexcitation via the substrate and are less susceptible to radiation damage than

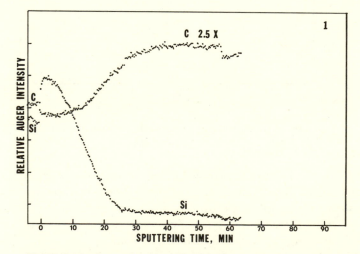

Figure 2.18. Auger-depth profiles for silicon films deposited on pyrolytic graphite.

monolayers on insulator surfaces. The surfaces of ionic crystals are sensitive to electron-beam-induced decomposition. Organic adsorbates are resistant to electron bombardment if they possess conjugated π electron systems (aromatic molecules, phthalocyanines), which apparently aids their deexcitation. Saturated organic adsorbates, the paraffins, for example, are readily desorbed by the incident electron beams.

It is necessary to carry out electron scattering experiments using beams of as low intensity as possible in order to minimize the possibility of radiation damage.

Thermal Desorption Spectroscopy (TDS)

Thermal desorption (flash desorption)[64] is the most widely used technique at present for studying the kinetics of desorption and for determining the heats of adsorption and the concentration of adsorbed molecules on crystal surfaces. When combined with mass spectrometry, the composition of the desorbed molecules is also determined. The sample is cleaned in ultrahigh vacuum and a gas is allowed to adsorb on the surface at known pressures while the surface is kept at a fixed temperature. Then the sample is heated at a controlled rate, and the pressure changes during the desorption of the molecules are recorded as a function of time and temperature. The pressure–temperature profile is usually referred to as the "desorption spectrum." The experiment is performed either in a closed system or in a flow system.

The desorption rate $F(t)$ of the species from a surface of area A (cm^2) is given by

$$\frac{A}{kV} F(t) = \frac{p - p_0}{\tau_0} + \frac{dp}{dt} \tag{8}$$

where p is the pressure rise from the steady-state pressure p_0 of the system prior to flashing, τ_0 is the mean residence time of gaseous species in the reaction vessel (τ_0 is defined as volume/pumping speed), V is the volume, and k is a constant.

In a flow system at high pumping speeds, $\tau_0 \to 0$ and Eq. (8) is reduced to

$$\frac{A}{kV} F(t) \approx \frac{p - p_0}{\tau_0} = \frac{\Delta p}{\tau_0} \tag{9}$$

This means that the rate is proportional to the pressure change in the system. Alternatively, for a nearly static system at low pumping speeds, τ_0

$\rightarrow \infty$. In this case, the duration of the flashing period is short compared to τ_0. Then the desorption rate is given by

$$F(t) = \frac{dp}{dt} \tag{10}$$

The desorption rate $F(t)$ is commonly expressed as

$$F(t) = \nu f(\sigma) \exp\left(-\frac{E_{des}}{RT}\right) \tag{11}$$

where ν is the preexponential factor, $f(\sigma)$ is an adsorbate concentration-dependent function. The various procedures for determining these parameters are well described in the literature.

Analysis using the temperature where the desorption peak is obtained is given by Redhead.[65] Assuming that ν and E_{des} are independent of the adsorbate concentration σ and t, E_{des} can be obtained for zero-, first-, and second-order desorption, respectively, as

$$E_0/R = \frac{\nu_0}{\sigma\alpha} \exp\left(-\frac{E}{RT_p}\right) \tag{12}$$

$$E_1/RT_p^2 = (\nu_1/\alpha) \exp\left(-\frac{E_1}{RT_p}\right) \tag{13}$$

$$E_2/RT_p^2 = (\nu_2\sigma/\alpha) \exp\left(-\frac{E_2}{RT_p}\right) \tag{14}$$

where T_p is the temperature at which a desorption peak is at a maximum and σ is the initial adsorbate concentration. Suffix 0, 1, or 2 denotes the zeroth-, first-, or second-order desorption processes. α is a constant of proportionality for the temperature rise with time, usually $T = T_0 + \alpha t$, that is, the temperature of the sample is raised linearly with time. As seen from the equations, T_p is independent of σ for the first-order desorption process. Alternatively, T_p is increased or is decreased with σ_0 for the zeroth- or second-order process, respectively. Eqs. (12), (13), and (14) allow one to determine the activation energy and the preexponential factor and also to distinguish between zeroth-, first-, and second-order desorption processes from the measurements of the dependence of the peak temperatures upon initial adsorbate concentrations and heating rate, α. Instead of a continuous temperature rise, a stepwise heating has also been used, especially for studies of temperature-programmed desorption. The advantage of this method is the more accurate evaluation of the rate constant for desorption at constant temperature and better resolution of multiple desorption peaks.

Recent determinations of the heats of chemisorption on single-crystal surfaces have utilized thermal desorption most frequently. The activation energy of desorption, E_{des}, for the desorption of the adsorbed species is equal to $E_a + \Delta H_{ads}$, where E_a is the activation energy of adsorption of the same species. Therefore, E_{des} cannot be equated with the heat of adsorption ΔH_{ads} as long as adsorption is an activated process. However, the adsorption of several gases such as O_2, CO, CO_2, H_2, and N_2 on transition metal surfaces is believed not to require activation energy. In this circumstance E_{des} is a good approximation for the heat of adsorption. A typical thermal desorption spectrum is shown in Figure 2.19.

Ellipsometry

Ellipsometry is a useful technique for measuring adsorption isotherms on a single-crystal surface.[66] The amplitude and phase of the polarized light reflected from the surface change upon adsorption of a gas even at below monolayer concentrations. These changes appear to be a linear function of the *optical thickness,* which is closely related to the average thickness of the adsorbed layer. By measuring the optical thickness as a function of gas pressure over a single-crystal surface, the adsorption isotherm can be obtained at different temperatures. Thus the heat of adsorption can be calculated. Ellipsometry can be readily combined in an up-to-date ultrahigh-vacuum system with other adsorption studies on well-defined surfaces.

The optical properties of an isotropic and absorbing medium (metal) are defined by two characteristic constants—refractive index (n) and adsorption index (k). These constants can be determined experimentally by measuring the change in polarization of reflected light.[67] This is done by resolving the electric vector of the incident polarized light into two components, E_p and E_s (parallel and perpendicular to the plane of incidence; see Figure 2.20); the state of polarization can then be conveniently defined by the phase difference (Δ) between the parallel and perpendicular components upon reflection:

$$\Delta = \delta_p - \delta_s \tag{15}$$

and the amplitude ratio

$$\tan \psi = \frac{E_s''/E_p''}{E_s/E_p} \qquad (E'' \text{ for reflected light}) \tag{16}$$

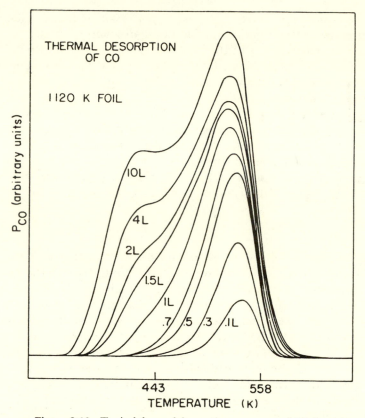

Figure 2.19. Typical thermal desorption spectrum of CO from a rhodium foil at different CO exposures (1 L = 1 Langmuir = 10^{-6} torr-sec).

The electromagnetic theory of light predicts that the two scattered components are retarded in phase and reduced in amplitude to different extents upon reflection from a metal. As Δ and ψ can be measured very accurately (with an error of not more than 0.02°) by means of an ellipsometer, the constants n and k of the pure metal surface can be determined.

In the presence of a surface layer in which the optical constants change from those of the surrounding medium to those of the bulk metal, the reflection formulas are modified and can be derived from Maxwell's equations. In the case of a thin, uniform, nonabsorbing, and isotropic film on an absorbing substrate, the linear relationships

$$\Delta = \bar{\Delta} + \alpha d \qquad (17)$$

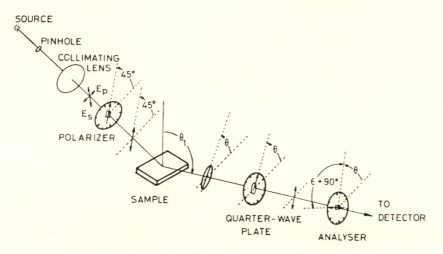

Figure 2.20. Scheme of the ellipsometry technique. Arrangement for measuring ϕ and Δ with linearly polarized light incident at 45°.

$$\psi = \bar{\psi} + \beta d \tag{18}$$

(where $\bar{\Delta}$ and $\bar{\psi}$ are the phase shift and amplitude for the film-free metal) between Δ and ψ for the film-covered metal surface and the film thickness d can be derived. The constants α and β in Eqs. (17) and (18) are functions of the complex refractive index $n_{cs} = n(1 + ik)$ of the substrate (here the pure solid surface), the refractive index of the film, the angle of incidence, the wavelength of the incident light, and the refractive index of the incident medium.

A computer program[68] determines the values of Δ and ψ for different film thicknesses d by using the complex refractive index of the substrate, the refractive index of the film, the refractive index of the incident medium, the angle of incidence, and the wavelength of the incident light. These values of Δ and ψ are then compared with the measured values of Δ and ψ, and the corresponding film thickness d is calculated.

Secondary-Ion Mass Spectroscopy (SIMS)

SIMS analyzes the mass and angular distribution of clusters of atoms ejected during ion bombardment of a surface, using incident ions with kinetic energies of the order of 1 keV; the ejected clusters are often ionized. Knowledge of the cluster masses allows the chemical composition of the surface to be investigated. For example, a Ni(100) surface saturated with oxygen and bombarded with 2000-eV Ar^+ ions produces mainly clusters of Ni^+, Ni_2^+, Ni_3^+, $O^\pm$, $O_2^\pm$, $NiO^\pm$, Ni_2O^+, NiO_2^-, and $Ni_2O_3^-$. The depth resolution of SIMS is of the order of a few atomic layers.

SIMS is very sensitive to the chemical state of the surface atoms and to the compounds present at the surface.[69] In addition, hydrogen and its compounds can be detected and isotopes can be discriminated between. When combined with a sensitivity of 10^{-4} to 10^{-6} monolayer and low primary current densities (on the order of 10^{-9} A/cm^2), corresponding to sputtering rates of one monolayer in several hours, it is clear that this technique is orders of magnitude more sensitive and less destructive than Auger electron spectroscopy. However, a quantitative surface compositional analysis requires comparison of the experimental yield of the individual clusters with corresponding yields obtained theoretically. This may be done by numerical simulation of the complex collision process,[28b] but the accuracy of the result cannot yet be ascertained. The accuracy of the compositional analysis depends to some extent on such poorly known factors as the interatomic potential, ionization cross sections, and quantum-mechanical corrections to a treatment based on classical trajectories.

Once the theory of the SIMS process is properly understood, this technique should be capable of analyzing some aspects of the detailed surface geometry, such as the registries of adsorbates on a substrate and whether molecules adsorb associatively or dissociatively.[70]

A typical SIMS spectrum is displayed in Figure 2.21. It should be noted that both positive and negative ions are detectable and convey useful information.

Ion-Scattering Spectroscopy (ISS)

In ISS, ions such as H^+, He^+, and Ar^+ are scattered off a surface and their energy distribution is observed.[71] During the scattering process the ions lose energy to the surface atoms. The collision process is usually so rapid (with kinetic energies of the order of 1 keV to 1 MeV) that a binary collision model is a good description of the situation. It is then easy to relate the energy loss ΔE to the mass M of the surface atoms involved:

$$\frac{E - \Delta E}{E} = \frac{1}{(1 + M/m)^2} \left\{ \cos \Theta \pm \left[\left(\frac{M}{m} \right)^2 - \sin^2 \Theta \right]^{1/2} \right\}^2$$

where E is the incident energy, m the ion mass, the θ the laboratory scattering angle. Once the masses M of the surface atoms are known, the chemical composition follows. A scheme of the experiment and a typical ion scattering spectrum are shown in Figure 2.22.

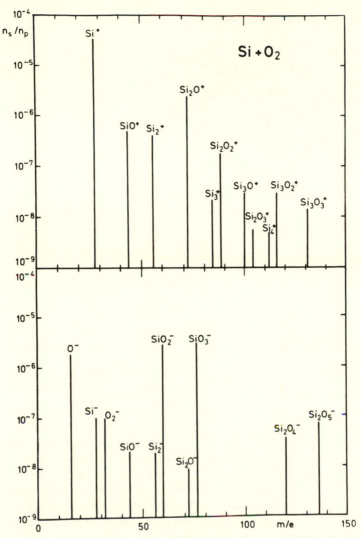

Secondary ion spectrum of a surface structure composed of two atomic species: Si(111) + 20 L O_2, $Si_mO_n^{\pm}$ emission

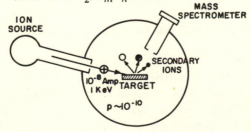

Figure 2.21. Scheme of the secondary-ion mass spectroscopy (SIMS) technique and typical SIMS positive- and negative-ion spectra.

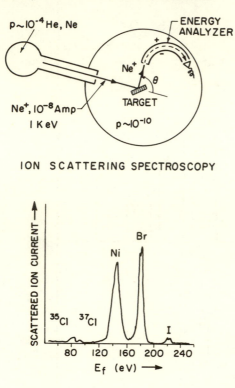

ION SCATTERING SPECTROSCOPY

Figure 2.22. Ion-scattering spectroscopy (ISS) technique and typical ISS spectrum.

Depending on energy and incidence direction (including channeling and blocking effects), the depth resolution of ISS can vary from a monolayer to about 300 Å. Quantitative evaluations of the chemical composition of surfaces may be achieved in special cases, but are generally hampered by uncertainties in the ion–atom scattering potential (especially at the lower energies), the possibility of multiple scatterings, and the ever-present question of the depth distribution of individual species.

Methods Sensitive to Vibrational Structure of Surfaces

Infrared Spectroscopies (IR) and Raman Spectroscopy

Absorption of infrared radiation by characteristic vibrations of atoms and molecules on the surface can be used to study the structure and bonding of these species. In the early stages of development of this technique it was used primarily in the transmission mode.[72a] Surface sensitivity was obtained by using small particles of high surface area distributed in porous and infrared transparent supports such as silica.[72b] Such a geometry is typical for catalysts, and this is one of the areas of surface science where IR found frequent application.[72c] By studies of the vibrational frequencies of adsorbates, the catalyst structure, adsorbate structure, and bonding could be investigated.[72d] Reaction cells were built that permitted monitoring the IR spectra during the chemical reactions, and surface reaction intermediates could be detected. One great advantage of infrared spectroscopy is that it may be carried out while the surface is subjected to high pressures or in the presence of liquids; electron spectroscopies cannot be used under such circumstances. However, IR studies of well-characterized, small-area single-crystal surfaces are difficult because of the very small cross section for interaction (about 10^6 times smaller than for vibrational excitation by electrons) and because the accessible energy losses are somewhat restricted so that only a limited set of vibrations can be detected.

To overcome these difficulties several new techniques of IR spectroscopy are being developed. Reflection absorption infrared spectroscopy (RAIR) proved to be a viable and exciting technique for studies of adsorbates on single crystal surfaces.[72e-h] Alternatively one can measure the changes in phase and amplitude of back-reflected, plane-polarized radiation. Infrared spectroscopy based on this effect is called infrared ellipsometry.[72i] Photoacoustic reflection-absorption spectroscopy is another technique that is being rapidly developed. When the sample is irradiated periodically with an infrared laser pulse it experiences temperature oscillations as a result of absorption of radiation.[72j] This pulsed heat wave causes pressure fluctuations in the gas above the surface that are detected by microphone or a piezoelectric transducer.[72k] Reflection Raman spectroscopy also takes advantage of the availability of high intensity and tunable lasers usually operating in the visible region. A high-resolution monochromator is used to separate the components of the scattered light that are detected by a photon counting system that is connected to a computer for data analysis and signal averaging.[72l] A major problem is background fluorescence that makes the detection of the weak scattering signals difficult.[72m] Raman spectra that were about 10^4 to 10^6 more intense than anticipated were observed for pyridine adsorbed on silver electrodes and from other adsorbed heterocyclic molecules.[72n] It appears that surface roughness plays an important role in generating what is called the enhanced or resonance Raman scattering effect.[72o]

The intense research activity in the field of optical spectroscopy from surfaces will yield many interesting and new techniques for surface studies in the near future.

High-Resolution Electron Energy Loss Spectroscopy (HREELS)

Electrons scattering off surfaces can lose energy in various ways. One of these involves excitation of the vibrational modes of atoms and molecules on the surface. The technique to detect vibrational excitation from surfaces by incident electrons is called high-resolution electron energy loss spectroscopy (HREELS).[73] It requires a beam of electrons with an energy of incidence of about 1 to 10 eV and with an energy spread of at most about ±10 meV (corresponding to about ±80 wave numbers). This energy spread is about an order of magnitude smaller than the energy spread of the electrons used presently for low-energy electron diffraction or detected in electron spectroscopies. This highly monochromatic electron beam, upon incidence on the surface, excites the vibrational mode of the various chemical bonds (M—H, M—C, M—O, C—C, C—H, where M is a substrate atom). These modes have frequencies in the range 100 to 300 meV (800 to 2400 wave numbers) and thus are readily detectable with the high-energy resolution of this instrument. The electrons are back-reflected from the surface with energies equal to

$$E_{\text{reflected}} = E_{\text{incident}} - E_{\text{vibration}}$$

and are detected by a suitable energy analyzer. The high-resolution electron energy loss spectrum from hydrogen and deuterium adsorbed on a tungsten (110) crystal face[74] is shown in Figure 2.23. Not only is hydrogen readily detectable at coverages much less than a monolayer, but isotope shifts caused by the different masses of H and D are also observed. The peaks are narrow and of high intensity and yield information on the location and structural symmetry of the sites where the surface atoms or molecules are located.

Adsorbed species with chemical bonds that are perpendicular to the surface plane are more readily detectable than are chemical bonds that are parallel to the surface. The sensitivity of this technique appears to be 0.1 to 1 percent of a monolayer. Thus the structure of the different adsorption sites that are filled up successfully with increasing coverage of the adsorbate can all be detected. A wealth of surface structural information is becoming available from the high-resolution electron loss spectra of adsorbed hydrocarbons and other molecules. A unique feature of this technique is that it is able to detect adsorbed hydrogen on the surface because of the high-frequency vibrational modes of this atom

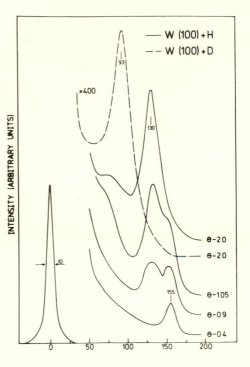

Figure 2.23. High-resolution electron energy loss spectra for H and D adsorbed atomically on W(100). The elastic peak is shown at left. The loss energy for hydrogen is plotted along the horizontal axis. The coverage varies from $\phi = 0.4$ to $\phi = 2.0$ (saturation), exhibiting a change in adsorption site, while the deuterium spectrum is shown at $\phi = 2.0$ only. (After H. Froitzheim, H. Ibach, and S. Lehwald, Phys. Rev. Lett. **36**, 1549 [1976].)

when it is bound to the surface or to other adsorbed atoms (C—H, O—H). Hydrogen cannot be readily detected by techniques such as LEED or other electron spectroscopies because of its small elastic and inelastic cross sections for scattering. This unique feature makes high-resolution electron energy loss spectroscopy an important tool for studying the surface chemistry of hydrogen, hydrocarbons, and other hydrogen-containing molecules.

At present, high-resolution electron energy loss spectroscopy and LEED appear to be the most powerful combination for surface-structure determinations. Although the theory of HREELS is not yet well developed, the vibrational spectra give an indication of the orientation of adsorbed molecules and the site symmetry of adsorbates. This, in turn, narrows the choices of trial geometries that have to be considered during surface-structure analysis using the dynamical theory of LEED.

Electron Tunneling Spectroscopy

This technique takes advantage of the electron tunneling phenomenon that occurs when two metal electrodes are separated by a very thin insulating oxide layer (one to several angstroms thick).[75] If a voltage is applied between the metals, electrons will tunnel through the insulator from the filled electronic states of one metal to the empty states of the other metal. The current will increase linearly with applied voltage.

If an adsorbed monolayer or multilayer of atoms or molecules is present at the oxide metal interface, the tunneling electrons lose energy to the vibrational excitation of the adsorbed species. By measuring the energy distribution of the tunneling electrons, the vibrational spectra of the adsorbate species is obtained. Inelastic tunneling spectroscopy yields the vibrational structure of the adsorbate, which is generally displayed by plotting the second derivative of the applied voltage with respect to the current, d^2V/dI^2, versus the applied voltage, V. A typical tunneling spectrum is shown in Figure 2.24. Here an Al–oxide–Pb junction was employed and the adsorbate was benzoic acid.[76] The metals are generally vaporized, the aluminum first because of the relative ease of prepa-

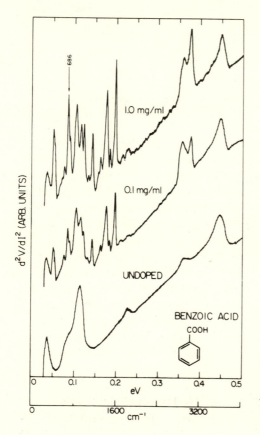

Figure 2.24. Electron tunneling spectrum from Al–oxide–Pb junctions with benzoic acid adsorbate. Doping solution strengths are indicated. The undoped junction data are included for comparison. (After Langan and Hansma, ref. 76.)

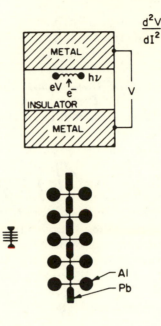

Figure 2.25. Idealized and actual views of tunneling spectroscopy samples.

ration of a thin oxide insulating layer on top of it. Then the adsorbate is introduced, and, finally, the second metal is vaporized on top of it. A typical sandwich geometry is shown in Figure 2.25.

The experiment must be carried out at 4.2 K or below to obtain appreciable tunnel current. The spectrum exhibits high sensitivity, as the electrons couple more strongly to molecular vibrations than do photons that are used in infrared spectroscopy. Another difficulty in addition to the need for low temperatures is the control of the oxide metal interface. Once the adsorbate is deposited, it is covered by a metal deposit that cannot be exchanged or removed. Nevertheless, this technique provides information about the solid–solid interface that is very difficult to obtain by any other means.

Modeling of Surfaces

Model Calculations

Cutting across the domains of the various techniques we have mentioned are the model calculations.[77] These are theoretical attempts to predict the structure of surfaces from first principles. The model calculations differ from the theories mentioned in conjunction with the ex-

89

perimental techniques listed above in that the former are not primarily designed to describe the interaction of a probe with a surface, although, obviously, much overlap exists. Thus the calculation of electronic states at surfaces seeks to describe from first principles a situation (the structure of the surface) that is analyzed experimentally by any of the techniques mentioned above, using external probes. But some of these techniques also involve the motion of electrons through the surface region. This motion, in turn, is clearly related to the electronic structure of the surface, and so the first-principles calculation and the surface-analysis technique may have and often do have much in common.

Model calculations are used in surface science by comparing their predictions with observed values of electronic-level energies, including densities of states (obtained mainly from UPS, XPS, and INS); of atomic and molecular binding energies (mainly from TDS); surface-structure determination (mostly by LEED); vibration frequencies (from IR and HREELS); work-function changes; and so on. Model calculations can be subdivided into those that take a semi-infinite or filmlike model of the surface and those that represent the surface by a cluster containing a finite number of atoms. The former are based on the methods of solid-state physics, while the latter originate in molecular physics. The surface, especially when atomic adsorbates are present, is an intermediate situation between the crystal interior (with its three-dimensional periodicities) and molecules (with their limited dimensions). The loss of some translational symmetry is troublesome for solid-state theories, while the essentially infinite size of surfaces is troublesome for molecular physics. In a first stage, therefore, model calculations were applied to idealized simple surfaces that could not be directly compared to real surfaces. However, progress has led to a situation in which more and more cases of agreement with experimental results from the various techniques of surface science are reported. These will be discussed as individual cases in Chapters 5 and 6.

Clusters

Recently, it has been recognized that similarities exist between, on the one hand, adsorbate-covered metal surfaces and, on the other hand, clusters consisting of a few (2 to about 12) metal atoms surrounded by various ligands, especially small attached molecules.[78] This opens up a new avenue for surface studies, since knowledge gained about clusters may possibly be extrapolated to surfaces. Clusters are the nearest approximation to a crystal surface that is presently available experimentally for comparison. Such clusters can be analyzed structurally by X-ray crystallography of clusters regularly arranged in single crystals. To what extent the analogy between clusters and surfaces is sufficiently close to make extrapolations from the former to the latter (or vice versa) is not clear yet because of the lack of directly comparable experimental data.

One promising class of structures concerns CO adsorption on metals and its metal carbonyl cluster counterparts. In such clusters CO is often found to attach itself molecularly to the metal frame, with the C end bonded to either one, two, or three metal atoms (terminal, edge, and face bonding, respectively). There are corresponding clear differences in bond lengths (not only in the metal–C bond but also in the CO bond) and in binding energies between the different bonding configurations. There are also clusters containing hydrocarbon ligands, which are useful analogies in the study of hydrocarbon adsorption. For example, the adsorption of ethylene (C_2H_4) and acetylene (C_2H_2) may be studied by comparison with the clusters $Co_3(CCH_3)(CO)_9$, $Ru_3(CCH_3)H_3(CO)_9$, Br_3CCH_3, $(Ph_3P)_2Pt(C_2Ph_2)$, $Co_2(C_2Ph_2)(CO)_6$, $Rh_2(h^5-C_5H_5)_2(CO)_2$ $(CF_3C_2CF_3)$, and $Os_3(C_2Ph_2)(CO)_{10}$.

Techniques for Studies of the Dynamics of Surface Processes

Several techniques have been developed recently to investigate the kinetics and mechanisms of surface reactions and energy transfer that take place at the solid–gas interface. One powerful method involves the scattering of atom or molecular beams from surfaces, or molecular beam surface scattering.[13,40] The scheme of the molecular beam experiment is shown in Figure 2.26. The well-collimated molecular beam of the reactant gas or gas mixture is scattered from the crystal surface, and the products that are desorbed at a given solid angle are detected by mass spectrometry. By rotation of the mass spectrometer around the sample, the angular distribution of the scattered products can be determined. If the incident molecular beam is chopped at well-defined frequencies, the time of flight of the incident molecules between the chopper and the detector is determined by phase-shift measurements. This information yields the residence time of the molecules on the surface. Chopping the product molecules that are desorbed from the surface also permits determination of their velocity. The experimental variables of the system are temperature, atomic structure and composition of the surface, and velocity and angle of incidence of the molecular beam. In reactive scattering experiments the mass spectrometer detects the product distribu-

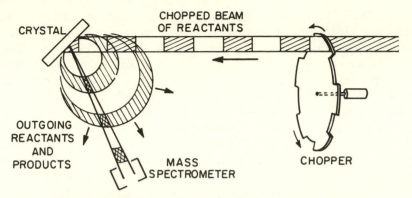

Figure 2.26. Scheme of the molecular beam–surface scattering experiment.

tion and rates of formation of product molecules (reaction probabilities) as a function of the system variables upon a single scattering. From the dependence of the reaction rate on the incident velocity or "beam temperature," the activation energy for adsorption is determined. From the surface-temperature dependence of the rate, the activation energy of the surface reaction is obtained. The surface residence time of the molecules, the kinetic energy, internal energy and the angular distribution of the products reveal the nature of energy transfer during the gas–surface interaction.

Reactive scattering studies may be divided into two groups:[79] (1) those in which the reactions take place on the surface that acts as a catalyst, for example $A_{2(gas)} + B_{2(gas)} \rightarrow 2AB_{(gas)}$; and (2) another group, which includes surface reactions where the surface atom is one of the reactants, for example $A_2 + S = SA_{2(gas)}$. Only a few of the multitudes of possible reactions of these two types have been investigated by molecular beam surface scattering. Several review papers are available that the reader can use to learn more about the technique and experimental results.

Catalytic reactions, however, are usually studied at high pressures where multiple collisions with the surface are possible and where the thermodynamics and kinetics of the surface reactions may be very different from those at low pressures that are appropriate conditions for molecular beam scattering. Higher-pressure studies involve a steady-state or time-dependent measurement of the reaction rates and product distribution, using catalyst surfaces with well-characterized atomic structure and composition. These are necessarily of small surface area, of the order of ~ 1 cm^2, usually single crystals or polycrystalline foils. Recently, new instruments have been developed in our laboratory[9] to permit *in situ* studies of the reactivity of small-area crystal surfaces at both low and high pressures. The scheme of one such apparatus is shown in Figure 2.27. At low pressures, in the range 10^{-7} to 10^{-4} torr, the reaction rate and product distribution are monitored by a quadrupole mass spectrometer, while the surface structure and composition are determined by low-energy electron diffraction and Auger electron spectroscopy, respectively. Then a small cup (total volume approximately 30 cm³) can be placed around the crystal sample to isolate it from the rest of the chamber and can be pressurized to over 100 atm if desired, using a mixture of gaseous reactants. The high-pressure reaction chamber is connected to a gas chromatograph, which serves to monitor both rate and product distributions in this circumstance. The structure and composition of the surface can be determined *in situ* by LEED and AES before and after the high-pressure experiment once the cup is removed. The crystal sample can be heated during both low- and high-pressure experiments.

In the chemical technology, small catalytic particles, 20 to 150 Å in size, which are dispersed on large-surface-area supports (usually oxides or oxihydrides of aluminum or silicon), are commonly utilized. The chemical reactions are carried out this way to achieve large contact areas and thus optimum reaction rates that are proportional to the surface area. The reactant pressures are in the range 1 to 100 atm. Studies of crystal surfaces at low pressures or at high pressures can be related to the behavior of conventional catalytic systems as follows.

First, the reaction is studied at low pressure on well-characterized small-area samples to establish correlations between reactivity and surface structure and surface composition. Then the same chemical reaction is studied at high pressure (1 to 100 atm) over small-area samples, and the pressure dependence of the reaction rate is determined over a range of nine orders of magnitude. Then the rates and product distributions that were determined at high pressures on single-crystal surfaces are compared with the reactivities of highly dispersed small-particle catalytic systems studied under similar experimental conditions. Our experiments indicate that small-surface (approximately 1 cm²) single-crystal catalysts can readily be used in studies as long as the reaction rates are greater than 10^{-6} product molecule/surface atom/sec. In the field of catalysis, the rate so defined is commonly called the "turnover number." Most important catalytic reactions (hydrogenation, dehydrogenation, oxidation, isomerization, dehydrocyclization, hydrogenolysis) usually have rates greater than the detection limit, even at low pressure. It

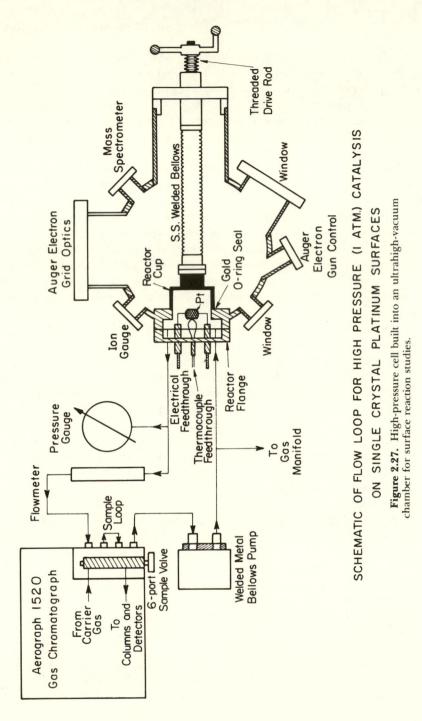

SCHEMATIC OF FLOW LOOP FOR HIGH PRESSURE (1 ATM.) CATALYSIS

ON SINGLE CRYSTAL PLATINUM SURFACES

Figure 2.27. High-pressure cell built into an ultrahigh-vacuum chamber for surface reaction studies.

should be noted that the high-pressure apparatus used to study small-surface-area single-crystal surfaces can also be used to study reactions at solid–liquid interfaces. In this circumstance an electrode may be placed in the high-pressure chamber together with the crystal surface that serves as the other electrode. Then liquids can be introduced into the chamber and an electrochemical reaction at the solid–liquid interface carried out. After the reaction the liquid is pumped out and the surface may be analyzed in high vacuum with the various surface diagnostic techniques. In this way the adsorption and reactions that take place at solid–gas and solid–liquid interfaces may be compared and correlated.

Another technique that provides rate constants and other kinetic parameters for elementary surface reactions involves thermal desorption studies while the chemical surface reactions are taking place.[80] The reactants are adsorbed on the surface, which is then heated at a controlled rate. The surface reaction takes place in a certain temperature regime, and the products desorb at well-defined temperatures (thermal desorption peak). The coverage and temperature dependence of the thermal desorption spectrum yield the rate parameters. In many experimental circumstances, isotope labeling and exchange during the surface reactions facilitate identification of the possible reaction intermediates.

References

1. J. Gland, Ph.D. thesis, University of California, Berkeley, 1976.
2. G.A. Somorjai and L.E. Firment, J. Chem. Phys. **66,** 2001 (1977).
3. G.A. Somorjai and L.E. Firment, J. Chem, Phys. **69,** 3940 (1972).
4. M. Kaminsky, ed. *Radiation Effects on Solid Surfaces,* American Chemical Society, Washington, D.C., 1976, pp. 1–30.
5. E.W. Müller and T.T. Tsong, *Field Ion Microscopy,* American Elsevier, New York, 1969.
6. J.H. Sinfeld, G.H. Via, and F.W. Lytle, J. Chem. Phys. **68**(4), 2009 (1978).
7. A.E. Barrington, *High Vacuum Engineering,* Prentice-Hall, Englewood Cliffs, N.J., 1963.
8. R.P. Redhead, J.P. Hobson, and E.V. Kornelsen, *The Physical Basis of Ultrahigh Vacuum,* Chapman & Hall, London, 1968.
9. D.W. Blakely, E. Kozak, B.A. Sexton, and G.A. Somorjai, J. Vac. Sci. Technol. **13,** 1091, (1976).
10. A.T. Hubbard, R.M. Ishikawa, and J. Katekari, J. Electroanal. Chem. **26,** 271 (1978).
11. H.H. Farrell and G.A. Somorjai, J. Chem. Phys. **20,** 215 (1971).
12. S.T. Ceyer and G.A. Somorjai, Annu. Rev. Phys. Chem. **28** (1977).
13. S.T. Ceyer, R.J. Gale, S.L. Bernasek, and G.A. Somorjai, J. Phys. Chem. **64,** 1934 (1976).
14. H.D. Hagstrum and G.E. Becker, J. Vac. Sci. Technol. **14,** 369 (1977).

15. G.A. Somorjai, in *Ceramic Microstructures 1976,* eds. R.M. Fulrath and J.A. Pask, Westview Press, Boulder, Colo., 1977.
16. L.L. Kesmodel and G.A. Somorjai, Am. Crystallogr. Assoc. **13,** 67 (1977).
17. H. Ibach, Phys. Rev. Lett. **24,** 1416 (1970).
18. S. Andersson, Solid State Commun. **21,** 75 (1977).
19. S.H. Overbury and G.A. Somorjai, Disc. Meet. Faraday Soc. **60,** 279 (1971).
20. (a) L.L. Kesmodel and G.A. Somorjai, Acc. Chem. Res. **9,** 392 (1976); (b) G.A. Somorjai, *Principles of Surface Chemistry,* Prentice-Hall, Englewood Cliffs, N.J., 1972; (c) G. Ertl and J. Küppers, *Low Energy Electrons and Surface Chemistry,* Verlag Chemie, Weinheim, 1979.
21. (a) M. Henzler, Surf. Sci. **19,** 159 (1970); (b) G.E. Laramore, J.E. Houston, and R.L. Park, J. Vac. Sci. Technol. **10,** 196 (1973).
22. (a) J.E. Houston and R.L. Park, Surf. Sci. **21,** 209 (1970); (b) G. Ertl and J. Küppers, Surf. Sci. **21,** 61 (1970); (c) H. Jagodzinski, D. Wolf, and W. Moritz, Surf. Sci. **77,** 223, 249, 265, 293 (1978).
23. G. Ertl and D. Schillinger, J. Chem. Phys. **69,** 479 (1978).
24. G.-C. Wang, T.-M. Lu, and M.G. Lagally, J. Chem. Phys. **69,** 479 (1978).
25. R.J. Behm, K. Christmann, and G. Ertl, Solid State Commun., **25,** 763 (1978).
26. T.J. Kaminska, L.L. Kesmodel, P.C. Stair, and G.A. Somorjai, Phys. Rev. **B11,** 623 (1975).
27. F. Jona, Surf. Sci. **68,** 204 (1977).
28. (a) J.B. Pendry, *Low Energy Electron Diffraction,* Academic Press, London, 1974; (b) M.A. Van Hove and S.Y. Tong, *Surface Crystallography by Low-Energy Electron Diffraction: Theory, Computation and Structural Results,* Springer-Verlag, Heidelberg, 1979.
29. (a) M. Lagally, T.C. Ngoc, and M.B. Webb, Phys. Rev. Lett. **26,** 1557 (1971); (b) C.W. Tucker and C.B. Duke, Surf. Sci. **29,** 237 (1972).
30. (a) D.L. Adams and U. Landman, Phys. Rev. Lett. **33,** 585 (1974); (b) P.I. Cohen, J. Unguris, and M.B. Webb, Surf. Sci. **58,** 429 (1976); (c) D.L. Adams and U. Landman, Phys. Rev. **B15,** 3775 (1977); (d) S.L. Cunningham, C.-M. Chan, and W.H. Weinberg, Phys. Rev. **B18,** 1537 (1978).
31. S.Y. Tong, M.A. Van Hove, and B.J. Mrstik, Proc. 7th Int. Vac. Congr. and 3rd Int. Conf. Solid Surf. Vienna, 2407 (1977).
32. (a) R.S. Zimmer and B.W. Holland, J. Phys. **C8,** 2395 (1975); (b) S.Y. Tong and M.A. Van Hove, Phys. Rev. **B16,** 1459 (1977).
33. R. Feder, Phys. Status Solidi **B62,** 135 (1974).
34. R.M. Goodman and G.A. Somorjai, J. Chem. Phys. **52,** 6325 (1970).
35. H.H. Farrell and G.A. Somorjai, Adv. Chem. Phys. **20,** 215 (1971).
36. (a) J.F. Menadue, Acta Crystallogr. **A28,** 1 (1972); (b) N. Masud and J.B. Pendry, J. Phys. **C9,** 1833 (1976); (c) N. Masud, C.G. Kinniburgh, and J. B. Pendry. J. Phys. **C10,** 1 (1977).
37. (a) A.V. Crewe, J. Wall, and J. Langmore, Science **168,** 133 (1970); (b) J.M. Cowley, *Diffraction Physics,* North-Holland, Amsterdam, 1975; (c) M.S. Isaacson, J. Langmore, N.W. Parker, D. Kopf, and M. Utlaut, Ultramicroscopy **1,** 359 (1976); (d) K. Yagi, K. Takayanagi, K. Kobayashi, N. Osakabe, Y. Tanishiro, and G. Honjo, *Electron Microscopy, 1978,* vol. 1 (Proc. 9th Int. Congr. on Electron Microsc., Toronto, 1978), ed. J.M. Sturgess, Microscopical Society of Canada, Toronto, 1978.

38. (a) E.W. Müller, Z. Phys. **136,** 131 (1951); (b) E.W. Müller, J. Appl. Phys. **27,** 474 (1956); (c) E.W. Müller, in *Advances in Electronics and Electron Physics,* vol. 13, ed. L. Marton, Academic Press, New York, 1960; (d) E.W. Müller and T.T. Tsong, *Field Ion Microscopy,* American Elsevier, New York, 1969; (e) G. Erlich, Surf. Sci. **63,** 422 (1977); (f) G. Ayrault and G. Ehrlich, J. Chem. Phys. **60,** 281 (1974).

39. (a) W. Heiland and E. Taglauer, Surf. Sci. **68,** 96 (1977); (b) F.W. Saris and J.F. Van der Veen, Proc. 7th Int. Vac. Congr. and 3rd Int. Conf. Solid Surf., Vienna, 2503 (1977); (c) L.C. Feldman, R.L. Kauffman, P.J. Silverman, R.A. Zuhr, and J.H. Barrett, Phys. Rev. Lett. **39,** 38 (1977); (d) J.F. Van der Veen, R.M. Tromp, R.G. Smeenk, and F.W. Saris, Surface Sci. **82,** 468 (1979).

40. S.T. Ceyer, R.J. Gale, S.L. Bernasek, and G.A. Somorjai, J. Chem. Phys. **64,** 1934 (1976).

41. B.R. Williams, J. Chem. Phys. **55,** 3220 (1971).

42. S.T. Ceyer and G.A. Somorjai, Annu. Rev. Phys. Chem. **28** (1977).

43. J.A. Meyers and D.P. Frankl, Surf. Sci. **51,** 61 (1975).

44. J.E. Lennard-Jones and K. Devonshire, Proc. Roy. Soc. (Lond.) **A158,** 253 (1938).

45. R.I. Masel, R.P. Merrill, and W.H. Miller, J. Vac. Sci. Technol. **13,** 355 (1976).

46. J.L. Beeby, J. Phys. **C7,** 1 (1975).

47. S.L. Bernasek and G.A. Somorjai, J. Vac. Sci. Technol. **12**(2), 655 (1975).

48. (a) P.A. Lee, Phys. Rev. **B13,** 5261 (1976); (b) P.H. Citrin, P. Eisenberger, and R.C. Hewitt, J. Vac. Sci. Technol. **15,** 449 (1978).

49. E.W. Plummer, in *Photoemission and the Electronic Properties of Surfaces,* eds. B. Feuerbacher, B. Fitton, and R.F. Willis, Wiley, London, 1979.

50. C.S. Fadley, in *Electron Spectroscopy: Theory, Techniques and Applications,* vol. 2, eds. C.R. Brundle and A.D. Baker, London Academic Press, 1978.

51. J.T. Yates, Jr., T.E. Madey, and N.E. Erickson, Surf. Sci. **43,** 257 (1974).

52. P.W. Palmberg, in *Electron Spectroscopy,* ed. D.A. Shirley, North-Holland, Amsterdam, 1972.

53. C.H. Li and S.Y. Tong, Phys. Rev. Lett. **40,** 46 (1978).

54. (a) D.E. Eastman and J.K. Cashion, Phys, Rev. Lett. **27,** 1520 (1971); (b) A.M. Bradshaw, L.S. Cederbaum, and W. Domcke, *Struct. Bonding,* **24** (1975); (c) E.W. Plummer, in *Photoemission and the Electronic Properties of Surfaces,* eds. B. Feuerbacher, B. Fitton, and R.F. Willis, Wiley, London, in press.

55. J.W. Gadzuk, Phys. Rev. **B14,** 2267 (1976).

56. (a) J.W. Gadzuk, Phys. Rev. **B10,** 5030 (1974); (b) J.B. Pendry, Surf. Sci. **57,** 679 (1976); (c) A. Liebsch, Phys. Rev. Lett. **38,** 248 (1977); (d) K. Jacobi, M. Scheffler, K. Kambe, and F. Forstmann, Solid State Commun. **22,** 17 (1977); (e) C.H. Li and S.Y. Tong, Phys. Rev. Letters **40,** 46 (1978).

57. (a) H.D. Hagstrum, Phys. Rev. **150,** 495 (1966); (b) G.E. Becker and H.D. Hagstrum, J. Vac. Sci. Technol. **10,** 31 (1973); (c) H.D. Hagstrum and G.E. Becker, J. Vac. Sci. Technol. **14,** 369 (1977).

58. (a) P.D. Johnson and T.A. Delchar, to be published; (b) S.W. Wang, G. Ertl, J. Küppers, and H. Conrad, Phys. Rev. Lett. **42,** 1082 (1979).

59. (a) C. Herring and M.H. Nichols, Rev. Mod. Phys. **21,** 185 (1949); (b) N.D.

Lang and W. Kohn, Phys. Rev. **B3**, 1215 (1971); (c) J. Topping, Proc. Roy. Soc. (Lond.) **A114**, 67 (1927).

60. S.H. Overbury and G.A. Somorjai, Disc. Meet. Faraday Soc. **60**, 279 (1975).
61. A. Jablonski, S.H. Overbury, and G.A. Somorjai, Surf. Sci. **65**, 578 (1977).
62. C.-A. Chang and W.J. Siekhaus, J. Appl. Phys. **46**(8), 3402 (1975).
63. G.A. Somorjai and E.J. Szalkowski, J. Chem. Phys. **61**, 2064 (1974).
64. R. Gomer, ed., *Interactions on Metal Surfaces,* Springer-Verlag, New York, 1975, pp. 102–142.
65. P.A. Redhead, Vacuum **12**, 203 (1962).
66. J.M. Morabito, Jr., R.H. Muller, R.F. Steiner, and G.A. Somorjai, Surf. Sci. **16**, 234 (1969).
67. A.C. Hall, Surf. Sci. **16**, 1 (1969).
68. C.A. Fenstermaker and F.L. McCrackin, Surf. Sci. **16**, 85 (1969).
69. A. Benninghoven, Surf. Sci. **35**, 427 (1973).
70. B.J. Garrison, N. Winograd, and D.E. Harrison, Jr., Phys. Rev. **B18**, 600 (1978).
71. (a) W. Heiland and E. Taglauer, Surf. Sci. **68**, 96 (1977); (b) F.W. Saris and J.F. van der Veen, Proc. 7th Int. Vac. Congr. and 3rd Int. Conf. Solid Surf., Vienna, 2503 (1977).
72. (a) M.L. Hair, *Infrared Spectroscopy in Surface Chemistry,* Marcel Dekker, New York, 1967; (b) G. Blyholder, in *Experimental Methods in Catalytic Research,* ed. R.B. Anderson, Academic Press, New York, 1968; (c) E.L. Force and A.T. Bell, J. Catal. **38**, 440 (1975); (d) R.P. Eischens and W.A. Pliskin, Adv. Catalysis **10**, 1 (1958); (e) R.G. Greenler, R. R. Rahn, and J.P. Schwartz, J. Catal. **23**, 42 (1971); (f) H.A. Pearce and N. Sheppard, Surf. Sci. **59**, 205 (1976); (g) J. Pritchard and T. Catterick, in *Experimental Methods in Catalytic Research,* ed. R.B. Anderson, Academic Press, New York, 1968; (h) A.M. Bradshaw and F.M. Hoffman, Surf. Sci. **72**, 513 (1978); (i) R.W. Stobie, B. Rao, and E.J. Digham, Surf. Sci. **56**, 334 (1976); (j) W.R. Harshbarger and M.B. Robin, A. Chem. Res. **6**, 329 (1973); (k) A. Rosenwaig, Opt. Commun. **7**, 305 (1973); (l) W.N. Delgass, G.L. Haller, R. Kellerman, and J.H. Lunsford, *Spectroscopy in Heterogeneous Catalysis,* Academic Press, New York, 1979; (m) B.A. Morrow, J. Phys. Chem. **81**, 2663 (1977); (n) D.L. Jeanmarie and R.P. van Dyke, J. Electroanal. Chem. **82**, 1 (1977); (o) B.A. Morrow, L.S.M. Lee, I.A. Cody, J. Phys. Chem. **80**, 2761 (1976).
73. (a) F.M. Propst and T.C. Piper, J. Vac. Sci. Technol. **4**, 53 (1967); (b) H. Ibach, Phys. Rev. Lett. **24**, 1416 (1970); (c) S. Andersson, Solid State Commun. **21**, 75 (1977).
74. H. Ibach, Phys. Rev. Lett. **24**, 1416 (1970).
75. P.K. Hansma, Phys. Rep. **30c**(2) (April 1977).
76. J.D. Langan and P.K. Hansma, Surf. Sci. **52**, 211 (1975).
77. (a) Overviews of theoretical methods in surface model calculations are given in chapters by T.B. Grimley and R.P. Messmer in *The Nature of the Surface Chemical Bond,* eds. T.N. Rhodin and G. Ertl, North-Holland, Amsterdam, 1978; (b) T.L. Einstein and J.R. Schrieffer, Phys. Rev. **B7**, 3529 (1973); (c) T.B. Grimley and C. Pisani, J. Phys. **C7**, 2831 (1974); (d) N.D. Lang and A.R. Williams, Phys. Rev. Lett. **34**, 531 (1975); (e) O. Gunnarson, H. Hjelmberg, and B.I. Lundqvist, Phys. Rev. Lett. **37**, 292 (1976); (f) S.W. Wang and W.H. Weinberg, Surf. Sci. in press; (g) D.J.M. Fassaert and A. van

der Avoird, Surf. Sci. **55**, 291, 313 (1976); (h) S.G. Louie, K.M. Ho, J.R. Chelikowsky, and M.L. Cohen, Phys. Rev. Lett. **37**, 1289 (1976); (i) R.V. Koasowski, Phys. Rev. Lett. **37**, 219 (1976); (j) J.A. Appelbaum, G.A. Baraff, and D.R. Hamann, Phys. Rev. **B14**, 588 (1976); (k) W.A. Harrison, Surf. Sci. **55**, 1 (1976); (l) D.W. Bullett and M.L. Cohen, J. Phys. Chem. **10**, 2083, 2101 (1977); (m) R.P. Messmer, S.K. Knudsen, K. H. Johnson, J.B. Diamond, and C.Y. Yang, Phys. Rev. **B13**, 1396 (1976); (n) I.P. Batra and O. Robaux, Surf. Sci. **49**, 653 (1975); (o) R.F. Marshall, R.J. Blint, and A.B. Kunz, Phys. Rev. **B13**, 333 (1976); (p) S.P. Walch and W.A. Goddard III, J. Am. Chem. Soc. **98**, 7908 (1976); (q) A.B. Anderson, J. Chem. Phys. **66**, 2173 (1977).
78. (a) P. Chini G. Longoni, and V.G. Albano, Adv. Organomet, Chem. **14**, 285 (1976); (b) E.L. Muetterties, T.N. Rhodin, E. Band, C.F. Brucker, and W.R. Pretzer, Chem. Rev. **79**, 91 (1979).
79. G.A. Somorjai, Acc. Chem. Res. **9**, 248 (1976).
80. R.J. Madix and J.A. Schwarz, Surf. Sci. **24**, 264 (1974).

3. Composition of Surfaces: Thermodynamic Guidelines and Experimental Results

Role of Surface Forces in the Control of Surface Composition

The development of surface diagnostic techniques that can determine the composition of the topmost layer of atoms with sensitivity of 1 percent ($\sim 10^{13}$ atoms/cm^2) or less has led to many surprising findings that have been recently uncovered by the surface and material scientists. Electron and ion scattering studies of the monolayer composition (mainly by AES and SIMS) revealed that the surface is always covered with impurities, even after extensive cleaning treatments. These are commonly carbon, sulfur, oxygen, silicon, and aluminum, which seem to segregate to the surface from the bulk, which stores a seemingly infinite supply of these frequently unwanted materials on our sensitive scale of surface analysis. Even after the complete elimination of these segregated impurities by exhaustive cleaning treatments, the surface composition of multicomponent alloy systems was found to be very different from the bulk composition.[1] There appeared to be a surface region, often several atomic layers thick, within which the composition varied until at greater depths the bulk composition was reached. Closer scrutiny of selected systems also revealed the existence of surface compounds. Bimetallic alloy clusters have been identified which exhibit the phase behavior typical of very small metallic particles with near-unity dispersion—very different from the phase diagrams of the bulk systems of the same metals.[2]

Since both the chemical and the mechanical properties of the condensed phase critically depend on the surface composition, investigations of the physical-chemical parameters that control the surface composition are of great importance. Perhaps the most important driving force for the segregation of impurities and for the change of composition of alloys and other multicomponent systems is the need to minimize the surface free energy of the condensed phase system. Let us therefore review the thermodynamic considerations that lead to changes in surface composition.

Consider a metal B that is dissolved at low concentrations in another metal, A.[3] Metal B may have a tendency to segregate to the surface of A if it forms a strong surface bond for one reason or another. Figure 3.1 shows schematically the various relative energies of vaporization, dissolution, and adsorption of B on A that lead to surface segregation. At any bulk concentration some of the B atoms will always be at the surface as a result of the surface–bulk equilibrium. By forming a strong surface chemical bond, the surface concentration of B could be greatly enhanced. This occurs if the heat of desorption of B from the surface of A is larger than the heat of vaporization of B. Since the binding energy of B on the various surfaces of A metal may markedly change as a function of crystal orientation, the extent of surface segregation can be strongly dependent on the surface structure of A.

It is also possible that B atoms form a weaker chemical bond with A atoms at the surface as compared with the bulk. In this circumstance the B atoms will be repelled from the surface and their surface concentration will be less than the expected equilibrium concentration.

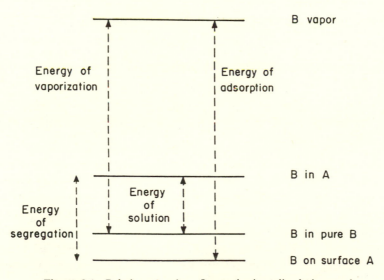

Figure 3.1. Relative energies of vaporization, dissolution, and adsorption of metal B on metal A.

The thermodynamic relationships that express the changes of surface free energies as a function of surface composition can be obtained by using the Gibbs adsorption equation,[4]

$$dγ = -S^s \, dT - \sum_i Γ_i \, dμ_i \tag{1}$$

where changes in surface tension $dγ$ are related to the specific surface entropy, S^s (entropy per mole of surface atoms); the temperature, T; the excess number of moles of compound i at the surface, $Γ_i$; and the change of chemical potential $dμ_i$ associated with placing the extra atoms of the ith component on the surface. For an isothermal system we have

$$dγ = -\sum_i Γ_i \, dμ_i \tag{2}$$

or

$$Γ_i = -\left(\frac{∂γ}{∂μ_i}\right)_{T,μ_j} \qquad j \neq i \tag{3}$$

For a dilute binary system exhibiting ideal solution behavior,

$$μ_2{}^b = μ_2{}^{o,b} + RT \ln X_2{}^b \tag{4}$$

where $μ_2{}^b$ and $μ_2{}^{o,b}$ are the chemical potentials of the second component in the bulk solution and pure component, respectively, and $X_2{}^b$ is the mole fraction in the bulk. Using Eq. (4), we can write

$$Γ_2 = -\frac{1}{RT}\left(\frac{∂γ}{∂\ln X_2{}^b}\right)_{T,X} \tag{5}$$

If we measure the variation of the bulk atom fraction of component 2 as a function of temperature at a fixed surface concentration, the segregation isostere is obtained,

$$\left(\frac{d\ln X_2{}^b}{dT}\right)_{Γ_2} = -\frac{ΔH_{segr}}{RT^2} \tag{6}$$

where $ΔH_{segr}$ is the heat of segregation of component 2. For a given bulk composition in equilibrium, any change of bulk composition with temperature is due to surface segregation. The variation of surface composition is readily measurable by the various surface-sensitive techniques. Thus it is useful to rewrite Eq. (6) as

$$\frac{X_2^s}{X_1^s} = \frac{X_2^b}{X_1^b} \exp \left(- \frac{\Delta H_{\text{segr}}}{RT} \right) \tag{7}$$

where X_2^s and X_1^s are the atom fractions of the two components at the surface, respectively.[5]

By plotting the surface atom fraction ratio as a function of X_2^b for a variety of bulk alloy compositions at a constant temperature, an isotherm is obtained, as shown in Figure 3.2. This is a Langmuir isotherm type (to be discussed in Chapter 5) called the McLean isotherm,[3] and it was first applied to describe grain-boundary segregation in metals. By plotting the bulk atom fraction ratio as a function of the reciprocal temperature for a constant surface composition, ΔH_{segr} is obtained as shown in Figure 3.3. If a positive slope is obtained, it indicates an exothermic heat of segregation whose magnitude is determined from such a plot. This simple model assumes that the heat of segregation remains unchanged with increasing coverage and that segregation stops when the monolayer surface coverage of B is reached. Clearly, Eq. (7) will have to be modified if these conditions do not hold.

The equilibrium surface concentration is established relatively slowly, since the migration of atoms from the bulk to the surface, or vice versa, is diffusion-limited. The time required to reach the equilibrium surface composition depends on the rate of bulk diffusion of the segregating constituents.[3] For segregation to a planar surface, the surface concentration as a function of time is given by the equation

$$\frac{X_2^s(t)}{X_2^s(t = \infty)} = 1 - \exp \left(\frac{Dt}{\alpha^2 d^2} \right) \operatorname{erfc} \left(\frac{Dt}{\alpha^2 d^2} \right)^{1/2} \tag{8}$$

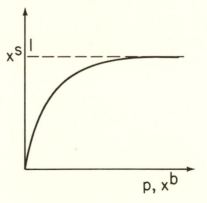

Figure 3.2. McLean isotherm: the surface excess as a function of bulk atom fraction.

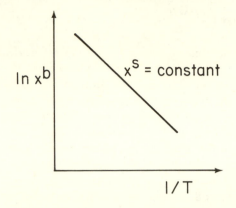

$\ln x^b$

$x^S = $ constant

$1/T$

Figure 3.3. Obtaining ΔH_{segr} by plotting the logarithm of the bulk atom fraction as a function of the reciprocal temperature for a constant surface composition.

where $X_2^s(t)$ and $X_2^s(t = \infty)$ are the surface atom fractions of the segregating component at time t and in equilibrium, respectively; α is the ratio of equilibrium surface and bulk atom fractions $[X_2^s(t = \infty)/X_2^b]$; D is the diffusivity of 2 in the bulk; and d is the thickness of the segregated surface region (assumed to be a monolayer).

The driving force for surface segregation is the difference in the binding energies between the two metal atoms, A–B, and the binding energies in the pure components, A–A and B–B. The same change in chemical bonding gives rise to a change in the surface tension of the binary system as compared to the surface tension for the pure constituents. Assuming ideal solution behavior, Eq. (4) can be rewritten[1] to express the equality of the chemical potentials in the bulk and at the surface,

$$\mu_2^b = \mu_2^{\circ,b} + RT \ln X_2^b = \mu_2^{\circ,s} + RT \ln X_2^s - \gamma a_2 \qquad (9)$$

where a_2 is the surface area covered by 1 mole of the second component of the binary system. For a pure, one-component system, $X_2^s = X_2^b = 1$, we have

$$\mu_2^{\circ,s} - \mu_2^{\circ,b} = \gamma_2 a_2 \qquad (10)$$

where γ_2 is the surface tension of the pure substance 2. Using Eq. (10) in rewriting Eq. (9) and assuming that $a_1 = a_2 = a$, then for a two-component system we have the equations

$$\gamma a = \gamma_1 a + RT \ln X_1^s - RT \ln X_1^b$$

and

$$\gamma a = \gamma_2 a + RT \ln X_2^s - RT \ln X_2^b$$

This can be rewritten[1] as

$$\frac{X_2^s}{X_1^s} = \frac{X_2^b}{X_1^b} \exp\left[\frac{(\gamma_1 - \gamma_2)a}{RT}\right] \tag{11}$$

which is the final result for the monolayer ideal solution model, where γ_2 and γ_1 are the surface tensions of the pure components 2 and 1. It is readily seen that the constituent with the lower surface tension will have a higher concentration at the surface.

The surface tension of solids and liquids can be measured by a variety of techniques or obtained by correlation with other thermodynamic properties. For metals, excellent correlations between γ and the heat of sublimation ΔH_{subl} can be established. Since sublimation as well as the creation of a unit area of surface are processes that give rise to the ΔH_{subl} and γ, respectively, a correlation between these two parameters is expected if the bond energies can be estimated by the addition of nearest-neighbor bonds. For example, in an fcc solid, each bulk atom has 12 nearest neighbors ($Z = 12$), whereas in the (111) surface (highest-atomic-density crystal plane) three of these are missing.[6] Thus 12 bonds are broken when an atom is moved from the solid into the vapor phase, whereas only 3 bonds are removed when the surface is created. Converting γ (energy/cm^2) to a molar quantity by multiplying it with a (area/mole), we have the relation $\gamma a = \gamma_m = \frac{3}{12}\Delta H_{subl} = 0.25\Delta H_{subl}$. The experimental data for metals actually lead to the result[7] $\gamma_m \approx 0.16\Delta H_{subl}$. By utilizing this experimental correlation, we can rewrite Eq. (11) to yield[8]

$$\frac{X_2^s}{X_1^s} = \frac{X_2^b}{X_1^b} \exp\left[\frac{0.16\,(\Delta H_{1\,subl} - \Delta H_{2\,subl})a}{RT}\right] \tag{12}$$

The metal constituent with the lower heat of sublimation will accumulate at the surface in excess. This relationship with small modifications has been used to predict surface enrichment at alloy surfaces. It should be noted that no simple γ–ΔH_{subl} correlation exists for oxide or organic solid surfaces as the one found for metal surfaces.

Equations (11) and (12) also predict that the surface composition of ideal solutions should be an exponential function of temperature. Thus while the bulk composition of multicomponent systems is not much affected by temperature, the surface concentration of the constituents may change markedly. According to these equations, the surface and bulk compositions should approach the same atom fraction ratios at high temperatures.

The surface segregation of one of the constituents becomes more pronounced the larger the difference in surface tensions between the components that make up the solution. Surface segregation is expected to be prevalent for metal solutions, since metals have the highest surface tensions. In addition, surface segregation can be readily detected for oxides as well as for organic solutions which are systems with smaller surface tensions.

Metallic alloys are not ideal solutions, since they generally have some finite heat of mixing. In the derivation of Eq. (11) this heat of mixing was ignored by assuming that the bond energy between unlike atoms, $E_{1,2}$, is equal to the average of the bond enthalpies between like atoms, namely, by assuming that [1,5]

$$E_{1,2} = \frac{E_{1,1} + E_{2,2}}{2}$$

If we assume that this is not the case, but instead define a regular solution parameter, Ω, as

$$\Omega = N_0 Z \left(E_{1,2} - \frac{E_{11} + E_{22}}{2} \right) \tag{13}$$

where N_0 is Avogadro's number and Z is the bulk coordination number, the regular solution parameter, Ω, is directly related to the heat of mixing, ΔH_m, by the relation

$$\Omega = \frac{\Delta H_m}{X_1^b (1 - X_1^b)} \tag{14}$$

Therefore, from heat-of-mixing data, which in many cases are readily available, the parameter Ω and the bond energy $E_{1,2}$ can be estimated.

The surface composition in the regular solution monolayer approximation is given by[7]

$$\frac{X_2^s}{X_1^s} = \frac{X_2^b}{X_1^s} \exp \left[\frac{(\gamma_1 - \gamma_2)a}{RT} \right] \exp \left\{ \frac{\Omega(l + m)}{RT} [(X_1^b)^2 - (X_2^b)^2] \right.$$

$$\left. + \frac{\Omega l}{RT} [(X_2^s)^2 - (X_1^s)^2] \right\} \tag{15}$$

where l is the fraction of nearest neighbors to an atom in the plane and m is the fraction of nearest neighbors below the layer containing the atom. For example, for an atom with $Z = 12$ nearest neighbors (three above, three below, and six in the same plane), $l = \frac{6}{12} = 0.5$ and $m = \frac{3}{12} =$

0.25. In this approximation the surface composition becomes a fairly strong function of the heat of mixing, its sign, and its magnitude, in addition to the surface-tension difference and temperature.[8]

Surface Composition of Alloys from Model Calculations

Let us view a few examples of how this equation predicts surface segregation. For Au–Ag alloys the quantity Ω is constant to within 17 percent throughout the entire composition range, which is about as close to regular behavior as is found in any metallic alloys.[9] When Eq. (15) is applied to the Au–Ag system, the result is shown in Figure 3.4a and 3.4b. The calculation was carried out for an fcc(111) face, and surface energy data were used instead of heats of sublimation. The agreement between the experimental data and model calculations is quite good.

It should be noted that surface composition is strongly temperature-dependent. As the temperature is increased, the surface excess concentration of the segregating constituent should diminish exponentially if the models expressed by Eq. (11), (12), and (15) are obeyed. This effect is readily discernible by comparing Figures 3.4a and 3.4b, where the surface excesses of silver in Au–Ag are plotted at two different temperatures.

The most artificial aspect of this model, which is expressed in Eq. (15), is the monolayer approximation, and in fact it is unnecessary to restrict all layers deeper than the top monolayer to have the bulk composition. Williams and Nason[10] presented a four-layer model in which the top four layers were allowed to have a composition different from the bulk, while the fifth and deeper layer had the bulk composition. The results of the derivation are four coupled equations relating the surface composition of the four layers to T, Ω, ΔH_{subl}, X^b, and the crystal-structure parameters. To demonstrate, the results of these types of calculations for the Au–Ag system and the liquid Pb–In system are shown in Figures 3.4a, 3.4b, and 3.5, respectively. The surface enrichment diminishes rapidly with depth into the surface, as might be expected for this model, which considers only nearest-neighbor bonding. Further, if as in Ag–Au, $\Omega < 0$ [which by Eq. (14) implies attractive interactions between unlike atoms], then there is a reversal in enrichment in adjacent layers. That is, Ag enrichment occurs in the first layer, but Ag depletion takes place in the second layer. This represents a tendency toward ordering of the alloy. For Pb–In, where $\Omega > 0$, the attraction between like atoms is on the average greater than between unlike atoms, so that Pb, the component with the lowest surface energy, clusters at the surface. If $\Omega = 0$, the

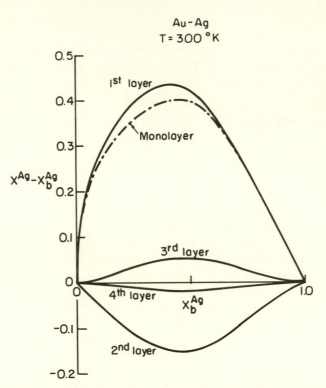

Figure 3.4a. Surface excess of Ag as a function of bulk composition at 300 K.

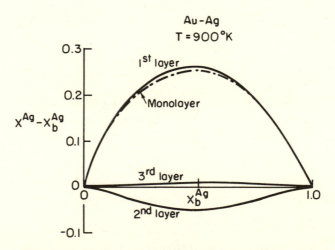

Figure 3.4b. Surface excess of Ag as a function of bulk composition at 900 K.

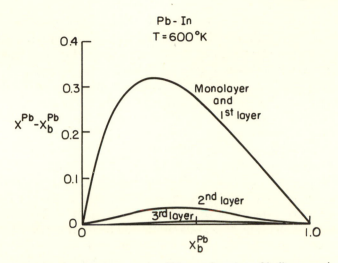

Figure 3.5. Surface excess of Pb as a function of bulk composition at 600 K.

depth distribution collapses to only a single-monolayer type of segregation.

Although Williams only considered four layers at a vacuum–solid interface, a variety of other theoretical treatments have dealt with up to an infinite number of layers. The model calculations have usually been applied to regular liquid solutions at a solid–liquid or liquid–vapor interface, although in defining the liquid phase the liquid takes on a quasi-crystalline aspect in that it is partitioned into layers. In these cases the liquid is usually taken to be semiinfinite.

There are irregularities at solid surfaces—steps and kinks at which atoms have fewer nearest neighbors than in the (111) plane of an fcc solid. According to the thermodynamic models discussed so far, surface segregation should be different at these sites. These surface irregularities have no analogy in liquids, and they can be of great importance in a variety of surface phenomena, ranging from crystal growth to heterogeneous catalysis. Burton et al.[11] approached this problem using the regular solution approximation and using appropriate value for l and m, the fraction of in-plane and out-of-plane nearest neighbors, respectively. The lower the coordination of a particular site, the greater is the tendency for segregation. Thus in alloy microclusters, sites at edges and corners are more enriched in the segregating species than are sites on flat terraces.

Burton et al.[11] also studied thin films of various thicknesses for alloys

109

in which there is a miscibility gap ($\Omega > 0$). In this case the films were thin enough that segregation to the surface led to a nonnegligible depletion of the bulk. These are the first calculations to take this mass balance into account, since other theories have assumed an infinite bulk. The results are shown in Figure 3.6 for parameters appropriate for a 50 atom percent Au–Ni alloy. The results indicate enrichment of Au at both surfaces of the film in all cases except the two-layer film. The Au–Ni bulk phase diagram has a miscibility gap below $T_c = 1000$ K. For $T > T_c$, the segregation of Au takes place only in the surface region, with a core that approaches the bulk composition as the film thickness increases. For $T < T_c$, the films exhibit phase separation, with the Au-rich phase accumulating at the surface and the Ni-rich phase accumulating at the center.

This structure, consisting of a film with an outer shell of one phase and an inner shell of another, was suggested by Sachtler and Jongpier[12] in their explanation of experimental findings in Cu–Ni alloys. Their "cherry" model suggests that during simultaneous deposition of thin films at temperatures $T \geqslant T_c$, the two phases will nucleate and grow separately. As the film equilibrates, the more easily diffusing species will diffuse to coat the crystallites of the other phase. The resulting "cherry" has a shell enriched in the more easily diffusing species, which is also usually the more surface-active species, since rates of diffusion usually increase with a decrease in bonding strength.

In the theories presented up to now, the driving force for segregation has been the fact that the surface is a region of reduced atomic coordination. In solids there is another driving force for segregation, the reduction of strain. McLean[13] has pointed out that solute atoms which differ in size from the solvent lattice atoms create a strain in the lattice. At a grain boundary there are open sites where more space is available to the

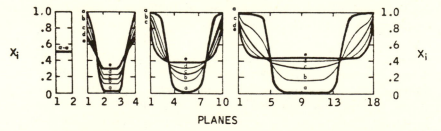

PLANES

Figure 3.6. Calculated composition, X_i, for the various planes of 50 percent Ni–Au (average composition) thin films with (111) surfaces. Data are shown for film thicknesses of 2 to 18 planes and various temperatures. Temperature of the miscibility gap is 1100 K. (After J.J. Burton et al., ref. 11.)

atoms. By migrating to these sites a solute can reduce the strain energy. McLean used the ideal solution model and arrived at an expression identical to Eq. (11) except that the argument of the exponential involves a difference between the strain energy caused by the solute atom located at the grain boundary and one located within the bulk. His expression is

$$\frac{X_2^s}{X_1^s} = \frac{X_2^b}{X_1^b} \exp\left(-\frac{Q}{RT}\right) \tag{16}$$

where

$$Q = -\frac{24\pi K G r_1 r_2 (r_2 - r_1)^2}{3K r_1 + 4G r_1} \tag{17}$$

where K is the bulk shear modulus of the solute, G the shear modulus of the solvent, and r_1 and r_2 the appropriate radii for the solvent and solute, respectively. The atomic radius is frequently obtained from the atomic volume (Seitz radius), although it can also be readily obtained from crystallographic data.

Complete treatment of the equilibrium surface composition must involve the minimization of the total free energy of the multicomponent regular solution.[8] To this end, the contributions of atom interactions and surface and solute strain energies all have to be included in the calculation. To the first approximation, all these effects can be combined into a unified formalism by setting the heat of segregation, ΔH_{segr}, equal to the exponent of the right-hand side of Eqs. (15) and (17):

$$\Delta H_{segr} = (\gamma_1 - \gamma_2)a + \Omega(l + m)[(X_1^b)^2] + \Omega l[(X_2^s)^2 - (X_1^s)^2]$$

$$- \frac{24\pi K G r_1 r_2 (r_1 - r_2)^2}{3K r_2 + 4G r_1} \tag{18}$$

The specific surface entropy term should also include the contribution of alloy surface interaction, surface forces, and strain.

Empirical Correlation

Burton and Machlin[14] have proposed a qualitative method for identifying the segregating component in binary alloy systems. For dilute alloys, they suggested, the solid–liquid equilibrium phase diagram reveals which component is in excess at the surface. If the solid–liquid distribution coefficient is less than unity [i.e., $(X_{solid}/X_{liquid}) < 1$], the solute segregates to the surface. If $(X_{solid}/X_{liquid}) > 1$, the solvent will segregate.

These cases are shown in Figures 3.7 and 3.8. They rationalize this correlation by pointing out several similarities between a surface and the liquid phase (lower coordination and more tendency toward disorder). Although the agreement between experiment and their qualitative prediction based on the phase diagram is good for many systems, the suggested correlation should be considered only as a guide to be used in the absence of suitable experimental data.

How the Surface Composition Is Measured

At present, two experimental techniques are used most frequently for quantitative determination of the surface composition—Auger electron spectroscopy (AES) and ion-scattering spectrometry (ISS). These techniques have been described in some detail in Chapter 2. We shall review here only those properties that aid surface chemical analysis. Most workers in the field assume that the peak-to-peak height of an Auger transition obtained in the usual derivative mode is proportional to the surface atom fraction of the element giving rise to electronic transition in the binary alloy or other multicomponent system. The pure component is used in general as the reference state to calibrate the peak heights. Matrix effects due to the changing strength of electron back-scattering as one element with a different atomic number is substituted for another make absolute intensity measurements ambiguous. To avoid these dif-

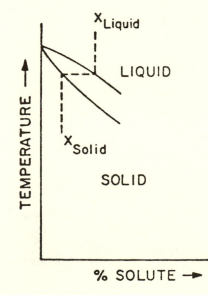

Figure 3.7. Characteristics of phase diagrams in the dilute solid solution region, which lead to the segregation of solute. (After Burton and Machlin, ref. 14.)

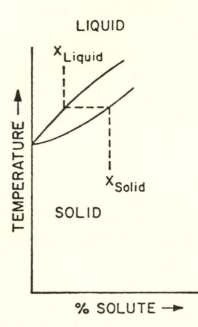

LIQUID

X_{Liquid}

X_{Solid}

TEMPERATURE →

SOLID

% SOLUTE →

Figure 3.8. Characteristics of phase diagrams in the dilute solid solution region, which lead to segregation of solvent. (After Burton and Machlin, ref. 14.)

ficulties, Auger peak ratios of the elements are given and equated with ratios of atom fractions. In this way the uncertainties due to the matrix effects from alloy to alloy are minimized.[15]

Auger electron spectroscopy yields the surface composition in the top one to seven layers; to obtain the composition of the topmost layers, the signal component due to the layers below must be subtracted. This procedure causes considerable difficulty if the aim is the precise, quantitative determination of the monolayer composition, especially for Auger peaks at higher energies (larger than 100 eV). ISS is a high-sensitivity probe that detects only those atoms that are in the topmost layer and provides, in principle, a true measure of the surface composition. For this reason ISS may also be used to calibrate Auger peak intensities. The drawback of the ion-scattering method is that it progressively erodes the surface that is being analyzed by sputtering, thereby introducing problems of interpretation. To overcome this difficulty, the ion beam is allowed to scan the surface while data are taken to minimize the radiation damage, and ion currents as low as possible are employed. In Figure 3.9 the surface atom fraction of gold, determined by AES and ISS studies, is plotted as a function of the bulk atom fraction for the Ag-Au system. This is the experimentally determined surface phase diagram for this regular solution system. The surface-layer compositions obtained by the two different surface-sensitive techniques are in good agreement.

113

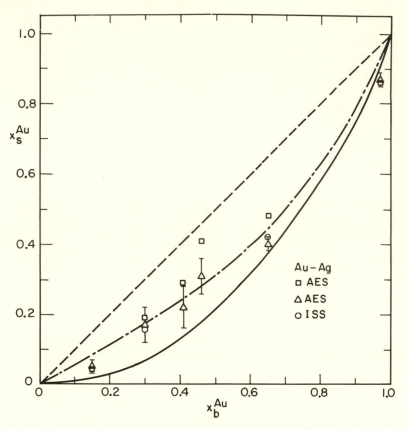

Figure 3.9. Surface phase diagram of Au–Ag alloy.

The solid line gives the calculated surface composition using the regular solution model and the dashed line indicates the curve that would be obtained in the absence of surface enrichment. The regular solution model appears to overestimate somewhat the surface segregation in this case, although the surface is clearly enriched in silver.

Although AES and ISS are readily employed for studies of samples of high conductivity, they are more difficult to use for studies of insulated surfaces. Charging of the surface changes the energy of the backscattering electrons, and the poor thermal conductivity results in localized heating. Since the surface composition, unlike the bulk composition, is temperature-dependent [see Eq. (5)], this may cause difficulties of analysis and interpretation. Different methods of discharging, using auxiliary electron guns, can be employed to minimize the buildup of

Table 3.1. Surface composition of alloys: experimental results and predictions of the regular solution and unified segregation models

Alloy systems	Phase diagram	Segregating constituent		
		Predicted		
		Regular solution	Unified	Experimental
Ag–Pd[1-4]	simple	Ag	Ag	Ag
Ag–Au[5-9]	simple	Ag	Ag	Ag
Au–Pd[3,10]	simple	Au	Au	Au
Ni–Pd[11]	simple	Pd	Pd	Pd
Fe–Cr[12]	low-T phase	Cr	Cr	Cr
Au–Cu[13-18]	low-T ordered phases	Cu	Au	Au, none, or Cu, depending on composition
Cu–Ni[19-34]	low-T miscibility gap	Cu	Cu	Cu
Au–Ni[35,36]	miscibility gap	Au	Au	Au
Au–Pt[37,38]	miscibility gap	Au	Au	Au
Pb–In[39,40]	intermediate phase	Pb	Pb	Pb
Au–In[41]	complex	In	In	In
Al–Cu[42,43]	complex	Al	Al	Al
Pt–Sn[44,45]	complex	Sn	Sn	Sn
Fe–Sn[46]	complex	Sn	Sn	Sn
Au–Sn[47]	complex	Sn	Sn	Sn

1. R. Bouwman, G.H.M. Lippits, and W.M.H. Sachtler, J. Catal. **25**, 350 (1972).
2. K. Christmann and G. Ertl, Surf. Sci. **33**, 254 (1972).
3. B.J. Wood and H. Wise, Surf. Sci. **52**, 151 (1975).
4. H.J. Mathieu and D.H. Landolt, Surf. Sci. **52**, 228 (1975).
5. S.H. Overbury and G.A. Somorjai, Surf. Sci. **55**, 209 (1976).
6. S.C. Fain and J.M. McDavid, Phys. Rev. **B9**, 5099 (1974).
7. G.C. Nelson, J. Vac. Sci. Technol. **13**, 512 (1976).
8. W. Farber and P. Braun, Vak.-Tech. **23**, 239 (1974).
9. W. Farber, G. Betz, and P. Braun, Nucl. Instr. Meth. **132**, 351, (1976).
10. A. Jablonski, S.H. Overbury, and G.A. Somorjai, Surf. Sci. **65**, 578 (1977).
11. C.T.H. Stoddart, R.L. Moss, and D. Pope, Surf. Sci. **53**, 241 (1975).
12. C. Leygraf, G. Hultquist, and S. Ekelund, Surf. Sci. **46**, 157 (1974).
13. P. Palmberg and T.N.Rhodin, J. Appl. Phys. **39**, 2425 (1968).
14. H.C. Potter and J.M. Blakely, J. Vac. Sci. Technol. **12**, 635 (1975).
15. V.S. Sundarem, R.S. Alben, and W.D. Robertson, Surf. Sci. **46**, 653 (1974).
16. W. Farber and P. Braun, Surf. Sci. **41**, 195 (1974).
17. R.A. Van Santen, L.H. Toneman, and R. Bouwman, Surf. Sci. **47**, 64 (1975).
18. J.M. McDavid and S.C. Fain, Surf, Sci. **52**, 161 (1975).
19. W.M.H. Sachtler and G.J.H. Dorgelo, J. Catal. **4**, 654 (1965).
20. L. Elford, F. Muller, and O. Kubaschewski. Ber. Bunsenges. Phys. Chem. **73**, 601 (1969).
21. D.T. Qunto, V.S. Sundarem, and W.D. Robertson, Surf. Sci. **28**, 504 (1971).
22. G. Ertl and J. Küppers, J. Vac. Sci. Technol. **9**, 829 (1971).
23. G. Ertl and J. Küppers, Surf. Sci. **24**, 104 (1971).
24. K. Nakayama, M. Ono, and H. Shimizu, J. Vac. Sci. Technol. **9**, 749 (1972).
25. Y. Takasu and H. Shimizu, J. Catal. **29**, 479 (1973).
26. M.L. Tarng and G.K. Wehner, J. Appl. Phys. **42**, 2449 (1971).
27. M. Ono, Y. Takasu, K. Nakayama, and T. Yamashina, Surf. Sci. **26**, 313 (1971).
28. H. Shimizu, M. Ono, and K. Nakayama, Surf. Sci. **36**, 817 (1973).
29. Y. Takasu, H. Konno, and T. Yamashina, Surf. Sci. **45**, 321 (1974).
30. C.R. Helms, J. Catal. **36**, 114 (1975).
31. C.R. Helms and K.Y. Yu, J. Vac. Sci. Technol. **12**, 276 (1975).
32. H.H. Brongersma and T.M. Buck, Surf. Sci. **53**, 649 (1975).
33. C.R. Helms, K.Y. Yu, and W.E. Spicer, Surf. Sci. **52**, 217 (1975).

(*Continued*)

115

Table 3.1.—Continued

34. J.J. Burton and E. Hyman, J. Catal. **37**, 114 (1975).
35. F.L. Williams and M. Boudart, J. Catal. **30**, 438 (1973).
36. J.J. Burton, C.R. Helms, and R.S. Polizzotti, J. Vac. Sci. Technol. **13**, 204 (1976).
37. R. Bouwman and W.M.H. Sachtler, J. Catal. **19**, 127 (1970).
38. F.J. Luijers, R.P. Dessing, and W.M.H. Sachtler, J. Catal. **33**, 316 (1974).
39. S. Berglund and G.A. Somorjai, J. Chem. Phys. **59**, 5537 (1973).
40. N.J. Chou, S.K. Lahiri, R. Hammer and K.L. Komarek, J. Chem. Phys. **63**, 2758 (1975).
41. S. Thomas, Appl. Phys. Lett. **24**, 1 (1974).
42. J. Ferrante, ACTA Metall. **19**, 743 (1971).
43. J. Ferrante, Scripta Metall. **5**, 1129 (1971).
44. R. Bouwman, L.J. Toneman, and A.A. Holscher, Surf. Sci. **35**, 8 (1973).
45. R. Bouwman and P. Biloen, Surf. Sci. **41**, 348 (1974).
46. M.P. Seah, Surf. Sci. **40**, 595 (1973).
47. S.H. Overbury and G.A. Somorjai, J. Chem. Phys. **66**, 3181 (1977).

space charge. The use of metal foils in the back of the thinly cut sample, or metal grids placed on top of the sample, can also minimize charging and heating effect in some cases.

Experimental Data and Correlation to the Various Models Predicting Surface Enrichment

Most of the experimental data were obtained for binary alloy systems. Many of these obey the regular solution model, and thus the surface enrichment may be similar to that calculated from Eq. (15). Indeed, careful AES or ISS studies indicate the surface segregation of the same constituents that was predicted by either the regular solution or the unified regular solution model. Table 3.1 lists the systems that were studied and the segregating components that were experimentally observed and also predicted. The agreement is certainly satisfactory. By using the unified model [Eq. (18)], the heat of segregation can be computed[8] for all of these systems and compared to the experimental ΔH_{segr}. Again, the agreement is satisfactory. It should be noted that the Burton–Machlin equilibrium diagram,[14] which yields a qualitative correlation, also predicts well which element is likely to segregate to the surface. Thus it appears that for binary-metal-alloy systems that exhibit regular solution behavior there are reliable methods to predict surface composition.

Surface Composition of Systems with Complex Phase Diagrams: The Au–Sn System

Most of the multicomponent systems studied so far were binary alloys that obey the regular solution model. The question arises whether similar thermodynamic arguments can be used for systems with complex phase diagrams that also form high-binding-energy compounds, to predict their surface composition. The Au–Sn system is a suitable example of this type of alloy.[16]

Sn has a much smaller surface free energy than that of Au (685 ergs/cm² for Sn versus about 1400 ergs/cm² for Au near their melting points),[17,18] and on this basis Sn may be expected to segregate to the surface. In addition, there is a considerable atomic size difference between pure Au and pure Sn, so that strain effects may also be expected to lead to segregation in some instances. The bulk phase diagram given in Figure 3.10 indicates complex bulk behavior for this system, with at least three distinct phases forming between the composition of 50.0 atom percent Au and pure Au. The bulk structures of each of these phases have been well characterized. The δ phase is a very strongly ordered phase, having a hexagonal type of structure (B8, isotypic with NiAs) which exists over a very narrow range of composition at 50.0 atom percent Au. The ζ phase is simply a hexagonal close-packed lattice (a3,

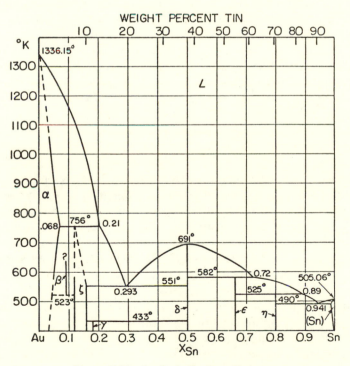

Figure 3.10. Au–Sn bulk-phase diagram.

isotypic with Mg), with Au and Sn randomly distributed in the lattice. This phase forms at compositions between 83.5 and 89 atom percent Au. The primary α phase is simply the face-centered cubic Au lattice, with the Sn solute substitutionally mixed in. The maximum solubility of Sn in Au is 7 atom percent Sn.

The following questions arise. Is there a segregation of one of the components to the surface of a one-phase alloy such as the δ, the ζ, or the α phase alloys, and if there is, why does segregation occur and to what extent? If there are two phases present, are both phases present at the surface in amounts expected from the lever rule, or is there phase redistribution at the surface which causes one phase to coat the surface?

Alloys throughout the composition range from 50 to 99 atom percent Au were prepared and studied by Auger electron spectroscopy, with the goal of answering these questions. Let us discuss the surface composition of each.

δ Phase (50 atom percent Au)

The Auger data seem to indicate that the surface composition is identical to that of the bulk for this alloy. This seems surprising at first, since because of the lower surface energy of pure Sn, it is expected that Sn should segregate to the surface. This contradiction can be explained by consideration of the bulk structure of the δ-phase alloy. This alloy is very strongly ordered and remains so up to the melting point. If surface segregation is to take place, this order must be defeated by interchanging Au and Sn atoms. For this alloy, then, it appears that the ordered lattice is stable enough to be energetically favored over the disordered state that would be brought about by surface segregation.

In work carried out by Bouwman et al.,[19] the surface composition of Pt–Sn alloy, which has the same structure as the δ-phase Au–Sn alloy, was studied, and segregation of Sn to the surface was noted. For this alloy, apparently, the tendency for Sn to segregate to the surface is strong enough to bring about the disordering. For the Au–Cu system, which forms ordered compounds at about 25, 50, and 75 atom percent Au, the long-range ordering is weak enough that it breaks up before the melting point is reached. Fain and McDavid[20] found surface segregation of Au, the component believed to have the lower surface energy, for films of a wide bulk composition range.

It would be desirable to be able to predict these results by comparing quantitatively the tendency for the alloy to remain ordered and the tendency for segregation to occur. The driving force for *segregation* may be expected to be related to the difference between the surface energy of the pure components, or to the difference in their heats of sublimation.

The tendency of the alloy to *order* should be related to ΔH_m, the heat of mixing of the alloy, or, better, to $\Omega = \Delta H_m / X_1^b X_2^b$, where X_1^b and X_2^b are the bulk atom fractions of the two components in the alloy. The ratio $|H_{sub}^A - H_{sub}^B|/|\Omega|$ may be of use in predicting whether segregation takes place. This ratio is 1.05 for Au–Sn that remains ordered at the surface. For Pt–Sn the heat of mixing at 298 K is -58.6 kJ/g-atom,[21] which gives a value for this ratio of 1.13. This larger ratio indicates that for Pt–Sn the tendency toward segregation is somewhat greater than for the AuSn alloy, in agreement with the experimental results.

It is evident from this discussion that the strength of bonding between unlike atom pairs in the alloy is of importance in predicting the surface composition in case of more complex phase diagrams. Qualitative information about this effect can be obtained from the alloy heats of mixing, but much theoretical and experimental work is necessary before this effect will be fully understood.

ζ Phase (86.7 atom percent Au)

For the ζ-phase alloy, contrary to the δ-phase alloy, there is no long-range ordering in the bulk, and all lattice points are equivalent whether occupied by Au or Sn. For this alloy the value of $|H_{sub}^A - H_{sub}^B|/|\Omega|$ ranges from 2.07 to 2.21 at 273 K, which is larger than for the δ-phase alloys of Au–Sn and Pt–Sn. In this circumstance it is expected to be easier for Sn to segregate to the surface of the alloy. The Auger intensity data for this sample are consistent with there being only about 43 atom percent Au in the surface monolayer. It should be pointed out that this composition does not exist in the bulk phase diagram, and the layer may be thought of as a distinct surface phase. It would be of interest to study the structure of this surface by a variety of surface-sensitive techniques (low-energy electron diffraction, for example).

There are possibly two driving forces causing this segregation. The first is the lower surface tension of pure Sn as compared to pure Au. Second, the effects of lattice strain might also be important, since Sn is located in a lattice where the nearest-neighbor distance (2.91 Å) is smaller than that in pure Sn (3.01 Å). Segregation of Sn to the surface sites where there are fewer nearest neighbors can relieve this strain. Any theory that predicts the surface composition of alloys must take both effects into account.

α Phase

The Auger data for this alloy phase indicate strong segregation of Sn to the surface. Here again, the driving force for segregation is probably due partly to the lower surface free energy of Sn. In this case, however, possibly the more important driving force is the relief of lattice strain caused by the large Sn atoms dissolved substitutionally in the Au lattice. Burton's phase-diagram correlation would also predict segregation of Sn in this alloy phase.

Surface-Phase Diagram

It is of interest to use these data to construct a surface-phase diagram. Surface-phase diagrams of binary alloys can be quite complicated because there are so many variables. If more than the top monolayer has a composition different from the bulk, one variable is required for each layer. In addition, these compositions may depend upon the crystal face and will certainly depend upon temperature, at least over a wide temperature range. For the Au–Ag system[9] the monolayer approximation will be assumed and the surface monolayer composition will be that of the average crystal faces, since the experimental data are for polycrystalline samples. The surface-phase diagram then becomes a plot of $X^s_{Au}(X^b_{Au}, T)$. In Figure 3.11, a portion of a constant-temperature slice of the surface-phase diagram for the Au–Sn system is given.[5,16] The temperature would vary along the y-axis perpendicular to the plane of the figure. The bulk structure is given along the x-axis and is a single line cut from the bulk-phase diagram shown in Figure 3.12. The entire bulk-phase diagram would be contained in the xy plane. In Figure 3.12 the surface monolayer composition is plotted along the z-axis as a function of bulk composition at a temperature of 150°C.

In the surface monolayer two surface phases are present between bulk composition of $X^b_{Au} = 0.50$ and $X^b_{Au} = 0.835$. These are labeled δ_s and ζ_x since they derive from the bulk structure and δ and ζ but have undetermined structures. The dashed–dotted lines indicate that both phases are present but that each phase has its own surface composition, which is invariant throughout the bulk composition range. The solid diagonal line between the two phase boundaries gives the average surface composition and indicates that the lever rule is likely to be obeyed at the surface in this two-phase field. Through the narrow ζ-phase region the surface composition is shown as constant at the value determined for the 86.7 atom percent Au alloy. A composition variation is shown for the α phase.

This phase diagram summarized the experimentally determined Au–Sn surface compositions in a compact way.[15] Such surface-phase diagrams have been constructed for simpler systems, such as for the Au–Ag system and the Au–Cu system. In the future, surface-phase diagrams will undoubtedly become more common and will improve as more

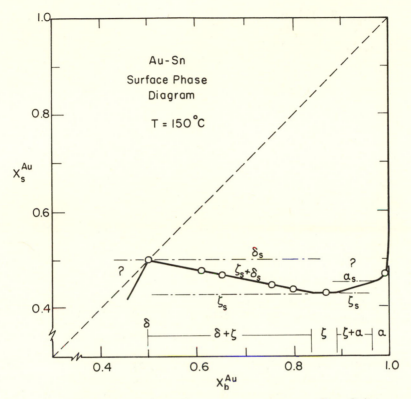

Figure 3.11. Portion of a constant-temperature slice of the Au–Sn surface-phase diagram.

information about the structure, composition, and temperature dependence of various surface phases is obtained.

From studies of regular solution type and complex binary-alloy systems, three physical parameters can be identified that control the surface composition. These are (1) the heat of mixing ΔH_m of the alloy, which reflects the strength of the chemical bond between the two constituents; (2) the relative surface tension of the pure components; and (3) the lattice strain energy, which is due to the mismatch of the atomic sizes of the constituents. A large negative value of ΔH_m, as compared to the decrease of total surface free energy achieved by surface segregation due to the surface-tension differences or the strain energy term, stabilizes the bulklike composition at the surface. Conversely, small relative values of ΔH_m, as compared to the surface-tension or strain terms, leads to surface compositions that are different from the bulk

121

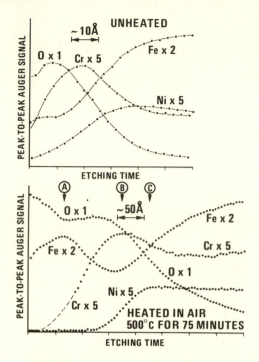

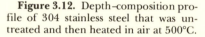

Figure 3.12. Depth–composition profile of 304 stainless steel that was untreated and then heated in air at 500°C.

stoichiometry. The ratio $|\Delta H_{\mathrm{subl}(1)} - \Delta H_{\mathrm{subl}(2)}|/|\Omega|$, where $\Omega = \Delta H_m/X_1 X_2$ is the regular solution parameter, may be of use in predicting whether surface segregation takes place in an alloy. From studies of the Pt–Sn and Au–Sn systems, it appears that if this ratio is 1.0 or larger, surface segregation occurs. For values of this ratio that are smaller than unity, surface segregation may not take place.

Small but chemically significant deviations from stoichiometry have been observed frequently in the surface layers of compounds with high heats of formation. Ionic solids (alkali halides and LiH) exhibit an excess of one of the ions.[22] Compound semiconductors, as well as oxides, show detectable nonstoichiometry when heated at elevated temperatures (discussed in Chapter 4). Perhaps one of the most important consequences of large vacancy concentrations in the surface is the appearance of unusual oxidation states. Large concentrations of Ti^{3+} appear to be stabilized in the TiO_2 and $SrTiO_3$ surface layers,[23,24] and there is evidence for the presence of Al^+ or Al^{2+} formal oxidation-state ions in the Al_2O_3 crystal surfaces.[25] These oxidation states are stabilized only in the surface environment and have unique chemical and electrical properties.

Bimetallic Clusters

An interesting phenomenon occurs for small particles of multicomponent systems. In the limit of very small particle size, when all the atoms are located on the surface (dispersion = surface atoms/total number of atoms ≈ unity), any variation in composition between the surface and bulk should disappear. There is evidence, however, for the formation of surface alloy systems with phase diagrams that are very different from the bulk phase diagram.[26] Bimetallic systems such as Ru–Cu and Ir–Au, which exhibit negligible solubility in the solid state, become miscible and form solid solutions when deposited as small particles with near-unity dispersion.[27] The exploration of the electronic structure, thermodynamic properties, and chemistry of high-dispersion multicomponent systems is a fertile area of investigation in surface science.

Effect of the Ambient Condition on the Surface Composition

So far we have discussed the surface composition of multicomponent systems that are in equilibrium with their vapor or in which clean surface–bulk equilibrium is obtained in ultrahigh vacuum. In most circumstances, however, the surface is covered with a monolayer of adsorbates that frequently form strong chemical bonds with the surface atoms. This solid–gas interaction can markedly change the surface composition in some cases. For example, carbon monoxide, when adsorbed on the surface of a Ag–Pd alloy, forms much stronger bonds with Pd. While the clean surface is enriched with Ag, in the presence of CO, Pd is attracted to the surface, where it arrives by surface diffusion to form strong carbonyl bonds.[28] When the adsorbed CO is removed, the composition returns to its original Ag-enriched state. Of course, nonvolatile adsorbates, such as carbon or sulfur, may have a similar influence on the surface composition as long as their bonding to the various constituents of the multicomponent system is different.

Adsorbates should therefore be viewed as an additional component of the multicomponent system. A strongly interacting adsorbate converts a binary system to a ternary system. As a result, the surface composition may markedly change with changing ambient conditions. To demonstrate this with a practical example, in Figure 3.12 the depth–composition profiles of 304 stainless steel are shown. One sample was heated in air at 500°C for 75 min, while the other was not subjected to heavy oxidation treatment. The samples were ion-sputtered, layer by layer, and the composition was monitored by AES during the sputtering to obtain the depth–composition profile. The 304 steel is enriched with Cr near the surface region, which appears to be composed of chromium oxide (top 10 layers). After oxidation, iron migrates to the surface, which becomes predominantly iron oxide in the top five or six atomic layers, and Cr oxide is no longer detectable near the surface.

The mechanical properties of solids, embrittlement, and crack propagation, among others, depend markedly on the surface composition. These studies indicate that the surface composition, and thus the mechanical properties of structural steels, may change drastically when the ambient conditions are changed from those of reducing to those of oxidizing environments.

References

1. S. Overbury and G.A. Somorjai, Discussions of the Faraday Soc. **60**, 279 (1975).
2. J. Sinfelt, Science **195**(4279), 641 (1977).
3. J. Blakely, Mater. Sci. Cen. Rep. 2891, Cornell University, Ithaca, N.Y., 1977.
4. G.N. Lewis and M. Randall, *Thermodynamics*, McGraw-Hill, New York, 1961, p. 161.
5. J. Blakely and J. Shelton, *Surface Physics of Materials*, vol. 1, Academic Press, New York, 1975, p. 199.
6. A. Jablonski, Adv. Colloid Interface Sci. **8**, 213 (1977).
7. S. Overbury, P. Bertrand, and G.A. Somorjai, Chem. Rev. **75**(5), 550 (1975).
8. P. Wynblatt and R. Ku, Surf. Sci. **65**, 520 (1977).
9. P. Hultgren et al., *Selected Values of the Thermodynamic Properties of Binary Alloys*, American Society for Metals, Metals Park, Ohio, 1973.
10. F.L. Williams and D. Nason, Surf. Sci. **45**, 377 (1974).
11. J.J. Burton, E. Hyman, and D.A. Fedak, J. Catal. **37**, 106 (1975).
12. W.M.H. Sachtler and R. Jongpier, J. Catal. **4**, 665 (1965).
13. D. McLean, *Grain Boundaries in Metals*, Oxford University Press, London, 1957.
14. J.J. Burton and E.S. Machlin, Phys. Rev. Lett. **37**, 1433 (1976).
15. S.H. Overbury, Ph.D. thesis, University of California, Berkeley, 1976, pp 73–78.
16. Ibid., pp. 116–176.
17. E.D. Greenhill and S.R. McDonald, Nature **171**, 37 (1953).
18. J.S. Vermaak and D.K. Wilsdorf, J. Phys. Chem. **72**, 4150 (1968).
19. R. Bouwman, G.H.M. Lippits, and W.M.H. Sachtler, J. Catal. **25**, 350 (1972).
20. S.C. Fain and J.M. McDavid, Bull. Am. Phys. Soc. **19**, 357 (1974).
21. R. Bouwman, L.H. Toneman, and A.A. Holscher, Surf. Sci. **35**, 8 (1973).
22. T.E. Gallon, I.A. Higginbotham, M.Prutton, and H. Tokutaka, Surf. Sci. **21**, 224 (1970).
23. G.A. Somorjai, Y.W. Chung, and W.J. Lo, Surf. Sci. **64**, 588 (1977).

24. W.J. Lo, and G.A. Somorjai, Phys. Rev. **B17,** 4942 (1978).
25. T.M. French and G.A. Somorjai, J. Phys. Chem. **74,** 2489 (1970).
26. J. Sinfelt, Science **195**(4279), 643 (1977).
27. Ibid., p. 645.
28. R. Bouwman, G.H.M. Lippits, and W.M.H. Sachtler, J. Catal. **25,** 350 (1972).

4. Structure of Clean Surfaces

Results of Surface Crystallography Studies

The structural heterogeneity and the varied composition of solid surfaces introduce a great deal of complexity into surface studies of all types. It is therefore essential that we also study surfaces that are less complex. For studies of surface structure, we could start with one face of a single crystal of a monatomic solid where most of the atoms are in their identical equilibrium positions in the well-ordered surface. The surface should be clean, since adsorbates or impurities may affect the location of atoms in the substrate surface. We should then introduce surface irregularities, adatoms, steps, and kinks systematically and study their atomic structure. Our investigations may then be extended to clean diatomic and polyatomic solid surfaces, where possible variations of surface composition that lead to nonstoichiometry may add to the structural complexity. When the atomic surface structures of many clean crystal faces of monatomic and polyatomic solids have been determined this way, we should be able to evolve a sound physical picture of the atomic surface structure of small particles or polycrystalline foils that are composed of all these crystal faces. Using the structural information obtained on clean crystal surfaces as a reference state, we are in a position to study the structure and chemical bonding of adsorbates of all types.

In this chapter we discuss our present understanding of the atomic structure of clean solid surfaces. First, we review what is known about the atomic surface structure of monatomic solids of the simplest, most uniform types. This review is followed by a discussion of the surface structure of similar low-Miller-index surfaces of polyatomic solids. Then we discuss the surface structures of molecular solids, followed by the surface structures of the structurally more complex high-Miller-index surfaces, which have different surface irregularities. Finally, we discuss the surface structures of compounds formed by solid-state reactions at the surface and phase transformations that occur at the surface.

Most of our present information on the structures of clean solid surfaces comes from low-energy electron diffraction studies. Therefore, we shall focus mostly on the results obtained by this technique. Several important structural changes that occur at surfaces are detectable by low-energy diffraction and surface crystallography.[1] These are:

1. Relaxation. The surface atoms seek new equilibrium positions that change the interlayer distance between the first and second layers of atoms or ions. Contraction is generally observed; expansion is also theoretically possible but has not yet been detected by surface crystallography. The relaxation of the first layer of atoms toward the second layer changes the bond angles but does not affect the number of nearest neighbors (coordination number) or the rotational symmetry of the surface atoms. Thus the surface unit cell remains the same as that for the "ideal" surface structure obtained by the projection of the bulk X-ray unit cell to that surface.

2. Reconstruction. The surface atoms seek new equilibrium positions that change not only the bond angles but also the rotational symmetry and the number of nearest neighbors. In this circumstance the surface unit cell becomes different from that predicted by the projection of the bulk X-ray unit cell. This reconstructed surface may maintain the same structure over a wide temperature range or may reconstruct again as the surface temperature changes.

3. Relaxation or reconstruction induced by changes of surface composition. For polyatomic solids the surface composition may be very different from that in the bulk. Changes in the surface stoichiometry induce relaxation of the surface atoms. In addition, new oxidation states may be stabilized in the surface layer as the surface composition varies. The new oxidation-state atoms are either smaller or larger than the other surface atoms and therefore may cause reconstruction.

Before discussing the various examples that demonstrate these surface structural changes, let us review the notations that were developed to identify surface structures of different types.

Conversion of the Diffraction Pattern to a Surface Structure; Notation of the Surface Structures on Low-Miller-Index Crystal Planes

LEED diffraction patterns represent the reciprocal lattice of the surface; the diffraction pattern must be inverted to real space in order to obtain the real space surface structure. In this section we see how this conversion is performed. First, the relationship between the reciprocal and real lattices of the substrate will be shown; then, determination of adsorbate surface structures from the LEED patterns will be discussed.

The diffraction pattern or reciprocal lattice has translational periodicity which is given by the vector **T***, which has the form

$$\mathbf{T}^* = n^*\mathbf{a}^* = m^*\mathbf{b}^* \tag{1}$$

where n^* and m^* are integers and **a*** and **b*** are the vectors of the reciprocal unit cell. **T*** is related to the translational lattice vector of the real lattice, **T**,

$$\mathbf{T} = n\mathbf{a} = m\mathbf{b} \tag{2}$$

where n and m are integers and **a** and **b** are the vectors of the primitive surface mesh. The reciprocal unit cell vectors **a*** and **b*** are related to the real-space unit-cell vectors **a** and **b** by the following equations:

$$\mathbf{a}^* = \frac{\mathbf{b} \times \mathbf{z}}{\mathbf{a} \cdot \mathbf{b} \times \mathbf{z}} \tag{3a}$$

$$\mathbf{b}^* = \frac{\mathbf{z} \times \mathbf{a}}{\mathbf{a} \cdot \mathbf{b} \times \mathbf{z}} \tag{3b}$$

where **z** is the surface normal. The relationship between the reciprocal and real space vectors for a two-dimensional hexagonal lattice is shown in Figure 4.1.

Reconstruction of the clean surface or adsorption of a gas on a surface usually results in a change in the diffraction pattern corresponding to

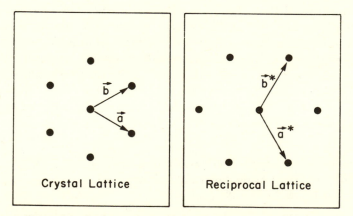

Figure 4.1. Real-space vectors **a** and **b** and reciprocal space vectors **a*** and **b*** of a two-dimensional hexagonal lattice.

the appearance of a new surface periodicity. This is illustrated in Figure 4.2, which shows a diffraction pattern of a clean Pt(111) surface and the diffraction pattern formed after the adsorption of an ordered layer of adsorbate. Figure 4.3 shows the unit cell responsible for the diffraction patterns in Figure 4.2 superimposed on a model of the Pt(111) surface. No information concerning the location of the adsorbate molecules within this unit mesh and relative to the substrate atom positions is indicated. This information can be obtained only from analysis of the diffraction spot intensities.

To make the transition from the diffraction pattern in Figure 4.2 to the surface structure in Figure 4.3, the adsorbate surface reciprocal mesh is referenced to the substrate reciprocal mesh. This is done by a visual inspection of the diffraction pattern, in which the differences in spot intensities are neglected and only the positions of the diffraction beams are considered.

For the general case, the relationship of adsorbate reciprocal mesh to the substrate reciprocal is given by the equations

$$\mathbf{a}^{*\prime} = m^{*}_{11}\, \mathbf{a}^{*} + m^{*}_{12}\, \mathbf{b}^{*} \tag{4a}$$

$$\mathbf{b}^{*\prime} = m^{*}_{21}\, \mathbf{a}^{*} + m^{*}_{22}\, \mathbf{b}^{*} \tag{4b}$$

where $\mathbf{a}^{*\prime}$ and $\mathbf{b}^{*\prime}$ are the vectors of the primitive adsorbate reciprocal mesh and the coefficients m^{*}_{11}, m^{*}_{12}, m^{*}_{21}, and m^{*}_{22} define the matrix

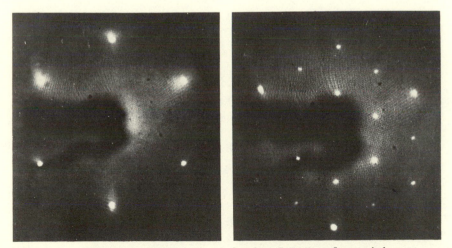

Figure 4.2. LEED patterns of a clean Pt(111) surface and the same surface with an ordered adsorbate. For both diffraction patterns, the incident beam energy is 68 eV.

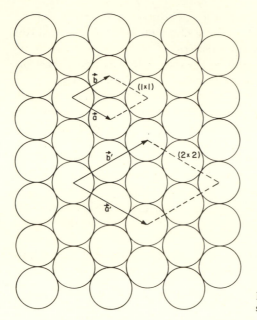

Figure 4.3. Real-space unit cells of Pt(111)-(1 × 1) and Pt(111)-(2 × 2) surface structures.

$$M^* = \begin{pmatrix} m^*_{11} & m^*_{12} \\ m^*_{21} & m^*_{22} \end{pmatrix}$$

In real space the adsorbate mesh is related to the substrate mesh by the equations

$$\mathbf{a}' = m_{11}\,\mathbf{a} + m_{12}\,\mathbf{b} \tag{5a}$$

$$\mathbf{b}' = m_{21}\,\mathbf{a} + m_{22}\,\mathbf{b} \tag{5b}$$

where $\mathbf{a}'$ and $\mathbf{b}'$ are the vectors of the primitive adsorbate mesh and the coefficients m_{11}, m_{12}, m_{21}, and m_{22} define the matrix

$$M = \begin{pmatrix} m_{11} & m_{12} \\ m_{21} & m_{22} \end{pmatrix}$$

The coefficients of the two matrices M and M^* are related by the following equations:

$$m_{11} = \frac{m^*_{22}}{m^*_{11}\,m^*_{22} - m^*_{21}\,m^*_{12}} \tag{6a}$$

$$m_{12} = \frac{-m^*_{21}}{m^*_{11}\,m^*_{22} - m^*_{21}\,m^*_{12}} \tag{6b}$$

$$m_{21} = \frac{-m^*_{12}}{m^*_{11}\, m^*_{22} - m^*_{21}\, m^*_{12}} \qquad (6c)$$

$$m_{22} = \frac{m^*_{11}}{m^*_{11}\, m^*_{22} - m^*_{21}\, m^*_{12}} \qquad (6d)$$

so that if either M or M^* is known, the other may be readily calculated. In LEED experiments, M^* is determined by visual inspection of the diffraction pattern and then transformed to give M, which defines the surface structure in real space.

For the case of ordered adsorption on Pt(111), visual inspection of the LEED patterns in Figure 4.2 gives

$$M^* = \begin{pmatrix} \frac{1}{2} & 0 \\ 0 & \frac{1}{2} \end{pmatrix}$$

By employing Eqs. (6a) through (6d), the matrix M is found to be

$$\begin{pmatrix} 2 & 0 \\ 0 & 2 \end{pmatrix}$$

so $\mathbf{a}' = 2\mathbf{a}$ and $\mathbf{b}' = 2\mathbf{b}$, as depicted in Figure 4.3.

In addition to the matrix method of denoting surface structures, another system, originally proposed by Wood,[2] is also used. Whereas the matrix notation can be applied to any system, Wood's notation can only be used when the angle between the adsorbate vectors $\mathbf{a}'$ and $\mathbf{b}'$ is the same as the angle between the substrate vectors $\mathbf{a}$ and $\mathbf{b}$. If this condition is met, the surface structure is labeled using the general form $p(n \times m) R\Phi°$ or $c(n \times m) R\Phi°$, depending on whether the unit cell is primitive or centered. In Wood's notation the adsorbate unit cell is related to the substrate unit mesh by the scale factors n and m, where

$$|\mathbf{a}'| = n\, |\mathbf{a}| \qquad (7a)$$

$$|\mathbf{b}'| = m\, |\mathbf{b}| \qquad (7b)$$

$R\Phi°$ indicates a rotation of the adsorbate unit mesh by $\Phi°$ from the substrate unit mesh. For $\Phi = 0$, the $R\Phi°$ label is omitted, so the surface structure in Figure 4.3 is labeled as $p(2 \times 2)$ or simply (2×2). The label for the total system refers to the type of substrate, the surface structure formed by the adsorbate, and the adsorbate. The platinum–acetylene adsorbate system shown in Figure 4.3 would be labeled Pt(111)-$\begin{pmatrix} 2 & 0 \\ 0 & 2 \end{pmatrix}$-$C_2H_2$ in matrix notation and as Pt(111)-$p(2 \times 2)$-C_2H_2 in Wood's notation. Wood's notation is more commonly used, and the matrix notation is

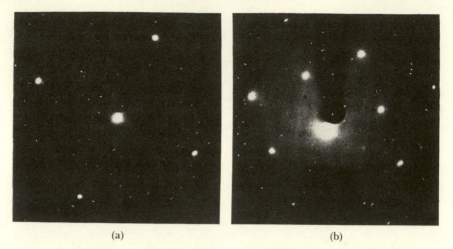

(a)   (b)

Figure 4.4. LEED patterns of (a) clean Rh(100) at 74 eV, and (b) oxygen-covered Rh(100) at 85 eV.

usually applied only to systems where the angle between the adsorbate vectors differs from the angle between substrate vectors.

An example of an adsorbate that has a centered unit cell is shown in Figures 4.4 and 4.5. In Figure 4.4 diffraction patterns are shown from a clean Rh(100) surface and from a Rh(100) surface after exposure to oxygen. By visual inspection it can be seen that

$$M^* = \begin{pmatrix} \frac{1}{2} & -\frac{1}{2} \\ \frac{1}{2} & \frac{1}{2} \end{pmatrix}$$

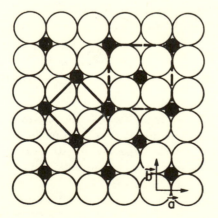

Figure 4.5. Real space unit cells of the $(2 \times 2)R45°-O$ (solid lines) and $c(2 \times 2)-O$ (dashed lines) on the Rh(100) surface.

so using Eqs. (6a) through (6d) yields

$$M = \begin{pmatrix} 1 & 1 \\ -1 & 1 \end{pmatrix}$$

M defines the primitive unit cell of the adsorbate, which is drawn with solid lines in Figure 4.5. This unit cell is labeled $(\sqrt{2} \times \sqrt{2})R45°$ in Wood's notation. Since the centered unit cell drawn in with dotted lines in Figure 4.5 also describes the adsorbate unit cell, another way of label-

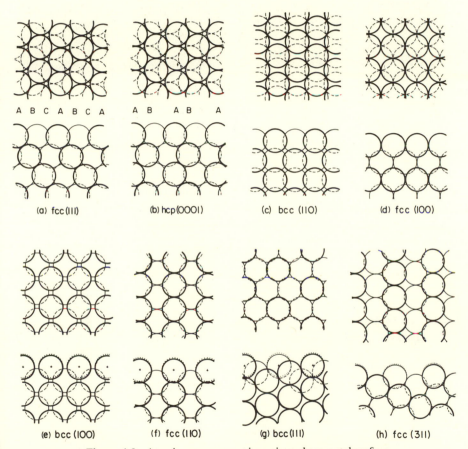

Figure 4.6. Atomic arrangement in various clean metal surfaces. In each of panels a to h, the top and bottom sketches give top and side views, respectively. Thin-lined atoms are behind the plane of thick-lined atoms. Dotted lines represent atoms in unrelaxed (ideal bulk) positions.

ing this structure would be $c(2 \times 2)$. The total system is labeled as Rh(100)-$\left(\begin{smallmatrix} 1 & 1 \\ -1 & 1 \end{smallmatrix}\right)$-0, Rh(100)-$(\sqrt{2} \times \sqrt{2})R45°$-0, or Rh(100)-$c(2 \times 2)$-0. Of these three labels, only the first two refer to the primitive unit cell of the oxygen surface structure.

The structures of the low-Miller-index surfaces in the face-centered cubic (fcc), body-centered cubic (bcc), and hexagonal close-packed (hcp) crystal structures are shown in Figure 4.6. The unreconstructed surface has a surface unit cell that is predicted by the projection of the bulk X-ray unit cell onto that surface. That unit cell is denoted as $p(1 \times 1)$ by Wood's notation, where p stands for "primitive" (as opposed to c for "centered"). The same surface unit mesh is denoted by $\left(\begin{smallmatrix} 1 & 0 \\ 0 & 1 \end{smallmatrix}\right)$ in the more general

Table 4.1. Wood and matrix notation for a variety of superlattices on low-Miller-index crystal surfaces

Substrate	Overlayer unit cell	
	Wood notation	Matrix notation
fcc(100), bcc(100)	$p(1 \times 1)$	$\begin{pmatrix} 1 & 0 \\ 0 & 1 \end{pmatrix}$
	$c(2 \times 2) = (\sqrt{2} \times \sqrt{2})R45°$	$\begin{pmatrix} 1 & -1 \\ 1 & 1 \end{pmatrix}$
	$p(2 \times 1)$	$\begin{pmatrix} 2 & 0 \\ 0 & 1 \end{pmatrix}$
	$p(1 \times 2)$	$\begin{pmatrix} 1 & 0 \\ 0 & 2 \end{pmatrix}$
	$p(2 \times 2)$	$\begin{pmatrix} 2 & 0 \\ 0 & 2 \end{pmatrix}$
	$(2\sqrt{2} \times \sqrt{2})R45°$	$\begin{pmatrix} 2 & 2 \\ -1 & 1 \end{pmatrix}$
fcc(111) (60°between basis vectors)	$p(2 \times 1)$	$\begin{pmatrix} 2 & 0 \\ 0 & 1 \end{pmatrix}$
	$p(2 \times 2)$	$\begin{pmatrix} 2 & 0 \\ 0 & 2 \end{pmatrix}$
	$(\sqrt{3} \times \sqrt{3})R30°$	$\begin{pmatrix} 1 & 1 \\ -1 & 2 \end{pmatrix}$
fcc(110)	$p(2 \times 1)$	$\begin{pmatrix} 2 & 0 \\ 0 & 1 \end{pmatrix}$
	$p(3 \times 1)$	$\begin{pmatrix} 3 & 0 \\ 0 & 1 \end{pmatrix}$
	$c(2 \times 2)$	$\begin{pmatrix} 1 & -1 \\ 1 & 1 \end{pmatrix}$
bcc(110)	$p(2 \times 1)$	$\begin{pmatrix} 2 & 0 \\ 0 & 1 \end{pmatrix}$

matrix notation. In Table 4.1 several surface structures that are more commonly detected on low-Miller-index surfaces are listed both by their matrix and by the Wood notation.

Surface Structures of Clean Low-Miller-Index Metal Surfaces

Several metal surfaces have been extensively studied by a detailed intensity analysis of the LEED beam. Indeed, the study of metal surfaces provided the testing ground for LEED multiple-scattering theories and placed surface crystallography on a firm foundation.

In Table 4.2 the clean surface structures of solids that have been subjected to surface crystallography studies are listed. The location of atoms in the topmost layer is indicated whether they remain unchanged (bulklike) or display relaxation or reconstruction.

Thermodynamically favored surfaces are those with densely packed planes of atoms exposed (Fig. 4.6). In conventional crystallographic terms, these are the low-Miller-index planes [e.g., the familiar (100) and (111) planes of a face-centered cubic lattice] and the bcc (110) plane. The surface unit cells of a low index face of many clean metal surfaces have generally been found to be those expected from the projection of the bulk X-ray unit cell on to the surface, referred to as (1 × 1), and the uppermost layer z-spacing is equal to the bulk value to within about 5 percent. These include the (111) crystal faces of face-centered cubic, aluminum,[3] platinum,[4] nickel,[5] and rhodium,[6] and the (0001) crystal faces of hcp cadmium[7] and beryllium.[8]

However, the Al(110) surface shows a 5 to 15 percent contraction,[9] the Mo(100) surface a 11 to 12 percent contraction,[10] and the W(100) surface a 6 percent contraction[11] of the upper-layer z-spacing with respect to the bulk, while retaining the (1 × 1) surface unit cell. In general, crystal planes having relatively less dense packing of atoms [fcc(110), bcc(100), etc.] will be more prone to relaxation compared to the most densely packed plane of a given crystal surface. This is consistent with the removal of a larger number of nearest-neighbor atoms in forming a surface of the less densely packed planes. To minimize the surface free energy in these cases, a relocation of surface atoms from bulk positions is, therefore, quite likely.

The physical or chemical origin of these contractions can also be explained in different terms. First, one can imagine the electron cloud attempting to smooth its surface (as if there were an electron gas surface tension), thereby producing electrostatic forces that draw the surface

135

Table 4.2. Results of surface crystallography studies for clean surfaces

Monatomic solid surfaces fcc(111)	Surface bond-length relaxation (%)	Method	References	Comments
Ag(111)	0	LEED	2	
		LEED	3	
Al(111)	−1 to +1.5	LEED	1	
	−1 [−3]	LEED and Fourier transf.	1e	
Au(111)	0	LEED	2b	reconstructions often observed
Co(111)	0	LEED	4	high-temperature phase
Cu(111)	−1.3 to 0	LEED	5	
Ir(111)	−0.8	Fourier transf.	6	
Ni(111)	0	LEED	7	
		LEED and averaging	8	
Pt(111)	0	LEED and LEED	9	
		MEIS, HEIS	10	
Rh(111)	−0.3 ± 0.6	LEED	11	
hcp(0001)				
Be(0001)	0	LEED	12	
Cd(0001)	0	LEED	13	
Co(0001)	0	LEED	4	low-temperature phase
Ti(0001)	−0.5	LEED	14	
Zn(0001)	−0.5	LEED and averaging	15	
bcc(110)				
Fe(110)	0	LEED	16	
Na(110)	0	LEED	17	
W(110)	0	LEED and averaging	18	
		LEED	19	

fcc(100)				
Ag(100)	0	LEED	22	
Al(100)	0	LEED	1b, 13, 20	
	0	MEED and LEED	21	
Au(100)	0	LEED	23	metastable surface
Co(100)	−1.5	LEED	24	
Cu(100)	0	LEED and averaging	5a, 25, 26	
	0	LEED and Fourier transf.	27	
	0		1c	
Ni(100)	0	LEED	7a, 7c, 28	
	0	LEED and Fourier transf.	1e	
Pt(100)	0	LEED	29	metastable surface
Rh(100)	0	LEED	30	
Xe	0	LEED	31	
bcc(100)				
Fe(100)	−0.7 to −1.5	LEED	32	less contraction when clean
Mo	−4	LEED	33	
W(100)	−2 to −4	LEED	19, 34	
	0 to −2.5	HEIS	35	
(100)c(2 × 2)	−6	LEED	36	low-temperature reconstruction—zigzag rows of touching top-layer W atoms
fcc(110)				
Ag(110)	−2 to −3	LEED	39	
	−6 to −10	LEED and Fourier transf.	37	
Al(110)	−3 to −4.5	LEED	20	
	−9 to −15	LEED and Fourier transf.	37	
		MEED	21	
		model calc.	38	

(continued)

137

Table 4.2.—continued

Monatomic solid surfaces fcc(111)	Surface bond-length relaxation (%)	Method	References	Comments
Ir(110) (2 × 1)	−2	LEED	40	missing row reconstruction, second-layer atoms pressed sideways somewhat
(110) (1 × 1)	−2.5	LEED	41	quarter-monolayer of randomly positioned O prevents reconstruction
Ni(110)	−1.5	LEED	7a, 28	
	−1.2	LEED and Fourier transf.	37	
	−1.2	MEIS	42	
	+0.3	MEIS	42	one-third monolayer of randomly positioned O present
Rh(110)	−0.9	LEED	43	
bcc(111) Fe(111)	−1.5	LEED	44	
fcc(311) Cu(311)	−1	LEED	45	
Compound surfaces CoO(111)	−5	LEED	46	O-terminated polar face of NaCl structure—top-layer contraction −5% (−15%)

Surface		Method	Description	Ref.
GaAs (100)	0	LEED	bulk structure with As termination; no relaxation	47
(110)	−2.5 −3.6	LEED	zincblende structure with top Ga and As atoms rotated, respectively, into and out of surface (keeping about constant mutual bond length, rotated by projected angle of 27°; Ga and As back bonds contracted by −2.5% and −3.6%, respectively)	48
		Model calc.		49
(110) + (1 × 1)As	0	LEED	substrate has unrelaxed bulk structure; As bonded to surface Ga as in bulk (possible small bond-angle change)	47a
$MgO(100)$	0	LEED	unrelaxed bulk NaCl structure within ±5%	50
$MoS_2(0001)$	−1.6 −3	LEED	layer compound cleaved between two three-plane layers; top contraction by −1.6% (−4.7%); first van der Waals spacing contracted (−3%)	51
$Na_2O(111)$	0	LEED	fluorite structure terminated between two Na layers; no relaxation	52
$NbSe_2(0001)$	−0.6 −1.4	LEED	as $MoS_2(0001)$, but top contraction by −0.2% (−0.6%); first van der Waals spacing contracted (−1.4%)	51
$NiO(100)$	0	LEED	unrelaxed bulk NaCl structure within ±5%	53

(*continued*)

Table 4.2.—continued

Monatomic solid surfaces fcc(111)	Surface bond-length relaxation (%)	Method	References	Comments
Si(111)"(1 × 1)"	−2	LEED	54	bulk structure with −2% (−15%) top contraction
		model calc.	55	bulk structure with −1% (−6%) top contraction
		cluster	56a	bulk structure with −4% (−30%) top contraction
(111)p(2 × 1)	−1	LEED	54	top layer contracted −1% (−8%) buckled ±3% (±22%); second-layer spacing contraction −10% (−10%)
		cluster model calc.	56	qualitatively as above, but buckled ±10% (±50%); no second-layer spacing contraction
(111) (7 × 7) buckled		LEED	57	top double layer may tend toward one planar (graphitic) layer, buckled and coincident with substrate with period (7 × 7)
(100)p(2 × 1) atom pairing		LEED	58–60	top atom pairing (Schlier–Farnsworth model) with elastic relaxations down several layers
		model calc.	61	
TiS$_2$(0001)	−1.7	LEED	62	as MoS$_2$, but top contraction by −1.7% (−5%); first van der Waals spacing contracted (−5%)
TiSe$_2$(0001)	+1.7	LEED	62	as MoS$_2$, but top expansion by +1.7% (+5%); first van der Waals spacing contracted (−5%)
ZnO(0001)	−3	LEED	63	unreconstructed Zn-terminated wurtzite structure with top contraction by −3% (−25%)
(10$\bar{1}$0)	0	LEED	64–66	unreconstructed wurtzite structure, top Zn and O pulled into surface somewhat

ZnSe(110)	−2.5	LEED	67	zincblende structure reconstructed as GaAs(110) qualitatively as above
		model calc.	49a	

1. (a) D.S. Boudreaux and V. Hoffstein, Phys. Rev. **B3**, 2447 (1971); (b) D.W. Jepsen, P.M. Marcus, and F. Jona, Phys. Rev. **B6**, 3684 (1972); (c) A.E. Laramore and C.B. Duke, Phys. Rev. **B5**, 267 (1972); (d) M.R. Martin and G.A. Somorjai, Phys. Rev. **B7**, 3607 (1973); (e) D.L. Adams and U. Landman, Phys. Rev. **B15**, 3775 (1977).

2. (a) F. Forstmann, Jap. J. Appl. Phys., Suppl **2**, Pt 2; (b) F. Soria, J.L. Sacedon, P.M. Echenique, and D. Titterington, Surf. Sci. **68**, 448 (1977).

3. T.C. Ngoc, M.G. Lagally, and M.B. Webb, Surf. Sci. **35**, 117 (1973).

4. B.W. Lee, L. Alsenz, S. Ignatiev, and M.A. Van Hove, Phys. Rev. **B17**, 1510 (1978).

5. (a) G.E. Laramore, Phys. Rev. **B9**, 1304 (1974); (b) P.R. Watson, F.R. Shepherd, D.C. Frost, and K.A.R. Mitchell, Surf. Sci. **72**, 562 (1974).

6. C.-M. Chan, S.L. Cunningham, M.A. Van Hove, W.H. Weinberg, and S.P. Withrow, Surf. Sci. **66**, 294 (1977).

7. (a) J.E. Demuth, P.M. Marcus, and D.W. Jepsen, Phys. Rev. **B11**, 1460 (1975); (b) R. Feder, Phys. Rev. **B15**, 1751 (1977); (c) A.E. Laramore, Phys. Rev. **B8**, 515 (1973).

8. T.C. Ngoc, M.G. Lagally, and M.B. Webb, Surf. Sci. **35**, 117 (1973).

9. (a) L.L. Kesmodel and G.A. Somorjai, Phys. Rev. **B11**, 630 (1975); (b) L.L. Kesmodel, P.C. Stair, and G.A. Somorjai, Surf. Sci. **64**, 342 (1977).

10. (a) E. Bogh and I. Stensgaard, Proc. 7th Int. Vac. Congr. and 3rd Int. Conf. Solid Surf., Vienna, A-2757 (1977); (b) J.F. Vander Veen, R.G. Smeenk, and F.W. Saris, Proc. 7th Int. Vac. Congr. and 3rd Int. Conf. Solid Surf., Vienna, 2515 (1977).

11. D.C. Frost, K.A.R. Mitchell, F.Z. Shepherd, and P.R. Watson, Proc. 7th Int. Vac. Congr. and 3rd Int. Conf. Solid Surf., Vienna, A-2725 (1977).

12. J.A. Strozier and R.O. Jones, Phys. Rev. **B3**, 3228 (1971).

13. H.D. Shih, F. Jona, D.W. Jepsen, and P.M. Marcus, Commun. Phys. **1**, 25 (1976).

14. H.D. Shih, F. Jona, D.W. Jepsen, and P.M. Marcus, J. Phys. **C9**, 1405 (1976).

15. W.N. Unertl and H.V. Thapliyal, J. Vac. Sci. Technol. **12**, 263 (1975).

16. R. Feder and G. Gafner, Surf. Sci. **57**, 45 (1976).

17. (a) S. Andersson, J.B. Pendry, and P.M. Echenique, Surf. Sci. **65**, 539 (1977); (b) P.M. Echenique, J. Phys. **C9**, 3193 (1976).

18. M.A. Lagally, J.C. Buchholz, and A.C. Wang, J. Vac. Sci. Technol. **12**, 213 (1975).

19. (a) R. Feder, Phys. Status Solidi **B62**, 135 (1974); (b) M.A. Van Hove and S.Y. Tong, Surf. Sci. **54**, 91 (1976).

20. R.H. Tait, S.Y. Tong, and T.N. Rhodin, Phys. Rev. Lett. **28**, 553 (1972).

21. N. Masud, C.G. Kinnuburgh, and J.B. Pendry, J. Phys. **C10**, 1 (1977).

22. (a) D.W. Jepsen, P.M. Marcus, and F. Jona, Phys. Rev. **B8**, 5523 (1973); (b) W. Moritz, Doctoral thesis, University of Munich, 1976.

23. R. Feder, Surf. Sci. **68**, 229 (1977).

24. M. Maglietta, E. Zanazzi, and F. Jona, Bull. Am. Phys. Soc. **22**, 355 (1977).

25. (a) G. Capart, Surf. Sci. **26**, 429 (1971); (b) P.M. Marcus, D.W. Jepsen, and F. Jona, Surf. Sci. **31**, 180 (1972); (c) J.B. Pendry, J. Phys. **C4**, 2514 (1971).

26. J.B. Pendry, *Low Energy Electron Diffraction*, Academic Press, London, 1974.

27. A.A. Klieman and J.M. Burkstrand, Surf. Sci. **50**, 493 (1975).

28. R.H. Tait, S.Y. Tong, and T.N. Rhodin, Phys. Rev. Lett. **28**, 553 (1972).

29. R. Feder, Surf. Sci. **68**, 229 (1977).

30. K.A.R. Mitchell, F.R. Shepherd, P.R. Watson, and D.C. Frost, Surf. Sci. **72**, 562 (1978).

31. A. Ignatiev, J.B. Pendry, and T.N. Rhodin, Phys. Rev. Lett. **26**, 129 (1971).

32. (a) R. Feder, Phys. Status Solidi **58**, K137 (1973); (b) K.O. Legg, F. Jona, D.W. Jepsen, and P.M. Marcus, J. Phys. **C10**, 937 (1977).

33. (a) L.J. Clarke, Proc. 7th Int. Vac. Congr. and 3rd Int. Conf. Solid Surf., Vienna, A-2725 (1977); (b) T.E. Felter, R.A. Barker, and P.J. Estrup, Phys. Rev. Lett. **38**, 1138 (1977); (c) A. Ignatiev, F. Jona, H.D. Shih, D.W. Jepsen, and P.M. Marcus, Phys. Rev. **B11**, 4787 (1975).

34. (a) M.K. Debe, D.A. King, and F.S. Marsh, Surf. Sci. **68**, 457 (1977); (b) R. Feder, Phys. Rev. Lett. **36**, 598 (1976); (c) R. Feder, Surf. Sci. **63**, 283 (1977); (d) J. Kirschner and R. Feder, Verh. Deutsch. Phys. Ges. **2**, 557 (1978); (e) B.W. Lee, A. Ignatiev, S.Y. Tong, and M.A. Van Hove, J. Vac. Sci. Technol. **14**, 291 (1977).

(continued)

Table 4.2.—continued

35. L.C. Feldman, R.L. Kauffman, R.J. Silverman, R.A. Suhr, and J.H. Barrett, Phys. Rev. Lett. **39**, 38 (1977).

36. (a) R.A. Barker, P.J. Estrup, F. Jona, and R.M. Marcus, Solid State Commun. **25**, 375 (1978); (b) M.K. Debe and D.A. King, Phys. Rev. Lett. **39**, 708 (1977); (c) M.K. Debe and D.A. King, J. Phys. **C10**, L303 (1977); (d) T.E. Felter, R.A. Barker, and P.J. Estrup, Phys. Rev. Lett. **38**, 1138 (1977).

37. C.-M. Chan, S.L. Cunningham, M.A. Van Hove, and W.H. Weinberg, Surf. Sci. **62**, 1 (1977).

38. M.W. Finnis and V. Heine, J. Phys. **F4**, 637 (1974),

39. M. Maglietta, E. Zanazzi, F. Jona, D.W. Jepsen, and P.M. Marcus, J. Phys. **C10**, 3287 (1977).

40. C.-M. Chan, S.L. Cunningham, K.L. Luke, M.A. Van Hove, W.H. Weinberg, and S.P. Withrow, Surf. Sci. **91**, 440 (1980).

41. C.-M. Chan, S.L. Cunningham, K.L. Luke, W.H. Weinberg, and S.P. Withrow, J. Vac. Sci. and Technol. **16**, 642 (1979).

42. J.F. Van Der Veen, R.A. Smeenk, R.M. Tromp, and F.W. Saris, Ned. Tijdschr, Vacuumtech. **2/3/4**, 284 (1978).

43. D.C. Frost, S. Hengrasmee, K.A.R. Mitchell, F.R. Shepherd, and P.R. Watson, Surf. Sci. **76**, L585 (1978).

44. H.D. Shih, F. Jona, D.W. Jepsen, and P.M. Marcus, Bull. Am. Phys. Soc. **22**, 257 (1977).

45. R.W. Streater, W.T. Moore, P.R. Watson, D.C. Frost, and K.A.R. Mitchell, Surf. Sci. **72**, 744 (1978).

46. A. Ignatiev, B.W. Lee, and M.A. Van Hove, Proc. 7th IVC and 3rd ICSS, Vienna (1977), p. 1733.

47. (a) B.J. Mrstik, M.A. Van Hove, and S.Y. Tong, Bull. Am. Phys. Soc. **23**, 391 (1978); (b) S.Y. Tong, M.A. Van Hove, and B.J. Mrstik, Proc. 7th IVC and 3rd ICSS, Vienna (1977), p. 2407.

48. (a) C.B. Duke, A.R. Lubinsky, B.W. Lee, and P.J. Mark, Vac. Sci. Technol. **13**, 761 (1976); (b) A.R. Lubinsky, C.B. Duke, B.W. Lee, and P. Mark, Phys. Rev. Lett. **36**, 1058 (1976); (c) P. Mark, G. Cisneros, M. Bonn, A. Kahn, C.B. Duke, A. Paton, and A.R. Lubinsky, J. Vac. Sci. Technol. **14**, 910 (1977); (d) S.Y. Tong, A.R. Lubinsky, B.J. Mrstik, and M.A. Van Hove, Phys. Rev. **B17**, 3303 (1978).

49. (a) J.E. Rowe, S.B. Christman, and G. Margaritondo, Phys. Rev. Lett. **35**, 1471 (1975); (b) W.E. Spicer, P.W. Chye, P.E. Gregory, T. Sukegawa, and I.A. Babaloba, J. Vac. Sci. Technol. **13**, 233 (1976); (c) J.R. Chelikowsky, S.G. Louis, and M.L. Cohen, Phys. Rev. **B14**, 4724 (1976); (d) C. Calandra, F. Manghi, and C.M. Bertoni, J. Phys. **C10**, 1911 (1977); (e) E.J. Mele and J.D. Joannopoulos, Phys. Rev. **B17**, 1816 (1978); (f) D.J. Chadi, Phys. Rev. Lett. **41**, 1062 (1978).

50. (a) C.B. Kinniburgh, J. Phys. **C8**, 2382 (1975); (b) C.G. Kinniburgh, J. Phys. **C9**, 2695 (1976).

51. (a) B.J. Mrstik, R. Kaplan, T.L. Reinecke, M. Van Hove, and S.Y. Tong, Phys. Rev. **B15**, 897 (1977); (b) B.J. Mrstik, R. Kaplan, T.L. Reinecke, M. Van Hove, and S.Y. Tong, Il Nuovo Cimento **38b**, 387 (1977).

52. S. Andersson, J.B. Pendry, and P.M. Echenique, Surf. Sci. **65**, 539 (1977).

53. C.G. Kinniburgh and J.A. Walker, Surf. Sci. **63**, 274 (1977).

54. H.D. Shih, F. Jona, D.W. Jepsen, and P.M. Marcus, Phys. Rev. Lett. **37**, 1622 (1975).

55. L.C. Snyder and Z. Wasserman, Surf. Sci. **77**, 52 (1978).

56. (a) W.S. Verwoerd and F.J. Kok, Ned. Tijdschr. Vacuumtech. **2/3/4**, 303 (1978); (b) D.J. Chadi, Phys. Rev. Lett. **41**, 1062 (1978).

57. (a) J.D. Levine, S.H. McFarlane, and P. Mark, Phys. Rev. **B16**, 5415 (1977); (b) P. Mark, J.D. Levine, and S.H. McFarlane, Phys. Rev. Lett. **38**, 1408 (1977).

58. J.A. Appelbaum and D.R. Hamann, Surf. Sci. **74**, 21 (1978).

59. (a) K.A.R. Mitchell and M.A. Van Hove, Surf. Sci. **75**, 147L (1978); (b) S.Y. Tong and A.L. Maldonado, Surf. Sci. **78**, 459 (1978).

60. S.Y. Tong, private communication.

61. J.A. Appelbaum, G.A. Baraff, and D.R. Hamann, Phys. Rev. Lett. **35**, 729 (1975); Phys. Rev. **B14**, 588 (1976).

62. B. Lau, B.J. Mrstik, S.Y. Tong, and M.A. Van Hove, Phys. Rev. **B15**, 897 (1977).

63. C.B. Duke, and A.R. Lubinsky, Surf. Sci. **50**, 605 (1975).

64. C.B. Duke, A.R. Lubinsky, S.C. Chang, B.W. Lee, and P. Mark, Phys. Rev. **B15**, 4865 (1977).

65. C.B. Duke, A.R. Lubinsky, B.W. Lee, and P. Mark, J. Vac. Sci. Technol. **13**, 761 (1976).

66. A.R. Lubinsky, C.B. Duke, S.C. Chang, B.W. Lee, and P. Mark, J. Vac. Sci. Technol. **13**, 189 (1976).

67. (a) C.B. Duke, A.R. Lubinsky, M. Bonn, G. Cisneros, and P. Mark, J. Vac. Sci. Technol. **14**, 294 (1977); (b) P. Mark, G. Cisneros, M. Bonn, A. Kahn, C.B. Duke, A. Paton, and A.R. Lubinsky, J. Vac. Sci. Technol. **14**, 910 (1977).

atoms toward the substrate. This effect should be stronger the less closely packed the surface is. Second, with fewer neighbors, the two-body repulsion energy is smaller, allowing greater atomic overlap and therefore more favorable bonding at shorter bond lengths. Third, for surface atoms the bonding electrons are partly shifted from the cut bonds to the remaining noncut bonds, thereby increasing the charge content of the latter and so reducing the bond length. On ionic crystal surfaces the asymmetries in the ionic electrostatic forces at surfaces may explain the contrast between similar bond-length contractions observed on $CoO(111)$[12] and the lack of observable contractions on $MgO(100)$[13] and $NiO(100)$.[14]

The foregoing descriptions of the origin of bond-length contractions at surfaces are consistent with the observations made when adsorbates are deposited on these surfaces. The shortened bond lengths are again systematically lengthened (sometimes to more than their bulk values) by the presence of adsorbates.

More dramatically, the (100) and (110) faces of iridium, platinum, and gold are reconstructed,[15–19] that is, the two-dimensional surface unit cell is different from that given by the termination of the bulk structure on the plane of interest.

The ideal unreconstructed Ir, Pt, and Au (100) surfaces have a square net of atoms. However, surface reconstruction that produces a superlattice basically five times longer in one direction than for this ideal surface is observed experimentally by low-energy electron diffraction. For Ir(100) the superlattice is more precisely denoted (5 × 1), while for Pt(100) the superlattice is described by the matrix notation $\left(\begin{smallmatrix} 5 & 1 \\ -1 & 14 \end{smallmatrix}\right)$, and for Au(100) the superlattice is denoted (5 × 20). The (100) surfaces of platinum, for example, exhibit the diffraction pattern illustrated in Figure 4.7. The spots from a nominal (1 × 1) surface occur at the corners of the squares, but there are extra or "fractional-order" spots in between indicative of domains of a $\left(\begin{smallmatrix} 5 & 1 \\ -1 & 14 \end{smallmatrix}\right)$ superstructure. This large unit cell can be constructed by the superposition of smaller unit cells, which are rationally related, producing the so-called "coincidence" structures.

The restructured (100) surfaces exhibit the same apparent surface structure in the temperature range from 300 K to the melting point. This structure has not yet been studied below 300 K. In the presence of various adsorbates (CO or olefinic hydrocarbons, for example), the surface snaps back to its bulklike, fourfold rotational symmetry surface unit cell structure in as little time as it takes to adsorb less than a monolayer of the gas. Upon recleaning, the reconstructed surface reappears. It is possible to prepare a clean surface with a metastable (1 × 1) bulklike surface

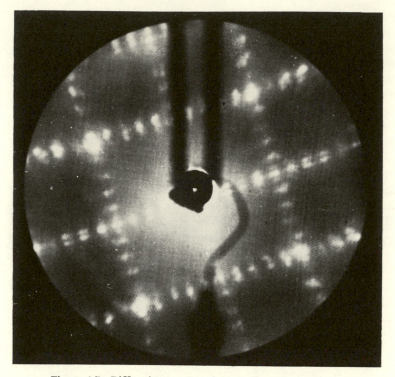

Figure 4.7. Diffraction pattern of the (5 × 1) type surface structure on the (100) crystal face of platinum at 124 eV.

structure as was shown for Pt and Ir (110) surfaces.[20,21] By bombardment using oxygen ions and subsequent careful annealing, a metastable (1 × 1) surface structure is produced. When this metastable surface is heated to about 600 K, it reconverts to its stable reconstructed (5 × 1) surface structure. Among the highest-atomic-density, lowest-surface-free-energy surfaces, only Au(111) has been reported to show signs of reconstruction.[22] Future studies are likely to further elucidate the relationship between the electronic structure and coordination number of surface atoms and their ease of relaxation and reconstruction.

Evidence from LEED intensity analyses[23] and laser simulation[24] of the diffraction pattern indicates that these three surfaces have a hexagonal close-packed top layer that fits in slightly different ways on the square-net substrate. For Ir(100) the hexagonal layer is uniaxially contracted (relative to bulk interatomic distances) by about 2 percent to obtain the fit over five substrate interatomic distances. This surface structure is shown in Figure 4.8. For Pt(100) an isotropic contraction of the hexagonal overlayer by about 2.5 percent, together with a rotation by 0.67°, seems to explain the data, and for Au(100) an isotropic contraction by about 4 percent without rotation is most likely. In addition, the hexagonal top layers appear to buckle by about 0.5 Å, owing to the irregular positioning of individual top-layer atoms with respect to the square-net substrate. It should be noted that the first time reconstruction on metal surfaces was reported was for the Pt(100)-(5 × 1) surface structure in 1965.[15a] There was considerable debate during the next several years as to whether the structures that were uncovered with increasing frequency were the properties of the clean surface or were induced only by the presence of contaminants. With the advent of widespread use of Auger electron spectroscopy, it has become clear that in most cases the reconstruction is the property of the clean metal surface. However, impurities on or below the surface can markedly influence the thermal stability of these clean surface structures.[18]

The (110) face of these materials (Pt, Ir, and Au) often exhibits a (2 × 1) reconstruction or, more generally, (n × 1) reconstructions [such as on Ir(110) with n = 2, 3, or 4], with sometimes a statistical distribution of the values of the integer n [as on Au(110), where $n = 2$ dominates]. Several models for these reconstructions have been suggested, but the "missing-

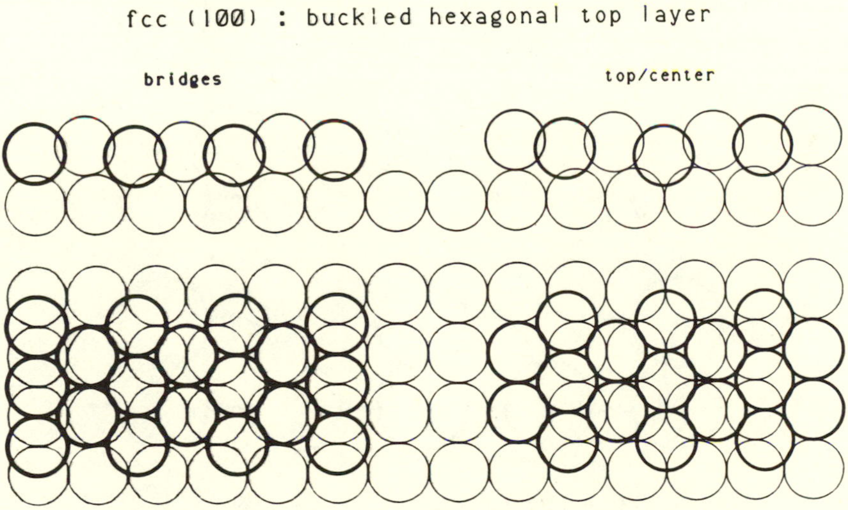

fcc (100) : buckled hexagonal top layer

bridges   top/center

Figure 4.8. Arrangement of atoms on the reconstructed (100) crystal faces of platinum, gold, and iridium. Side and top views are shown.

row" model seems to be the most promising in studies of Ir(110) (2 × 1) (Fig. 4.9) and Au(110)(n × 1) with random n. In this model[25] small facets of the hexagonal close-packed (111) face are built [note the analogy with the fcc(100) reconstructions], which is consistent with the knowledge that the (111) face of fcc metals is energetically the most favorable.

The (100) crystal faces of Mo and W exhibit a different kind of reconstruction, which is only detectable at low surface temperatures.[26] At 300 K these crystal faces have the same unit cell as expected from the projection of the bulk bcc X-ray unit cell to this surface. Upon cooling, a $c(2 \times 2)$ surface structure appears that exhibits increasing diffraction beam intensities with decreasing temperature. Adsorbates can inhibit this reconstruction just as in the case of the (100) surfaces of Au, Pt, and Ir at elevated temperatures.[26] The low-temperature reconstruction of the clean surface is reversible and reproducible as the temperature is cycled. The mechanism responsible for this could be a charge-density wave that induces a structural wave which can have a wavelength related to the lattice constant [as with W(100)] or not related to it [as with Mo(100)].

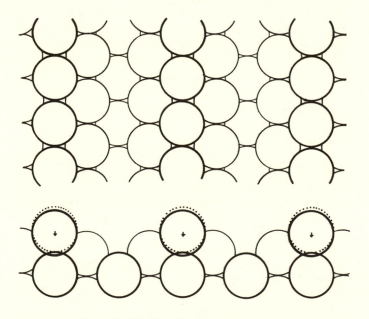

fcc (110) (2×1) missing row

Figure 4.9. Atomic arrangement in the missing-row model of the reconstructed iridium (110) crystal surface.

Hydrogen chemisorption at less than full coverage appears not to change the structure of the $W(100)c(2 \times 2)$ surface noticeably. Interestingly, chemisorption of hydrogen at room temperature on an unreconstructed $W(100)$ surface seems to generate the same $c(2 \times 2)$ reconstruction as that obtained by cooling.[26a] At full hydrogen coverage, the reconstruction disappears and W—W bond-length contractions seem to disappear as well.[27]

It is probable that reconstruction of low-Miller-Index metal surfaces at low or high temperatures will be uncovered for other bcc, fcc, or hcp solids as more clean crystal faces of these materials are investigated. The nature and cause of these surface-phase transformations are not well established at present. The ease of structural transition from metastable to stable structure upon adsorption or removal of adsorbates indicates the likelihood of electronic transitions that accompany reconstruction. Chemical-bonding arguments put forward by Brewer,[28,29] which are very successful in predicting alloy phase diagrams, can be used to rationalize the surface reconstruction of some of the metals. According to this theory, the formation of the solid from the metal atoms involves electron promotion to higher bound but unoccupied orbitals. The positive promotion energy is paid for by the negative lattice energy, as the energy of the system is decreased by forming many metal–metal bonds. At the surface there are fewer nearest neighbors, however, as compared to atoms in the bulk. Thus the electronic structure that is stabilized in the lower-symmetry-surface environment may be markedly different from that for the bulk metal. Since the surface atoms are surrounded by atoms only on one side and there is vacuum on the other side, they may change their coordination number by slight relocation with simultaneous changes of the electronic structure. A comprehensive theory of metal surface reconstruction would certainly aid future studies in this field.

Other metal surfaces that exhibit reconstruction include the bismuth[30] $(1 1\bar{2}0)$, antimony[30] $(1 1\bar{2}0)$, and tellurium[31] (0001) crystal surfaces.

Surface Structure of Alloys

There is considerable interest in alloy surfaces, due in part to their potential as efficient catalysts and in part to their important mechanical and corrosion-resistant properties. Order–disorder transformations, which are well characterized for many bulk alloy systems, may also be studied in the surface region. Studies of the surface structures by LEED have been carried out for alloys such as Cu–Au,[32] Cu–Al,[33] and Ag–Pd.[34] The appearance of superlattice beams in the LEED patterns for the

Cu₃Au(100) surface,[35] for example, indicates the presence of long-range order in the alloy surface, as in the bulk below the order–disorder transition temperature, T_c = 390°C. However, the temperature dependence of these beams seems to indicate a different behavior of the long-range-order parameter for the surface of this alloy as compared to the bulk. The surface segregation of gold was reported for the Cu–Au(100) surface.[35]

Structure of Low-Miller-Index Semiconductor Surfaces

Elemental Semiconductors

Several elemental semiconductor surfaces (Si, Ge) have been studied by LEED, and in some cases diffraction beam intensities have been analyzed. Whereas surface reconstruction is relatively rare for metals, it seems to be very common for semiconductors. In fact, surface transformations from one ordered structure to another take place frequently as the temperature is increased. In a general way, this behavior can be ascribed to the more localized, directional character of bonding in semiconductors as opposed to the delocalized bonding picture more appropriate for metals. At the surface the atoms seek equilibrium positions that optimize their bonding interaction. A complete set of LEED intensity data has been obtained for the (2 × 1) reconstructed surface of Si(100), and its surface structure has recently been solved by surface crystallography.[36] For Si(100)p(2 × 1) a long search has produced a structure derived from the Schlier–Farnsworth model,[36e] in which adjoining surface atoms (each with two unsatisfied "dangling bonds") simply bond together by bending over toward each other and pairing up dangling bonds (one dangling bond per surface atom remains unsatisfied). A substantial bond bending occurs, and this distortion propagates elastically through the lattice down to a few layers' depth, as shown in Figure 4.10. Adsorption of hydrogen up to a certain coverage onto this surface seems not to change the nature of this reconstruction. However, a higher coverage of hydrogen destroys the reconstruction and restores the bulk geometry at the silicon substrate surface.[37]

Let us consider the case of the reconstructed silicon (111) surface. Upon cleavage in ultrahigh vacuum, the LEED diffraction pattern shows diffraction beams indicative of (2 × 1) superlattice periodicity.[38] This structure is metastable and converts with increasing temperature to an apparent (1 × 1) structure around 400°C. Upon further heating in about the 600 to 700°C temperature range, the (1 × 1) structure converts to a

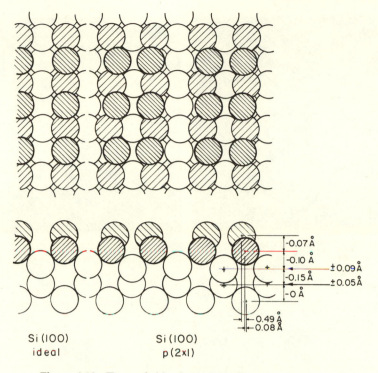

Figure 4.10. Top and side views of ideal bulklike Si(100) at the left, and Si(100) $p(2 \times 1)$ in the modified Schlier–Farnsworth model at the right. Layer-spacing contractions and intralayer atomic displacement relative to the bulk structure are given. Shading differentiates surface layers.

(7×7) superstructure. The beautiful diffraction patterns of this structure are shown in Figure 4.11. Rowe and Phillips[39] have argued that whereas a surface buckling model of the kind proposed by Haneman[40] provides a satisfactory explanation of the metastable (2×1) surface, a qualitatively different model, such as the one proposed by Lander[38c] involving ordered surface vacancies, is necessary to explain the properties of the annealed (7×7) surface. At present the evidence is inconclusive, and most experiments and theories have focused on the (2×1) structure. The essence of Haneman's model for this surface is as follows.[40] In the bulk material the Si atoms are tetrahedrally coordinated with an sp^3-hybrid bonding scheme. The surface atoms, however, have only three nearest neighbors, and the remaining dangling bond may have a tendency to become more p-like. If this happens, the back bonds

149

Figure 4.11. Low-energy electron diffraction patterns of the reconstructed Si(111) crystal face exhibiting a (7 × 7) surface structure taken at four different electron energies.

will tend toward sp^2 hybridization or trigonal bonding, which is essentially planar. These considerations suggest a movement of the surface atom toward the second plane of atoms (contraction of the back bonds), but this will, in turn, give rise to lateral forces on the second-layer atoms, forces that can be released if other atoms in the upper layer are slightly raised. The net result, of course, is a slight buckling or rumpling of the surface caused by the raising and lowering (0.1 to 0.2 Å) of alternate rows of surface atoms, thereby producing a (2 × 1) periodicity. A number of theoretical calculations for the electronic structure of the idealized

Si(111)-(1 × 1) surface have been reported, notably the initial self-consistent model due to Appelbaum et al.,[41] showing the partially occupied dangling-bond surface state band lying in the semiconductor band gap. Schlüter and Cohen[42] have subsequently considered the effect of the buckling model for (2 × 1) reconstruction and found that the dangling bond state is split with a transfer of charge from the inwardly relaxed atoms to the outwardly relaxed ones, the surface becoming partially ionic. The reconstruction of Ge and diamond low-Miller-index surfaces has also been reported.[43,44] However, these crystal faces have not been studied in great detail as has been the case for the Si low-Miller-index surfaces. It would be of considerable importance to study the surface structural changes that may occur at tin crystal faces. Tin exists in the tetragonal modification, and at low temperatures it converts to the diamondlike structure.

Compound Semiconductors

The tetrahedrally coordinated compound semiconductors have either wurtzite or zincblende structures. Only a few crystal surface structures have been determined by LEED structure analysis. However, the available data indicate the frequent presence of surface relaxation or reconstruction on the surfaces of these solids. The GaAs(110) crystal face is unreconstructed, but LEED intensity analysis by Lubinsky et al.[45] indicates about a 0.5-Å vertical displacement of the anions and cations in the surface layer, with the cations lying closer to the second layer of atoms under the surface. The proposed surface structure is shown in Figure 4.12. This surface has an equal number of gallium and arsenic atoms present at crystal stoichiometry.

Nonstoichiometry is apparently a major factor in the observed reconstruction of the polar faces of compound semiconductors such as GaAs(111) and ($\bar{1}\bar{1}\bar{1}$).[46] The (111) face, for example, would ideally have all Ga atoms at the surface bonded to As atoms immediately beneath the surface, whereas the reverse would be true of the ($\bar{1}\bar{1}\bar{1}$) face. However, the ($\bar{1}\bar{1}\bar{1}$) surface has been found to lose As at elevated temperatures and to form a ($\sqrt{19} \times \sqrt{19}$) surface structure, whereas the low-temperature (2 × 2) structure is arsenic-stabilized. Similarly, phosphorus is found to preferentially desorb at high temperatures from the GaP($\bar{1}\bar{1}\bar{1}$) surface.[47] Another zincblende structure material, ZnSe, appears to behave similarly to GaAs. Its nonpolar (110) crystal face shows no signs of reconstruction, only a small relaxation[48] of about 0.6 Å.

Among the wurtzite structure semiconductors, only ZnO[49] and CdS[50] have been studied. The nonpolar ZnO($11\bar{2}0$) crystal face is unrecon-

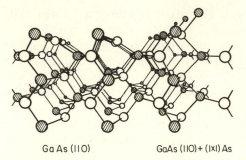

GaAs (110)   GaAs (110)+ (1x1) As

Figure 4.12. Perspective view looking along the surface of clean, reconstructed GaAs(110) at the left, and GaAs(110) + (1 × 1)As at the right. Open and shaded circles represent galium and arsenic atoms, respectively.

structed, whereas the polar ZnO(10$\bar{1}$0) surface exhibits marked relaxation. The uppermost Zn sublattice spacing is contracted by 0.3 Å and the oxygen sublattice spacing by 0.1 Å. This structural rearrangement provides evidence for the ionic surface relaxation predicted by Benson et al.[51]

A LEED intensity analysis has been reported for the (0001) crystal faces of the layered compounds MoS_2 and $NbSe_2$.[52,53] These interesting compounds consist of layers of covalently bonded atoms coupled to similar layers by weak van der Waals forces. Each layer has a transition-metal-layer atom sandwiched between planes of the sulfur or selenium atoms, respectively. In the case of MoS_2, the investigators found no evidence for surface reconstruction. However, assuming a 1.6 percent contraction of the top layer (relaxation), good agreement resulted between calculated and experimental I–V profiles. For $NbSe_2$ there is also a small contraction of the topmost atomic layer, but there is no reconstruction.

Structure of Ionic Crystal Surfaces

Ionic crystals are insulators consisting of a lattice of alternating positively and negatively charged ions (e.g., Na^+ and Cl^-) for which the bulk cohesive energy is due to Coulomb forces between ions. However, at the surfaces of these materials a net electric field arises from the ionic half-space beneath the surface, which, in turn, may polarize the ions in the surface layer. These polarization fields affect the anions and cations differently because of their different charge or size, and may cause considerable distortion at the surface. Definitive studies of the surface atomic structure of ionic materials have seldom been made by low-energy electron diffraction. However, McRae and Caldwell[54] did find

LEED evidence for relaxation. Their studies indicate a distortion of the (100) surface of LiF, indicating that the top Li and F sublayers do not lie in the same plane (the surface is periodically buckled). This result is qualitatively consistent with the theoretical predictions of Benson and Claxton[55] and has been further investigated with LEED intensity calculations.

A number of studies have pointed to possible nonstoichiometry of alkali-halide crystal surfaces upon cleavage.[56,57] These surfaces may also become charged or damaged under electron beam exposure. In general, there is preferential desorption of the halogen atom from the surface by the electron beam with rather high efficiency and associated formation of F or M color centers. Some of these effects can be minimized by working at elevated temperatures to increase conductivity and to permit rapid diffusion of ions from the bulk to recombine with vacancies at the surface. The elementary theoretical models of the surface structure outlined above may have to be modified to include the possibility of varying degrees of nonstoichiometry at the surface. LEED studies of the LiH(100) crystal face indicated[58] that this surface loses hydrogen preferentially and the surface becomes covered with Li. During these chemical changes, however, the surface diffraction pattern remained characteristic of a (1 × 1) unreconstructed surface structure.

Structure of Oxide Surfaces

The interaction of oxygen with metals to produce various surface oxides is of considerable chemical and technological interest. However, relatively few structural studies have been carried out by LEED. Changes in chemical composition have been related to the formation of new surface unit cells, as evidenced for the (0001) surface of α-alumina (Al_2O_3), where reconstruction at elevated temperatures in vacuum was associated with loss of oxygen.[59] The observed transformation from a (1 × 1) to a ($\sqrt{31}$ × $\sqrt{31}$) unit cell could be reversed by oxidation of the surface in 10^{-4} torr of oxygen at 1000 to 1200°C. The reconstructed surface has been interpreted in terms of a reduced oxide surface layer containing Al^+ or Al^{2+} formal charge ions.[59] This is one of several examples in which the surface environment stabilized a reduced oxidation state. Owing to a change in the ion size and also as a result of nonstoichiometry, restructuring occurs in a reversible manner with respect to the bulk unit cell. Fiermans and Vennik[60] have studied the transformation of a $V_2O_5(010)$ surface to one characteristic of

$V_{12}O_{26}(010)$ under the influence of the electron beam. Upon preferential removal of oxygen, the authors found that the transformation proceeds by domain formation and that two different intermediate superstructures of (4×1) and (1×2) periodicity were involved, depending on the degree of sample nonstoichiometry. Reconstruction that is induced by nonstoichiometry has been found for the (100) faces of the rutile structure of TiO_2.[61] This surface, upon heating, loses oxygen, and a series of surface structures of (1×3), (1×5), and (1×7) unit cells form. Upon heating the (1×7) surface structure in oxygen, the surface reverts to its more stoichiometric (1×3) surface structure.[61] These surface structural changes are associated with the loss of oxygen in the surface layers and with the formation of ordered oxygen vacancy structures. Simultaneously, the presence of Ti^{3+}, a lower formal oxidation state of Ti ions can be detected at the surface by electron loss spectroscopy and ultraviolet photoelectron spectroscopy. The presence of lower-oxidation-state ions also changes the chemical activity of the (100) surface, as shown by studies of the chemisorption of water, hydrogen, and oxygen. $BaTiO_3$[62] and $SrTiO_3$[63] crystal faces have also been studied. These ternary oxides show interesting structural and compositional variations, as the function of temperature and surface preparation. The "stoichiometric" $SrTiO_3$ crystal surfaces contain large concentrations of Ti^{3+}, lower formal oxidation-state ions indicating partial reduction in the surface layer. The surface composition varies reversibly with temperature. On heating, the Ti^{3+} surface concentration changes markedly, but upon cooling, it returns to its original value. Upon high-temperature ion bombardment (at 900 K using inert gas ions), the surface composition changes entirely. A strontium oxide surface compound forms which is stable between 300 K and 1100 K. This surface compound may only be removed by chemical etching or by ion bombardment at 300 K. The chemical activity of these surfaces depends very strongly on both surface stoichiometry and the surface oxidation states of the various elements.

Surface-structure analysis is lacking for oxide surfaces at present. In more recent work leading to quantitative structural determinations, Legg et al. have reported LEED intensity data for the (100) surface of MgO.[64] An analysis by Kinniburgh[65] of modern LEED intensity data indicates that the MgO(100) crystal face is unreconstructed and shows no evidence of relaxation in the topmost surface layer. Aberdam and coworkers have observed a $(\sqrt{3} \times \sqrt{3})$ structure on the (001) crystal face of barium titanate ($BaTiO_3$) prepared by different heat treatments.[62] The surface rearrangement is considered to be due to the ordering of vacancies at the surface similar to that observed in $SrTiO_3$. However, structure analyses have not yet been carried out for these surface structures.

Structure of Molecular Crystal Surfaces

Molecular crystals constitute a large and important group of materials that include most organic solids. Only very recently have the surface structures of some of these materials been investigated on an atomic scale by LEED. Ice and naphthalene have been grown by vapor deposition[66] on a Pt(111) substrate, and observation of LEED diffraction patterns has facilitated studies of the surface morphologies as a function of substrate structure, temperature, and exposure. The ice structure was obtained by exposing a clean Pt(111) surface to a water vapor flux of 10^{14} molecules/cm^2/sec at substrate temperatures of from 125 to 155 K for several minutes. The diffraction pattern observed is almost identical to that of domains of a Pt(111)-$(\sqrt{3} \times \sqrt{3})R30°$ surface structure rotated 60° to each other; the domains are of the order of 30 Å in linear dimensions. The pattern is probably due to domains of the (111) face of fcc ice grown parallel to the Pt(111) surface. Similarly, ordered surface structures of naphthalene were grown between 105 and 200 K, and the observed diffraction pattern is that expected from the monoclinic naphthalene crystals growing with (001) planes parallel to the Pt(111) surface. Several other materials were studied similarly, including benzene, trioxane, n-octane, cyclohexane, and methanol.[67]

Ordered films of Cu, Fe, and metal-free phthalocyanines have recently been grown by vapor deposition on Cu(100), Cu(111), and Pt(111) substrates (monolayer to 500 Å film thickness) and studied by conventional LEED techniques.[68] The diffraction patterns are consistent with a relatively large surface unit cell containing one phthalocyanine molecule, with the plane of the molecule parallel to the surface plane. The first layer of molecules is chemically bonded to the substrate. It appears that the central metal atom of the molecule plays only a limited role in this bonding. Other materials that have been studied using similar vapor deposition techniques are amino acids; glycine, tryptophan, and alanine, grown on metal substrates.[69]

Each of the studies described above has indicated that growth of an ordered monolayer phase is essential to ordered growth of the film, and that suitable matching of metal substrate and molecular crystal is of considerable importance in such studies. It appears that for many molecular crystals the structure of the ordered monolayer controls the growth and orientation of subsequent layers that are deposited in growing the molecular single-crystal thin film. The control of the monolayer that establishes the orientation and growth pattern of subsequent layers is called pseudomorphism, a phenomenon that appears to be important and predominant for large molecular crystals grown by vapor deposition on ordered substrates.

General problems encountered in LEED studies of molecular crystals are sample damage and space charging under electron beam exposure. The vapor pressure of the sample must also be temperature-controlled to allow study under ultrahigh-vacuum conditions. The charging effect can be largely avoided by vapor growth of suitably thin films on conducting substrates. This procedure does, however, rather severely limit the types of surfaces that can be studied, and it is anticipated that LEED systems utilizing much lower beam currents can remedy the charging problem as well as alleviate the problem of electron beam damage. These advances would then allow study of a wide variety of molecular crystal surfaces of higher molecular weight, which could be prepared by suitable cleavage or cutting of bulk crystals.

Structure of Inert-Gas Crystal Surfaces

Thin single crystal films of Xe, Kr, and Ar have been grown at low temperatures on ordered crystal surfaces. The first of these studies was reported by Lander and Morrison,[70a] who have grown Xe crystals on graphite single-crystal surfaces. Xenon forms an ordered monolayer surface structure and grows on top of the basal plane of graphite with a (111) surface orientation. Argon and xenon crystals were grown on Nb,[71] and Kr on graphite surfaces.[72] In all cases studied so far, the growing inert-gas crystal maintains its (111) surface orientation regardless of substrate orientation on all the substrates. It appears that for these systems the attractive adsorbate–adsorbate interaction controls the growth morphology.[70b-d] This is in contrast to the growth behavior of most molecular crystal surfaces, where strong substrate–adsorbate interaction frequently locks in the monolayer surface structure and all subsequent layers assume the same orientation as the monolayer while they grow.

Stepped, High-Miller-Index Surfaces

Low-energy electron diffraction studies have been applied, mostly, to the study of the surface structure of close-packed faces of solids of low Miller index. These surfaces are chosen for structural investigations because they have the lowest surface free energy, and they are therefore stable with respect to rearrangement or disordering up to or near the melting point. Studies of surfaces of high Miller index and higher surface free energy are important in their own right. They are known to play important roles during evaporation, condensation, and melting. Steps and kinks are sites where atoms break away as an initial step leading to desorption or where atoms migrate to upon condensation to be incorporated into the crystal lattice. Since the addition or removal of atoms to or from a kink site leaves the surface structure of the solid unchanged, theories of crystal growth,[73] evaporation,[74] and the kinetics of melting[75] have identified the significance of these lower-coordination-number sites in controlling the rate processes associated with phase changes.

Recent studies of chemisorption and catalytic reactions using single-crystal surfaces revealed different binding energies and enhanced chemical activity at steps and kinks on high-Miller-index transition metal surfaces as compared to low-Miller-index surfaces.[76] Adsorption of diatomic and polyatomic molecules frequently leads to dissociation with higher probability at steps and kinks than on atomic terraces.[76] It is therefore important to elucidate their atomic structure and stability under a variety of experimental conditions in the presence of reactive and inert gases and in vacuum.

The earliest diffraction observation from a high index surface is probably that of niobium.[77] The first detailed study of a stepped surface of this type is that of Ellis and Schwoebel.[78] They examined a uranium dioxide, UO_2, crystal cut at $11.4°$ from the (111) plane in the (112) zone. Heating this sample at 1100 K in ultrahigh vacuum for 1 hr produced a diffraction pattern resembling that from a $UO_2(111)$ crystal face except that the diffraction spots were elongated and appeared to be split into multiplets. Heating at 1200 K in 10^{-7} torr oxygen generated a pattern with each (111) beam resolved into a well-defined doublet at certain electron energies. This behavior, with doublets appearing in place of single spots, is characteristic of terrace geometry and has been reported for all the stepped surfaces examined. Figure 4.13 shows diffraction patterns from four different high-Miller-index surfaces of platinum. The appearance of doublet diffraction spots is readily detectable. Low-energy electron diffraction investigations of copper,[79] germanium,[80] gallium arsenide,[81] and platinum high-Miller-index planes[82] indicate that the surfaces consist of low-index planes separated by steps often one atom in height. The ordered stepped surfaces displayed varying degrees of thermal stability.

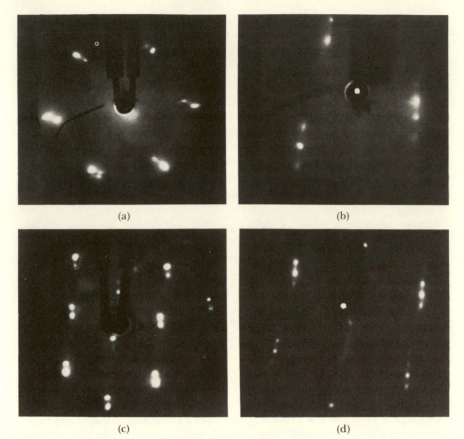

(a)

(b)

(c)

(d)

Figure 4.13. LEED patterns of the (a) Pt(755), (b) Pt(679), (c) Pt(544), and (d) Pt(533) stepped surfaces.

Figure 4.14 shows one crystallographic zone of a face-centered cubic crystal. The corners of the triangle represent the three lowest-Miller-index, most-stable crystal faces, (111), (100), and (110). When higher-Miller-index crystal faces of fcc metals are cut at angles that would place them along the sides of the stereographic triangle, they generally exhibit a one-atom-height step-terrace configuration. Both step and terrace orientations are of low Miller index, (111), (100), or (110), for these surfaces, and the terrace width depends on the angle of cut with respect to the flat low-index surfaces; the steeper the angle, the narrower the terrace. When surfaces are prepared whose Miller indices fall inside the triangle, they also exhibit a step-terrace structure, except that the steps

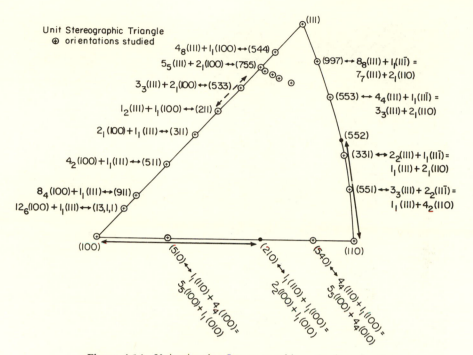

Figure 4.14. Unit triangle of stereographic projection, showing location of various platinum surfaces that were studied. The crystal faces are denoted by their Miller index and also by the microfacet notation.

are now of high Miller index. As a result, the steps have a high density of ledges or kinks that, under proper conditions, also order. In Figure 4.15, several stepped surfaces are schematically displayed to indicate how variable the terrace widths and step orientations are. In Figure 4.16, several crystal faces are shown that have different concentrations of ordered kinks along the steps. By suitable preparation, crystal faces with well-controlled step and kink concentrations can be prepared, and the densities of these irregularities can be systematically varied, if desired. The terrace width is calculated from the doublet separation in the diffraction pattern, while the step height is obtained from the variation of the intensity maximum of the doublet diffraction beam features with electron energy.[83] Let us consider the analysis of these diffraction patterns.

Several approaches are available in the literature, all kinematic and all yielding the same results. Henzler,[84] extending the derivation by Ellis and Schwoebel,[85] has shown that the scattered intensity, I, at an angle ϕ

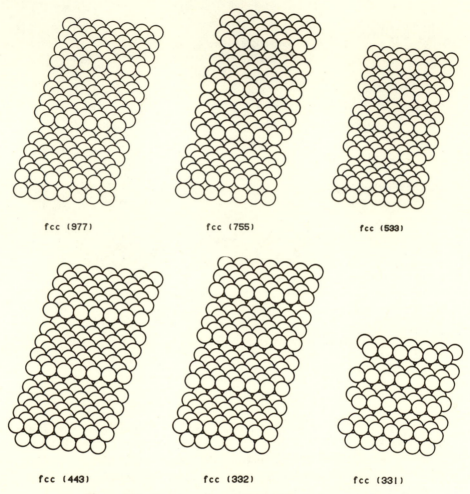

fcc (977)   fcc (755)   fcc (533)

fcc (443)   fcc (332)   fcc (331)

Figure 4.15. Structure of several high-Miller-index stepped surfaces with different terrace widths and step orientations.

with electron beam incidence normal to the terraces is given by

$$I - \text{constant} = \frac{\sin^2 \left[\tfrac{1}{2}k \cdot a(N + 1) \sin \phi \right]}{\sin^2 \left(\tfrac{1}{2}k \cdot a \sin \phi \right)}$$

$$\times \sum_{i=-\infty}^{+\infty} \delta \left[\tfrac{1}{2}k(N \cdot a + g) \sin \phi + \tfrac{1}{2}kd \, (1 + \cos \phi) - i\pi \right] \qquad (8)$$

where the terrace has $(N + 1)$ rows, $k = 2\pi/\lambda$, a is the separation of the

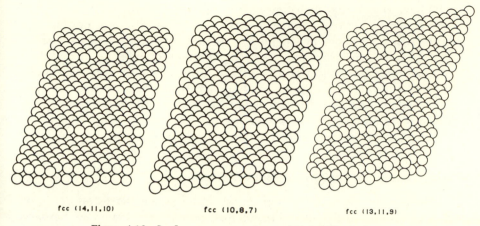

fcc (14,11,10)   fcc (10,8,7)   fcc (13,11,9)

Figure 4.16. Surface structures of several high-Miller-index surfaces with deferring kink concentrations in the steps.

atomic rows, d the step height, and g the horizontal shift of one terrace compared to that below it. The first term is the intensity distribution for a grating of $(N + 1)$ slits, and the maxima are given by the Bragg equation

$$\tfrac{1}{2}k \cdot a \sin \phi = n\pi \qquad (9)$$

The second term in the sum of δ functions with a separation $\Delta\phi$ given (near $\phi = 0$) by $\Delta\phi = \lambda/(Na + g)$, in other words, dependent only on the width and the displacement of the terraces. When two delta functions fall on a maximum of the intensity curve, a doublet arises, and when only one delta function falls on the maximum of the intensity function, a singlet is observed. The delta functions converge toward the specular reflection of the high-index plane. The spot pattern itself, however, converges toward a (00) spot of the terrace plane. It has been shown that the separation of the doublet is inversely proportional to $(Na + g)$, the terrace width, which is therefore easily determined. Also, the step height can be found from

$$V_{00} \text{ (singlet max)} = \frac{150}{4d^2}s^2 \qquad (10)$$

where V_{00} are the voltages where a singlet of maximum intensity is observed, d is the step height, and s is an integer. This method has been applied to the determination of step height by Henzler[81] and by Joyner et al.[82,86] The diffraction patterns to be expected from stepped surfaces have also been examined, using laser simulation, by Campbell and Ellis,[87]

161

who have shown that the single scattering model of the diffraction pattern is potentially very informative. The terrace width does not have to be very precise to obtain satisfactory diffraction patterns. Houston and Park,[88] in a theoretical study, have shown that there may be a great deal of uncertainty in the step width. All that is needed is that on an average the step width be well defined to obtain a diffraction pattern of satisfactory quality. That is, if the diffraction pattern indicates that the terrace width is six atoms wide, this does not rule out the presence of a large number of terraces four, five, seven, or eight atoms wide.

Nomenclature of High-Miller-Index Surfaces

The atomic structures of high-Miller-index surfaces are composed of terraces, steps, and kinks. The terrace and kink concentrations vary widely. The structure of these surfaces cannot be readily visualized, however, from their Miller indices. Inversely, it is difficult to determine the Miller indices from the atomic surface structures with known average concentrations of terraces, steps, and kinks.

A notation has been developed that permits decomposition of high Miller indices to several simple low-Miller-index vectors.[89] For example, the stepped (755) surface is given by (755) = 5(111) + 2(100), where the (755) surface is composed of five-atom-wide (111) orientation terraces and (100) orientation steps one atom in height. Thus the complex surface structure is assembled from microfacets that identify the terrace, step, and kink site orientations, if present, and are thereby more representative of the real surface structure. A kinked platinum surface, for example, the Pt(10, 8, 7), is described by this nomenclature as Pt(S)-[7(111) + (110) + 2(100)], where S denotes a stepped or kinked surface. This surface is composed of terraces of (111) orientation that are about seven atoms wide, separated by single-atom-height steps of (110) orientation. The steps have kinks that are two atoms apart in the step and are of (100) orientation. In Table 4.3 and Figure 4.14, the Miller indices and the corresponding microfacet notations are listed for many stepped and kinked surface structures. The rules of conversion from one notation to the other and the methods to obtain the relative surface concentrations of the different sites and the angles of inclination with respect to the low-Miller-index terrace orientation are described in detail elsewhere.[89b]

Structure and Stability of High-Miller-Index Surfaces

Early studies by field-ion microscopy and electron microscopy have disclosed that many stepped surfaces exhibit high thermal stability. Be-

Table 4.3. List of surfaces of fcc crystals, giving the Miller indices, the microfacet notation, and the step notation. The step notation gives the number of atoms across a terrace, including the first and last atoms. This number is one larger than the number of unit cells across a terrace, as used in the microfacet notation.

$(hk\ell)$	Microfacet notation	Step notation
$(n + 1, n + 1, n - 1)$	$n_n (111) + 1_1 (11\bar{1})$	$(n + 1)(111) \times (111)$
(110)	$(\frac{1}{2})_1 (111) + (\frac{1}{2})_1 (11\bar{1})$	$2(111) \times (111)$
(331)	$2_2 (111) + 1_1 (11\bar{1})$	$3(111) \times (111)$
(221)	$(\frac{3}{2})_3 (111) + (\frac{1}{2})_1 (11\bar{1})$	$4(111) \times (111)$
$(n + 2, n, n)$	$n_n (111) + 2_1 (100)$	$(n + 1)(111) \times (100)$
(311)	$1_1 (111) + 2_1 (100)$	$2(111) \times (100)$ or $2(100) \times (111)$
(211)	$1_2 (111) + 1_1 (100)$	$3(111) \times (100)$
(533)	$3_3 (111) + 2_1 (100)$	$4(111) \times (100)$
$(n + 3, n + 1, n)$	$n_{2n} (111) + 1_1 (310)$	$n(111) \times (310)$
(421)	$1_2 (111) + 1_1 (310)$	$1(111) \times (310)$
(532)	$2_4 (111) + 1_1 (310)$	$2(111) \times (310)$
(643)	$3_6 (111) + 1_1 (310)$	$3(111) \times (310)$
$(2n + 1, 1, -1)$	$2n_n (100) + 1_1 (11\bar{1})$	$(n + 1)(100) \times (111)$
$(3, 1, -1)$	$2_1 (100) + 1_1 (11\bar{1})$	$2(100) \times (111)$ or $2(111) \times (100)$
$(5, 1, -1)$	$4_2 (100) + 1_1 (11\bar{1})$	$3(100) \times (111)$
$(n + 1, 1, 0)$	$n_n (100) + 1_1 (110)$	$(n + 1)(100) \times (110)$
(210)	$1_1 (100) + 1_1 (110)$	$2(100) \times (110)$ or $2(110) \times (100)$
(310)	$2_2 (100) + 1_1 (110)$	$3(100) \times (110)$
(410)	$3_3 (100) + 1_1 (110)$	$4(100) \times (110)$
$(2n + 1, 2n + 1, 1)$	$2n_n (110) + 1_1 (111)$	$(n + 1)(110) \times (111)$
(331)	$2_1 (110) + 1_1 (111)$	$2(110) \times (111)$
(551)	$4_2 (110) + 1_1 (111)$	$3(110) \times (111)$
(771)	$6_3 (110) + 1_1 (111)$	$4(110) \times (111)$
$(n + 1, n, 0)$	$n_n (110) + 1_1 (100)$	$(n + 1)(110) \times (100)$
(210)	$1_1 (110) + 1_1 (100)$	$2(110) \times (100)$ or $2(100) \times (110)$
(320)	$2_2 (110) + 1_1 (100)$	$3(110) \times (100)$
(430)	$3_3 (110) + 1_1 (100)$	$4(110) \times (100)$

cause of the lower work functions associated with edges, they are readily detectable by FIM. Well-ordered step structures can be prepared by field evaporation, and these structures exhibit one-atom-height steps on the iridium and tungsten tips that were used for these investigations. Using gold decoration and replica techniques, the step structure that formed during growth and evaporation processes on alkali halide surfaces has been revealed by TEM. The steps were uniform and one lattice plane in height on cleaved (100) orientation NaCl surfaces.

The ordered one-atom-height step-periodic terrace configuration appears to be the stable surface structure of many high-Miller-index surfaces. In a series of studies, Blakely and Somorjai[90] investigated the structure and stability of 22 high-Miller-index crystal faces of platinum. Most of the crystal faces were stable, when clean, in their one-atom-height step-ordered terrace configurations. When the surfaces are disordered by ion bombardment or by heat treatment near the melting point, annealing at lower temperatures in vacuum regenerates the stepped ordered surface structure. However, in the presence of a monolayer of adsorbed carbon or oxygen, many of the stepped surfaces undergo restructuring. The type of structural rearrangements that take place are shown in Figure 4.17. The step height and the terrace width may double, or faceting may take place, whereby the step orientation becomes as prominent as that of the terrace, giving rise to new diffraction features recognizable by LEED. Upon removing the impurities from the surface,

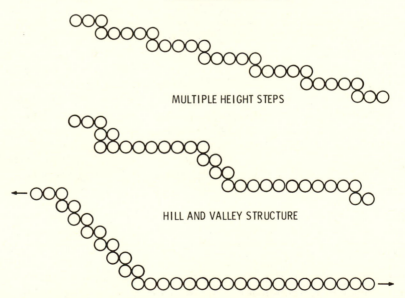

MONATOMIC HEIGHT STEPS

MULTIPLE HEIGHT STEPS

HILL AND VALLEY STRUCTURE

Figure 4.17. Schematic representation of surfaces exhibiting one-atom step height configuration, multiple-height step structure, and hill-and-valley configuration consisting of large facet planes. Reconstruction from one type to another may occur on adsorption.

the one-atom-height step-ordered terrace surface structure is usually regenerated.

Interestingly, the type of restructuring that takes place in the presence of carbon is generally very different from the structural rearrangement in the presence of oxygen.[91] In Figure 4.18, the platinum surface structures are shown on stereographic projections when clean, carbon-covered, or oxygen-covered. The different restructuring effects of the donor and acceptor impurities on the clean metal surface structures are striking. Monatomic height steps appear only occasionally on oxygen- and carbon-covered surfaces. Multiple-height steps and multiple-width terraces occur vicinal to the (111) in oxygen and between the (011) and (012) in all conditions of the experiment. The "hill-and-valley" configuration or faceted surfaces occur predominantly on graphite-covered samples. The studies also show the remarkable stability of many of the highly stepped surfaces.

The impurity-induced changes of surface structure have an important influence on sintering, densification of solids, or structure-sensitive surface reactions. It would also be important to be able to generalize these findings to other metal surfaces. However, in the absence of more experimental data, this would certainly be premature. More information on the structural stability of clean high-Miller-index surfaces of various solids and the effects of impurities on their structural stability is needed.

The driving force for this surface reconstruction in the presence of adsorbates appears to be the difference in chemical bonding of the adsorbates to the different crystal faces of platinum, which alters the relative surface free energies of the various crystal faces. Surfaces that have the lowest surface free energies, and thus the greatest stability when clean, become less stable than other crystal faces when covered with adsorbates. As a result, there is a thermodynamic driving force for restructuring.

Why do high-Miller-index surfaces exhibit the ordered step-terrace configuration? A repulsive interaction between steps would keep them apart and evenly spaced. An attractive interaction would cause step coalescence and faceting. Equilibrium theories that evaluate the structure of stepped or vicinal surfaces that have the lowest surface free energies have been proposed by Herring,[91] and by Gruber and Mullins.[92] These models fit qualitatively the observed anizotrophy of the surface tension for several vicinal surfaces with relatively low step densities. For terraces narrower than five atoms wide, their models break down, probably from neglecting interactions between steps.

Ellis and Schwoebel[93] have proposed a kinetic model for step motion

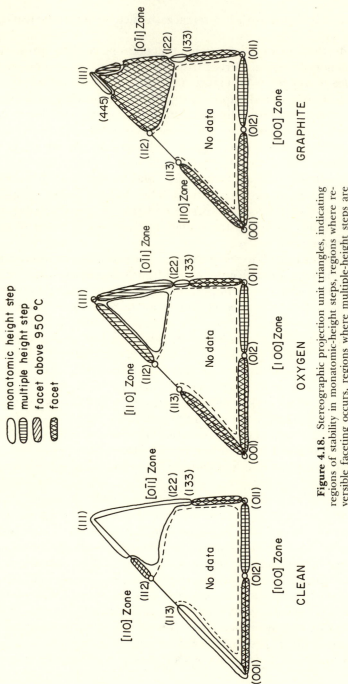

Figure 4.18. Stereographic projection unit triangles, indicating regions of stability in monatomic-height steps, regions where reversible faceting occurs, regions where multiple-height steps are stable, regions where steps are unstable, and the hill-and-valley forms for clean platinum surfaces, platinum surfaces heated in an oxygen ambient, and platinum surfaces covered with a monolayer of graphitic carbon.

and stepped surface stability which treats the probability of a step capturing a diffusing atom. The model considers a one-dimensional lattice with higher potential-energy barriers in place of steps. The key feature is that the probabilities for capture and detachment of an atom at a step depend on the direction of approach to the step. The atoms that diffuse across the terrace may be reflected at the top of the next step and be forced to turn back. This dynamic theory of stepped surface stability is somewhat more flexible than the equilibrium theories and, using realistic interaction potentials, can be related to the equilibrium theories that explain the stability.

Although most of the structural studies of stepped surfaces have concentrated on metals, several investigations of stepped semiconductor crystal faces have been reported. Henzler and Clabes[94] observed that the step structures of cleaned Ge and Si surfaces disappeared on heat treatment to about half the melting temperatures as a result of faceting. Olshanetsky[95] found stable, ordered step structures on Ge surfaces. Some of the crystal surfaces exhibited double step heights. Double-height steps were also detected on stepped ZnO surfaces.[96]

Relaxation of the interlayer spacing at step edges was found also by these investigators on germanium and silicon vicinal surfaces.[97] The interlayer spacing was 4 percent less than the calculated values based on bulklike interlayer distances. Calculations by Tsang and Falicov[98] for stepped ionic crystal surfaces also predict a relaxation of ionic positions that would minimize the polarization at these low-coordination-number sites.

Adsorbates at step and kink sites on transition-metal surfaces frequently have different binding energies than those at sites on the flat, low-Miller-index surfaces. This gives rise to preferential segregation of carbon on stepped nickel surfaces. Hydrogen exhibits stronger binding at kink and step sites on platinum than on the Pt(111) crystal face.[99] The different binding energies associated with adsorbates at surface irregularities markedly influence the chemical properties of transition-metal surfaces. Catalytic activity and selectivity are influenced by them as well as the rate of activation or "poisoning," a very important phenomenon in chemical technology.[100] These properties of high-Miller-index surfaces are discussed in more detail in subsequent chapters.

An interesting technique to measure surface diffusion rates on single-crystal surfaces that takes advantage of the relatively larger surface free energies of high-Miller-index surfaces as compared to low-index crystal faces was developed by Bonzel.[101] By suitable chemical etching, a sinusoidal crystal surface was produced that diffracts a laser beam and exhibits a LEED diffraction pattern characteristic of stepped surfaces with a broad distribution of terrace width. Upon heating, the surface flattens by diffusion of the surface atoms, which changes the amplitude of the sinusoidal profile with corresponding changes in the diffraction features. From these time-dependent changes at various crystal temperatures, the diffusion rates and the activation energies for surface diffusion can be determined. This technique, together with diffusion studies on field-emission tips utilizing field-ion microscopy,[102] appear to be ideally suited for studies of the important phenomena of surface diffusion on well-characterized surfaces.

Studies of Antiferromagnetic Phase Transformations by LEED

While studying the surface structure of the freshly cleaved (100) crystal face of NiO, Palmberg et al.[103] detected a $c(2 \times 2)$ surface structure. NiO is antiferromagnetic and has a Neel temperature of 525 K. The $c(2 \times 2)$ surface structure was detectable only below the Neel temperature and at low electron energies, 30 to 70 eV. The intensity of the fractional-order diffraction beams was about 1 to 3 percent of the integer-order beam intensities.

Spin ordering in the antiferromagnetic NiO leads to the formation of a magnetic unit cell that is twice as large as the atomic unit cell in one crystallographic direction on the (100) crystal face. After repeating these experiments in several laboratories,[104,105] there is little doubt that LEED can indeed detect the presence of the antiferromagnetic surface structure. Calculations indicate that the intensity of this exchange scattering of low-energy electrons is strong enough to be detectable with the usual LEED geometry.[106]

The $c(2 \times 2)$ antiferromagnetic cell on the Ni(100) surface is detectable only when the surface is clean. Impurities, adsorbed on carbon or sulfur, must severely modify the exchange interaction to render the surface structure undetectable.

It is hoped that LEED studies in the near future will also explore the nature of antiferromagnetic surface structures of other surfaces of NiO and other materials.

Recently, the emission of spin-polarized electrons from solids has been investigated by Siegmann and his co-workers,[107] who also developed polarized low-energy electron diffraction (PLEED). The surface magnetization of ferromagnetic Ni(110) has been successfully studied by PLEED.[108]

Surface-Atoms Vibrations: The Debye–Waller Factor

In addition to its primary use in giving information on the structures of surfaces and adsorbates, LEED may also serve as a probe of the surface vibrational amplitudes. We mention here the most elementary model, based on single scattering of electrons from identical atoms at a surface. Just as in X-ray diffraction, the intensities of the diffraction spots decay exponentially with the mean-square amplitude of the atomic vibrations:[109]

$$I = I_0 \exp\left(- \sum_i \Delta \mathbf{k}_i^2 <u_i^2> \right) \tag{11}$$

In this expression I_0 is the intensity from a rigid lattice of scatterers, $\Delta \mathbf{k}$ the momentum transfer of the scattered electron, and $<u_i^2>$ the mean-square amplitude of vibration of the atom from its equilibrium position.

This equation expresses the fact that only those atoms that are in their equilibrium positions in the ordered structure will contribute to the intensity of the Bragg peaks. The electrons that are scattered from atoms displaced from their equilibrium positions will contribute to the diffuse background intensity. The vibration of the atoms does not change the width of the diffraction beams, only their intensity. Increasing the crystal temperature increases the amount of time the atoms spend away from their equilibrium position; thus the intensity of the diffraction beams falls off as the temperature is raised. The exponent $2\mathcal{M} = \Sigma \, \Delta \mathbf{k}_i^2 <\mu_i^2>$ is commonly called the Debye–Waller factor.

Using the Debye approximation in the high-temperature limit ($T > \Theta_D/2\pi$), we have

$$<u_i^2> = \frac{3h^2T}{Mk_B\Theta_D^2} \tag{12}$$

where Θ_D is the Debye temperature and M the atomic mass. We may specialize to the case of specular reflection, in which only the amplitude of vibration perpendicular to the surface enters Eq. (11). Using Eqs. (11) and (12) and noting that $\Delta \mathbf{k}_z = 2\sqrt{2mE/\hbar^2} \cos \phi$ (where ϕ is the incident beam angle measured from the surface normal), it is easy to show that

$$\log\left(\frac{I}{I_0}\right) = \frac{-24mE \cos^2 \phi}{Mk_B\Theta_D^2(z)} \times T \tag{13}$$

This equation implies that a plot of the logarithm of the intensity at a given energy as a function of temperature is a straight line, the slope of which yields $\Theta_D(z)$, a measure of the surface vibration amplitude perpendicular to the surface.

In reality, the electron beam penetration varies as a function of energy, so that Eq. (13) provides, at any given energy, an *effective Debye temperature*, which is some average of the surface and bulk layers. In empirical fashion, however, one may arrive at a *surface Debye temperature* from the low-energy limit of this effective Debye temperature. Due caution must be exercised in interpreting results with this method, however, since the electron penetration is always finite and does not increase monotonically with energy. Furthermore, multiple-scattering processes are very important at selected energies, and a much more elaborate approach is needed to correctly interpret the effect of these processes on the temperature dependence of the diffracted intensity.

The effective surface Debye temperatures of various crystal surfaces have been determined by the approximate method outlined above.[110] These values are listed in Table 4.4. In most cases the surface vibration amplitude appears to be roughly twice the amplitude in the bulk.

Table 4.4. Surface and bulk root-mean-square displacement ratios and Debye temperatures for several metals

	$\dfrac{<u^2\perp>^{1/2}\text{(surface)}}{<u^2>^{1/2}\text{(bulk)}}$	Θ(surface) (K)	Θ(bulk) (K)
Pb(110), (111)[a,b]	2.43 (1.84)	37 (49)	90
Bi(0001), (01$\bar{1}$2)[b]	2.42	48	116
Pd(100), (111)[a]	1.95	142	273
Ag(100),[c] (110),[c] (111)[d]	2.16 (1.48)	104 (152)	225
Pt(100), (110), (111)[e]	2.12	110	234
Ni(110)[f]	1.77	220	390
Ir(100)[g]	1.63	175	285
Cr(110)[h]	1.80	333	600
Nb(110)[i]	2.65	106	281
V(100)[j]	1.52	250	380
Rh(100), (111)[k]	1.35	260	350

a. R.M. Goodman, H.H. Farrell, and G.A. Somorjai, J. Chem. Phys. **48,** 1046 (1968).
b. R.M. Goodman and G.A. Somorjai, J. Chem. Phys. **52,** 6325 (1970).
c. J.M. Morabito, Jr., R.F. Steiger, and G.A. Somorjai, Phys. Rev. **179,** 638 (1969).
d. E.R. Jones, J.T. McKinney, and M.B. Webb, Phys. Rev. **151,** 476 (1966).
e. H.B. Lyon and G.A. Somorjai, J. Chem. Phys. **44,** 3707 (1966).
f. A.U. McRae, Surf. Sci. **2,** 522 (1964).
g. R.M. Goodman, Ph.D. dissertation, University of California, Berkeley, 1969.
h. R. Kaplan and G.A. Somorjai, Solid State Commun. **9,** 505 (1971).
i. D. Tabor and J. Wilson, Surf. Sci. **20,** 203 (1970).
j. D.J. Cheng, R.F. Wallis, C. Megerle, and G.A. Somorjai, Phys. Rev. **B12,** 5599 (1975).
k. D.G. Castner, Ph.D. dissertation, University of California, Berkeley, 1979.

Adsorbates should have a marked influence on surface-atom vibrations, since they change the bonding environment with respect to that on the clean surface. The adsorption of oxygen on tungsten increases the surface Debye temperature[111] with respect to the bulk value due to the stronger W—O bond as compared to the W—W bond. Studies of surface-atom vibrations in the presence of adsorbates should provide information on the nature of the surface bond. However, this bonding information is perhaps more directly available from high-resolution electron energy loss (HREELS) studies than from monitored changes of the surface Debye–Waller factor.

References

1. L.L. Kesmodel and G.A. Somorjai, Trans. Am. Crystallogr. Assoc. **13**, 67 (1977).
2. E.A. Wood, J. Appl. Phys. **35**, 1306 (1964).
3. (a) D.S. Boudreaux and V. Hoffstein, Phys. Rev. **B3**, 2447 (1971); (b) D.W. Jepsen, P.M. Marcus, and F. Jona, Phys. Rev. **B6**, 3684 (1972); (c) G.E. Laramore and C.B. Duke, Phys. Rev. **B5**, 267 (1972); (d) M.T. Martin and G.A. Somorjai, Phys. Rev. **B7**, 3607 (1973); (e) D.L. Adams and U. Landman, Phys. Rev. **B15**, 3775 (1977).
4. (a) L.L. Kesmodel and G.A. Somorjai, Phys. Rev. **B11**, 633 (1975); (b) L.L. Kesmodel, P.C. Stair, and G.A. Somorjai, Surf. Sci. **63**, 342 (1977).
5. (a) J.E. Demuth, P.M. Marcus, and D.W. Jepsen, Phys. Rev. **B11**, 1460 (1975); (b) R. Feder, Phys. Rev. **B15**, 1751 (1977); (c) G.E. Laramore, Phys. Rev. **B8**, 515 (1973).
6. D.C. Frost, K.A.R. Mitchell, F.R. Shepherd, and P.R. Watson, Proc. 7th Int. Vac. Congr. and 3rd Int. Conf. Solid Surf., Vienna (1977).
7. H.D. Shih, F. Jona, D.W. Jepsen, and P.M. Marcus, Commun. Phys. **1**, 25 (1976).
8. J.A. Strozier and R.O. Jones, Phys. Rev. **B3**, 3228 (1971).
9. R.H. Tait, S.Y. Tong, and T.N. Rhodin, Phys. Rev. Lett. **28**, 553 (1972).
10. (a) L.J. Clarke, Proc. 7th Int. Vac. Congr. and 3rd Int. Conf. Solid Surf., Vienna, A2725 (1977); (b) T.E. Felter, R.A. Barker, and P.J. Estrup, Phys. Rev. Lett. **38**, 1138 (1977); (c) A. Ignatiev, F. Jona, H.D. Shih, D.W. Jepsen, and P.M. Marcus, Phys. Rev. **B11**, 4787 (1975).
11. (a) M.K. Debe, D.A. King, and F.S. Marsh, Surf. Sci. **68**, 437 (1977); (b) R. Feder, Phys. Rev. Lett. **36**, 598 (1976); (c) R. Feder, Surf. Sci. **63**, 293, (1977); (d) J. Kirschner and R. Feder, Verh. Deutsch. Phys. Ges. **2**, 557 (1978); (e) B.W. Lee, A. Ignatiev, S.Y. Tong, and M.A. Van Hove, J. Vac. Sci. Technol. **14**, 291 (1977); (f) R. Feder, Phys. Status Solidi **B62**, 135 (1974); (g) M.A. Van Hove and S.Y. Tong, Surf. Sci. **54**, 91 (1976).
12. A. Ignatiev, B.W. Lee, and M.A. Van Hove, Proc. 7th Int. Vac. Congr. and 3rd Int. Conf. Solid Surf., Vienna, 1733 (1977).
13. (a) C.G. Kinniburgh, J. Phys. **C2**, 2382 (1975); (b) C.G. Kinniburgh, J. Phys. **C9**, 2695 (1976).
14. C.G. Kinniburgh and J.A. Walker, Surf. Sci. **63**, 274 (1977).

15. (a) S. Hagstrom, H.B. Lyon, and G.A. Somorjai, Phys. Rev. Lett. **15**, 491 (1965); (b) H.B. Lyon and G.A. Somorjai, J. Chem Phys. **46**, 2539 (1967); (c) L. L. Kesmodel and G.A. Somorjai, MTP Int. Rev. Sci. **7**, 1 (1975).

16. (a) J.T. Grant, Surf. Sci. **18**, 282 (1969); (b) A. Ignatiev, A. V. Jones, and T. N. Rhodin, Surf. Sci. **30**, 573 (1972).

17. (a) D.G. Fedak and N.A. Gjostein, Surf. Sci. **8**, 77 (1967); (b) P.W. Balmberg and T.N. Rhodin, Phys. Rev. **161**, 586 (1967).

18. M. Salmerón and G.A. Somorjai, Surf. Sci. **91**, 373 (1980).

19. C.M. Chan, S.L. Cunningham, K.L. Luke, M.A. Van Hove, W.H. Weinberg, and S.R. Withrow, J. Vac. Sci. Technol. **16**, 642 (1979).

20. H.P. Bonzel, C.R. Helms, and S. Kelemen, Phys. Rev. Lett. **35**, 1237 (1975).

21. T.N. Rhodin and G. Brodén, Surf. Sci. **60**, 466 (1976).

22. D.M. Zehner, B.R. Appleton, T.S. Noggle, J.W. Miller, J.H. Barrett, L.H. Jenkins, and E.O. Schorr, J. Vac. Sci. Technol. **12**, 454 (1975).

23. MA. Van Hove, R.J. Koestner, P.C. Stair, J.P. Biberian, L.L. Kesmodel, and G.A. Somorjai, Surf. Sci., to be published.

24. M.A. Van Hove, R.J. Koestner, P.C. Stair, J.P. Biberian, L.L. Kesmodel, and GA. Somorjai, Surf. Sci., to be published.

25. C.M. Chan, S. L. Cunningham, K.L. Luke, W.H. Weinberg, and S.P. Withrow, to be published.

26. (a) P.A. Barker, P.S. Estrup, F. Jona, and P.M. Marcus, Solid State Commun. **25**, 375 (1978); (b) M. K. Debe and D. A. King, Phys. Rev. Lett. **39**, 708 (1977); (c) M.K. Debe and D.A. King, J. Phys. **C10**, L303 (1977); (d) T.E. Felter, R.A. Barker, and P.J. Estrup, Phys. Rev. Lett. **38**, 1138 (1977).

27. A. Ignatiev, B.W. Lee, and M.A. Van Hove, to be published.

28. L.Brewer, Science **161**, 115 (1968).

29. L. Brewer, in *Electronic Structure and Alloy Chemistry,* ed. P. A. Beck, Wiley, New York, 1963.

30. F. Jona, Surf. Sci. **8**, 57 (1967).

31. S. Andersson, I. Marklund, and D. Andersson, in *The Structure and Chemistry of Solid Surfaces,* ed. G.A. Somorjai, Wiley, New York, 1969.

32. J.M. McDavid and S. C. Fain, Surf. Sci. **52**, 161 (1975).

33. (a) J. Ferrante, Acta Metall. **19**, 743 (1971); (b) J. Ferrante, Scripta Metall. **5**, 1129 (1971).

34. R. Bouwman, G.J.M. Lipitts, and W.M.H. Sachtler, J. Catal. **25**, 350 (1972).

35. R.A. Van Santen, L.H. Toneman, and R. Bouwman, Surf. Sci. **47**, 64 (1975).

36. (a) J.A. Appelbaum and D.R. Hamann, Surf. Sci. **74**, 21 (1978); (b) K.A.R. Mitchell and M.A. Van Hove, Surf. Sci. **75**, 147L (1978); (c) S.Y. Tong and A.L. Maldonado, Surf. Sci. **78**, 459 (1978); (d) J.A. Appelbaum, G.A. Baraff, and D.R. Hamann, Phys. Rev. Lett. **35**, 729 (1975); Phys. Rev. **B14**, 588 (1976); (e) R.E. Schlier and H.E. Farnsworth, *Semiconductor Surface Physics,* Univ. of Pennsylvania Press, Philadelphia, 1957, p. 3.

37. S.J. White, D.P. Woodruff, B.W. Holland, and R.S. Zimmer, Surf. Sci. **68**, 457 (1977).

38. (a) J.D. Levine, S.H. McFarlane, and P. Mark, Phys. Rev. **B16**, 5415 (1977); (b) P. Mark, J.D. Levine and S.H. McFarlane, Phys. Rev. Lett. **38**, 1408 (1977); (c) J.J. Lander, Progr. in Solid State Chem. **2** (1965).

39. D. Rowe and R. Phillips, Phys. Rev. Lett. **32**, 1315 (1974).
40. D. Haneman, Phys. Rev. **121**, 1093 (1961).
41. J.A. Appelbaum, G.A. Baraff, and D.R. Hamann, Phys. Rev. Lett. **35**, 729 (1975).
42. G. Schlüter and H. Cohen, Phys. Rev. Lett. **34**, 1385 (1975).
43. (a) J.J. Lander and J. Morrison, J. Appl. Phys. **34**, 1411 (1963); (b) G. Rovida et al., Surf. Sci. **14**, 93 (1969).
44. (a) J.B. Marsh and H.E. Farnsworth, Surf. Sci. **1**, 3 (1964); (b) J.J. Lander and J. Morrison, Surf. Sci. **4**, 241 (1966).
45. (a) C.B. Duke, A.R. Lubinsky, B.W. Lee, and P. Mark, J. Vac. Sci. Technol. **13**, 761 (1976); (b) A.R. Lubinsky, C.B. Duke, B.W. Lee, and P. Mark, Phys. Rev. Lett. **36**, 1058 (1976).
46. (a) A.Y. Cho, Appl. Phys. **41**, 2780 (1970); (b) A.U. MacRae and G.W. Gobeli, J. Appl. Phys. **35**, 1629 (1964); (c) D.L. Heron and D. Haneman, Surf. Sci. **21**, 12 (1970).
47. H.H. Brongersma and P.M. Mul, Surf. Sci. **35**, 393 (1973).
48. (a) C.B. Duke, A.R. Lubinsky, M. Bonn, G. Cieneros, and P. Mark, J. Vac. Sci. Technol. **14**, 294 (1977); (b) P. Mark. G. Cieneros, M. Bonn, A. Kahn, C.B. Duke, A. Paton, and A.R. Lubinsky, J. Vac. Sci. Technol. **14**, 210 (1977).
49. (a) C.B. Duke and A.R. Lubinsky, Surf. Sci. **50**, 605 (1975); (b) C.B. Duke, A.R. Lubinsky, S.C. Change, B. W. Lee, and P. Mark, Phys. Rev. **B15**, 4865 (1977); (c) C.B. Duke, A.R. Lubinsky, B.W. Lee, and P. Mark, J. Vac. Sci. Technol. **13**, 761 (1976); (d) A.R. Lubinsky, C.B. Duke, S.C. Chang, B.W. Lee, and P. Mark, J. Vac. Sci. Technol. **13**, 189 (1966).
50. B.D. Campbell, G.A. Haque, and H.F. Farnsworth, in *The Structure and Chemistry of Solid Surfaces*, ed. G.A. Somorjai, Wiley, New York, 1969.
51. G.C. Benson, P.I. Freeman, and E. Dempsey, J. Chem. Phys. **39**, 302 (1963).
52. (a) B.J. Mrstik, R. Kaplan, T.L. Reinecke, M.A. Van Hove, and S.Y. Tong, Phys. Rev. **B15**, 897 (1977); (b) B.J. Mrstik, P. Kaplan, T.L. Reinecke, M.A. Van Hove, and S.Y. Tong, Nuovo Cimento **38B**, 387 (1977).
53. M.A. Van Hove, S.Y. Tong, and M.H. Elconin, Surf. Sci. **64**, 85 (1977).
54. E.G. McRae and C.W. Caldwell, Jr., Surf. Sci. **2**, 509 (1964).
55. G.C. Benson and T.A. Claxton, J. Chem. Phys. **48**, 1356 (1968).
56. See, for example, T.E. Gallon, I.G. Higginbotham, M. Prutton, and H. Tokutaka, Surf. Sci. **21**, 224 (1970), and references therein.
57. Ibid., p. 226, for table.
58. C.E. Holcombe, Jr., and G.L. Powell, Surf. Sci. **30**, 561 (1972).
59. T.M. French and G.A. Somorjai, J. Phys. Chem. **74**, 2489 (1970).
60. L. Fiermans and J. Vennik, Surf. Sci. **18**, 317 (1969).
61. Y.W. Chung, W. Lo, and G.A. Somorjai, Surf. Sci. **65**, 419 (1977).
62. A. Aberdam and C. Gaub, Surf. Sci. **27**, 571 (1971); 559, (1971).
63. W.J. Lo and G.A. Somorjai, Phys. Rev. **B17**, 4942 (1978).
64. K.O. Legg, M. Prutton, and C. Kinniburgh, J. Phys. **C7**, 4236 (1974).
65. (a) C.G. Kinniburgh, J. Phys. **C8**, 2482 (1975); (b) C.G. Kinniburgh, J. Phys. **C9**, 2695 (1976).
66. L. Firment and G.A. Somorjai, Surf. Sci. **55**, 413 (1976).
67. L. E. Firment and G.A. Somorjai, J. Chem. Phys. **66**, 3901 (1977).

68. J.C. Buchholz, J. Chem. Phys. **66,** 573 (1977).
69. L.L. Atanasoska, J.C. Buchholz, and G.A. Somorjai, Surf. Sci. **72,** 189 (1978).
70. (a) J.J. Lander and J. Morrison, Surf. Sci. **6,** 1 (1967); (b) J. Suzanne, J.P. Holcombe, and M. Bienfait, Surf. Sci. **40,** 414 (1973); **44,** 141; **47,** 204 (1975); (c) J. Suzanne, J.P. Holcombe, and M. Bienfait, J. Cryst. Growth **31,** 87 (1975); (d) J. Suzanne, J.P. Holcombe, M. Bienfait, and P.M. Asri, Solid State Commun. **15,** 1585 (1974).
71. H.H. Farrell, M. Strongin, and J.M. Dickey, Phys. Rev. **B6,** 4703 (1972).
72. M.D. Chinn and S.C. Fain, Jr., Phys. Rev. Lett. **39,** 146 (1977).
73. See, for example, N. Cabrera, Disc. Faraday Soc. **28,** 16 (1959).
74. See, for example, J.P. Hirth, and G.M. Pound, *Condensation and Evaporation,* Pergammon Press, Oxford, 1963, pp. 135–64, and references therein.
75. (a) J.P. Stark, Acta Met. **13,** 1181 (1965); (b) I.N. Stranski, W. Gaus, and H. Rau, Ber. Buusenges, Physick Chem. **67,** 965 (1963).
76. G.A. Somorjai, Adv. Catal. **26,** 1 (1977).
77. T.W. Haas, Surf. Sci. **5,** 345 (1966).
78. W.P. Ellis and R.L. Schwoebel, Surf. Sci. **11,** 82 (1968).
79. J. Perdereau and G.E. Rhead, Surf. Sci. **24,** 555 (1971).
80. M. Henzler, Surf. Sci. **19,** 159 (1970).
81. M. Henzler, Surf. Sci. **22,** 12 (1970).
82. R.W. Joyner, B. Lang, and G.A. Somorjai, Surf. Sci. **30,** 440 (1972).
83. R.W. Joyner, B. Lang, and G.A. Somorjai, Surf. Sci. **30,** 454 (1972).
84. M. Henzler, Surf. Sci. **19,** 159 (1970).
85. W.P. Ellis and R.L. Schwoebel, Surf. Sci. **11,** 82 (1968).
86. R.W. Joyner and G.A. Somorjai, in *Surface and Defect Properties of Solids,* ed. M.W. Roberts and J.M. Thomas, vol. 2, London, The Chem. Soc. Publ., 1973, p. 1.
87. B.D. Campbell and W.P. Ellis, Surf. Sci. **10,** 118 (1968).
88. J.E. Houston and R.L. Park, Surf. Sci. **26,** 269 (1971).
89. (a) B. Lang, R. Joyner, and G.A. Somorjai, Surf. Sci. **30,** 440 (1972); 454 (1972); (b) M.A. Van Hove and G.A. Somorjai, Surf. Sci., to be published, 1980.
90. D.W. Blakely and G.A. Somorjai, Surf. Sci. **65,** 419 (1977).
91. See D.W. Blakely, Ph.D. dissertation, University of California, Berkeley (1976).
92. E.E. Gruber and W.W. Mullins, J. Phys. Chem. Solids **28,** 875 (1967).
93. W.P. Ellis and R.L. Schwoebel, Surf. Sci. **11,** 82 (1968).
94. M. Henzler and J. Clabes, Proc. 2nd Int. Conf. Solid Surf., J. Appl. Phys., Suppl. **2,** Pt. 2, 389 (1974).
95. B.Z. Olshanetsky, S.M. Repinsky, and A.A. Shklayaev, Surf. Sci. **69,** 205 (1971).
96. M. Henzler, Surf. Sci. **36,** 109 (1973).
97. M. Henzler and J. Clabes, Proc. 2nd Int. Conf. Solid Surf., J. Appl. Phys., Suppl. **2,** Pt. 2, 389 (1974).
98. Y.W. Tsang and L.M. Falicov, J. Phys. **C9,** 51 (1976).
99. M. Davis, Surf. Sci. **91,** 73 (1980).
100. G.A. Somorjai, J. Catal. **27,** 453 (1972).

101. (a) H.P. Bonzel, in *Structure and Properties of Metal Surfaces,* ed. S. Shimodaira, Maruzen, Tokyo, 1973, p. 248; (b) H.P. Bonzel and E.E. Lotta, Surf. Sci. **76,** 275 (1978).
102. G. Erlich, in *Surface Science Recent Progress and Perspectives,* ed. T.S. Jayadenaah and R. Vanselow, CRC Press, Cleveland, Ohio, 1974.
103. P.W. Palmberg, R.E. DeWames, and L.A. Vredevoe, Phys. Rev. Lett. **21,** 682 (1968).
104. K. Hayakawa, K. Namikawa, and S. Miyake, J. Phys. Soc. Jap. **31,** 1408 (1971).
105. M. Prutton, W.D. Doyle, and K. Legg, unpublished.
106. (a) R.E. DeWames and D. Falicov, Phys. Status, Solidi **39,** 445 (1970); (b) T. Suzuki, Japan J. App. Phys. **17,** 320 (1978).
107. M. Campagna, D.T. Pierce, F. Meier, K. Sattler, and H.C. Siegmann, Adv. in Electronics and Electron Phys. **41,** 113 (1976).
108. R.J. Celotta, D.T. Pierce, G.C. Wang, S.D. Bader, and G.P. Felcher, Phys. Rev. Lett. **43,** 728 (1979).
109. J. James, in *The Optical Principles of the Diffraction of X-Rays,* G. Bell & Sons, London, 1958, pp. 1–92.
110. R.M. Goodman, J. Chem. Phys. **52,** 6325 (1970).
111. P.J. Estrup, in *The Structure and Chemistry of Solid Surfaces,* ed. G.A. Somorjai, Wiley, New York, 1969, p. 191.

5. Structure of Adsorbed Monolayers on Solid Surfaces

Introduction

An atom or molecule that approaches the surface of a solid always experiences a net attractive potential.[1] As a result, there is a finite probability that it will be trapped on the surface and the phenomenon that we call adsorption will occur. Under the usual environmental conditions (about 1 atm and 300 K in the presence of oxygen, nitrogen, water vapor, and assorted hydrocarbons), all solid surfaces are covered with a monolayer of adsorbate and the buildup of multiple adsorbate layers is often detectable. The constant presence of the adsorbate layer influences all the chemical, mechanical, and electronic surface properties. Adhesion, lubrication, the onset of chemical corrosion, or photoconductivity are just a few of the many macroscopic surface processes that are controlled by the various properties of a monolayer of adsorbates.

In this chapter, we review the various experimental parameters that can be used to characterize the adsorbate layer in the submonolayer to few-monolayer range. Then we discuss the principles of ordering of the adsorbate layer, since one of the most exciting observations of low-energy electron diffraction studies is the predominance of ordering within these layers. We list the ordered adsorbate layer structures and summarize what can be learned about the nature of their bonding from the available structural data. This will be done separately for the many (~1000) surfaces whose two-dimensional unit cells are known in terms of shape, size, and orientation, and for the fewer (~100) surfaces for which, additionally, the contents of the unit cell are known, such as adsorption site and bond lengths. Many types of adsorption will be covered, including atomic and molecular adsorption, and coadsorption for metallic, nonmetallic, and organic adsorbates.

Principles of Monolayer Adsorption

Consider a uniform surface with a number n_0 of equivalent adsorption sites. The ratio of the number of adsorbed atoms or molecules, n, to n_0 is defined as the coverage, $\theta = n/n_0$. The coverage in the monolayer is usually less than or equal to unity for a uniform surface. For a heterogeneous surface that exhibits multiple binding sites (more than one site per substrate unit cell), small adsorbate atoms may build up coverages somewhat greater than unity. We shall, however, ignore this possibility for the present.

When adsorption occurs on a clean surface, heat is liberated during the formation of the surface bond. The heat of adsorption, ΔH_{ads}, associated with the layer of adsorbates reveals the strength of interaction between atoms and molecules in the monolayer and the surface on which they are adsorbed. These two macroscopic, experimentally measurable parameters, θ and ΔH_{ads}, usually well characterize the adsorbed monolayer, and the form of their interdependence often reveals the nature of bonding in the adsorbed layer.

Atoms or molecules may impinge on a surface from the gas phase, where they establish a surface concentration $[n_A]_s$ (molecules/cm^2). Let us assume that only one type of species of concentration $[n_A]_g$ (molecules/cm^3) exists in the gas phase, so that the adsorption process can be written

$$A_{(gas)} \underset{k'}{\overset{k}{\rightleftarrows}} A_{(surface)}$$

and the net rate of adsorption may be expressed as

$$F(\text{molecules/cm}^2/\text{sec}) = k[n_A]_g - k'[n_A]_s \tag{1}$$

where k and k' are the rate constants for adsorption and desorption, respectively. Starting with a nearly clean surface far from equilibrium, the rate of desorption may be taken as zero and Eq. (1) can be simplified to

$$F(\text{molecules/cm}^2/\text{sec}) = k[n_A]_g \tag{2}$$

where

$$k = \alpha \left(\frac{RT}{2\pi M_A} \right)^{1/2} \quad \text{cm/sec}$$

α is the adsorption coefficient, M_A the molecular weight of the impinging molecules, R the gas constant, and T the temperature. The surface concentration $[n_A]_s$ (molecules/cm^2) under these conditions is the product of the incident flux, F, and the surface residence time, τ (sec):

$$[n_A]_s = F\tau \tag{3}$$

If the incident molecules stay on the surface long enough to achieve thermal equilibrium with the surface atoms, τ has a form $\tau = \tau_0 e^{\Delta H_{ads}/RT}$, where τ_0 is related to the average vibrational frequency associated with the immobile adsorbate. The value of τ_0 may be markedly different if the adsorbate possesses one or two translational degrees of freedom along the surface.

The heat of adsorption, ΔH_{ads}, defined as the binding energy of the adsorbed species, is always positive. Clearly, the larger ΔH_{ads} and the lower the temperature, T, the longer is the residence time. For a given incident flux, larger ΔH_{ads} and lower temperatures yield higher coverages. Replacing the vapor density by the pressure, using the ideal gas law $[n_A]_g = N_A P/RT$ (where N_A is Avogadro's number), we can rewrite Eq. (3) as

$$[n_A] \text{ (molecules/cm}^2) = \frac{\alpha P N_A}{\sqrt{2\pi M_A RT}} \tau_0 \exp\left(\frac{\Delta H_{ads}}{RT}\right)$$

$$= 3.52 \times 10^{22} \frac{P(\text{torr})}{\sqrt{M_A T}} \tau_0 \exp\left(\frac{\Delta H_{ads}}{RT}\right) \tag{4}$$

From the knowledge of P, T and ΔH_{ads}, $[n_A]_s$ can be estimated. For example, assuming that $\tau_0 = 10^{-12}$ sec and $\alpha = 1$, $\Delta H_{ads} = 2$ kcal/mol, and $T = 300$ K, the surface concentration of argon at $P = 10^{-6}$ torr is immeasurable, $[n_A]_s \approx 10^4$ molecules/cm^2 (one monolayer is about 10^{15} molecules/cm^2). It is still a fraction of a monolayer at 1 atm. However, at $T = 100$ K, the surface is saturated with a monolayer of argon at 1 atm ($\sim 10^{15}$ molecules/cm^2). For a higher value of ΔH_{ads}, say 15 kcal/mol, the surface is covered with a measurable quantity (1 to 100 percent of a monolayer) of gas at 300 K, even at 10^{-6} torr. Gas–surface systems characterized by weak interactions ($\Delta H_{ads} < 15$ kcal/mol accompanied by short residence times), which require adsorption studies to be carried out at low T and at relatively high pressure (~ 1 atm), are called physical adsorption systems. Adsorbates that are characterized by stronger chemical interactions ($\Delta H \geq 15$ kcal/mol), where near-monolayer adsorption commences even at 300 K and at low pressures ($\leq 10^{-6}$ torr), are called chemisorbed systems. Although these traditional names imply two distinct types of adsorption, the various gas–surface systems exhibit a gradual change from the physisorption to the chemisorption regime.

The coverage, θ, may be varied by changing the pressure over the surface while maintaining a well-chosen constant temperature. The θ versus $P(T)$ curve so obtained for any given gas–surface system is called the adsorption isotherm. The simplest adsorption isotherm is obtained from Eq. (4), which we can rewrite as

$$\theta = k''P \tag{5}$$

with

$$k'' = \frac{1}{n_0} \frac{\alpha N_A}{\sqrt{2\pi M_A RT}} \tau_0 \exp\left(\frac{\Delta H_{ads}}{RT}\right)$$

Thus the coverage is proportional to the first power of the pressure at a given temperature provided that we have an unlimited number of adsorption sites on the surface and that ΔH_{ads}, which reflects the nature of the gas–surface interaction, does not change as the coverage is changing. Langmuir[2] has derived a different adsorption isotherm, which has become very useful in describing many adsorption processes that terminate when a monolayer coverage is reached. He assumed that any gas molecule that strikes an adsorbed molecule must reflect from the surface, whereas it adsorbs when impinging on the bare surface. If $[n_0]$ is the surface concentration on a completely covered surface, the concentration of surface sites available for adsorption after building up an adsorbate concentration $[n_A]_s$ is $[n_0] - [n_A]_s$. In equilibrium the desorption rate that is given by $k'[n_A]_s$ is equal to the rate of adsorption $k([n_0] - [n_A]_s)P$

$$k'[n_A]_s = k([n_0] - [n_A]_s)P \tag{6}$$

which can be rearranged to yield the Langmuir adsorption isotherm,

$$\theta = \frac{bP}{1+bP} \tag{7}$$

where $b = k/k'$. Typical adsorption isotherms that obey Eq. (5) and (7) are shown in Figures 5.1 and 5.2, respectively. It should be noted that a linear Langmuir plot can be obtained by plotting $1/[n_A]_s$ against $1/P$, where the slope is $1/b[n_0]$ and the intercept is $1/[n_0]$, as seen after rearrangement of Eq. (7). The adsorption isotherms are used primarily to determine the surface area of porous solids and the heats of adsorption. The isotherms yield the amount of gas adsorbed. By multiplying with the area occupied per molecule that is determined independently, the total surface area is determined. For example, the area per molecule is 16.2 Å^2 for N_2 and 25.6 Å^2 for krypton on a large variety of surfaces. The heat of adsorp-

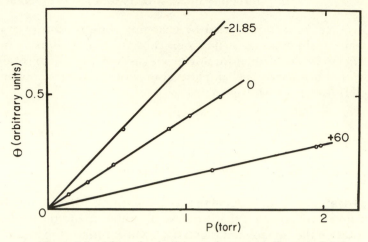

Figure 5.1. Adsorption isotherms of argon on silica gels (labels on curves indicate different temperatures in degrees Celsius).

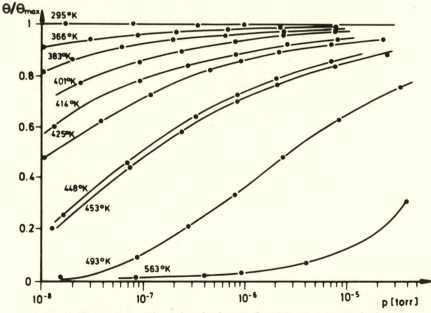

Figure 5.2. Adsorption isotherms for CO on Pt(111) single-crystal surfaces (after G. Ertl, M. Neumann, and K. M. Streit, Surf. Sci. **64,** 393 [1977]).

tion is obtained from adsorption isotherms measured at different temperatures using the Clausius–Clapeyron equation,

$$\left[\frac{d \ln P}{d(1/T)} \right]_{\theta=\text{const}} = -\frac{\Delta H_{\text{ads}}}{R} \tag{8}$$

Principles of Ordering of Adsorbed Monolayers

Causes of Ordering

Once a molecule lands on the solid surface, it may slide along the surface plane or remain bound at a specific site during much of its surface residence time. As long as ΔH_{ads} and the activation energy for bulk diffusion $\Delta E_{D\,(\text{bulk})}^{*}$ are high enough as compared with kT ($\geq 10\,kT$), we are assured of a residence time that is long enough to permit thermal equilibration among the adsorbates and between the adsorbate and substrate atoms (that is, adsorption). Ordering, however, depends primarily on the depth of the potential-energy barrier that keeps an atom or molecule from hopping to a neighboring site along the surface. The activation energy for surface diffusion, $\Delta E_{D(\text{surface})}^{*}$, is an experimental parameter that is of the magnitude of this potential-energy barrier. $\Delta E_{D(\text{surface})}^{*}$ may be obtained for self-diffusion or for the diffusion of adsorbates on well-characterized surfaces by several techniques. Among them, field-ion microscopy[3] and sinusoidal-wave analysis[4] are the most prominent at present. The $\Delta E_{D(\text{surface})}^{*}$ for Ar, W adatoms, and O atoms on tungsten surfaces are 2, 15, and 10 kcal, respectively. For small values of $\Delta E_{D(\text{surface})}^{*}$, ordering is restricted to low temperatures, since the adsorbate atoms become very mobile as the temperature is increased. For higher values of $\Delta E_{D(\text{surface})}^{*}$, ordering cannot commence at low temperatures, since the adsorbate atoms need to have a considerable mean free path along the surface to find their equilibrium position once they land on the surface at a different location. Of course, if the temperature is too high, the adsorbed atoms or molecules desorb or vaporize.

It should be noted that in the limit of the very high heats of chemisorption, one may form surface compounds: oxides or carbides, for example. In this circumstance, ordering of the new surface phase may require the relocation of the substrate atoms as well as the adsorbate atoms. Such chemisorption-induced reconstructions have been observed for several systems, and their presence makes the conditions necessary for ordering in the surface layer very difficult to analyze. Some of these systems will be discussed later in the chapter.

The interatomic forces responsible for the binding of adsorbates at surfaces and for the ordering of overlayers are of various types. The binding of adsorbates to substrates is frequently due to the strong covalent chemical forces, resulting from the presence of electron orbitals overlapping both the substrate and the adsorbate. Some adatoms (notably the rare gases) and many molecules will stick only weakly to substrates. The binding force is then predominantly due to the van der Waals interaction, and we have physisorption.

The binding forces have components perpendicular and parallel to the surface. The perpendicular component is primarily responsible for the binding energy (heat of adsorption), while the parallel component often determines the binding site along the surface. The binding site may, however, also be affected by adsorbate–adsorbate interactions, which are also responsible for any ordering within an overlayer. These interactions may be arbitrarily subdivided into direct adsorbate–adsorbate interactions (not involving the substrate at all) and substrate-mediated interactions: the latter are complicated many-atom interactions. Dipole–dipole interactions are an example of such interactions, involving the exact charge distribution of the adsorbed particles, the shape of the electrostatic dipolar fields at the surface, and, of course, self-consistency requirements, since dipolar charge distributions are themselves affected by nearby dipoles.

The adsorbate–adsorbate interactions may be repulsive; they always are repulsive at sufficiently small adsorbate–adsorbate separations. They may be attractive at larger separations, giving rise to the possibility of island formation. They may be oscillatory, moving back and forth between attractive and repulsive as a function of adsorbate–adsorbate separation, with a period of several angstroms giving rise, for example, to non-close-packed islands.[5] Such is the case of oxygen adsorbed on W(110), for instance.[6] And they are usually anisotropic, differing according to the orientation of the lines connecting pairs of adsorbates, since the single-crystal substrate surface is inherently anisotropic. Additional anisotropy occurs with many-adsorbate interactions [as observed for oxygen on W(110)[6]], as one can easily illustrate for a single adsorbate near a cluster of two adsorbates: it may produce a three-in-line cluster or an L-shaped cluster. This particular form of interaction is ideally suited for studies with field-ion microscopy, especially by observing the diffusion of such clusters along surfaces.[7] However, the analysis of such observations is only in its early stages.

Except for the strong repulsion at close separations, which prevents adsorbates from penetrating each other, the adsorbate–adsorbate interactions are usually weak as compared to the adsorbate–substrate interactions, even when one considers only the components of the forces parallel to the surface. Thus, in the case of chemisorption, where the adsorbate–substrate interaction dominates, one finds that the adsorbates usually choose an adsorption site that is independent of the coverage and of the overlayer arrangement (the positions the other adsorbates choose). As will be discussed in more detail later in the chapter, this adsorption site is usually that location which provides the largest number of nearest substrate neighbors, which is indeed independent of the position of other adsorbates. Adsorbates with these properties normally do not accept close packing: the substrate controls the overlayer geometry and imposes a unique adsorption site. Close packing of an adsorbate layer is, however, often observed with other adsorbates. Then the overlayer chooses its own lattice (normally a hexagonal close-packed arrangement) with its own lattice constant independent of the substrate lattice: formation of so-called "incommensurate" lattices. In this case no unique adsorption site exists; each adsorbate is differently situated with respect to the substrate. This situation is especially common in the physisorption of rare gases, with their relatively weak adsorbate–substrate interactions, which therefore allows the adsorbate–adsorbate interactions to play the dominant role in determining the overlayer geometry. Sometimes, the substrate imposes a particular orientation on the overlayer lattice in this circumstance.

The chemisorption case is exemplified by oxygen and sulfur on metals; the physisorption case by krypton and xenon on metals and graphite. Intermediate cases do exist; for example, although undissociated CO on metals is not physisorbed but chemisorbed, it nevertheless seems in many cases to be able to produce incommensurate close-packed hexagonal overlayers. Also, some metal surfaces [for example, Pt(100), Ir(100), and Au(100)] reconstruct into different lattices, exhibiting the effect of adsorbate–adsorbate interactions (here the adsorbate is just another metal atom of the same species as in the substrate).

As will be seen in a later section, the variety of possible ordered surface structures is immense. This is a reflection of the different magnitudes of the various forces responsible for the bonding and the ordering. When one realizes that each of these bonding forces varies in three dimensions, often drastically, it is not surprising that a very large number of combinations, and hence of structures, is possible. It may be true that every conceivable two-dimensional ordering arrangement is possible in nature on surfaces, even with simple adsorbates on simple surfaces.

The theory of the binding of single adsorbates to substrates is understood to some extent today and will be discussed later in the chapter, but the theory of adsorbate–adsorbate interactions, especially of large-scale ordering, is in its infancy.

Degree of Ordering

A perfectly ordered surface represents the energetically most favorable surface configuration. However, no real surface is perfectly ordered. There are several reasons for this. First, some thermal energy is always available to make an adsorbate jump into an energetically less favorable configuration; for example, adsorbate atoms in an ordered overlayer can jump out of registry. Even at zero temperature the zero-point motion gives rise to disorder in the form of vibrations about the atomic equilibrium positions. Second, in no experiment is the surface allowed to reach the asymptotic equilibrium; some forms of disorder have characteristic half-lives of the order of many hours. Thus an adatom trapped interstitially in a normally unoccupied site of a $c(2 \times 2)$ overlayer arrangement on a square-lattice substrate will, at low enough temperatures, have very little chance of migrating to a proper unoccupied site prescribed by the $c(2 \times 2)$ lattice, since such a site may be located at a considerable distance. Other examples of long half-life disorder are steps in the surface (if they are undesirable), bulk defects extending to the substrate surface, and, of course, impurities.

Some forms of disorder common in adsorbed layers are islands of clustered adsorbates leaving patches of bare substrate; domains in which different patches of the overlayer have identical structure but do not match at their junction because of an error in registry (an error in relative positioning parallel to the surface); and periodicity errors, in which individual adsorbates do not fit in the periodic arrangement of the surrounding adsorbates. These periodicity errors subdivide into those that involve inequivalent sites (such as an adsorbate choosing a twofold bridge site while the overlayer as a whole involves only fourfold hollow sites); those that retain unique adsorption sites, but improper ones; cases of individual disorder (one adsorbate in a wrong position); and cases of collective disorder (such as phonons and liquid layers).

Although perfect order is never present, perfect disorder also does not exist at surfaces. In the liquid (or the gaseous) state of overlayers on surfaces, the adsorbates cannot pass through each other; this gives rise to a limited amount of short-range order. Additionally, there is always some nonzero parallel component of the substrate–adsorbate interaction that will make the adsorbates spend more of their time at one type of location than at others; this, also, is a form of ordering.

Surface-sensitive diffraction techniques, especially LEED, can, in principle, detect any kind of ordering or disordering at a surface. The exact state of a surface at any moment can be represented by a Fourier series that describes the surface in terms of all possible periodicities (Fourier components). Each different periodicity present in an overlayer produces diffraction into a well-defined direction specified by the period and the orientation of the particular periodicity. The intensity of the diffraction measures the amount of order with that periodicity (this intensity is modulated by the surface structure perpendicular to the surface and cannot be taken to be a direct measure of the amount of order without proper care). Therefore, diffraction methods allow one to easily filter out many forms of disorder. In LEED, one may analyze just the sharp spots observed on a screen and thereby filter out all disorder that has periodicities defined by points between those spots. Nondiffraction methods do not have this kind of disorder-filtering capability: they usually average over all information, whether from the ordered or disordered part of the surface. Nondiffraction methods may sometimes have other types of disorder filtering, however. If disorder produces features at different energies of a spectrum, for example, such energies might be screened out. Thus in high-resolution electron energy loss spectroscopy, undesired surface adsorbates might produce resonance levels at different energy losses, which can then be ignored.

As an example of the analysis of the ordering of an overlayer of adsorbates, we may take the question of detecting island formation. LEED provides a means for identifying when island formation takes place,[5,6] although it does not always give a definitive answer. To monitor island formation, the presence of adsorbate-induced extra spots in the diffraction is necessary. Thus the adsorbate must produce a superlattice, and we assume this case in the following discussion.

To recognize island formation, one takes advantage of the difference between coherent and incoherent diffraction from a set of N identical scatterers. If the waves scattered off the individual scatterers are incoherent in their phases, the observed intensity will be proportional to N (addition of intensities). If, however, these scattered waves are coherent, the intensity will be proportional to N^2 (addition of amplitudes). Incoherency occurs either when the incident wave arrives with incoherent phases at different scatterers, which occurs in practice for scatterers separated by at least the "coherence length" of the incident beam, or when the scatterers themselves are located incoherently (are disordered).

The key to the detection of island formation is the coherence length (also called instrument response width or transfer width) of the incident electron beam, typically 100 Å. If the coherence length were much smaller or much larger than 100 Å, one would not obtain information about island formation from LEED. If the coherence length were variable, this degree of freedom would be valuable to study island formation on different scales (to a limited extent it is variable, namely, by changing the angle of incidence or the electron energy, but more flexibility would be useful). At low coverages, if islands form that are smaller than the coherence length and also farther apart than the coherence length, a diffraction pattern characteristic of an island is produced: each of the extra spots has a sharpness inversely proportional to the island diameter. The extra-spot intensity is then proportional to the square of the coverage, if one assumes that additional adsorbates will attach themselves to islands. If, instead, they initiate new islands (still far apart), the intensity will increase linearly with coverage; with such island birth the spot sharpness is constant. In reality, both island growth and island birth can take place simultaneously in varying proportions, depending on such factors as the surface mobility of the adsorbate and the binding energy of adsorbates to islands; then the extra-spot intensities will vary with a law between the first and second powers of the coverage. The extra-spot sharpness will simultaneously be less than inversely proportional to the island diameter, that is, more constant. In contrast, away from the island, adsorption at low coverages gives no extra spots, but a weak, diffuse background.

However, these relations change as soon as the coverage becomes sufficiently high that either the island diameter is at least equal to the coherence length or the island–island distance is at most equal to the coherence length. With an island diameter at least equal to the coherence length, the extra-spot sharpness saturates at a value determined by the coherence length, while the extra-spot intensities become linear with coverage. With an island–island distance at most equal to the coherence length, the extra spots remain relatively diffuse with increasing coverage (due to a relatively constant and small island size), while these spots weaken again due to antiphase domains (in some cases extra-spot splitting occurs). For comparisons, nonisland adsorption at these higher coverages produces either no order at all (no extra spots) or else spots with a sharpness determined by the ordering distance and with an intensity quadratic in the coverage.

Complications in actual studies along the lines described above come from uncertainties in the question of island growth versus island birth and a lack of understanding of the factors determining these. Also, the range of the ordering forces plays a role that should be explored more systematically than hitherto. And, of course, the bonding configuration may change with coverage, causing a change in intensities not related to the effects described above.

Generally, it is difficult to obtain experimental information about the exact form of disorder present on any actual surface; much work remains to be done in this direction.

Effect of Temperature on the Ordering of
Adsorbed Monolayers

In Figure 5.3 the influence of temperature on the ordering of C_4 to C_3 saturated hydrocarbon molecules on the Pt(111) crystal face is shown. At the highest temperatures, adsorption may not take place, since under the exposure conditions the rate of desorption is greater than the rate of condensation of the vapor molecules. As the temperature is decreased, the surface coverage increases and ordering becomes possible. First, one-dimensional lines of molecules form, then upon further dropping of the temperature ordered two-dimensional surface structures form. Not surprisingly, the temperatures at which these ordering transitions occur depend on the molecular weights of the hydrocarbons, which also control their vapor pressure, their heats of adsorption, and their activation energies for surface diffusion. As the temperature is further decreased, multilayer adsorption may occur and epitaxial growth of crystalline thin films of hydrocarbon commences.

Figure 5.3 clearly demonstrates the controlling effect of temperature on the ordering and the nature of ordering of the adsorbed monolayer. Although changing the pressure at a given temperature may be used to vary the coverage by small amounts and thereby change the surface structures in some cases, the variation of temperature has a much more drastic effect on ordering. All the important ordering parameters (the rates of desorption, surface, and bulk diffusion) are exponential functions of the temperature.

An example of the control of ordering by surface diffusion is shown in Figure 5.4. Naphthalene forms a poorly ordered structure when adsorbed on a Pt(111) crystal face at 300 K. Upon heating the almost glassy layer to 450 K, a well-ordered (6 × 6) surface structure forms. For large molecules, surface diffusion plays a visibly important role in ordering, as detected by several investigations.

Temperature also markedly influences chemical bonding to surfaces. There are adsorption states that can only be populated if the molecule overcomes a small potential-energy barrier. The various bond-breaking

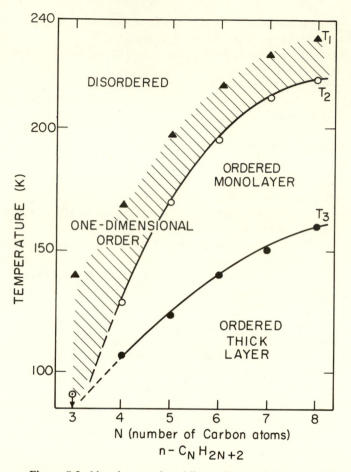

Figure 5.3. Monolayer and multilayer phases of the n-paraffins C_3 to C_8 on Pt(111) and the temperatures at which they are observed at 10^{-7} torr.

processes are similarly activated. The adsorption of most reactive molecules on chemically active solid surfaces takes place without bond breaking at sufficiently low temperatures. As the temperature is increased, bond breaking occurs sequentially until the molecule is atomized. Thus the chemical nature of the molecular fragments will be different at various temperatures. There is almost always a temperature range, however, for the ordering of intact molecules in chemically reactive adsorbate–substrate systems. It appears that for these systems ordering

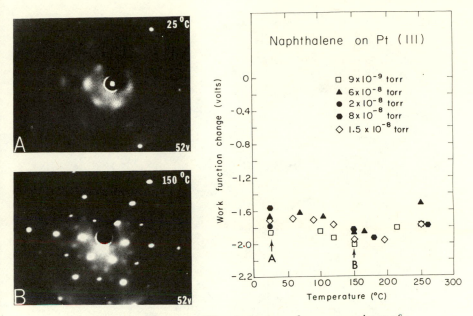

Figure 5.4. Electron diffraction pattern from a monolayer of naphthalene on Pt(111) for an electron energy of 52 eV as a function of temperature. Sharp spots correspond to good ordering. The work-function changes obtained at different pressures are also shown as a function of temperature.

is restricted to low temperatures below 150 K, and consideration of surface mobility becomes, perhaps, secondary.

Effect of Surface Irregularities on Ordering

When a solid surface is viewed under the optical or the electron microscope, it almost always exhibits a large degree of roughness on the macroscopic scale. There are protruding hills several hundred atomic layers in height and discontinuities that separate relatively smooth terraces. A typical electron microscope picture of an etched platinum single-crystal surface is shown in Figure 5.5. On the atomic scale, however, the surface appears to be much smoother. The very high quality low-energy electron diffraction pattern commonly observed from most cleaned and annealed solid surfaces must require the presence of domains of ordered atoms larger than 100 to 200 Å in diameter. The coherent scattering of electrons that yields the sharp, small, and high-intensity diffraction spots can occur only if the size of the scattering areas is larger than the electron

189

Figure 5.5. Scanning electron microscopy (SEM) photograph of an etched Pt(100) surface tilted at 45° to the incident electron beam to enhance the picture contrast.

coherence length. Were the ordered domains smaller, a broadening of the diffraction spots would be observed, which is in fact what happens if the surface is roughened by ion bombardment, for example. Another technique, field-ion microscopy (FIM), which can display the surface topography of a small tip of $\sim 10^{-4}$ cm diameter with atomic resolution, also shows the large degree of atomic order that is possible at surfaces.

As long as nucleation is an important part of the adsorbate ordering process, surface roughness is likely to play an important role in preparing ordered surface structures. It is observed frequently that the ease of ordering and the quality of the ordered surface structures of adsorbates change from one substrate sample to another. There is often great "improvement" in the ordering characteristics right after ion-bombardment cleaning and brief thermal annealing of the substrate surface; then ordering becomes poorer as the substrate is annealed and thereby ordered more and more. The transformation temperature or pressure at which one adsorbate surface structure converts into another can also be affected by the presence of uncontrolled surface irregularities. Although the surface structures of adsorbates are, by and large, reproducible from sample to sample and laboratory to laboratory, the uncertainties in the experimental conditions necessary to form the ordered surface structures are caused most frequently by uncontrolled surface defects. Other causes that could influence ordering are the presence of small amounts of surface impurities that block nucleation sites or interfere with the kinetics of ordering or impurities below the surface that are pulled to the surface during adsorption and ordering.

It is much easier to investigate the effect of surface irregularities on ordering using stepped crystal surfaces. Unlike the case of uncontrolled surface defects on a (111) face of an fcc metal, for example, steps are readily detectable by LEED or FIM. They are likely to be ordered with a well-defined periodicity, and surfaces can be prepared in such a way that steps are the predominant highest-concentration surface defects.

The influence of atomic height steps on the ordering of adsorbate structures has been investigated in several studies.[8] In general, the smaller the ordered terrace between steps, the stronger the effect of steps on ordering. The ordering of small molecular adsorbates on a high-Miller-index (755) Rh surface was largely unaffected by the presence of steps.[8a] However, ordering was influenced by steps on the larger step-density Rh(331) crystal face.[8a] Nitrogen and carbon layers were observed to extend over several terraces on stepped copper surfaces,[8b] and the ordering of Ar and Kr was unaffected by the presence of steps on copper and silver (211) crystal faces.[8c] Just as in the case of uncontrolled irregularities, steps can markedly affect the nucleation of ordered domains. It is frequently observed on W and Pt stepped surfaces that when two or three equivalent ordered domains may form in the absence of steps, only one of the ordered domains grows in the presence of steps. Oxygen surface structures exhibit this phenomenon as well and have been studied in the greatest detail.

In many cases, ordering is no longer observable in the presence of steps. Ordered carbonaceous layers form on the Ir(111) crystal face, while ordering is absent on the stepped iridium surface. Ordering is absent on stepped Pt surfaces for most molecules that would order on the low-Miller-index (111) or (100) surfaces.

In some cases the step sites have different chemistry: they break chemical bonds, thereby producing new chemical species on the surface. This happens, for example, during NO adsorption on a stepped platinum surface.[9] In this circumstance the step effect on ordering is through the new types of chemistry introduced by the presence of steps. Hydrocarbons dissociate readily at stepped surfaces of platinum or nickel, whereas this occurs much more slowly on the low-Miller-index surfaces in the absence of a large concentration of steps.[10] As a result, ordered hydrocarbon surface structures cannot be formed on the stepped surfaces of these metals, whereas they can be produced on the low-Miller-index surfaces.

There is also a great deal of evidence for increased sticking probability at stepped surfaces. The change in the magnitude of the adsorption probability ranges from 20 percent to orders of magnitude. Also, several studies have revealed increased binding energies at step sites: 10 to 20 percent increases in binding energies at steps on Ni and Pt surfaces are common. Both the increased adsorption probability and the binding energies at steps may strongly affect the kinetics of ordering. Thus there are many reasons for the different ordering characteristics of adsorbed monolayers in the presence of surface irregularities.

Unit Cells of Adsorbed Monolayers

The unit cells of adsorbate layers are primarily a function of the coverage: as the coverage varies, many adsorbates produce complete series of successive different unit cells. The coverage is defined here in such a way that unit coverage, $\theta = 1$, occurs when the adsorbate occupies all equivalent adsorption sites.

One can correlate the coverage, θ, with certain features of the unit cells that the adsorbates can adopt on surfaces. Let us define S to be the area of the substrate unit cell.

When $1/\theta$ is an integer n, there are n substrate unit cells per adsorbate, and a superlattice with unit cell area nS can occur. Thus for $\theta = \frac{1}{2}$, a superlattice with unit cell area $2S$ may exist, examples of which are designated $p(2 \times 1)$ and $c(2 \times 2)$ and are illustrated in Figure 5.6.

When $1/\theta$ is a rational number m/n (m and n indivisible integers), there are m substrate unit cells per set of n adsorbates. A superlattice with unit-cell area mS can form, with the superlattice unit cell containing n arbitrarily positioned adsorbates. It may happen in this case that the adsorbates between themselves (ignoring the substrate) form a structure that has a smaller unit cell than the superlattice unit cell. One must then distinguish between the overlayer unit cell (defined in the absence of a substrate) and the so-called "coincidence" unit cell (describing the combined substrate–overlayer system). An example is shown in Figure 5.7 for the case of Pd(100) + $(2\sqrt{2} \times \sqrt{2})$R45°−2CO, which has two molecules per coincidence unit cell.[11] Note that this adsorbate has man-

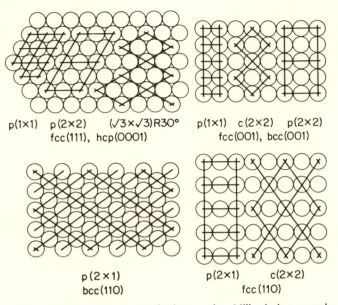

Figure 5.6. Common superlattices on low-Miller-index crystal surfaces. The Wood notation is used.

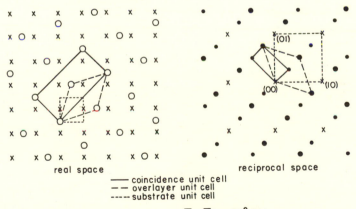

Figure 5.7. Real space (left) and reciprocal space (right) applicable to Pd(100) + (2 × 2)R45°−2CO. Substrate (overlayer) atoms and diffraction spots are indicated by crosses (circles). Large filled circles represent kinematically produced spots; small filled circles represent multiple diffraction spots.

aged to combine bridge sites with an approximately hexagonal arrangement.

When $1/\theta$ is an irrational number, the overlayer lattice bears, in general, no relationship to the substrate lattice; the surface unit cell becomes infinite, and the unit cell areas become incommensurate. This case corresponds to totally independent lattices, as is approximated by physisorbed systems.

In practice, the distinction between rational and irrational values of $1/\theta$ is unimportant, because LEED cannot distinguish between unit cells larger than the coherence distance of the electron beam (~ 100 Å). It is customary to designate as incommensurate any overlayer that produces a coincidence unit cell larger than the LEED coherence distance. In fact, a truly incommensurate overlayer is impossible, since it could only occur in the limit of vanishing adsorbate–substrate forces parallel to the surface.

Review of Ordered Adsorbate Structures

Introduction

While only about 40 surface structures have been analyzed by methods of surface crystallography in order to determine the precise location of adsorbed atoms or molecules, nearly 1000 ordered surface structures of adsorbates have been reported. It appears that almost any adsorbate monolayer may be ordered to form at least one, and frequently several, structures under appropriate conditions of gas exposure and temperature. The proper experimental conditions achieve a balance among the various surface forces (such as heats of adsorption, activation energies of surface, and bulk diffusion), which facilitates the formation of large ordered domains that yield sharp diffraction features. The ordering of adsorbed monolayers is a very sensitive function of temperature. For example, the lowering of the temperature of rhodium single crystals from 300 to 270 K greatly increases the size of the ordered domains of CO, O_2, and other adsorbates, which, in turn, visibly improves the quality of the diffraction patterns. Similar observations are commonly reported for other adsorbed monolayer systems as well. The adsorbate ordering obviously also depends strongly on coverage, since a particular periodic arrangement of adsorbates at one coverage cannot freely accommodate a change in coverage. Among other, similar examples, Pb deposited on Au(100) produces $c(2 \times 2)$, $c(7\sqrt{2} \times \sqrt{2})R45°$, $c(3\sqrt{2} \times$

$\sqrt{2}$)R45°, and $c(6 \times 2)$ arrangements as the coverage is varied. (The unit-cell notation is discussed in Chapter 4.)

In this section we present comprehensive tabulations of the observed ordered structures for any adsorbate on any substrate. For most of these cases the surface structure has not been analyzed beyond the implications of unit-cell shape, size, and orientation. Many of these structures are good candidates for a structural analysis of binding sites, bond lengths, and bond angles. It is hoped that the list of geometrically analyzed structures will grow rapidly so as to present an expanding base for the extraction of fundamental laws governing the adsorption phenomenon.

Low-Miller-index surfaces of metallic single crystals are the most commonly used substrates in LEED investigations. The reasons for their widespread use are that they have the lowest surface free energy of the various metal crystal faces and therefore are the most stable, have the highest rotational symmetry, and are the most densely packed. Also, in the case of transition metals and semiconductors, they are chemically less reactive than are the higher-Miller-index crystal faces.

The metal substrates used in the LEED experiments have either face-centered cubic (fcc), body-centered cubic (bcc), or hexagonal close-packed (hcp) crystal structures. For the cubic metals the (111), (100), and (110) planes are the low-Miller-index surfaces, and they have threefold, fourfold, and twofold rotational symmetry, respectively. The top layer of a (111) surface actually has sixfold symmetry, but the rotational symmetry of the top two layers together is threefold. Since the near-surface region can influence where gases adsorb on the surface and the LEED intensities exhibit threefold rotational symmetry at normal incidence, the (111) surface will be considered to have threefold rotational symmetry. Although most of the adsorption studies have been carried out on fcc and bcc crystals, several studies have been reported on hcp crystals. For hcp metals, the basal or (0001) plane is the surface most frequently studied by LEED investigations, and it is the most densely packed plane, having threefold rotational symmetry.

Metals on Metals

Table 5.1 lists the surface structures of ordered metal monolayers adsorbed on metal surfaces. For each substrate the crystallographic structure, the distance between nearest neighbors, and the heat of sublimation (that is, proportional to the surface free energy) are given. For each metal adsorbate the identical information is provided together with

Table 5.1. Adsorption properties of metal monolayers on metal substrates. The clean substrate properties are also given for comparison. Substrates are ordered by lattice type (fcc, bcc, hcp, cubic, diamond, and rhombic). The structures, nearest-neighbor distances, and heats of vaporization refer to the bulk material of the substrate or the adsorbate. VD, ID, and S stand for vapor deposition, ion beam deposition, and surface segregation, respectively. TD, WF, and TED stand for thermal desorption, work-function measurements, and transmission electron diffraction, respectively.

Substrate metal	Adsorbed metal	Structure	Nearest-neighbor distance	Heat of vaporization (kcal/g-atom)	Deposition technique	Substrate orientation	Technique of investigation	Surface structures observed	References
Rh	—	fcc	2.69	127	—	—	—	—	—
Rh	Fe	bcc	2.48	85	VD	(100)	TED	Fe(100) and Fe(110) $\parallel$ Rh (100)	13
Ir	—	fcc	2.71	160	—	—	—	—	—
Ir	Cr	bcc	2.50	73	VD	(111)	LEED-AES-WF	hexagonal	14
Ir	Au	fcc	2.88	82	VD	(111)	LEED-AES-WF	$\begin{pmatrix} 1 & 0 \\ 0 & 1 \end{pmatrix}$	14
Ni	—	fcc	2.49	91	—	—	—	—	—
Ni	Na	bcc	3.66	24	ID/VD	(100)	LEED-WF	$\begin{pmatrix} 1 & 1 \\ 1 & 1 \end{pmatrix}$	13, 15–20
Ni					ID	(111)	LEED	hexagonal	16, 19, 21
Ni					ID	(110)	LEED	disordered structures, hexagonal	16, 19, 21
Ni	K	bcc	4.52	19	ID/VD	(100)	LEED-WF/LEED	$\begin{pmatrix} 4 & 0 \\ 0 & 2 \end{pmatrix}$/hexagonal	15, 18, 22
Ni					ID	(110)	LEED	disordered structures	16
Ni	C$_s$	bcc	5.23	16	ID	(100)	LEED-WF/LEED	$\begin{pmatrix} 2 & 0 \\ 0 & 2 \end{pmatrix}$/hexagonal	15, 23
Ni					ID	(110)	LEED	disordered structures	16
Ni	Ba	bcc	4.35	36	VD	(100)	LEED-WF	disordered overlayer	15
Ni	Cr	bcc	2.50	73	VD	(100)	TED	$\begin{pmatrix} 1 & 0 \\ 0 & 1 \end{pmatrix}$	23
Ni	Mo	bcc	2.72	128	S	(111)	LEED	$\begin{pmatrix} 5 & 0 \\ 0 & 5 \end{pmatrix}, \begin{pmatrix} 4 & 0 \\ 0 & 4 \end{pmatrix}, \begin{pmatrix} 2 & 0 \\ 5 & 10 \end{pmatrix}, \begin{pmatrix} 1 & 0 \\ 5 & 10 \end{pmatrix}$	8
Ni	Fe	bcc	2.48	85	VD	(100)	TED	(110) Fe $\parallel$ (100) Ni	13
Ni	Co	hcp	2.50	93	VD	(100)	TED	$\begin{pmatrix} 1 & 0 \\ 0 & 1 \end{pmatrix}$	25
Ni	Cu	fcc	2.56	73	VD	(100)	RHEED-AES	$\begin{pmatrix} 1 & 0 \\ 0 & 1 \end{pmatrix}$	26

Substrate	Deposit	Structure	a		Method	Face	Technique	Orientation matrix	Ref.
Ni	Ag	fcc	2.89	61	VD	(111)	LEED/AES	$\begin{pmatrix} 6 & 0 \\ 0 & 6 \end{pmatrix}$	27, 28
Ni	Au	fcc	2.88	82	S	(111)	LEED-AES/LEED	$\begin{pmatrix} 6 & 0 \\ 0 & 6 \end{pmatrix}, \begin{pmatrix} 13 & 0 \\ 0 & 13 \end{pmatrix} / \begin{pmatrix} 6 & 0 \\ 0 & 6 \end{pmatrix}$	29, 30
Ni	Pb	fcc	3.50	42	VD	(100)	LEED	$\begin{pmatrix} 1 & \bar{1} \\ 1 & 1 \end{pmatrix}, \begin{pmatrix} \frac{1}{2} & \frac{1}{2} \\ 2 & 3 \end{pmatrix}$	31
Ni		fcc			VD	(111)	LEED	$\begin{pmatrix} 1 & \bar{1} \\ 1 & 2 \end{pmatrix}, \begin{pmatrix} 7 & 0 \\ 0 & 7 \end{pmatrix}, \begin{pmatrix} 13 & 0 \\ 0 & 13 \end{pmatrix}, \begin{pmatrix} 3 & 0 \\ 0 & 3 \end{pmatrix} / \text{hexagonal}$ rotated $\pm 3°$, $\begin{pmatrix} 3 & 0 \\ 0 & 3 \end{pmatrix}, \begin{pmatrix} 1 & 1 \\ 1 & 2 \end{pmatrix}$	31, 32
					VD	(110)	LEED	$\begin{pmatrix} 1 & \bar{1} \\ 1 & 1 \end{pmatrix}, \begin{pmatrix} 3 & 0 \\ 0 & 1 \end{pmatrix}, \begin{pmatrix} 4 & 0 \\ 0 & 1 \end{pmatrix}, \begin{pmatrix} 5 & 0 \\ 0 & 1 \end{pmatrix}$	31
Pd	—	fcc	2.75	90	—	—	—	—	—
Pd	Fe	bcc	2.48	85	VD	(100)	TED	$\begin{pmatrix} 1 & 0 \\ 0 & 1 \end{pmatrix}$ (100) Fe ∥ (100) Pd and (110) Fe ∥ (100) Pd	13
Pd	Ni	fcc	2.49	91	VD	(100)	TED	$\begin{pmatrix} 1 & 0 \\ 0 & 1 \end{pmatrix}$	33
Pd	Ag	fcc	2.89	61	VD	(100)	LEED	$\begin{pmatrix} 1 & 0 \\ 0 & 1 \end{pmatrix}$	34
Pd	Au	fcc	2.88	82	VD	(100)	LEED/TED	$\begin{pmatrix} 1 & 0 \\ 0 & 1 \end{pmatrix}$	34–36
Pd					VD	(111)	TED	$\begin{pmatrix} 1 & 0 \\ 0 & 1 \end{pmatrix}$	36
Cu	—	fcc	2.56	73	—	—	—	—	—
Cu	Fe	bcc	2.48	85	VD	(100)	TED	$\begin{pmatrix} 1 & 0 \\ 0 & 1 \end{pmatrix}$	13, 37, 38
Cu					VD	(111)	LEED-AES	$\begin{pmatrix} 1 & 0 \\ 0 & 1 \end{pmatrix}$	39
Cu	Co	hcp	2.50	93	VD	(100)	TED	$\begin{pmatrix} 1 & 0 \\ 0 & 1 \end{pmatrix}$	40, 41

(continued)

Table 5.1.—Continued

Substrate metal	Adsorbed metal	Structure	Nearest-neighbor distance	Heat of vaporization (kcal/g-atom)	Deposition technique	Substrate orientation	Technique of investigation	Surface structures observed	References
Cu	Ni	fcc	2.49	91	VD	(100)	TED	$\begin{pmatrix}1&0\\0&1\end{pmatrix}$	42
					VD	(111)	LEED/RHEED	$\begin{pmatrix}1&0\\0&1\end{pmatrix}$	43–45
Cu	Ag	fcc	2.89	61	VD	(100)	LEED	$\begin{pmatrix}2&0\\1&5\end{pmatrix}$	34, 46
					VD	(111)	LEED/RHEED, TED	$\begin{pmatrix}8&0\\0&8\end{pmatrix}$/three-dimensional crystals	45–50
Cu	Au	fcc	2.88	82	VD	(100)	LEED	$\begin{pmatrix}1&1\\1&1\end{pmatrix}$, $\begin{pmatrix}2&0\\1&7\end{pmatrix}$	34, 51
					VD	(111)	LEED-AES/RHEED	$\begin{pmatrix}2/3&2/3\\2/3&4/3\end{pmatrix}$, $\begin{pmatrix}2&0\\0&2\end{pmatrix}$/three-dimensional crystals	47, 52, 53
					VD	(110)	LEED-AES	$\begin{pmatrix}1&0\\1/2&3/2\end{pmatrix}$, $\begin{pmatrix}1&0\\0&2\end{pmatrix}$, $\begin{pmatrix}2&0\\0&2\end{pmatrix}$. complex structures	52
Cu	Sn	diamond	2.81	70	S	(100)	LEED-AES	$\begin{pmatrix}2&0\\0&2\end{pmatrix}$	54
					S	(111)	LEED-AES	$\begin{pmatrix}1&1\\1&2\end{pmatrix}$	54
Cu	Pb	fcc	3.50	42	VD	(100)	LEED/LEED-AES-TD/TED	$\begin{pmatrix}2&2\\2&2\end{pmatrix}$, $\begin{pmatrix}1&1\\2&3\end{pmatrix}$	55–59
					VD	(111)	LEED/LEED-AES-TD	$\begin{pmatrix}4&0\\0&4\end{pmatrix}$	55, 58
					VD	(110)	LEED/LEED-AES	$\begin{pmatrix}1&1\\1&1\end{pmatrix}$, $\begin{pmatrix}5&0\\0&1\end{pmatrix}$/$\begin{pmatrix}1&1\\1&1\end{pmatrix}$, $\begin{pmatrix}4&0\\0&1\end{pmatrix}$, $\begin{pmatrix}5&0\\0&1\end{pmatrix}$	55, 56
					VD	(711)	LEED-AES/LEED-AES-TD	$\begin{pmatrix}4&0\\0&1\end{pmatrix}$	56, 58
					VD	(511)	LEED-AES	$\begin{pmatrix}4&0\\0&1\end{pmatrix}$	56

Cu	Bi	rhombic	3.07		VD	(311)	LEED-AES	$\left(\begin{smallmatrix}\frac{3}{2} & 1\\ 2 & 1\end{smallmatrix}\right)$	58
					VD	(211)	LEED-AES-TD	$\left(\begin{smallmatrix}4 & 0\\ 0 & 2\end{smallmatrix}\right)$	58
				43	VD	(100)	LEED/LEED-AES	$\left(\begin{smallmatrix}2 & 0\\ 0 & 2\end{smallmatrix}\right),\left(\begin{smallmatrix}1 & 1\\ 1 & 1\end{smallmatrix}\right),\left(\begin{smallmatrix}\frac{1}{4} & 1\\ \frac{1}{4} & 5\end{smallmatrix}\right),\left(\begin{smallmatrix}\frac{5}{4} & 4\\ & 5\end{smallmatrix}\right)$	57, 60
Ag	—	fcc	2.89	61	VD	(111)	LEED	$\left(\begin{smallmatrix}\frac{1}{1} & 1\\ 1 & 2\end{smallmatrix}\right),\left(\begin{smallmatrix}2 & 1\\ 0 & 2\end{smallmatrix}\right),\left(\begin{smallmatrix}2 & 3\\ 1 & 2\end{smallmatrix}\right)$	—
Ag	Na	bcc	3.66	24	VD	(111)	LEED-AES-TD	$\left(\begin{smallmatrix}1 & 0\\ 0 & 1\end{smallmatrix}\right)$	62
Ag	Rb	bcc	4.84	18	VD	(110)	LEED-AES-TD	$\left(\begin{smallmatrix}1 & 0\\ 0 & 1\end{smallmatrix}\right)$	63
Ag		bcc			VD	(111)	LEED-AES-TD	$\left(\begin{smallmatrix}1 & 0\\ 0 & 1\end{smallmatrix}\right)$	64
Ag	Mg	hcp	3.20	32	VD	(111)	TED	disordered overlayer	65
Ag	Cr	bcc	2.50	73	VD	(111)	TED	disordered overlayer	65
Ag	Co	hcp	2.50	93	VD	(111)	TED	disordered overlayer	65
Ag	Ni	fcc	2.49	91	VD	(100)	TED	$\left(\begin{smallmatrix}1 & 0\\ 0 & 1\end{smallmatrix}\right)$	66
Ag	Pd	fcc	2.75	90	VD	(111)	TED/RHEED	hexagonal overlayer	65, 67
Ag	Cu	fcc	2.56	73	VD	(111)	TED	disordered overlayer	65
Ag		fcc			VD	(100)	TED	hexagonal overlayer	66
Ag	Au	fcc	2.88	82	VD	(111)	RHEED/TED	$\left(\begin{smallmatrix}1 & 0\\ 0 & 1\end{smallmatrix}\right)$	68–70
Ag		fcc			VD	(100)	TED	hexagonal overlayer	35
Ag		fcc			VD	(111)	LEED-AES/TED	$\left(\begin{smallmatrix}1 & 0\\ 0 & 1\end{smallmatrix}\right)$	65, 71, 72
Ag	Zn	hcp	2.66	27	VD	(111)	TED	no condensation	65
Ag	Cd	hcp	2.98	24	VD	(111)	TED	no condensation	65
Ag	Al	fcc	2.86	68	VD	(111)	TED	disordered overlayer	65
Ag	Tl	hcp	3.46	39	VD	(111)	TED	hexagonal overlayer	65
Ag	Sn	diam	2.81	70	VD	(111)	TED	disordered overlayer	65

(continued)

Table 5.1.—Continued

Substrate metal	Adsorbed metal	Structure	Nearest-neighbor distance	Heat of vaporization (kcal/g-atom)	Deposition technique	Substrate orientation	Technique of investigation	Surface structures observed	References
Ag	Pb	fcc	3.50	42	VD	(111)	TED	hexagonal overlayer	65, 73
Ag	Sb	rhomb	2.91	47	VD	(111)	TED	disordered overlayer	65
Ag	Bi	rhomb	3.07	43	VD	(111)	TED	disordered overlayer	65
Au	—	fcc	2.88	82	—	—	—	—	—
Au	Na	bcc	3.66	24	VD	(100)	LEED	series of structures, hexagonal	74
Au	Cr	bcc	2.50	73	VD	(111)	LEED-AES-WF	hexagonal	75
Au	Fe	bcc	2.48	85	VD	(100)	TED	$\begin{pmatrix}1&0\\0&1\end{pmatrix}$	76–78
Au					VD	(111)	TED	$\begin{pmatrix}1&0\\0&1\end{pmatrix}$	76, 78–80
Au	Pd	fcc	2.75	90	VD	(100)	LEED/TED	$\begin{pmatrix}1&0\\0&1\end{pmatrix}$	34, 81
Au					VD	(111)	TED	$\begin{pmatrix}1&0\\0&1\end{pmatrix}$	36, 82
Au	Pt	fcc	2.77	122	VD	(100)	LEED-AES/TED	$\begin{pmatrix}1&0\\0&1\end{pmatrix}$	3, 4
Au					VD	(111)	TED	$\begin{pmatrix}1&0\\0&1\end{pmatrix}$	83
Au	Cu	fcc	2.56	73	VD	(100)	LEED	$\begin{pmatrix}1&0\\0&1\end{pmatrix}$	34
Au					VD	(111)	RHEED	extra lines	84, 85
Au	Ag	fcc	2.89	61	VD	(100)	LEED/TED	$\begin{pmatrix}1&0\\0&1\end{pmatrix}$	34, 39, 86, 87
Au						(111)	LEED-AES	$\begin{pmatrix}1&0\\0&1\end{pmatrix}$	71
Au	Pb	fcc	3.50	42	VD	(100)	LEED-AES	$\begin{pmatrix}1&1\\1&1\end{pmatrix}, \begin{pmatrix}1&1\\3&4\end{pmatrix}, \begin{pmatrix}1&1\\1&2\end{pmatrix},$ $\begin{pmatrix}2&0\\1&3\end{pmatrix}$	5, 6
Au					VD	(111)	LEED-AES	hexagonal rotated $\pm 5°$ $\begin{pmatrix}1&0\\0&3\end{pmatrix}, \begin{pmatrix}1&0\\0&1\end{pmatrix}, \begin{pmatrix}7&0\\0&1\end{pmatrix},$	6, 88
Au					VD	(110)	LEED-AES	$\begin{pmatrix}7&0\\0&3\end{pmatrix}, \begin{pmatrix}4&0\\0&4\end{pmatrix}$	6, 88

Substrate	Overlayer	Structure	d (Å)	No.	Mode	Orientation	Method	Epitaxial relationship	Ref
Au					VD	(11,1,1)	LEED-AES	$\begin{pmatrix}1&\bar1\\\bar1&1\end{pmatrix}$, $\begin{pmatrix}2&0\\\bar1&3\end{pmatrix}$	6
					VD	(911)	LEED-AES	$\begin{pmatrix}1&\bar1\\\bar1&1\end{pmatrix}$, $\begin{pmatrix}2&0\\\bar1&3\end{pmatrix}$	6
					VD	(711)	LEED-AES	$\begin{pmatrix}1&\bar1\\\bar1&1\end{pmatrix}$, $\begin{pmatrix}2&0\\\bar1&3\end{pmatrix}$	6
					VD	(511)	LEED-AES	$\begin{pmatrix}1&\bar1\\\bar1&1\end{pmatrix}$, $\begin{pmatrix}2&0\\\bar1&3\end{pmatrix}$	6
					VD	(311)	LEED-AES	$\begin{pmatrix}5&0\\0&3\end{pmatrix}$	89
					VD	(320)	LEED-AES	$\begin{pmatrix}3&0\\0&3\end{pmatrix}$	89
					VD	(210)	LEED-AES	$\begin{pmatrix}1&0\\0&1\end{pmatrix}$	90
	Bi	rhombic	3.07	43	VD	(100)	LEED	$\begin{pmatrix}2&0\\\bar1&2\end{pmatrix}$	91
						(111)	LEED	$\begin{pmatrix}10&10\\10&20\end{pmatrix}$	91
						(110)	LEED	$\begin{pmatrix}1&\bar1\\\bar1&1\end{pmatrix}$, $\begin{pmatrix}2&1\\\bar1&1\end{pmatrix}$, $\begin{pmatrix}2&0\\0&1\end{pmatrix}$	91
Al	—	fcc	2.86	68	—	—	—	—	—
Al	Na	bcc	3.66	24	ID	(100)	LEED-AES-WF	$\begin{pmatrix}1&\bar1\\\bar1&1\end{pmatrix}$, $\begin{pmatrix}2&0\\\bar1&4\end{pmatrix}$	92–94
Al					ID	(111)	LEED-AES-WF	$\begin{pmatrix}1&\bar1\\\bar1&2\end{pmatrix}$, $\begin{pmatrix}2&0\\0&1\end{pmatrix}$	93
Al	Mn	cubic	2.24	54	VD	(111)	LEED-AES	$\begin{pmatrix}6&0\\\bar1&2\end{pmatrix}$, hexagonal rotated ± 9°	95
Al	Fe	bcc	2.48	85	VD	(100)	TED	poor epitaxy	13
Al	Ni	fcc	2.49	91	VD	(111)	TED	$\begin{pmatrix}1&1\\\bar1&2\end{pmatrix}$	96
Al	Sn	diamond	2.81	70	VD	(100)	LEED-AES	$\begin{pmatrix}2&0\\\bar1&3\end{pmatrix}$	97
Al					VD	(111)	LEED-AES	hexagonal rotated ± 9°	98
Al	Pb	fcc	3.50	42	VD	(100)	LEED-AES	$\begin{pmatrix}2&0\\\bar1&2\end{pmatrix}$	97
Al					VD	(111)	LEED-AES	hexagonal rotated ± 9°	98
Nb	—	bcc	2.86	172	—	—	—	—	—

(continued)

Table 5.1.—Continued

Substrate metal	Adsorbed metal	Structure	Nearest-neighbor distance	Heat of vaporization (kcal/g-atom)	Deposition technique	Substrate orientation	Technique of investigation	Surface structures observed	References
Nb	Sn	diamond	2.81	70	VD	(110)	LEED	disordered structures, $\begin{pmatrix} 3 & 0 \\ 0 & 1 \end{pmatrix}$	99
Ta	—	bcc	2.86	180	—	—	—		—
Ta	Au	fcc	2.88	82	VD	(100)	LEED	split $\begin{pmatrix} 1 & 1 \\ 1 & 2 \end{pmatrix}$	100
Ta	Al	fcc	2.86	68	VD	(110)	LEED	hexagonal, square	101
Ta	Th	fcc	3.60	137	VD	(100)	LEED-WF	$\begin{pmatrix} 1 & 1 \\ 1 & \bar{1} \end{pmatrix}$, $\begin{pmatrix} 1 & 0 \\ 0 & 1 \end{pmatrix}$	102
Mo	—	bcc	2.72	128	—	—	—		—
Mo	Na	bcc	3.66	24	ID	(110)	LEED-AES	no ordered structure	103
Mo	K	bcc	4.52	19	ID	(110)	LEED-AES	hexagonal	103
Mo	Rb	bcc	4.84	18	ID	(110)	LEED-AES/AES	hexagonal	103, 104
Mo	Cs	bcc	5.23	16	ID	(110)	LEED-AES	hexagonal	103
Mo	Ag	fcc	2.89	61	VD	(100)	SEM-AES/LEED-AES	(100) Ag ∥ (100) Mo and (011) Ag ∥ (001) Mo	105, 106
Mo	Al	fcc	2.86	68	VD	(110)	LEED-AES	hexagonal	107
Mo	Sn	rhombic	2.81	70	VD	(100)	LEED-AES	$\begin{pmatrix} 1 & 1 \\ 1 & \bar{1} \end{pmatrix}$, $\begin{pmatrix} 1 & 0 \\ 0 & 2 \end{pmatrix}$	108
W	—	bcc	2.74	185	—	—	—		—
W	Li	bcc	3.02	32	VD	(110)	LEED-WF	$\begin{pmatrix} \frac{1}{2} & \frac{5}{2} \end{pmatrix}$, $\begin{pmatrix} 2 & 0 \\ 0 & 2 \end{pmatrix}$, $\begin{pmatrix} 1 & 1 \\ \bar{1} & 2 \end{pmatrix}$	109–111
					VD	(112)	LEED-WF	$\begin{pmatrix} 4 & 0 \\ 0 & 1 \end{pmatrix}$, $\begin{pmatrix} 3 & 0 \\ 0 & 1 \end{pmatrix}$, $\begin{pmatrix} 2 & 0 \\ 0 & 1 \end{pmatrix}$, incoherent, $\begin{pmatrix} 1 & 0 \\ 0 & 1 \end{pmatrix}$	110, 112
W	Na	bcc	3.66	24	VD	(100)	RHEED-TD	$\begin{pmatrix} 1 & 1 \\ 1 & \bar{1} \end{pmatrix}$	113
					VD	(110)	LEED-WF	$\begin{pmatrix} \frac{1}{2} & \frac{5}{2} \end{pmatrix}$, $\begin{pmatrix} 2 & 0 \\ 0 & 2 \end{pmatrix}$, $\begin{pmatrix} 1 & 1 \\ \bar{1} & 2 \end{pmatrix}$, $\begin{pmatrix} 1 & 1 \\ 0 & 8 \end{pmatrix}$, $\begin{pmatrix} 1 & 1 \\ 0 & 5 \end{pmatrix}$, hexagonal	7

W									
W	K	bcc	4.52		ID	(112)	LEED	$\begin{pmatrix} 2 & 0 \\ 0 & 1 \end{pmatrix}$, compressed $\begin{pmatrix} 2 & 0 \\ 0 & 1 \end{pmatrix}$	114
W	Rb	bcc	4.84	19	VD	(100)	RHEED	$\begin{pmatrix} 1 & 1 \\ 1 & 1 \end{pmatrix}$	115
W	Cs	bcc	5.23	18	ID	(100)	LEED-AES	$\begin{pmatrix} 1 & -1 \\ 1 & 1 \end{pmatrix}$, $\begin{pmatrix} 2 & 0 \\ 0 & 2 \end{pmatrix}$, hexagonal	116
W				16	ID–VD	(100)	LEED-AES/LEED-WF	$\begin{pmatrix} 1 & -1 \\ 1 & 1 \end{pmatrix}$, $\begin{pmatrix} 2 & 0 \\ 0 & 2 \end{pmatrix}$, split $\begin{pmatrix} 2 & 0 \\ 0 & 2 \end{pmatrix}$/ $\begin{pmatrix} 2 & 0 \\ 0 & 2 \end{pmatrix}$, $\begin{pmatrix} 1 & 1 \\ 1 & 1 \end{pmatrix}$, hexagonal	117–119
W					VD	(110)	LEED/LEED-AES	disordered hexagonal, hexagonal	117, 120, 121
W	Be	hcp	2.22	74	VD	(110)	LEED	$\begin{pmatrix} 1 & 0 \\ 0 & 9 \end{pmatrix}$, $\begin{pmatrix} 1 & 0 \\ 0 & 1 \end{pmatrix}$, $\begin{pmatrix} 9 & 0 \\ 1 & 1 \end{pmatrix}$	122
W	Sr	fcc	4.30	34	VD	(110)	LEED-WF	$\begin{pmatrix} \frac{3}{2} & \frac{3}{2} \\ \frac{3}{2} & 5 \end{pmatrix}$, $\begin{pmatrix} 2 & 2 \\ 0 & 6 \end{pmatrix}$, $\begin{pmatrix} 2 & 2 \\ 1 & 6 \end{pmatrix}$, $\begin{pmatrix} 1 & 1 \\ 3 & 0 \end{pmatrix}$, hexagonal	123
W	Ba	bcc	4.35	36	VD	(100)	LEED-WF	$\begin{pmatrix} 2 & 0 \\ 8 & 2 \end{pmatrix}$, split $\begin{pmatrix} \frac{1}{2} & \frac{1}{2} \\ \frac{1}{2} & 2 \end{pmatrix}$, $\begin{pmatrix} \frac{1}{2} & \frac{1}{2} \\ \frac{1}{2} & 1 \end{pmatrix}$, $\begin{pmatrix} 1 & 1 \\ 1 & 1 \end{pmatrix}$	124
W					VD	(110)	LEED-WF	disordered hexagonal, hexagonal, $\begin{pmatrix} 2 & 2 \\ 0 & 6 \end{pmatrix}$, $\begin{pmatrix} 2 & 2 \\ 0 & 5 \end{pmatrix}$, $\begin{pmatrix} 3 & 3 \\ 1 & 5 \end{pmatrix}$, hexagonal compact	125
W	Sc	hcp	3.25	81	VD	(110)	LEED-WF	$\begin{pmatrix} 1 & 1 \\ 0 & 3 \end{pmatrix}$, $\begin{pmatrix} 2 & 2 \\ 0 & 8 \end{pmatrix}$	126
W	Y	hcp	3.55	93	VD	(110)	LEED-WF	hexagonal	127, 128
W	Zr	hcp	3.17	122	VD	(100)	LEED-RHEED	$\begin{pmatrix} 1 & 0 \\ 0 & 1 \end{pmatrix}$, $\begin{pmatrix} 2 & 0 \\ 1 & 2 \end{pmatrix}$	129
W	Fe	bcc	2.48	85	VD	(110)	LEED	three-dimensional crystals	10
W	Pd	fcc	2.75	90	VD	(110)	LEED-AES	$\begin{pmatrix} 1 & 0 \\ 0 & 3 \end{pmatrix}$, hexagonal	130
W	Cu	fcc	2.56	73	VD	(100)	LEED-AES-TD	$\begin{pmatrix} 2 & 0 \\ 0 & 1 \end{pmatrix}$, $\begin{pmatrix} 1 & 1 \\ 1 & 1 \end{pmatrix}$	131
W					VD	(110)	LEED/LEED-AES-WF-TD	hexagonal structures	131, 132, 133

(continued)

Table 5.1.—Continued

Substrate metal	Adsorbed metal	Structure	Nearest-neighbor distance	Heat of vaporization (kcal/g-atom)	Deposition technique	Substrate orientation	Technique of investigation	Surface structures observed	References
W	Ag	fcc	2.89	61	VD	(100)	LEED-AES-WF-TD	$\begin{pmatrix}2&0\\0&1\end{pmatrix}$, $\begin{pmatrix}1&1\\1&\bar1\end{pmatrix}$, $\begin{pmatrix}1&0\\0&1\end{pmatrix}$	134
W					VD	(110)	LEED-AES-WF-TD	hexagonal structures	134, 135
W	Au	fcc	2.88	82	VD	(100)	LEED-AES-WF-TD	$\begin{pmatrix}2&0\\0&1\end{pmatrix}$, $\begin{pmatrix}\frac{2}{1}&0\\2&2\end{pmatrix}$, $\begin{pmatrix}1&0\\0&1\end{pmatrix}$	134
W					VD	(110)	LEED-AES-WF-TD	hexagonal structures	134, 136
W	Hg	rhomb	3.01	14	VD	(100)	LEED-AES	$\begin{pmatrix}1&0\\0&1\end{pmatrix}$	137
W	Pb	fcc	3.50	42	VD	(100)	LEED/LEED-AES-WF-TD	disordered $\begin{pmatrix}2&0\\0&2\end{pmatrix}$, split $\begin{pmatrix}1&1\\1&\bar1\end{pmatrix}$	138, 139
								$\begin{pmatrix}1&1\\1&\bar1\end{pmatrix}$, $\begin{pmatrix}\frac{1}{2}&1\\2&2\end{pmatrix}$, hexagonal/ $\begin{pmatrix}2&0\\0&2\end{pmatrix}$	
					VD	(110)	LEED-AES-WF-TD	$\begin{pmatrix}2&0\\1&2\end{pmatrix}$, $\begin{pmatrix}1&1\\1&\bar1\end{pmatrix}$, $\begin{pmatrix}1&0\\0&1\end{pmatrix}$ split $\begin{pmatrix}3&0\\1&1\end{pmatrix}$, $\begin{pmatrix}3&0\\1&1\end{pmatrix}$	139, 140
W	Sb	rhomb	2.91	47	VD	(100)	RHEED	$\begin{pmatrix}2&0\\0&2\end{pmatrix}$, $\begin{pmatrix}1&1\\1&\bar1\end{pmatrix}$, $\begin{pmatrix}1&0\\0&1\end{pmatrix}$	141, 142
W					VD	(110)	RHEED/LEED-WF	$\begin{pmatrix}1&\bar1\\0&4\end{pmatrix}$, $\begin{pmatrix}\frac{2}{1}&0\\3&1\end{pmatrix}$, $\begin{pmatrix}4&0\\1&1\end{pmatrix}$	141, 143
W					VD	(112)	RHEED	$\begin{pmatrix}2&0\\0&1\end{pmatrix}$, $\begin{pmatrix}1&0\\0&1\end{pmatrix}$	141
W	Th	fcc	3.60	137	VD	(100)	LEED/LEED-AES-WF	$\begin{pmatrix}1&1\\1&\bar1\end{pmatrix}$, $\begin{pmatrix}1&0\\1&1\end{pmatrix}/\begin{pmatrix}1&\bar1\\1&1\end{pmatrix}$, $\begin{pmatrix}1&0\\0&1\end{pmatrix}$, $\begin{pmatrix}2&0\\1&3\end{pmatrix}$, hexagonal	144–146, 147, 148
Ti	—	hcp	2.89	106	—	—	—	—	—
Ti	Cu	fcc	2.56	73	VD	(0001)	LEED	extra spots	1
Ti	Cd	hcp	2.98	24	VD	(0001)	LEED	$\begin{pmatrix}1&0\\0&1\end{pmatrix}$	2
Re	—	hcp	2.74	152	—	—	—	—	—

Re	Ba	bcc	4.35	36	VD	(0001)	LEED-WF	$\begin{pmatrix} 2 & 0 \\ 0 & 2 \end{pmatrix}$, hexagonal	11
Zn	—	hcp	2.66	27	—	—	—		—
Źn	Cu	fcc	2.56	73	VD	(0001)	LEED	$\begin{pmatrix} 1 & 0 \\ 0 & 1 \end{pmatrix}$	9
Sb	—	rhombic	2.91	62	—	—	—		—
Sb	Fe	bcc	2.48	85	VD	(0001)	TED	$\begin{pmatrix} 1 & 0 \\ 0 & 1 \end{pmatrix}$	12

1. R.E. Schlier and H.E. Farnsworth, J. Phys. Chem. Solids **6**, 271 (1958).
2. H.D. Shih, F. Jona, D.W. Jepsen, and P.M. Marcus, Phys. Rev. **B15**, 5550 (1977); H.D. Shih, F. Jona, D.W. Jepsen, and P.M. Marcus, Phys. Rev. **B15**, 5561 (1977); H.D. Shih, F. Jona, D.W. Jepsen, and P.M. Marcus, Commun. Phys. **1**, 25 (1976).
3. J.P. Biberian and G.A. Somorjai, Phys. Rev. Lett., to be published (1980).
4. J.W. Matthews and W.A. Jesser, Acta Metall. **15**, 595 (1967); J.W. Matthews, Phil. Mag. **13**, 1207 (1966).
5. J.P. Biberian and G.E. Rhead, J. Phys. **F3**, 675 (1973); J.P. Biberian and M. Huber, Surf. Sci. **55**, 259 (1976); A.K. Green, S. Prigge, and E. Bauer, Thin Solid Films, **52**, 163 (1978).
6. J.P. Biberian, Surf. Sci. **74**, 437 (1978).
7. V.K. Nedvedev, A.G. Nauvomets, and A.G. Fedorus, Sov. Phys. Solid State **12**, 301 (1970); A.G. Naumovets and A.G. Fedorus, JETP Lett. **10**, 6 (1969).
8. L.G. Feinstein and E. Blanc, Surf. Sci. **18**, 350 (1969); T. Edmonds and J.J. McCarroll, Surf. Sci. **24**, 353 (1971).
9. I. Abbati, L. Braicovich, C.M. Bertoni, C. Calandra, and F. Manghi, Phys. Rev. Lett. **40**, 469 (1978); J. Abbati and L. Braicovich, Proc. 7th Vac. Congr. and 3rd Int. Conf. Solid Surf., Vienna, 1117 (1977).
10. A.J. Melmed and J.J. McCarroll, Surf. Sci. **19**, 243 (1970).
11. D.A. Gorodetskii and A.N. Knysh, Surf. Sci. **40**, 636 (1973); D.A. Gorodetskii and A.N. Knysh, Surf. Sci. **40**, 651 (1973).
12. T. Shigematsu, S. Hine, and T. Takada, J. Cryst. Growth **43**, 531 (1978).
13. D.C. Hothersall, Phil. Mag. **15**, 1023 (1967).
14. R.E. Thomas and G.A. Haas, J. Appl. Phys. **43**, 4900 (1972).
15. S. Andersson and B. Kasemo, Surf. Sci. **32**, 78 (1972).
16. R.L. Gerlach and T.N. Rhodin, Surf. Sci. **17**, 32 (1969).
17. R.L. Gerlach and J.B. Pendry, J. Phys. **C6**, 601 (1973).
18. S. Andersson and U. Jostell, Surf. Sci. **46**, 625 (1974).
19. R.L. Gerlach and T.N. Rhodin, in *The Structure and Chemistry of Solid Surfaces*, ed. G.A. Somorjai, Wiley, New York, 1969, p. 55.
20. S. Andersson and J.B. Pendry, J. Phys. **C5**, L41 (1972).
21. R.L. Gerlach and T.N. Rhodin, Surf. Sci. **10**, 446 (1968).
22. S. Andersson and U. Jostell, Solid State Commun. **13**, 829 (1973); S. Andersson and U. Jostell, Solid State Commun. **13**, 833 (1973).
23. C.A. Papageorgopoulos and J.M. Chen, Surf. Sci. **52**, 40 (1975); C.A. Papageorgopoulos and J.M. Chen, Surf. Sci. **52**, 53 (1975).
24. W.A. Jesser and J.W. Matthews, Phil. Mag. **17**, 475 (1968).
25. W.A. Jesser and J.W. Matthews, Acta Metall. **16**, 1307 (1968).
26. A. Chambers and D.C. Jackson, Phil. Mag. **31**, 1357 (1975).
27. L.G. Feinstein, E. Blanc, and D. Dufayard, Surf. Sci. **19**, 269 (1970).

(continued)

Table 5.1.—Continued

28. D.C. Jackson, T.E. Gallon, and A. Chambers, Surf. Sci. **36**, 381 (1973).

29. J.J. Burton, C.R. Helms, and R.S. Polizzotti, Surf. Sci. **57**, 425 (1976); J.J. Burton, C.R. Helms, and R.S. Polizzotti, J. Chem. Phys. **65**, 1089 (1976); J.J. Burton, C.R. Helms, and R.S. Polizzotti, J. Vac. Sci. Technol. **13**, 204 (1976).

30. J.R. Wolfe and H.W. Weart, *The Structure and Chemistry of Solid Surfaces*, ed. G.A. Somorjai, Wiley, New York, 1968, p. 32.

31. J. Perdereau and I. Szymerska, Surf. Sci. **32**, 247 (1972).

32. E. Alkhoury Nemen, R.C. Cinti, and T.T.A. Nguyen, Surf. Sci. **30**, 697 (1972).

33. J.W. Matthews, Thin Solid Films, **12**, 243 (1972).

34. P.W. Palmberg and T.N. Rhodin, J. Chem. Phys. **49**, 134 (1968).

35. J.W. Matthews, Phil. Mag. **13**, 1207 (1966).

36. K. Yagi, K. Takanayagi, K. Kobayashi, and G. Honjo, J. Cryst. Growth **9**, 84 (1971).

37. W.A. Jesser and J.W. Matthews, Phil. Mag. **17**, 595 (1968).

38. W.A. Jesser and J.W. Matthews, Phil. Mag. **15**, 1097 (1967).

39. U. Gradmann, W. Kummerle, and P. Tillmanns, Thin Solid Films **34**, 249 (1976).

40. W.A. Jesser and J.W. Matthews, Phil. Mag. **17**, 461 (1968).

41. A.I. Fedorenko and R. Vincent, Phil. Mag. **24**, 55 (1971).

42. R. Kuntze, A. Chambers, and M. Prutton, Thin Solid Films **4**, 47 (1969).

43. C.A. Haque and H.E. Farnsworth, Surf. Sci. **4**, 195 (1966).

44. U. Gradmann, Surf. Sci. **13**, 498 (1969).

45. U. Gradmann, Ann. Phys. **13**, 213 (1964); U. Gradmann, Ann. Phys. **17**, 91 (1966).

46. E. Bauer, Surf. Sci. **7**, 351 (1967).

47. R.W. Vook, C.T. Horng, and J.E. Macur, J. Cryst. Growth **31**, 353 (1975).

48. R.W. Vook and C.T. Horng, Phil. Mag. **33**, 843 (1976).

49. C.T. Horng and R.W. Vook, Surf. Sci. **54**, 309 (1976).

50. U. Gradmann, Phys. Kondens. Mater. **3**, 91 (1964).

51. P.W. Palmberg and T.N. Rhodin, J. Appl. Phys. **39**, 2425 (1968).

52. Y. Fujinaga, Surf. Sci. **64**, 751 (1977).

53. R.W. Vook and J.E. Macur, Thin Solid Films **32**, 199 (1976).

54. J. Erlewein and S. Hofmann, Surf. Sci. **68**, 71 (1977).

55. J. Henrion and G.E. Rhead, Surf. Sci. **29**, 20 (1972).

56. A. Sepulveda and G.E. Rhead, Surf. Sci. **66**, 436 (1977).

57. C. Argile and G.E. Rhead, Surf. Sci. **78**, 115 (1978).

58. M.G. Barthes and G.E. Rhead, Surf. Sci. **80**, 421 (1979).

59. K. Reichelt and F. Müller, J. Cryst. Growth **21**, 323 (1974).

60. F. Delamare and G.E. Rhead, Surf. Sci. **35**, 172 (1973).

61. F. Delamare and G.E. Rhead, Surf. Sci. **35**, 185 (1973).

62. P.J. Goddard, J. West, and R.M. Lambert, Surf. Sci. **71**, 447 (1978).

63. R.A. Marbrow and R.M. Lambert, Surf. Sci. **61**, 329 (1976).

64. P.J. Goddard and R.M. Lambert, Surf. Sci. **79**, 93 (1979).

65. R.C. Newman, Phil. Mag. **2**, 750 (1957).

66. L.A. Bruce and H. Jaeger, Phil. Mag. **36**, 1331 (1977).

67. C. Gonzalez, Acta Metall. **15**, 1373 (1967).

68. C.T. Horng and R.W. Vook, J. Vac. Sci. Technol. **11**, 140 (1974).

69. E. Grünbaum, G. Kremer, and C. Reymond, J. Vac. Sci. Technol. **6**, 475 (1969).
70. R.C. Newman and D.W. Pashley, Phil. Mag. **46**, 927 (1955).
71. F. Soria, J.L. Sacedon, P.M. Echenique, and D. Titterington, Surf. Sci. **68**, 448 (1977).
72. M. Klaua and H. Bethge, J. Cryst. Growth **3,4**, 188 (1968).
73. E. Grünbaum, Proc. Phys. Soc. (Lond.) **72**, 459 (1958).
74. E. Bauer, Structure et Propriétes des Solides, CNRS, Paris 1969.
75. R.E. Thomas and G.A. Haas, J. Appl. Phys. **43**, 4900 (1972).
76. E.F. Wassermann and H.P. Jablonski, Surf. Sci. **22**, 69 (1970).
77. D.C. Hothersall, Phil. Mag. **15**, 1023 (1967).
78. P. Gueguen, C. Camoin, and M. Gillet, Thin Solid Films **26**, 107 (1975).
79. G. Honjo, K. Takayanagi, K. Kobayashi, and K. Yagi, J. Cryst. Growth **42**, 98 (1977).
80. P. Gueguen, M. Cahareau, and M. Gillet, Thin Solid Films **16**, 27 (1973).
81. D. Cherns and M.J. Stowell, Thin Solid Films **29**, 107 (1975).
82. D. Cherns and M.J. Stowell, Thin Solid Films **29**, 127 (1975).
83. W.A. Jesser, J.W. Matthews, and D. Kuhlmann–Wilsdorf, Appl. Phys. Lett. **9**, 176 (1966).
84. J.E. Macur and R.W. Vook, 32nd Annu. Proc. Electron Microsc. Soc., St. Louis, Mo. 1974, ed. C.J. Arceneaux.
85. J.E. Macur, 33rd Annu. Proc. Electron Microsc. Soc. Amer., Las Vegas, Nev. 1975, ed. B.W. Bailey, p. 98.
86. H.E. Farnsworth, Phys. Rev. **40**, 684 (1932).
87. J.W. Matthews, Phys. Thin Films **4**, 137 (1967).
88. J. Perdereau, J.P. Biberian, and G.E. Rhead, J. Phys. **F4**, 798 (1974).
89. M.G. Barthes and G.E. Rhead, Surf. Sci., **85**, L211 (1979).
90. M.G. Barthes, Thesis, University of Paris, 1978.
91. A. Sepulveda and G.E. Rhead, Surf. Sci. **49**, 669 (1975).
92. B.A. Hutchins, T.N. Rhodin, and J.E. Demuth, Surf. Sci. **54**, 419 (1976).
93. J.O. Porteus, Surf. Sci. **41**, 515 (1974).
94. M. Van Hove, S.Y. Tong, and N. Stoner, Surf. Sci. **54**, 259 (1976).
95. I.A.S. Edwards and H.R. Thirsk, Surf. Sci. **39**, 245 (1973).
96. G. Dorey, Thin Solid Films **5**, 69 (1970).
97. C. Argile and G.E. Rhead, Surf. Sci. **78**, 125 (1978).
98. C. Argile, Thesis, University of Paris, 1978.
99. A.G. Jackson and M.P. Hooker, *The Structure and Chemistry of Solid Surfaces*, ed. G.A. Somorjai, Wiley, New York, 1969, p. 73.
100. A.G. Elliot, Surf. Sci. **51**, 489 (1975); J.P. Biberian, Surf. Sci. **59**, 307 (1976).
101. T.W. Haas, A.G. Jackson, and M.P. Hooker, J. Appl. Phys. **38**, 4998 (1967); A.G. Jackson, M.P. Hooker, and T.W. Haas, Surf. Sci. **10**, 308 (1968).
102. J.H. Pollard and W.E. Danforth, *The Structure and Chemistry of Solid Surfaces*, ed. G.A. Somorjai, Wiley, New York, 1969, p. 71; J.H. Pollard and W.E. Danforth, J. Appl. Phys. **39**, 4019 (1968).
103. S. Thomas and T.W. Hass, J. Vac. Sci. Technol. **9**, 840 (1972).
104. S. Thomas and T.W. Haas, Surf. Sci. **28**, 632 (1971).
105. K. Hartig, A.P. Janssen, and J.A. Venables, Surf. Sci. **74**, 69 (1978).
106. K. Hartig, Thesis, Ruhr-Universität, Bochum.
107. A.G. Jackson and M.P. Hooker, Surf. Sci. **28**, 373 (1971).
108. A.G. Jackson and M.P. Hooker, Surf. Sci. **27**, 197 (1971).
109. D.A. Gorodetsky, Yu. P. Melnik, and A.A. Yasko, Ukr. Fiz. Zh. **12**, 649 (1967).

(*continued*)

Table 5.1.—Continued

110. V.K. Medvedev and T.P. Smereka, Sov. Phys. Solid State **16**, 1046 (1974).
111. A.G. Naumovets and A.G. Fedorus, Sov. Phys. JETP **41**, 587 (1975).
112. V.K. Medvedev, A.G. Naumovets, and T.P. Smereka, Surf. Sci. **34**, 368 (1973).
113. A. Mlynczak and R. Niedermayer, Thin Solid Films **28**, 37 (1975).
114. J.M. Chen and C.A. Papageorgopoulos, Surf. Sci. **21**, 377 (1970).
115. P.W. Steinhage and H. Mayer, Thin Solid Films, **28**, 131 (1975).
116. S. Thomas and T.W. Haas, J. Vac. Sci. Technol. **10**, 218 (1973).
117. A.U. MacRae, K. Müller, J.J. Lander, and J. Morrison, Surf. Sci. **15**, 483 (1969).
118. C.A. Papageorgopoulos and J.M. Chen, Surf. Sci. **39**, 283 (1973).
119. V.B. Voronin, A.G. Nauvomets, and A.G. Fedorus, JETP Lett. **15**, 370 (1972); C.S. Wang, J. Appl. Phys. **48**, 1477 (1977).
120. A.G. Fedorus and A.G. Naumovets, Surf. Sci. **21**, 426 (1970).
121. A.G. Fedorus and A.G. Naumovets, Sov. Phys. Solid State **12**, 232 (1970).
122. H. Niehus, Thesis, Clausthal, 1975.
123. O.V. Kanash, A.G. Neumovets, and A.G. Fedorus, Sov. Phys. JETP **40**, 903 (1974).
124. D.A. Gorodetskii and Yu. P. Mel'nik Akad. Nauk SSSR **33**, 4 (1969).
125. D.A. Gorodetskii and Yu. P. Mel'nik, Surf. Sci. **62**, 647 (1977); D.A. Gorodetskii, A.D. Gorchinskii, V.I. Maksimenko, and Yu. P. Melnik, Sov. Phys. Solid State **18**, 691 (1976); D.A. Gorodetskii, A.M. Kornev, and Yu. P. Mel'nik, Izv. Akad. Nauk Ser. Fiz. **28**, 1337 (1964).
126. V.B. Voronin and A.G. Naumovets, Ukr. Fiz. Zh. **13**, 1389 (1968); V.B. Voronin, Sov. Phys. Solid State **9**, 1758 (1968); D.A. Gorodetskii and A.A. Yas'ko, Sov. Phys. Solid State **10**, 1812 (1969).
127. D.A. Gorodetskii, A.A. Yas'ko, and S.A. Shevlyakov, Izv. Akad. Nauk SSSR, Ser. Fiz. **35**, 436 (1971).
128. V.B. Voronin and A.G. Naumovets, Izv. Akad. Nauk SSSR Ser. Fiz. **35**, 325 (1971).
129. G.E. Hill, J. Marklund, and J. Martinson, Surf. Sci. **24**, 435 (1971).
130. D. Paraschkevov, W. Schlenk, R.P. Bajpai, and E. Bauer, Proc. 7th Int. Vac. Congr. and 3rd Int. Conf. Solid Surf., Vienna, 1977, p. 1737.
131. E. Bauer, H. Poppa, G. Todd, and F. Bonczek, J. Appl. Phys. **45**, 5164 (1974).
132. N.J. Taylor, Surf. Sci. **4**, 161 (1966).
133. A.R. Moss and B.H. Blott, Surf. Sci. **17**, 240 (1969).
134. E. Bauer, H. Poppa, G. Todd, and P.R. Davis, J. Appl. Phys. **48**, 3773 (1977).
135. J.B. Hudson and C.M. Lo, Surf. Sci. **36**, 141 (1973).
136. P.D. Augustus and J.P. Jones, Surf. Sci. **64**, 713 (1977).
137. R.G. Jones and D.L. Perry, Surf. Sci. **71**, 59 (1978).
138. D.A.Gorodetskii and A.A. Yas'ko, Sov. Phys. Solid State **14**, 636 (1972).
139. E. Bauer, H. Poppa, and G. Todd, Thin Solid Films **28**, 19 (1975).
140. D.A. Gorodetskii and A.A. Yas'ko, Sov. Phys. Solid State **11**, 640 (1969).
141. B.J. Hopkins and G.D. Watts, Surf. Sci. **47**, 195 (1975).
142. B.J. Hopkins and G.D. Watts, Surf. Sci. **45**, 77 (1974).
143. D.A. Gorodetskii and A.A. Yas'ko, Sov. Phys. Solid State **13**, 1085 (1971).
144. P.J. Estrup, J. Anderson, and W.E. Danforth, Surf. Sci. **4**, 286 (1966).
145. P.J. Estrup and J. Anderson, Surf. Sci. **7**, 255 (1967).
146. P.J. Estrup and J. Anderson, Surf. Sci. **8**, 101 (1967).
147. J.H. Pollard, Surf. Sci. **20**, 269 (1970).
148. J. Anderson, P.J. Estrup, and W.E. Danforth, Appl. Phys. Lett. **7**, 122 (1965).

the technique of deposition and all the ordered surface structures that form with increasing coverage.

One of the striking results of these studies revealed by inspection of Table 5.1 is the predominance of the formation of ordered monolayers regardless of the relative magnitudes of the surface free energies of the substrate and adsorbate metals. Surface thermodynamic considerations would predict monolayer formation only when the total surface free energy is minimized in this way (during the deposition of a metal of lower surface free energy on a metal substrate of higher surface free energy). If these circumstances are not met, the growth of three-dimensional crystallites is predicted (when the adsorbate surface free energy is greater than that of the substrate) to minimize the total surface energy. However, the experimental data indicate that regardless of the surface-free-energy differences (for example, even for Mo on Ni, Pt on Au, and Cu on Zn), ordered monolayer deposits form.

There is one exception to the formation of ordered monolayers: Fe on W forms three-dimensional crystallites even though surface thermodynamic considerations would predict monolayer formation.

At low adsorbate coverages, the surface structure of the deposited metal is determined by the substrate periodicity. Under these conditions the adsorbate–substrate interaction is predominant. At higher coverages, the adsorbate may continue to follow the substrate periodicity or may form coincidence structures with new periodicities that are unrelated to the substrate periodicity. The ordering geometry of large-radius metallic adatoms (especially K, Rb, and Cs) shows relatively little dependence on the substrate lattice; they tend to form hexagonal close-packed layers on any metal substrate. It appears that for these systems, adsorbate–adsorbate interaction predominates during ordering.

The available data are inadequate to permit a detailed analysis of the various factors that control the ordering of metal monolayers on metal surfaces. It is probable that both the electronic interaction between the two metals and the relative atomic sizes are important in determining the nature of ordering in the monolayer.

Surface Structures of Small Molecules on Low- and
High-Miller-Index Surfaces

In Tables 5.2 to 5.7 we list the observed adsorbate surface structures (excluding metallic adsorbates, listed in Table 5.1). The substrates are classified according to the rotational symmetry of their surfaces: threefold in Table 5.2, fourfold in Table 5.3, and twofold in Table 5.4. Stepped surfaces are considered in Tables 5.5 and 5.6. Finally, struc-

Surface	Adsorbed gas	Surface structure	References
Ag(111)	O_2	(2×2)-O	1
		$(\sqrt{3} \times \sqrt{3})R30°$-O	1
		not adsorbed	2
		(4×4)-O	3, 4
	I_2	$(\sqrt{3} \times \sqrt{3})R30°$-I	5–7
	Cl_2	$(\sqrt{3} \times \sqrt{3})R30°$-Cl	8
		(10×10)-Cl	8
		AgCl(111)	9
	$C_2H_4Cl_2$	$(\sqrt{3} \times \sqrt{3})R30°$-Cl	10, 11
		(3×3)-Cl	10, 11
	Br_2	$(\sqrt{3} \times \sqrt{3})R30°$-Br	12
		(3×3)-Br	12
	Xe	hexagonal overlayer	13–17
	Kr	hexagonal overlayer	13
	$CO+O_2$	$(2 \times \sqrt{3})$-$(CO + O_2)$	18
	NO	disordered	19
Al(111)	O_2	(4×4)-O	20
Au(111)	O_2	oxide	21
		not adsorbed	22
		adsorbed	22
Be(0001)	O_2	disordered	23
	CO	disordered	23
	H_2	not adsorbed	23
	N_2	not adsorbed	23
C(111), diamond	O_2	adsorbed	24
		not adsorbed	25
	N_2	not adsorbed	25
	NH_3	not adsorbed	25
	H_2S	not adsorbed	25
	H_2	(1×1)-H	26
	P	$(\sqrt{3} \times \sqrt{3})R30°$-P	26
C(0001), graphite	Xe	$(\sqrt{3} \times \sqrt{3})R30°$-Xe	27
	Kr	$(\sqrt{3} \times \sqrt{3})R30°$-Kr	28–30
CdS(0001)	O_2	disordered	31
Co(0001)	CO	$(\sqrt{3} \times \sqrt{3})R30°$-CO	32
		hexagonal overlayer	32
Cr(111)	O_2	$(\sqrt{3} \times \sqrt{3})R30°$-O	33
Cu(111)	O_2	disordered	34–36
		(7×7)-O	34, 37
		$(\sqrt{3} \times \sqrt{3})R30°$-O	34, 37
		(2×2)-O	34, 37, 38
		(3×3)-O	37
		$(11 \times 5)R5°$-O	39
		$(2 \times 2)R30°$-O	38, 40
		hexagonal	41
	CO	not adsorbed	42
		$(\sqrt{3} \times \sqrt{3})R30°$	43, 44
		$(\sqrt{7/3} \times \sqrt{7/3})R49.1$	43, 44
		$(3/2 \times 3/2)$	44

Table 5.2.—Continued

Surface	Adsorbed gas	Surface structure	References
	Cl_2	$(\sqrt{3} \times \sqrt{3})R30°$-Cl	8
		$(6\sqrt{3} \times 6\sqrt{3})R30°$-Cl	8
		$(12\sqrt{3} \times 12\sqrt{3})R30°$-Cl	8
		$(4\sqrt{7} \times 4\sqrt{7})R19.2°$-Cl	8
	H_2	not adsorbed	34
	H_2S	$(\sqrt{3} \times \sqrt{3})R30°$-S	45
		adsorbed	45
	Xe	$(\sqrt{3} \times \sqrt{3})R30°$-Xe	16
Cu/Ni(111)	CO	disordered	44
Fe(111)	O_2	(6×6)-O	46
		(5×5)-O	46
		(4×4)-O	46
		$(2\sqrt{7} \times 2\sqrt{7})R19.1°$-O	46
		$(2\sqrt{3} \times 2\sqrt{3})R30°$-O	46
	NH_3	disordered	47
		(3×3)-N	47
		$(\sqrt{19} \times \sqrt{19})R23.4°$-N	47
		$(\sqrt{21} \times \sqrt{21})R10.9°$-N	47
	H_2	adsorbed	48
Ge(111)	O_2	disordered	49, 50
		(1×1)	51, 52
	P	(1×1)-P	51
	H_2S	(2×2)-S	53
		(2×1)-S	54
	H_2Se	(2×2)-Se	53
	H_2O	(1×1)-H_2O	55, 56
	I_2	(1×1)-I	51
Ir(111)	O_2	(2×2)-O	57–62
		(2×1)-O	60
		Ir oxide	59
	CO	$(\sqrt{3} \times \sqrt{3})R30°$-CO	18, 58, 60, 61, 63, 64
		$(2\sqrt{3} \times 2\sqrt{3})R30°$-CO	58, 60, 61, 63, 64
	H_2O	not adsorbed	61
	H_2	adsorbed	65
	NO	(2×2)-NO	66
Mo(111)	O_2	(211) facets	67, 68
		(110) facets	68
		(4×2)-O	69
	H_2S	$c(4 \times 2)$-H_2S	70
		$MoS_2(0001)$	70
Nb(111)	O_2	(2×2)-O	71
		(1×1)-O	71
Ni(111)	O_2	(2×2)-O	72–81
		$(\sqrt{3} \times \sqrt{3})R30°$-O	72, 78, 82
		$(\sqrt{3} \times \sqrt{21})$-O	75
		NiO(111)	74–77, 83
	CO	$(\sqrt{3} \times \sqrt{3})R30°$-CO	77, 78, 84, 85
		hexagonal overlayer	85
		(2×2)-CO	73

(continued)

Table 5.2.—Continued

Surface	Adsorbed gas	Surface structure	References
		$(\sqrt{3} \times \sqrt{3})R30°$-O	82
		$(2 \times \sqrt{3})$-CO	82
		$(\sqrt{39} \times \sqrt{39})$-C	18, 82
		disordered	81
		$(\sqrt{7} \times \sqrt{7})R19.1°$	78, 79
		$c(4 \times 2)$	78, 79
	CO_2	(2×2)-CO_2	82
		$(\sqrt{3} \times \sqrt{3})R30°$-O	82
		$(2 \times \sqrt{3})$-CO_2	82
		$(\sqrt{39} \times \sqrt{39})$-C	18, 82
	H_2	(1×1)-H	73, 86, 87, 88, 89,
		(2×2)-H	90
		disordered	
	NO	$c(4 \times 2)$-NO	76
		hexagonal overlayer	76
		(2×2)-O	76
		(6×2)-N	76
	H_2S	(2×2)-S	80, 81, 91–94
		$(\sqrt{3} \times \sqrt{3})R30°$-S	91, 92
		(5×5)-S	91
		adsorbed	91
	H_2Se	(2×2)-Se	95
		$(\sqrt{3} \times \sqrt{3})R30°$-Se	95
	Cl_2	$(\sqrt{3} \times \sqrt{3})R30°$-Cl	96
		$\begin{pmatrix} 2 & 1 \\ 4 & 7 \end{pmatrix}$-Cl	96
	N_2	not adsorbed	97
Pd(111)	O_2	(2×2)-O	98
		$(\sqrt{3} \times \sqrt{3})R30°$-O	98
		(2×2)-PdO	98
	NO	$c(4 \times 2)$-NO	99
		(2×2)-NO	99
	CO	$(\sqrt{3} \times \sqrt{3})R30°$-CO	100, 101
		hexagonal overlayer	100
		$c(4 \times 2)$-CO	101
	H_2	(1×1)-H	102, 103
Pt(111)	O_2	(2×2)-O	104–110
		$(\sqrt{3} \times \sqrt{3})R30°$-O	107, 108, 110
		not adsorbed	111
		$(4\sqrt{3} \times 4\sqrt{3})R30°$-O	107, 108
		$PtO_2(0001)$	107, 108
		(3×15)-O	110
	CO	$(\sqrt{3} \times \sqrt{3})R30°$-CO	112
		$c(4 \times 2)$-CO	111–114
		hexagonal overlayer	112
		(2×2)-CO	114
	H_2	not adsorbed	114
		adsorbed	115, 116
	H_2+O_2	$(\sqrt{3} \times \sqrt{3})R30°$	105
	NO	disordered	117
	H_2O	$(\sqrt{3} \times \sqrt{3})R30°$-$H_2O$	118, 119
		$H_2O(111)$	119

Table 5.2. —Continued

Surface	Adsorbed gas	Surface structure	References
	S_2	(2×2)-S	120–123
		$(\sqrt{3} \times \sqrt{3})R30°$-S	120–123
		$\begin{pmatrix} 4 & -1 \\ -1 & 2 \end{pmatrix}$-S	120, 121
		hexagonal	122
	N	disordered	124
Re(0001)	O_2	(2×2)-O	125–127
	CO	not adsorbed	126
		(2×2)-CO	125
		disordered	128
		$(2 \times \sqrt{3})$	128
	H_2	not adsorbed	126
	N_2	not adsorbed	126
Rh(111)	O_2	(2×2)-O	129, 130
	CO	$(\sqrt{3} \times \sqrt{3})R30°$-CO	130
		(2×2)-CO	129, 130
	CO_2	$(\sqrt{3} \times \sqrt{3})R30°$-CO	130
		(2×2)-CO	130
	H_2	adsorbed	130
	NO	$c(4 \times 2)$-NO	130
		(2×2)-NO	130
Ru(0001)	O_2	(2×2)-O	129, 131, 132
	CO	$(\sqrt{3} \times \sqrt{3})R30°$-CO	129, 132, 133
		(2×2)-CO	129, 132
	CO_2	$(\sqrt{3} \times \sqrt{3})R30°$-$CO_2$	129
		(2×2)-CO_2	129
	H_2	$(1 \times)$-H	134
	N_2	adsorbed	134
	NH_3	(2×2)-NH_3	134, 135
		$(\sqrt{3} \times \sqrt{3})R30°$-$NH_3$	134
Si(111)	O_2	disordered	49, 52, 136
	N_2	(8×8)-N	136
	P	$(6\sqrt{3} \times 6\sqrt{3})$-P	137, 138
		(1×1)-P	137
		$(2\sqrt{3} \times 2\sqrt{3})$-P	137
		(4×4)-P	138
	Cl_2	disordered	139
		(7×7)-Cl	139, 140
		(1×1)-Cl	139, 140
	I_2	(1×1)-I	138
	H_2	(1×1)-H	141
		(7×7)-H	141
	NH_3	(8×8)-N	142
	PH_3	(7×7)-P	143
		(1×1)-P	143
		$(6\sqrt{3} \times 6\sqrt{3})$-P	143
		$(2\sqrt{3} \times 2\sqrt{3})$-P	143
Ti(0001)	O_2	(1×1)-O	53

(continued)

Table 5.2.—Continued

Surface	Adsorbed gas	Surface structure	References
	CO	(1×1)-CO	50, 144
		(2×2)-CO	144
	N_2	(1×1)-N	145, 146
		$(\sqrt{3} \times \sqrt{3})R30°$-N	145, 146
Th(111)	O_2	disordered	147
		$ThO_2(111)$	147
	CO	disordered	147
		$ThO_2(111)$	147
$UO_2(111)$	O_2	(3×3)-O	148
		$(2\sqrt{3} \times 2\sqrt{3})R30°$-O	148
W(111)	O_2	disordered	149
		(211) facets	150
Zn(0001)	O_2	(1×1)-O	151
		ZnO(0001)	152
(000$\bar{1}$)	O_2	$(\sqrt{3} \times \sqrt{3})R30°$-O	151

1. K. Muller, Z. Naturforsch., **20A,** 153 (1965).
2. H.A. Engelhardt and D. Menzel, Surf. Sci. **57,** 591 (1976).
3. H. Albers, W.J.J. Van der Wal, and G.A. Bootsma, Surf. Sci. **68,** 47 (1977).
4. G. Rovida, F. Pratesi, M. Maglietta, and E. Ferroni, Surf. Sci. **43,** 230 (1974).
5. F. Forstmann, W. Berndt, and P. Büttner, Phys. Rev. Lett. **30,** 17 (1973).
6. W. Berndt, Proc. 2nd Int. Conf. Solid Surf., 653 (1974).
7. F. Forstmann, Proc. 2nd Int. Conf. Solid Surf., 657 (1974).
8. P.J. Goddard and R.M. Lambert, Surf. Sci. **67,** 180 (1977).
9. Y. Tu and J.M. Blakely, J. Vac. Sci. Technol. **15,** 563 (1978).
10. G. Rovida, F. Pratesi, M. Maglietta, and E. Ferroni, Proc. 2nd. Int. Conf. Solid Surf. 117 (1974).
11. G. Rovida and F. Pratesi, Surf. Sci. **51,** 270 (1975).
12. P.J. Goddard, K. Schwaha, and R.M. Lambert, Surf. Sci. **71,** 351 (1978).
13. R.H. Roberts and J. Pritchard, Surf. Sci. **54,** 687 (1976).
14. N. Stoner, M.A. Van Hove, S.Y. Tong, and M.B. Webb, Phys. Rev. Lett. **40,** 243 (1978).
15. G. McElhiney, H. Papp, and J. Pritchard, Surf. Sci. **54,** 617 (1976).
16. M.A. Chesters, M. Hussain, and J. Pritchard, Surf. Sci. **36,** 161 (1973).
17. P.I. Cohen, J. Unguris, and M.B. Webb, Surf. Sci. **58,** 429 (1976).
18. T. Edmonds and R.C. Pitkethly, Surf. Sci. **17,** 450 (1969).
19. P.J. Goddard, J. West, and R.M. Lambert, Surf. Sci. **71,** 447 (1978).
20. S.M. Bedair and H.P. Smith, Jr., J. Appl. Phys. **42,** 3616 (1971).
21. M.A. Chesters and G.A. Somorjai, Surf. Sci. **52,** 21 (1975).
22. D.M. Zehner and J.F. Wendelken, Proc. 7th Int. Vac. Cong. and 3rd Int. Conf. Solid Surf., Vienna, 51 (1977).
23. R.O. Adams, in *The Structure and Chemistry of Solid Surfaces,* ed. G.A. Somorjai. Wiley, New York, 1969.
24. J.B. Marsh and H.E. Farnsworth, Surf. Sci. **1,** 3 (1964).
25. P.G. Lurie and J.M. Wilson, Surf. Sci. **65,** 453 (1977).
26. J.J. Lander and J. Morrison, Surf. Sci. **4,** 241 (1966).
27. J. Suzanne, J.P. Coulomb, and M. Bienfait, Surf. Sci. **40,** 414 (1973).
28. M.D. Chinn and S.C. Fain, Jr., J. Vac. Sci. Technol. **14,** 314 (1977).
29. H.M. Kramer and J. Suzanne, Surf. Sci. **54,** 549 (1976).
30. M.D. Chinn and S.C. Fain, Jr., Phys. Rev. Lett. **39,** 146 (1977).
31. B.D. Campbell, C.A. Haque, and H.E. Farnsworth, *The Structure and Chemistry of Solid Surfaces,* ed. G.A. Somorjai, Wiley, New York, 1969.
32. M.E. Bridge, C.M. Comrie, and R.M. Lambert, Surf. Sci. **67,** 393 (1977).
33. C. Jardin and P. Michel, Surf. Sci. **71,** 575 (1978).
34. G. Ertl, Surf. Sci. **6,** 208 (1967).
35. F.H.P.M. Habraken, E.P. Kieffer, and G.A. Bootsma, Proc. 7th Int. Vac. Congr. and 3rd Int. Conf. Solid Surf., Vienna, 887 (1977).
36. L. McDonnel and D.P. Woodruff, Surf. Sci. **46,** 505 (1974).

Table 5.2.—Continued

37. N. Takahashi et al., C.R. Acad. Sci. Paris **269B**, 618 (1969).
38. I. Marklund, S. Andersson, and J. Martinsson, Ark. Phys. **37**, 127 (1968).
39. G.W. Simmons, D.F. Mitchell, and K.R. Lawless, Surf. Sci. **8**, 130 (1967).
40. K. Okado, T. Halsushika, H. Tomita, S. Motov, and N. Takalashi, Shinku **13**(11), 371 (1970).
41. A. Oustry, L. Lafourcade, and A. Escaut, Surf. Sci. **40**, 545 (1973).
42. G. Ertl, Surf. Sci. **7**, 309 (1967).
43. J. Kessler and F. Thieme, Surf. Sci. **67**, 405 (1977).
44. C. Benndorf, K.H. Gressman, and F. Thieme, Surf. Sci. **61**, 646 (1976).
45. J.L. Domange and J. Oudar, Surf. Sci. **11**, 124 (1968).
46. S. Nakanishi and T. Horiguchi, Proc. 7th Int. Vac. Congr. and 3rd Int. Conf. Solid Surf., Vienna, A2727 (1977).
47. M. Grunze, F. Bozso, G. Ertl, and M. Weiss, Appl. Surf. Sci. **1**, 241 (1978).
48. F. Bozso, G. Ertl, M. Grunze, and M. Weiss, Appl. Surf. Sci. **1**, 103, (1978).
49. R.E. Schlier and H.E. Farnsworth, J. Chem. Phys. **30**, 917 (1959).
50. H.E. Farnsworth, R.E. Schlier, T.H. George, and R.M. Buerger, J. Appl. Phys. **29**, 1150 (1958).
51. J.J. Lander and J. Morrison, J. Appl. Phys. **34**, 1411 (1963).
52. G. Rovida et al., Surf. Sci. **14**, 93 (1969).
53. A.J. van Bommel and F. Meyer, Surf. Sci. **6**, 391 (1967).
54. B.Z. Olshanetsky, S.M. Repinsky, and A.A. Shklyaev, Surf. Sci. **64**, 224 (1977).
55. M. Henzler and J. Töpler, Surf. Sci. **40**, 388 (1973).
56. S. Sinharoy and M. Henzler, Surf. Sci. **51**, 75 (1975).
57. J.T. Grant. Surf. Sci. **25**, 451 (1971).
58. V.P. Ivanov, G.K. Boreskov, V.I. Savchenko, W.F. Egelhoff, Jr., and W.H. Weinberg, J. Catal. **48**, 269 (1977).
59. H. Conrad, J. Küppers, F. Nitschké, and A. Plagge, Surf. Sci. **69**, 668 (1977).
60. D.I. Hagen, B.E. Nieuwenhuys, G. Rovda, and G.A. Somorjai, Surf. Sci. **57**, 632 (1976).
61. J. Küppers and A. Plagge, J. Vac. Sci. Technol. **13**, 259 (1976).
62. V.P. Ivanov, G.K. Boreskov, V.I. Savchenko, W.F. Egelhoff, Jr., and W.H. Weinberg, Surf. Sci. **61**, 207 (1976).
63. C.M. Comrie and W.H. Weinberg, J. Vac. Sci. Technol. **13**, 264 (1976).
64. C.M. Comrie and W.H. Weinberg, J. Chem. Phys. **64**, 250 (1976).
65. B.E. Nieuwenhuys, D.I. Hagen, G. Rovida, and G.A. Somorjai, Surf. Sci. **59**, 155 (1976).
66. J. Kanski and T.N. Rhodin, Surf. Sci. **65**, 63 (1977).
67. J. Ferrante and G.C. Barton, NASA Tech. Note D-4735, 1968.
68. L.J. Clark, Proc. 7th Int. Vac. Congr. and 3rd Int. Conf. Solid Surf., Vienna, A2725 (1977).
69. H.M. Kennett and A.E. Lee, Surf. Sci. **48**, 606 (1975).
70. J.M. Wilson, Surf. Sci. **59**, 315 (1976).
71. R. Pantel, M. Buhor, and J. Bardolle, Surf. Sci. **62**, 739 (1977).
72. A.U. MacRae, Surf. Sci. **1**, 319 (1964).
73. L.H. Germer, E.J. Schneiber, and C.D. Hartman, Phil. Mag. **5**, 222 (1960).
74. R.L. Park and H.E. Farnsworth, Appl. Phys. Lett. **3**, 167 (1963).
75. P. Legaré and G. Marie, J. Chim. Phys. Physsichim. Biol. **68**(7–8), 1206 (1971).
76. H. Conrad, G. Ertl, J. Küppers, and E.E. Latta, Surf. Sci. **50**, 296 (1975).
77. P.H. Holloway and J.B. Hudson, Surf. Sci. **43**, 141 (1974).
78. H. Conrad, G. Ertl, J. Küppers, and E.E. Latta, Surf. Sci. **57**, 475 (1976).
79. W. Erley, K. Besoche, and H. Wagner, J. Chem. Phys. **66**, 5269 (1977).
80. P.M. Marcus, J.E. Demuth, and D.W. Jepsen, Surf. Sci. **53**, 501 (1975).
81. J.E. Demuth and T.N. Rhodin, Surf. Sci. **45**, 249 (1974).
82. T. Edmonds and R.C. Pitkethly, Surf. Sci. **15**, 137 (1969).
83. A.U. MacRae, Science **139**, 379 (1963).
84. G. Ertl, Surf. Sci. **47**, 86 (1975).
85. K. Christmann, O. Schober, and G. Ertl, J. Chem. Phys. **60**, 4719 (1974).
86. J.C. Bertolini and G. Dalmai-Imelik, Coll. Int. CNRS, Paris, July 1969.
87. M.A. Van Hove, G. Ertl, W.H. Weinberg, K. Christmann, and H.J. Behm, Proc. 7th Int. Vac. Congr. and 3rd Int. Conf Solid Surf., Vienna, 2415 (1977).
88. H. Conrad, G. Ertl, J. Küppers, and E.E. Latta, Surf. Sci. **58**, 578 (1976).
89. G. Casalone, M.G. Cattania, M. Simonetta, and M. Tescari, Surf. Sci. **72**, 739 (1978).
90. K. Christmann, O. Schober, G. Ertl, and M. Neumann, J. Chem. Phys. **60**, 4528 (1978).
91. M. Perdereau and J. Oudar, Surf. Sci. **20**, 80 (1970).
92. T. Edmonds, J.J. McCarrol, and R.C. Pitkethly, J. Vac. Sci. Technol. **8**(1), 68 (1971).

(continued)

Table 5.2.—Continued

93. J.E. Demuth, D.W. Jepsen, and P.M. Marcus, Phys. Rev. Lett. **32,** 1182 (1974).
94. T.N. Rhodin and J.E. Demuth, Proc. 2nd Int. Conf. Solid Surf. 167 (1974).
95. A.E. Becker and H.D. Hagstrum, Surf. Sci. **30,** 505 (1972).
96. W. Erley and H. Wagner, Surf. Sci. **66,** 371 (1977).
97. D.L. Adams and L.H. Germer, Surf. Sci. **27,** 21 (1971).
98. H. Conrad, G. Ertl, J. Küppers, and E.E. Latta, Surf. Sci. **65,** 245 (1977).
99. H. Conrad, G. Ertl, J. Küppers, and E.E. Latta, Surf. Sci. **54,** 235 (1977).
100. H. Conrad, G. Ertl, J. Koch, and E.E. Latta, Surf. Sci. **43,** 462 (1974).
101. A.M. Bradshaw and F.M. Hoffman, Surf. Sci. **72,** 513 (1978).
102. K. Christmann, G. Ertl, and O. Schober, Surf. Sci. **40,** 61 (1973).
103. H. Conrad, G. Ertl, and E.E. Latta, Surf. Sci. **41,** 435 (1974).
104. C.W. Tucker, Jr., Surf. Sci. **2,** 516 (1964).
105. C.W. Tucker, Jr., J. Appl. Phys. **35,** 1897 (1964).
106. H.P. Bonzel and R. Ku, Surf. Sci. **40,** 85 (1973).
107. B. Carriere, J.P. Deville, G. Maire, and P. Legare, Surf. Sci. **58,** 578 (1976).
108. P. Legare, G. Maire, B. Cariere, and J.P. Deville, Surf. Sci. **68,** 348 (1977).
109. J.A. Joebstl. J. Vac. Sci. Technol. **12,** 347 (1975).
110. W.H. Weinberg, D.R. Monroe, V. Lampton, and R.P. Merrill, J. Vac. Sci. Technol. **14,** 444 (1977).
111. B. Lang, R.W. Joyner, and G.A. Somorjai, Surf. Sci. **30,** 454 (1972).
112. G. Ertl, M. Neumann, and K.M. Streit, Surf. Sci. **64,** 393 (1977).
113. S.L. Bernasek and G.A. Somorjai, J. Chem. Phys. **60,** 4552 (1974).
114. A.E. Morgan and G.A. Somorjai, J. Chem. Phys. **51,** 3309 (1969).
115. K. Christmann, G. Ertl, and T. Pignet, Surf. Sci. **54,** 365 (1976).
116. K. Baron, D.W. Blakely, and G.A. Somorjai, Surf. Sci. **41,** 45 (1974).
117. C.M. Comrie, W.H. Weinberg, and R.M. Lambert, Surf. Sci. **57,** 519 (1976).
118. L.E. Firment and G.A. Somorjai, J. Chem. Phys. **63,** 1037 (1975).
119. L.E. Firment and G.A. Somorjai, Surf. Sci. **55,** 413 (1976).
120. W. Heegemann, E. Bechtold, and K. Hayek, Proc. 2nd Int. Conf. Solid Surf. 185 (1974).
121. W. Heegemann, K.H. Meister, E. Bechtold, and K. Hayek, Surf. Sci. **49,** 161 (1975).
122. Y. Berthier, M. Perdereau, and J. Oudar, Surf. Sci. **44,** 281 (1974).
123. Y. Berthier, M. Perdereau, and J. Oudar, Surf. Sci. **36,** 225 (1973).
124. K. Schwaka and E. Bechtold, Surf. Sci. **66,** 383 (1977).
125. H.E. Farnsworth and D.M. Zehner, Surf. Sci. **17,** 7 (1969).
126. G.J. Dooley and T.W. Haas, Surf. Sci. **19,** 1 (1970).
127. D.A. Gorodetsky and A.N. Knysh, Surf. Sci. **40,** 651 (1973).
128. M. Housley, R. Ducros, G. Piquard, and A. Cassuto, Surf. Sci. **68,** 277 (1977).
129. J.T. Grant and T.W. Haas, Surf. Sci. **21,** 76 (1970).
130. D.G. Castner, B.A. Sexton, and G.A. Somorjai, Surf. Sci. **71,** 519 (1978).
131. T.E. Madey, H.A. Engelhardt, and D. Menzel, Surf. Sci. **48,** 304 (1975).
132. J.C. Fuggle, E. Umbach, P. Feulner, and D. Menzel, Surf. Sci. **64,** 69 (1977),
133. T.E. Madey and D. Menzel, Proc. 2nd Int. Conf. Solid Surf. 229 (1974).
134. L.R. Danielson, M.J. Dresser, E.E. Donaldson, and J.T. Dickinson, Surf. Sci. **71,** 599 (1978).
135. L.R. Danielson, M.J. Dresser, E.E. Donaldson, and D.R. Sandstrom, Surf. Sci. **71,** 615 (1978).
136. J.J. Lander and J. Morrison, J. Appl. Phys. **33,** 2089 (1962).
137. J. Perdereau and G.E. Rhead, Surf. Sci. **24,** 555 (1971).
138. P.W. Palmberg, Surf. Sci. **25,** 104 (1971).
139. D.L. Adams and L.H. Germer, Surf. Sci. **32,** 205 (1972).
140. K.C. Pandey, T. Sakurai, and H.D. Hagstrom, Phys. Rev. **B16,** 3648 (1977).
141. H. Ibach and J.E. Rowe, Surf. Sci. **43,** 481 (1974).
142. R. Heckinbottom and P.R. Wood, Surf. Sci. **36,** 594 (1973).
143. A.J. van Bommel and J.E. Crombeen, Surf. Sci. **36,** 773 (1973).
144. H.D. Shih, F. Jona, D.W. Jepsen, and P.M. Marcus, J. Vac. Sci. Technol. **15,** 596 (1978).
145. H.D. Shih, F. Jona, D.W. Jepsen, and P.M. Marcus, Phys. Rev. Lett. **36,** 798 (1976).
146. H.D. Shih, F. Jona, D.W. Jepsen, and P.M. Marcus, Surf. Sci. **60,** 445 (1976).
147. R. Bastasz, C.A. Colmenares, R.L. Smith, and G.A. Somorjai, Surf. Sci. **67,** 45 (1977).
148. W.P. Ellis, J. Chem. Phys. **48,** 5695 (1968).
149. T.E. Madey, J. Czyzewski, and J.T. Yates, Jr., Surf. Sci. **57,** 580 (1976).
150. N.J. Taylor, Surf. Sci. **2,** 544 (1964).
151. H. Van Hove and R. Leysen, Phys. Status Solidi **A9**(1), 361 (1972).
152. W.N. Unertl and J.M. Blakely, Surf. Sci. **69,** 23 (1973).

216

Surface	Adsorbed gas	Surface structure	References
Ag(100)	O_2	disordered	1
	$C_2H_4Cl_2$	$c(2 \times 2)$-Cl	2, 3
	Se	$c(2 \times 2)$-Se	4
Al(100)	O_2	disordered	5–7
Au(100)	H_2S	(2×2)-S	8
		$c(2 \times 2)$-S	8
		(6×6)-S	8
		$c(4 \times 4)$-S	8
	CO	disordered	9
	Xe	disordered	9
C(100), diamond	O_2	disordered	10
		not adsorbed	11
	N_2	not adsorbed	11
	NH_3	not adsorbed	11
	H_2S	not adsorbed	11
Co(100)	CO	$c(2 \times 2)$-CO	12
		(2×2)-C	12
	O_2	(2×2)-O	13
		$c(2 \times 2)$-O	13
Cr(100)	O_2	$c(2 \times 2)$-O	14
		$Cr_2O_3(310)$	15
Cu(100)	O_2	(1×1)-O	16, 17
		(2×1)-O	16–18
		$(2 \times 4)R45°$-O	19–22
		(2×3)-O	23
		$c(4 \times 4)$-O	23
		$c(2 \times 2)$-O	21, 24–30
		(2×2)	24
		$(2 \times 2\sqrt{2})R45°$	27, 29–31
		Hexagonal	27
		(410) facets	27
	CO	$c(2 \times 2)$-CO	32–34
		hexagonal overlayer	33–35
		(2×2)-C	32, 36
	N_2	(1×1)-N	37
		$c(2 \times 2)$-N	20, 22, 26, 38, 39
Mo(100)	O_2	disordered	40, 41
		$c(2 \times 2)$-O	40, 44
		$(\sqrt{5} \times \sqrt{5})R26°$-O	40, 41, 44–46
		(2×2)-O	40, 45, 46
		$c(4 \times 4)$-O	40, 45, 47
		(2×1)-O	45, 56
		(6×2)-O	44
		(3×1)-O	44
		(1×1)-O	44
	CO	disordered	41
		(1×1)-CO	41, 43, 47, 48

(continued)

Table 5.3.—Continued

Surface	Adsorbed gas	Surface structure	References
		$c(2 \times 2)$-CO	37, 43, 48
		(4×1)-CO	43
	H_2	$c(4 \times 2)$-H	49
		(1×1)-H	49
	N_2	(1×1)-N	41
		$c(2 \times 2)$-N	50
	H_2S	(1×1)-S	51
		$(\sqrt{5} \times \sqrt{5})$-S	51, 52
		$c(2 \times 2)$-S	51
		$MoS_2(100)$	52
NaCl(100)	Xe	hexagonal overlayer	53
Nb(100)	O_2	$c(2 \times 2)$-O	54, 55
		(1×1)-O	54, 55
		(3×10)-NbO_2	54
	N_2	(5×1)-N	54
Ni(100)	O_2	(2×2)-O	56–60, 63, 64
		$c(2 \times 2)$-O	54, 56, 59, 60, 63–67, 70–75, 78
		(2×1)-O	59
		NiO(100)	61–65
		NiO(111)	62, 63
	CO	$c(2 \times 2)$-CO	59, 68, 69, 79–83
		(2×2)-CO	84
		hexagonal overlayer	80, 82, 83
		(2×2)-C	59
	CO_2	(2×2)-O + $c(2 \times 2)$-CO	85
	N_2	not adsorbed	86, 87
	H_2	disordered	59, 88, 89
		$c(2 \times 2)$-H	82
	H_2S	adsorbed	90
		(2×2)-S	28, 31, 90, 91
	Te	(2×2)-Te	92
	Xe	hexagonal overlayer	93
Fe(100)	O_2	$c(2 \times 2)$-O	94–98
		(1×1)-O	97, 99, 100, 104
		FeO(100)	94, 96, 101–103
		FeO(111)	96
		FeO(110)	98
		disordered	103, 104
	CO	$c(2 \times 2)$-CO	105
	H_2S	$c(2 \times 2)$-S	104, 105
	H_2	adsorbs	107
	NH_3	disordered	106, 108
		$c(2 \times 2)$-N	106, 108
	H_2O	$c(2 \times 2)$	109
Fe/Cr(100)	O_2	$c(2 \times 2)$-O	110
		$c(4 \times 4)$-O	110
		oxide	111

Table 5.3.—Continued

Surface	Adsorbed gas	Surface structure	References
Ge(100)	O_2	disordered	112, 113
	I_2	(3×3)-I	114
Ir(100)	O_2	(2×1)-O	115, 116
		(5×1)-O	115, 117
	CO	$c(2 \times 2)$-CO	115, 116
		(2×2)-CO	115
		(1×1)-CO	112
	CO_2	$c(2 \times 2)$-CO_2	115, 116
		(2×2)-CO_2	115, 116
		(7×20)-CO_2	115
	NO	(1×1)-NO	117, 118
	H_2	adsorbed	114, 116
	Kr	(3×5)-Kr	118
		Kr(111)	118
	CO + H_2	$c(3 \times 3)$	82
	H_2S	(2×2)-S	59, 72, 119, 120, 121
		$c(2 \times 2)$-S	59, 72, 76, 77, 78, 120, 121, 122, 123
		(2×1)-S	59
		$c(2 \times 2)$-H_2S	123
	H_2Se	(2×2)-Se	59, 72
		$c(2 \times 2)$-Se	59, 72, 76, 124, 125, 128
		(2×1)-Se	59
		$c(4 \times 2)$-Se	125
	Te	(2×2)-Te	59, 72, 125
		$c(2 \times 2)$-Te	59, 72, 78, 121, 125
		(2×1)-Te	59
		$c(4 \times 2)$-Te	123, 125
	SO_2	$c(2 \times 2)$-SO_2	126, 127
		(2×2)-SO_2	126, 127
NiO(100)	H_2	adsorbed	128
		Ni(100)	128
	H_2S	Ni(100)-$c(2 \times 2)$-S	129
	Cl_2	disordered	64
Pd(100)	CO	disordered	130
		$c(4 \times 2)$-CO	130
		$c(2 \times 2)$-CO	137
		(2×4)R45°-CO	131–133
		hexagonal overlayer	131, 133
	Xe	hexagonal overlayer	134
Pt(100)	O_2	not adsorbed	135, 136
		adsorbed	136, 137
		$(2\sqrt{2} \times 2\sqrt{2})$R45°-O	138, 139
		$PtO_2(0001)$	138
		(5×1)-O	137
		(2×1)-O	137

(continued)

219

Table 5.3.—Continued

Surface	Adsorbed gas	Surface structure	References
	CO	$c(4 \times 2)$-CO	135, 140–144
		$(3\sqrt{2} \times \sqrt{2})$R45°-CO	140–142, 145
		$(\sqrt{2} \times \sqrt{5})$R45°-CO	142, 147
		(2×4)-CO	146
		(1×3)-CO	146
		(1×1)-CO	135, 136, 143, 144
		$c(2 \times 2)$-CO	136, 144
	H_2	adsorbed	136, 145
		(2×2)-H	141, 146
		not adsorbed	136
	$CO + H_2$	$c(2 \times 2)$-$(CO + H_2)$	141, 146
	NO	(1×1)-NO	147
		$c(4 \times 2)$-NO	148
	N	disordered	149
	H_2S	(2×2)-S	150, 151
		$c(2 \times 2)$-S	151, 152
	S_2	(2×2)-S	153, 154
		$c(2 \times 2)$-S	153, 154
Rh(100)	O_2	(2×2)-O	155
		$c(2 \times 2)$-O	155
		$c(2 \times 8)$-O	156
	CO	$c(2 \times 2)$-CO	155
		hexagonal overlayer	155
		(4×1)-CO	156
	CO_2	$c(2 \times 2)$-CO	155
		hexagonal overlayer	155
	H_2	adsorbed	155
	NO	$c(2 \times 2)$-NO	155
Si(100)	O_2	(1×1)-O	112, 113, 156, 157
		(111) facets	112, 113, 156, 157
	H_2	(1×1)-H	152, 158, 159
		(2×1)-H	160
	H	(1×1)-H	161
		(2×1)-H	161
	NH_3	(111) facets	162
	I_2	(3×3)-I	163
Sr(100)	O_2	SrO(100)	164
Ta(100)	O_2	$(2 \times 8/9)$-O	165
		$c(3 \times 1)$-O	165
		(4×1)-O	165
	CO	$c(3 \times 1)$-O	165
	CO_2	$c(3 \times 1)$-O	165
	NO	$c(3 \times 1)$-O	165
	N_2	adsorbed	165
Th(100)	O_2	disordered	166
		ThO_2	166
	CO	disordered	166

Table 5.3.—Continued

Surface	Adsorbed gas	Surface structure	References
V(100)	O_2	(1×1)-O	167
		(2×2)-O	167
	H_2	disordered	167
W(100)	O_2	disordered	169
		(4×1)-O	168–172, 175
		(2×2)-O	169–173
		(2×1)-O	168–176
		(3×3)-O	170, 172, 174
		$c(2 \times 2)$-O	172
		$c(8 \times 2)$-O	172
		(3×1)-O	172
		(1×1)-O	172
		(8×1)-O	172
		(4×4)-O	172, 174
		(110) facets	172
	CO	disordered	177
		$c(2 \times 2)$-CO	168, 177
	H_2	$c(2 \times 2)$-H	168, 178–181
		(2×5)-H	179
		(4×1)-H	179
		(1×1)-H	181
	CO_2	disordered	182
		(2×1)-O	182
		$c(2 \times 2)$-CO	182
	NO	(2×2)-NO	183
		(4×1)-NO	183
		(2×2)-O	183
		(4×1)-O	183
		(2×1)-O	183
	N_2	$c(2 \times 2)$-N	79, 184, 185
	NH_3	disordered	186
		$c(2 \times 2)$-NH_2	186
		(1×1)-NH_2	186
	N_2O	(1×1)-N_2O	187
		(4×1)-N_2O	187
	$CO + N_2$	(4×1)-$(CO + N_2)$	184

1. H.A. Engelhardt and D. Menzel, Surf. Sci. **57**, 591 (1976).
2. G. Rovida and F. Pratesi, Surf. Sci. **51**, 270 (1975).
3. E. Zanazzi, F. Jona, D.W. Jepsen, and P.M. Marcus, Phys. Rev. **B14**, 432 (1976).
4. A. Ignatiev, F. Jona, D.W. Jepsen, and P.M. Marcus, Surf. Sci. **40**, 439 (1973).
5. F. Jona, J. Phys. Chem. Solids **38**, 2155 (1967).
6. S.M. Bedair, F. Hoffman, and H. P. Smith, Jr., J. Appl. Phys. **39**, 4026 (1968).
7. H.H. Farrell, Ph.D. dissertation, University of California, Berkeley, 1969.
8. M. Kostelitz, J. L. Domange, and J. Oudar, Surf. Sci. **34**, 431 (1973).
9. G. McElhiney and J. Pritchard, Surf. Sci. **60**, 397 (1976).
10. J.B. Marsh and H.E. Farnsworth, Surf. Sci. **1**, 3 (1964).
11. M. Maglietta and G. Rovida, Surf. Sci. **71**, 495 (1978).
12. M. Maglietta and G. Rovida, Surf. Sci. **71**, 495 (1978).
13. G. Rovida and M. Maglietta, Proc. 7th Int. Vac. Congr. and 3rd Int. Conf. Solid Surf., Vienna, 963 (1977).

(continued)

Table 5.3.—Continued

14. K. Horn, M. Hussain, and J. Pritchard, Surf. Sci. **63,** 244 (1977).
15. S. Ekelund and C. Leygraf, Surf. Sci. **40,** 179 (1973).
16. A.W. Simmons, D.F. Mitchell, and K.R. Lawless, Surf. Sci. **8,** 130 (1967).
17. L.K. Jordan and E.J. Scheibner, Surf. Sci. **10,** 373 (1968).
18. L. Trepte, C. Menzel-Kopp, and E. Mensel, Surf. Sci. **8,** 223 (1967).
19. R.N. Lee and H.E. Farnsworth, Surf. Sci. **3,** 461 (1965).
20. A. Oustry, L. Lafourcade, and A. Escaut, Surf. Sci. **40,** 545 (1973).
21. E.G. McRae and C.W. Caldwell, Surf. Sci. **57,** 77 (1976).
22. K. Okado, T. Halsushika, H. Tomita, S. Motov, and N. Takalashi, Shinku **13**(11), 371 (1970).
23. L. McDonnell and D.P. Woodruff, Surf. Sci. **46,** 505 (1974).
24. L. McDonnell and D.P. Woodruff, and K.A.R. Mitchell, Surf. Sci. **45,** 1 (1977).
25. G.G. Tibbetts, J.M. Burkstrand, and J.C. Tracy, Phys. Rev. **B15,** 3652 (1977).
26. E. Legrand-Bonnyns and A. Ponslet, Surf. Sci. **53,** 675 (1975).
27. G.C. Tibbetts, J.M. Burkstrand, and J.C. Tracy, J. Vac. Sci. Technol. **13,** 362 (1976).
28. P. Hoffman, R. Unwin, W. Wyrobisch, and A.M. Bradshaw, Surf. Sci. **72,** 635 (1978).
29. U. Gerhardt and G. Franz-Moller, Proc. 7th Int. Vac. Congr. and 3rd Int. Conf. Solid Surf., Vienna, 897 (1977).
30. J.R. Noonan, D.M. Zehner, and L.H. Jenkins, Surf. Sci. **69,** 731 (1977).
31. R.W. Joyner, C.S. McKee, and M.W. Roberts, Surf. Sci. **26,** 303 (1971).
32. J.C. Tracy, J. Chem. Phys. **56**(6), 2748 (1971).
33. C.R. Brundle and K. Wandelt, Proc. 7th Int. Vac. Congr. and 3rd Int. Conf. Solid Surf., Vienna, 1171 (1977).
34. M.A. Chesters and J. Pritchard, Surf. Sci. **28,** 460 (1971).
35. G. Ertl, Surf. Sci. **7,** 309 (1967).
36. R.E. Schlier and H.E. Farnsworth, J. Appl. Phys. **25,** 1333 (1954).
37. J. Perdereau and G.E. Rhead, Surf. Sci. **24,** 555 (1971).
38. J.M. Burkstrand, G.G. Kleiman, G.G. Tibbetts, and J.C. Tracy, J. Vac. Sci. Technol. **13,** 291 (1976).
39. H.K.A. Kann and S. Feuerstein, J. Chem. Phys. **50,** 3618 (1969).
40. K. Hayek and H.E. Farnsworth, Surf. Sci. **10,** 429 (1968).
41. H.E. Farnsworth and K. Hayek, Suppl. Nuovo Cimento **5,** 2 (1967).
42. G.J. Dooley and T.W. Haas, J. Chem. Phys. **52,** 461 (1970).
43. R. Riwan, C. Guillot, and J. Paigne, Surf. Sci. **47,** 183 (1975).
44. L.J. Clark, Proc. 7th Int. Vac. Cong. and 3rd Int. Conf. Solid Surf., Vienna, A2725 (1977).
45. H.M. Kennett and A.E. Lee, Surf. Sci. **48,** 606 (1975).
46. L. Lecante, R. Riwan, and G. Guillot, Surf. Sci. **35,** 271 (1973).
47. C. Guillot, R. Riwan, and J. Lecante, Surf. Sci. **59,** 581 (1976).
48. C.J. Dooley and T.W. Haas, J. Chem. Phys. **52,** 993 (1970).
49. A. Ignatiev, F. Jona, D.W. Jepsen, and P.M. Marcus, Surf. Sci. **49,** 189 (1975).
50. D. Tabor and J.M. Wilson, J. Cryst. Growth **9,** 60 (1971).
51. J.M. Wilson, Surf. Sci. **53,** 330 (1975).
52. A. Glachant, J.P. Coulomb, and J.P. Biberian, Surf. Sci. **59,** 619 (1976).
53. H.H. Farrell and M. Strongin, Surf. Sci. **38,** 18 (1973).
54. R. Pantel, M. Bufor, and J. Bardolle, Surf. Sci. **62,** 739 (1977).
55. A.U. McRae, Surf. Sci. **1,** 319 (1964).
56. H.E. Farnsworth and J. Tuul, J. Phys. Chem. Solids **9,** 48 (1958).
57. J.W. May and L.H. Germer, Surf. Sci. **11,** 443 (1968).
58. J.E. Demuth and T.N. Rhodin, Surf. Sci. **45,** 249 (1974).
59. E.G. McRae and C.W. Caldwell, Surf. Sci. **57,** 63 (1976).
60. P.H. Holoway and J.B. Hudson, Surf. Sci. **43,** 123 (1974).
61. G. Dalmai-Imelik, J.C. Bertolini, and J. Rousseau, Surf. Sci. **63,** 67 (1977).
62. D.F. Mitchell, P.B. Sewell, and M. Cohen, Surf. Sci. **61,** 355 (1976).
63. F.P. Netzer and M. Prutton, Surf. Sci. **52,** 505 (1972).
64. A.U. McRae, Science **139,** 379 (1963).
65. R.E. Schlier and H.E. Farnsworth, Adv. Catal. **9,** 434 (1957).
66. L.H. Germer and C.D. Hartman, J. Appl. Phys. **31,** 2085 (1960).
67. H.E. Farnsworth and H.H. Madden, Jr., J. Appl. Phys. **32,** 1933 (1961).
68. R.L. Park and H.E. Farnsworth, J. Chem. Phys. **43,** 2351 (1965).
69. L.H. Germer, Adv. Catal. **13,** 191 (1962).
70. L.H. Germer, R. Stern, and A.W. McRae, *Metal Surfaces,* American Society for Metals, Metals Park, Ohio (1963), p. 287.
71. P.M. Marcus, J.E. Demuth, and D.W. Jepsen, Surf. Sci. **53,** 501 (1975).

Table 5.3.—Continued

72. H.H. Brongersma and J.B. Theeten, Surf. Sci. **54,** 519 (1976).
73. Y. Murata, S. Ohtani, and K. Terada, Proc. 2nd Int. Conf. Solid Surf., 837 (1974).
74. J.E. Demuth, D.W. Jepsen, and P.M. Marcus, J. Vac. Sci. Technol. **11,** 190 (1974).
75. T.N. Rhodin and J.E. Demuth, Proc. 2nd Int. Conf. Solid Surf., 167 (1974).
76. S. Andersson, B. Kasemo, J.B. Pendry, and M.A. Van Hove, Phys. Rev. Lett. **31,** 595 (1973).
77. J.E. Demuth, D.W. Petersen, and P.M. Marcus, Phys. Rev. Lett. **31,** 540 (1973).
78. M. Onchi and H.E. Farnsworth, Surf. Sci. **11,** 203 (1968).
79. J.C. Tracy, J. Chem. Phys. **56**(6), 2736 (1971).
80. S. Andersson and J.B. Pendry, Surf. Sci. **71,** 75 (1978).
81. S. Andersson, Proc. 3rd Int. Vac. Congr. and 7th Int. Conf. Solid Surf., Vienna, 1019 (1977).
82. K. Horn, A.M. Bradshaw, and K. Jacobi, Surf. Sci. **72,** 719 (1978).
83. R.A. Armstrong, *The Structure and Chemistry of Solid Surfaces,* ed. G.A. Somorjai, Wiley, New York, 1969.
84. M. Onchi and H.E. Farnsworth, Surf. Sci. **13,** 425 (1969).
85. H.H. Madden and H.E. Farnsworth, J. Chem. Phys. **34,** 1186 (1961).
86. J.W. May and L.H. Germer, *The Structure and Chemistry of Solid Surfaces,* ed. G.A. Somorjai, Wiley, New York, 1969.
87. K. Christmann, O. Schober, G. Ertl, and M. Neumann, J. Chem. Phys. **60,** 4528 (1974).
88. K. Christmann, G. Ertl, and O. Schober, Surf. Sci. **40,** 61 (1973).
89. J.L. Domange and J. Oudar, Surf. Sci. **11,** 124 (1968).
90. R.W. Joyner, C.S. McKee, and M.W. Roberts, Surf. Sci. **27,** 279 (1971).
91. A. Salwan and J. Rundgren, Surf. Sci. **53,** 523 (1975).
92. M.A. Chesters, M. Hussain, and J. Pritchard, Surf. Sci. **35,** 161 (1973).
93. A.J. Pignocco and G.E. Pellissier, J. Electronchem. Soc. **112,** 1188 (1965).
94. C. Leygraf and S. Ekelund, Surf. Sci. **40,** 609 (1973).
95. G.W. Simmons and D.J. Dwyer, Surf. Sci. **48,** 373 (1975).
96. C.F. Brucker and T.N. Rhodin, Surf. Sci. **57,** 523 (1976).
97. C. Brucker and T. Rhodin, J. Catal. **47,** 214 (1977).
98. P.B. Sewell, D.F. Mitchell, and M. Cohen, Surf. Sci. **33,** 535 (1972).
99. K.O. Legg, F. Jona, D.W. Jepsen, and P.M. Marcus, Phys. Rev. **B16,** 5271 (1977).
100. C. Leygraf and S. Ekelund, Surf. Sci. **40,** 609 (1973).
101. T. Horiguchi and S. Nakanishi, Proc. 2nd Int. Conf. Solid Surf., 89 (1974).
102. M. Watanabe, M. Miyarmara, T. Matsudaira, and M. Onchi, Proc. 2nd Int. Conf. Solid Surf., 501 (1974).
103. F. Jona, K. O. Legg, H.D. Shih, D.W. Jepsen, and P.M. Marcus, Phys. Rev. Lett. **40,** 1466 (1978).
104. K.O. Legg, F. Jona, D.W. Jepsen, and P.M. Marcus, Surf. Sci. **66,** 25 (1977).
105. F. Bozso, G. Ertl, M. Grunze, and M. Weiss, Appl. Surf. Sci. **1,** 103 (1978).
106. M. Grunze, F. Bozso, G. Ertl, and M. Weiss, Appl. Surf. Sci. **1,** 241 (1978).
107. D.J. Dwyer and G.W. Simmons, Surf. Sci. **63,** 617 (1977).
108. C. Leygraf, G. Hultquist, and S. Ekelund, Surf. Sci. **51,** 409 (1975).
109. C. Leygraf and G. Hultquist, Surf. Sci. **61,** 69 (1976).
110. R.E. Schlier and H.E. Farnsworth, J. Chem. Phys. **30,** 917 (1959).
111. H.E. Farnsworth, R.E. Schlier, T.H. George, and R.M. Buerger, J. Appl. Phys. **29,** 1150 (1958).
112. J.J. Lander and J. Morrison, J. Appl. Phys. **34,** 1411 (1963).
113. J.T. Grant, Surf. Sci. **18,** 228 (1969).
114. T.N. Rhodin and G. Brodén, Surf. Sci. **60,** 466 (1976).
115. G. Brodén and T.N. Rhodin, Solid State Commun. **18,** 105 (1976).
116. B.E. Nieuwenhuys and G.A. Somorjai, Surf. Sci. **72,** 8 (1978).
117. J. Kanski and T.N. Rhodin, Surf. Sci. **65,** 63 (1977).
118. A. Ignatiev, T.N. Rhodin, and S.Y. Tong, Surf. Sci. **42,** 37 (1974).
119. J.L. Perdereau and J. Oudar, Surf. Sci. **20,** 80 (1970).
120. T. Edmonds, J.J. McCarrol, and R.C. Pitkethly, J. Vac. Sci. Technol. **8**(1), 68 (1971).
121. J.E. Demuth, D.W. Jepsen, and P.M. Marcus, Surf. Sci. **45,** 733 (1974).
122. T. Matsudaira, M. Nichijima, and M. Onchi, Surf. Sci. **61,** 651 (1976).
123. H.D. Hagstrum and A.E. Becker, Phys. Rev. Lett. **22,** 1054 (1969).
124. H. Froitzheim and H.D. Hagstrum, J. Vac. Sci. Technol. **15,** 485 (1978).
125. G.E. Becker and H.D. Hagstrum, J. Vac. Sci. Technol. **11,** 234 (1974).
126. N.W. Tideswell and J.M. Ballingal, J. Vac. Sci. Technol. **7,** 496 (1970).
127. J.M. Rickard, M. Perdereau, and L.G. Dufour, Proc. 7th Int. Vac. Congr. and 3rd Int. Conf. Solid Surf., Vienna, 847 (1977).
128. A. Steinbrunn, P. Dumas, and J.C. Colson, Surf. Sci. **74,** 201 (1978).

(continued)

Table 5.3.—Continued

129. J.C. Tracy and P.W. Palmberg, J. Chem. Phys. **51**, 2854 (1971).
130. A.M. Bradshaw and F.M. Hoffman, Surf. Sci. **72**, 513 (1978).
131. R.L. Park and H.H. Madden, Surf. Sci. **11**, 188 (1968).
132. H. Conrad, G. Ertl, J. Koch, and E.E. Latta, Surf. Sci. **43**, 462 (1974).
133. H. Conrad, G. Ertl, J. Küppers, and E.E. Latta, Surf. Sci. **65**, 235 (1977).
134. P.W. Palmberg, Surf. Sci. **25**, 104 (1971).
135. B. Lang, R.W. Joyner, and G.A. Somorjai, Surf. Sci. **30**, 454 (1972).
136. C.R. Helms, H.P. Bonzel, and S. Kelemen, J. Chem. Phys. **65**, 1773 (1976).
137. G. Pirug, G. Brodén, and H.P. Bonzel, Proc. 7th Int. Vac. Congr. and 3rd Int. Conf. Solid Surf., Vienna, 907 (1977).
138. P. Légaré, G. Maire, B. Carrière, and J.P. Deville, Surf. Sci. **68**, 348 (1977).
139. B. Lang, P. Légaré, and G. Maire, Surf. Sci. **47**, 89 (1975).
140. A.E. Morgan and G.A. Somorjai, J. Chem. Phys. **51**, 3309 (1969).
141. A.E. Morgan and G.A. Somorjai, Surf. Sci. **12**, 405 (1968).
142. C. Burggraf and A. Mosser, C.R. Acad. Sci. **268B**, 1167 (1969).
143. G. Kneringer and F.P. Netzer, Surf. Sci. **49**, 125 (1975).
144. G. Brodén, G. Pirug, and H.P. Bonzel, Surf. Sci. **72**, 45 (1978).
145. F.P. Netzer and G. Kneringer, Surf. Sci. **51**, 526 (1975).
146. A.E. Morgan and G.A. Somorjai, Trans. Am. Crystallogr. Assoc. **4**, 59 (1968).
147. H.P. Bonzel and G. Pirug, Surf. Sci. **62**, 45 (1977).
148. H.P. Bonzel, G. Brodén, and G. Pirug, J. Catal. **53**, 96 (1978).
149. K. Schwaka and E. Bechtold, Surf. Sci. **66**, 383 (1977).
150. Y. Berthier, M. Perdereau, and J. Oudar, Surf. Sci. **36**, 225 (1973).
151. T.E. Fischer and S.R. Kelemen, Surf. Sci. **69**, 1 (1977).
152. T.E. Fischer and S.R. Kelemen, J. Vac. Sci. Technol. **15**, 607 (1978).
153. W. Heegemann, E. Bechtold, and K. Hayek, Proc. 2nd Int. Conf. Solid Surf., 195 (1974).
154. W. Heegemann, K.H. Meister, E. Bechtold, and K. Hayek, Surf. Sci. **49**, 161 (1975).
155. D.G. Castner, B.A. Sexton, and G.A. Somorjai, Surf. Sci. **71**, 159 (1978).
156. J.J. Lander and J. Morrison, J. Appl. Phys. **33**, 2089 (1962).
157. S.J. White and D.P. Woodruff, Surf. Sci. **63**, 254 (1977).
158. S.J. White, D.P. Woodruff, B.W. Holland, and R.S. Zimmer, Surf. Sci. **74**, 34 (1978).
159. S.J. White, D.P. Woodruff, B.W. Holland, and R.S. Zimmer, Surf. Sci. **68**, 457 (1977).
160. H. Ibach and J.E. Rowen, Surf. Sci. **43**, 81 (1974).
161. T. Sakurai and H.D. Hagstrum, Phys. Rev. **B14**, 1593 (1976).
162. R. Heckinbottom and P.R. Wood, Surf. Sci. **36**, 594 (1973).
163. J.J. Lander and J. Morrison, J. Chem. Phys. **37**, 729 (1962).
164. A.P. Janssen and R.C. Schoonmaker, Surf. Sci. **55**, 109 (1976).
165. M.A. Chesters, B.J. Hopkins, and M.R. Leggett, Surf. Sci. **43**, 1 (1974).
166. T.N. Taylor, C.A. Colmenares, R.L. Smith, and G.A. Somorjai, Surf. Sci. **54**, 317 (1976).
167. K.K. Vijai and P.F. Packman, J. Chem. Phys. **50**, 1343 (1969).
168. P.J. Estrup, *The Structure and Chemistry of Solid Surfaces*, ed. G.A. Somorjai, Wiley, New York, 1969.
169. B.J. Hopkins, G.D. Watt, and A.R. Jones, Surf. Sci. **52**, 715 (1975).
170. C.A. Papageorgopoulous and J.M. Chen, Surf. Sci. **39**, 313 (1973).
171. A.M. Bradshaw, D. Menzel, and M-Steinkilberg, Proc. 2nd Int. Conf. Solid Surf., 841 (1974).
172. E. Bauer, H. Poppa, and Y. Viswanath, Surf. Sci. **58**, 578 (1976).
173. S. Prigge, H. Niehus, and E. Bauer, Surf. Sci. **65**, 141 (1977).
174. J.L. Desplat, Proc. 2nd Int. Conf. Solid Surf., 177 (1974).
175. P.E. Luscher and F.M. Propst, J. Vac. Sci. Technol. **14**, 400 (1977).
176. J. Anderson and W.E. Danforth, J. Franklin Inst. **279**, 160 (1965).
177. J. Anderson and P.J. Estrup, J. Chem. Phys. **46**, 563 (1967).
178. P.W. Tamm and L.D. Schmidt, J. Chem. Phys. **51**, 5352 (1969).
179. P.J. Estrup and J. Anderson, J. Chem. Phys. **45**, 2254 (1955).
180. R. Jaeger and D. Menzel, Surf. Sci. **63**, 232 (1977).
181. C.A. Papageorgopoulos and J.M. Chen, Surf. Sci. **39**, 283 (1973).
182. B.J. Hopkins, A.R. Jones, and R.I. Winton, Surf. Sci. **57**, 266 (1976).
183. S. Usami and T. Nakagima, Proc. 2nd Int. Conf. Solid Surf., 237 (1974).
184. P.J. Estrup and J. Anderson, J. Chem. Phys. **46**, 567 (1967).
185. D.L. Adams and L.H. Germer, Surf. Sci. **27**, 21 (1971).
186. P.J. Estrup and J. Anderson, J. Chem. Phys. **49**, 523 (1968).
187. W.H. Weinberg and R.P. Merrill, Surf. Sci. **32**, 317 (1972).

Surface	Adsorbed gas	Surface structure	References
Ag(110)	O_2	(2×1)-O	1–5
		(3×1)-O	1–4
		(4×1)-O	1–3
		(5×1)-O	1, 2
		(6×1)-O	1, 2
		(7×1)-O	1
	NO	disordered	6
	$C_2H_4Cl_2$	(2×1)-Cl	7
		$c(4 \times 2)$-Cl	7
	Xe	hexagonal overlayer	8
Al(110)	O_2	(331) facets	9
		(111) facets	10
Au(100)	H_2S	(1×2)-S	11
		$c(4 \times 2)$-S	11
C(110), diamond	O_2	not adsorbed	12
	N_2	not adsorbed	12
	NH_3	not adsorbed	12
	H_2S	not adsorbed	12
Cr(110)	O_2	(3×1)-O	13
		(100) facets	13, 14
		$Cr_2O_3(0001)$	13, 14
Cu(110)	O_2	(2×1)-O	15–20
		$c(6 \times 2)$-O	16–20
		(5×3)-O	16, 21
	CO	ordered 1d	22
		(2×3)-CO	22
		(2×1)-CO	22
		hexagonal overlayer	23
	H_2	not adsorbed	15
	H_2O	disordered	22
	H_2S	$c(2 \times 3)$-S	24
		adsorbed	24
	Xe	$c(2 \times 2)$-Xe	25
		hexagonal overlayer	25
Cu/Ni(110)	O_2	(2×1)-O	26
	CO	(2×1)-CO	26
		(2×2)-CO	26
	H_2S	$c(2 \times 2)$-S	26
Fe(110)	O_2	$c(2 \times 2)$-O	27–29
		$c(3 \times 1)$-O	27–29
		(2×8)-O	30
		FeO(111)	27–29, 31
		(2×1)-O	32
	CO	$\begin{pmatrix} 3 & -2 \\ 0 & 4 \end{pmatrix}$-CO	33

(continued)

Table 5.4.—Continued

Surface	Adsorbed gas	Surface structure	References
	N_2	$\begin{pmatrix} 3 & -2 \\ 0 & 4 \end{pmatrix}$-$N_2$	33
	H_2	(2×1)-H	34
		(3×1)-H	34
		(1×1)-H	34
	H_2S	(2×4)-S	35
		(1×2)-S	35
Fe/Cr(110)	O_2	Cr_2O_3(0001)	36
		amorphous oxide	37
Ge(110)	O_2	disordered	38, 39
		(1×1)-O	38, 39
	H_2S	(10×5)-S	40
Ir(110)	O_2	(1×2)-O	41
	CO	(2×2)-CO	41, 42
		(4×2)-CO	42
	H_2	adsorbed	41
	N_2	not adsorbed	41
LaB_6(110)	O_2	(1×1)-O	43
Mo(110)	O_2	(2×2)-O	44–46
		(2×1)-O	44–46
		(1×1)-O	44, 45
		disordered	47
	CO	(1×1)-CO	44, 46
		$c(2 \times 2)$-CO	48
		disordered	49
	CO_2	disordered	48
	H_2	adsorbed	46
	N_2	(1×1)-N	44
	H_2S	(2×2)-S	50
		$c(2 \times 2)$-S	50
		(1×1)-S	50
		$c(1 \times 3)$-S	50
		$c(1 \times 5)$-S	50
		(1×3)-S	50
		$c(1 \times 7)$-S	50
		(1×4)-S	50
		(1×5)-S	50
		$c(1 \times 11)$-S	50
		$\begin{pmatrix} 2 & 2 \\ -1 & 1 \end{pmatrix}$-S	50
(211)	O_2	(2×1)-O	51
		(1×2)-O	51
		(1×3)-O	51
		$c(4 \times 2)$-O	51
	CO	disordered	51

Table 5.4.—Continued

Surface	Adsorbed gas	Surface structure	References
	H_2	(1×2)-H	51
	N_2	not adsorbed	51
Na(110)	O_2	NaO(111)	52
Nb(110)	O_2	(3×1)-O	53
		NbO(111)	54
		NbO(110)	54
		NbO(220)	54
		oxide	53
	CO	disordered	53
		(3×1)-O	53
	H_2	(1×1)-H	75
Ni(110)	O_2	(2×1)-O	29, 55–63
		(3×1)-O	55, 57, 59–67
		(5×1)-O	55, 60
		(9×4)-O	57, 65, 66
		NiO(100)	57, 59, 61, 63, 65–67
	CO	(1×1)-CO	55, 67
		adsorbed	63
		$c(2 \times 1)$-CO	64, 68, 71
		(2×1)-CO	63, 69, 70
		$c(2 \times 2)$-CO	71
		(4×2)-CO	71
	H_2	(1×2)-H	62, 63, 66, 72–77
	NO	(2×3)-N	78
		(2×1)-O	78
	H_2O	(2×1)-H_2O	74
	H_2S	$c(2 \times 2)$-S	62, 78–80
		(3×2)-S	78
	H_2Se	$c(2 \times 2)$-Se	81
	$CO + O_2$	(3×1)-$(CO + O_2)$	61
Pd(110)	O_2	(1×3)-O	82
		(1×2)-O	82
		$c(2 \times 4)$-O	82
	CO	(5×2)-CO	82
		(2×1)-CO	82, 83
		(4×2)-CO	83
		$c(2 \times 2)$-CO	83
	H_2	(1×2)-H	84
Pt(110)	O_2	(2×1)-O	85, 86
		(4×2)-O	85
		adsorbed	87
		$c(2 \times 2)$-O	86
		PtO(100)	86

(continued)

227

Table 5.4.—Continued

Surface	Adsorbed gas	Surface structure	References
	CO	(1×1)-CO	88, 89
		(2×1)-CO	90
	C_3O_2	(1×1)-C_3O_2	91
	NO	(1×1)-NO	89, 92
	CO + NO	(1×1)-(CO + NO)	89
	H_2S	$c(2 \times 6)$-S	93–95
		(2×3)-S	93–95
		(4×3)-S	93–95
		$c(2 \times 4)$-S	93–95
		(4×4)-S	93, 94
Rh(110)	O_2	disordered	96, 97
		$c(2 \times 4)$-O	96, 97
		$c(2 \times 8)$-O	96, 97
		(2×2)-O	96, 97
		(2×3)-O	96, 97
		(1×2)-O	96, 97
		(1×3)-O	96, 97
	CO	(2×1)-CO	98
		$c(2 \times 2)$-C	98
Ru($10\bar{1}0$)	O_2	$c(4 \times 2)$-O	99, 100
		(2×1)-O	99, 100
		$c(2 \times 6)$-O	99
		(7×1)-O	99
		$c(4 \times 8)$-O	99
	CO	disordered	100
	H_2	not adsorbed	100
	N_2	not adsorbed	100
	NO	$c(4 \times 2)$-(N + O)	99, 100
		(2×1)-(N + O)	99, 100
		(2×1)-O	100
		$c(4 \times 2)$-O	100
		$c(2 \times 6)$-O	99
		(7×1)-O	99
		$c(4 \times 8)$-O	99
		(2×1)-N	100
		$c(4 \times 2)$-N	100
(101)	O_2	$\begin{pmatrix} 1 & 1 \\ 3 & 0 \end{pmatrix}$-O	101
		$\begin{pmatrix} 2 & 1 \\ 5 & 0 \end{pmatrix}$-O	101
		$\begin{pmatrix} 4 & 1 \\ 9 & 0 \end{pmatrix}$-O	101
	CO	$\begin{pmatrix} 1 & 1 \\ 3 & 0 \end{pmatrix}$-CO	102
		$\begin{pmatrix} 0 & 1 \\ 2 & 0 \end{pmatrix}$-C	102
	NO	disordered	103

Table 5.4.—Continued

Surface	Adsorbed gas	Surface structure	References
Si(110)	H_2	(1×1)-H	104
(311)	NH_3	adsorbed	105
Ta(100)	O_2	(3×1)-O	106, 107
		oxide	106, 107
	CO	disordered	106, 107
		(3×1)-O	106, 107
	H_2	(1×1)-H	107
	N_2	not adsorbed	106
(211)	O_2	(3×1)-O	106, 107
		oxide	106, 107
	CO	disordered	106, 107
		(3×1)-O	107
	H_2	(1×1)-H	107
	N_2	disordered	107
		(311) facets	107
TiO_2(100)	O_2	disordered	108
	H_2O	disordered	108
V(110)	O_2	(3×1)-O	106
	CO	disordered	106
		(3×1)-O	106
W(110)	O_2	(2×1)-O	58, 109, 110, 120
		$c(2 \times 2)$-O	111–113, 121
		(2×2)-O	120, 121
		(1×1)-O	115–119, 121
		$c(14 \times 7)$-O	58, 109, 121
		$c(21 \times 7)$-O	121
		$c(48 \times 16)$-O	121
		WO_3(100)	122
		WO_3(111)	122
	CO	disordered	123
		$c(9 \times 5)$-CO	123
		(1×1)-CO	112
		$c(2 \times 2)$-CO	112
		(2×7)-CO	124
		$c(4 \times 1)$-CO	124
		(3×1)-CO	124
		(4×1)-CO	124
		(5×1)-CO	124, 125
		(2×1)-(C + O)	124, 125
		$c(9 \times 5)$-(C + O)	124
	CO + O_2	$c(11 \times 5)$-(CO + O_2)	126
	H_2	(2×1)-H	127
	I_2	(2×2)-I	128
		(2×1)-I	128

(continued)

Table 5.4.—Continued

Surface	Adsorbed gas	Surface structure	References
(211)	O_2	(2×1)-O	129–134
		(1×2)-O	129, 130, 134
		(1×1)-O	130, 131, 133, 134
		(1×3)-O	130
		(1×4)-O	130, 134
	CO	disordered	132
		$c(6 \times 4)$-CO	132
		(2×1)-CO	132
		$c(2 \times 4)$-CO	132
	H_2	(1×1)-H	135
	NH_3	$c(4 \times 2)$-NH_2	136
	$CO + O_2$	(1×1)-$(CO + O_2)$	132
		(1×2)-$(CO + O_2)$	132
(210)	CO	(2×1)-CO	137
		(1×1)-CO	137
	N_2	(2×1)-N	138
(310)	N_2	(2×1)-N	138
		$c(2 \times 2)$-N	138
$ZnO(10\bar{1}0)$	O_2	(1×1)-O	139

1. H.A. Engelhardt and D. Menzel, Surf. Sci. **57**, 591 (1976).
2. H.A. Engelhardt, A.M. Bradshaw, and D. Menzel, Surf. Sci. **40**, 410 (1973).
3. G. Rovida and F. Pratesi, Surf. Sci. **52**, 542 (1975).
4. W. Heiland, F. Iberl, E. Taglauer, and D. Menzel, Surf. Sci. **53**, 383 (1975).
5. E. Zanazzi, M. Maglietta, U. Bardi, F. Jona, D.W. Jepsen, and P.M. Marcus, Proc. 7th Int. Vac. Congr. and 3rd Int. Conf. Solid Surf., Vienna, 2447 (1977).
6. R.A. Marbrow and R.M. Lambert, Surf. Sci. **61**, 317 (1976).
7. G. Rovida and F. Pratesi, Surf. Sci. **51**, 270 (1975).
8. M.A. Chesters, M. Hussain, and J. Pritchard, Surf. Sci. **35**, 161 (1973).
9. S.M. Bedair and H.P. Smith, Jr., J. Appl. Phys. **42**, 3616 (1971).
10. H. Van Hove and R. Leysen, Phys. Status Solidi **A9**, 361 (1972).
11. M. Kostelitz, J.L. Domange, and J. Oudar, Surf. Sci. **34**, 431 (1973).
12. P.G. Lurie and J.M. Wilson, Surf. Sci. **65**, 453 (1977).
13. P. Michel and Ch. Jardin, Surf. Sci. **36**, 478 (1973).
14. Ekelund and C. Leygraf, Surf. Sci. **40**, 179 (1973).
15. G. Ertl, Surf. Sci. **6**, 208 (1967).
16. N. Takahashi et al., C.R. Acad. Sci., Paris **269B**, 618 (1969).
17. G.W. Simmons, D.F. Mitchell, and K.R. Lawless, Surf. Sci. **8**, 130 (1967).
18. L.K. Jordan and E.J. Scheibner, Surf. Sci. **10**, 373 (1968).
19. L. Trepte, C. Menzel-Kopp, and E. Menzel, Surf. Sci. **8**, 223 (1967).
20. A. Oustry, L. Lafourcade, and A. Escaut, Surf. Sci. **40**, 545 (1973).
21. I. Marklund, S. Anderson, and J. Martinsson, Ark. Phys. **37**, 127 (1968).
22. G. Ertl, Surf. Sci. **7**, 309 (1967).
23. K. Horn, M. Hussain, and J. Pritchard, Surf. Sci. **63**, 244 (1977).
24. D.L. Domange and J. Oudar, Surf. Sci. **11**, 124 (1968).
25. M.A. Chesters, M. Hussain, and J. Pritchard, Surf. Sci. **54**, 687 (1976).
26. G. Ertl and J. Küppers, Surf. Sci. **24**, 104 (1971).
27. F. Portele, Z. Naturforsch. **24A**, 1268 (1969).
28. G. Dalmai-Imelik and J.C. Bertolini, C.R. Acad. Sci. Paris **270**, 1079 (1970).
29. K. Moliere and F. Portele, *The Structure and Chemistry of Solid Surfaces,* ed. G.A. Somorjai, Wiley, New York, 1969.
30. A.J. Pignocco and G.E. Pellisier, Surf. Sci. **7**, 261 (1967).

Table 5.4.—Continued

31. C. Leygraf and S. Ekelund, Surf. Sci. **40,** 609 (1973).
32. A. Melmed and J.J. Carroll, J. Vac. Sci. Technol. **10,** 164 (1973).
33. G. Gafner and R. Feder, Surf. Sci. **57,** 37 (1976).
34. F. Bozso, G. Ertl, M. Grunze, and M. Weiss, Appl. Surf. Sci. **1,** 103 (1978).
35. D.H. Buckley, NASA Tech. Note D-4589 (1970).
36. C. Leygraf and G. Hultquist, Surf. Sci. **61,** 69 (1976).
37. C. Leygraf, G. Hultquist, and S. Elelund, Surf. Sci. **51,** 409 (1975).
38. R.E. Schlier and H.E. Farnsworth, J. Chem. Phys. **30,** 917 (1959).
39. H.E. Farnsworth, R.E. Schlier, J.H. George, and R.M. Buerger, J. Appl. Phys. **29,** 1150 (1958).
40. B.Z. Olshanetsky, S.M. Repinsky, and A.A. Shklyaev, Surf. Sci. **64,** 224 (1977).
41. B.E. Nieuwenhuys and G.A. Somorjai, Surf. Sci. **72,** 8 (1978).
42. J.L. Taylor and W.H. Weinberg, J. Vac. Sci. Technol. **15,** 590 (1978).
43. E.B. Bas, P. Hafner, and S. Klauser, Proc. 7th Int. Vac. Congr. and 3rd Int. Conf. Solid Surf., Vienna, 881 (1977).
44. K. Hayek and H.E. Farnsworth, Surf. Sci. **10,** 429 (1958).
45. H.E. Farnsworth and K. Hayek, Suppl. Nuovo Cimento **5,** 2 (1957).
46. T.W. Haas and A.G. Jackson, J. Chem. Phys. **44,** 2121 (1966).
47. T. Miura and Y. Tuzi, Proc. 2nd Int. Conf. Solid Surf., 85 (1974).
48. A.J. Pignosco and G.E. Pellisier, Surf. Sci. **7,** 261 (1967).
49. E. Gillet, J.C. Chiarena, and M. Gillet, Surf. Sci. **67,** 393 (1977).
50. L. Peralta, Y. Bertier, and J. Oudar, Surf. Sci. **55,** 199 (1976).
51. G.J. Dooley and T.W. Haas, J. Vac. Sci. Technol. **7,** 49 (1970).
52. S. Andersson, J.B. Pendry, and P.M. Echenique, Surf. Sci. **65,** 539 (1977).
53. T.W. Haas, A.G. Jackson, and M.P. Hooker, J. Chem. Phys. **46,** 3025 (1967).
54. R. Pantel, M. Bujor, and J. Bardolle, Surf. Sci. **62,** 739 (1977).
55. A.U. McRae, Surf. Sci. **1,** 319 (1964).
56. L.H. Germer, E.J. Schneiber, and C.D. Hartman, Phil. Mag. **5,** 222 (1960).
57. J.W. May and L.H. Germer, Surf. Sci. **11,** 443 (1968).
58. L.H. Germer, R. Stern, and A.U. MacRae, *Metal Surfaces,* American Society for Metals, Metals Park, Ohio, 1963, p. 287.
59. T.L. Park and H.E. Farnsworth, J. Appl. Phys. **35,** 2220 (1964).
60. L.H. Germer and A.U. MacRae, J. Appl. Phys. **33,** 2923 (1962).
61. R.L. Park and H.E. Farnsworth, J. Chem. Phys. **40,** 2354 (1964).
62. L.H. Germer, J.W. May, and R.J. Szostak, Surf. Sci. **7,** 430 (1967).
63. J.E. Demuth and T.N. Rhodin, Surf. Sci. **45,** 249 (1974).
64. J. Küppers, Surf. Sci. **36,** 53 (1973).
65. D.F. Mitchell and P.B. Sewell, Proc. 7th Int. Vac. Congr. and 3rd Int. Conf. Solid Surf., Vienna, 963 (1977).
66. D.F. Mitchell, P.B. Sewell, and M. Cohen, Surf. Sci. **69,** 310 (1977).
67. A.G. Jackson and M.P. Hooker, Surf. Sci. **6,** 297 (1967).
68. H.H. Madden, J. Küppers, and G. Ertl, J. Chem. Phys. **58,** 3401 (1973).
69. H.H. Madden and G. Ertl, Surf. Sci. **35,** 211 (1973).
70. H.H. Madden, J. Küppers, and G. Ertl, J. Vac. Sci. Technol. **11,** 190 (1974).
71. T.N. Taylor and P.J. Estrup, J. Vac. Sci. Technol. **10,** 26 (1973).
72. C.A. Hague and H.E. Farnsworth, Surf. Sci. **1,** 378 (1964).
73. J.W. May and L.H. Germer, *The Structure and Chemistry of Solid Surfaces,* ed. G.A. Somorjai, Wiley, New York, 1969.
74. L.H. Germer and A.U. MacRae, Proc. Natl. Acad. Sci. U.S. **48,** 997 (1962).
75. T.W. Haas, J. Appl. Phys. **39,** 5854 (1968).
76. K. Christmann, O. Schober, G. Ertl, and M. Neumann, J. Chem. Phys. **60,** 4528 (1974).
77. T.N. Taylor and P.J. Estrup, J. Vac. Sci. Technol. **11,** 244 (1974).
78. M. Perdereau and J. Oudar, Surf. Sci. **20,** 80 (1970).
79. J.E. Demuth, D.W. Jepsen, and P.M. Marcus, Phys. Rev. Lett **32,** 1182 (1974).
80. T.N. Rhodin and J.E. Demuth, Proc. 2nd Int. Conf. Solid Surf., 167 (1974).
81. G.E. Becker and H.D. Hagstrum, Surf. Sci. **30,** 505 (1972).
82. G. Ertl and P. Rau, Surf. Sci. **15,** 443 (1969).
83. H. Conrad, G. Ertl, J. Koch, and E.E. Latta, Surf. Sci. **43,** 462 (1974).

(continued)

Table 5.4.—Continued

84. H. Conrad, G. Ertl, and E.E. Latta, Surf. Sci. **41**, 435 (1974).
85. C.W. Tucker, Jr., J. Appl. Phys. **35**, 1897 (1964).
86. R. Ducros and R.P. Merrill, Surf. Sci. **55**, 227 (1976).
87. M. Wilf and P.T. Dawson, Surf. Sci. **65**, 399 (1977).
88. H.P. Bonzel and R. Ku, Surf. Sci. **33**, 91 (1972).
89. R.M. Lambert and C.M. Comrie, Surf. Sci. **46**, 61 (1974).
90. R.M. Lambert, Surf. Sci. **49**, 325 (1974).
91. P.D. Reed and R.M. Lambert, Surf. Sci. **57**, 485 (1976).
92. C.M. Comrie, W.H. Weinberg, and R.M. Lambert, Surf. Sci. **57**, 619 (1976).
93. Y. Berthier, M. Perdereau, and J. Oudar, Surf. Sci. **36**, 225 (1973).
94. Y. Berthier, J. Oudar, and M. Huber, Surf. Sci. **65**, 361 (1977).
95. H.P. Bonzel and R. Ku, J. Chem. Phys. **58**, 4617 (1973).
96. C.W. Tucker, Jr., J. Appl. Phys. **38**, 2696 (1967).
97. C.W. Tucker, Jr., J. Appl. Phys. **37**, 4147 (1966).
98. R.A. Marbrow and R.M. Lambert, Surf. Sci. **67**, 489 (1977).
99. T.W. Orent and R.S. Hansen, Surf. Sci. **67**, 325 (1977).
100. R. Ku, N.A. Gjostein, and H.P. Bonzel, Surf. Sci. **64**, 465 (1977).
101. P.D. Reed, C.M. Comrie, and R.M. Lambert, Surf. Sci. **64**, 603 (1977).
102. P.D. Reed, C.M. Comrie, and R.M. Lambert, Surf. Sci. **59**, 33 (1976).
103. P.D. Reed, C.M. Comrie, and R.M. Lambert, Surf. Sci. **72**, 423 (1978).
104. T. Sakurai and H.D. Hagstrum, J. Vac. Sci. Technol. **13**, 907 (1976).
105. R. Heckingbottom and P.R. Wood, Surf. Sci. **36**, 594 (1973).
106. T.W. Haas, A.G. Jackson, and M.P. Hooker, J. Chem. Phys. **46**, 3025 (1967).
107. T.W. Haas, *The Structure and Chemistry of Solid Surfaces,* ed. G.A. Somorjai, Wiley, New York, 1969.
108. W.J. Lo, Y.W. Chung, and G.A. Somorjai, Surf. Sci. **71**, 199 (1978).
109. L.H. Germer, Phys. Today, July 1964, p. 19.
110. M.A. VanHove, S.Y. Tong, and M.H. Elconin, Surf. Sci. **64**, 85 (1977).
111. G.C. Wang, T.M. Lu, and M.G. Lagally, Proc. 7th Int. Vac. Congr. and 3rd Int. Conf. Solid Surf., Vienna, A2726 (1974).
112. J.M. Baker and D.E. Eastman, J. Vac. Sci. Technol. **10**, 223 (1973).
113. J.C. Buchholz and M.G. Lagally, J. Vac. Sci. Technol. **11**, 194 (1974).
114. J.C. Buchholz and M.G. Lagally, Phys. Rev. Lett. **35**, 442 (1975).
115. K. Besocke and S. Berger, Proc. 7th Int. Vac. Congr. and 3rd Int. Conf. Solid Surf., Vienna, 893 (1977).
116. T.E. Madey and J.T. Yates, Surf. Sci. **63**, 203 (1977).
117. T. Engel, H. Niehus, and E. Bauer, Surf. Sci. **52**, 237 (1975).
118. J.C. Buchholz, G.C. Wang, and M.G. Lagally, Surf. Sci. **49**, 508 (1975).
119. M.A. Van Hove and S.Y. Tong, Phys. Rev. Lett **35**, 1092 (1975).
120. E. Bauer and T. Engel, Surf. Sci. **71**, 695 (1978).
121. L.H. Germer and J.W. May, Surf. Sci. **4**, 452 (1966).
122. N.R. Avery, Surf. Sci. **41**, 533 (1974).
123. J.W. May and L.H. Germer, J. Chem. Phys. **44**, 2895 (1966).
124. Ch. Steinbruchel and R. Gomer, Surf. Sci. **67**, 21 (1977).
125. Ch. Steinbruchel and R. Gomer, J. Vac. Sci. Technol. **14**, 484 (1977).
126. J.W. May, L.H. Germer, and C.C. Chang, J. Chem. Phys. **45**, 2383 (1966).
127. J.K. Matsik, Surf. Sci. **29**, 324 (1972).
128. N.R. Avery, Surf. Sci. **43**, 191 (1974).
129. N.J. Taylor, Surf. Sci. **2**, 544 (1964).
130. C.C. Chang and L.H. Germer, Surf. Sci. **8**, 115 (1967).
131. T.C. Tracy and J.M. Blakeley, *The Structure and Chemistry of Solid Surfaces,* ed. G.A. Somorjai, Wiley, New York, 1969.
132. C.C. Chang, J. Electrochem. Soc. **115**, 354 (1968).
133. G. Ertl and M. Plancher, Surf. Sci. **48**, 364 (1975).
134. B.J. Hopkins and G.D. Watts, Surf. Sci. **44**, 237 (1974).
135. D.L. Adams et al., Surf. Sci. **22**, 45 (1970).
136. J.W. May, R.J. Szostak, and L.H. Germer, Surf. Sci. **15**, 37 (1969).
137. D.L. Adams and L.H. Germer, Surf. Sci. **32**, 205 (1972).
138. D.L. Adams and L.H. Germer, Surf. Sci. **27**, 21 (1971).
139. W. Göpel, Surf. Sci. **62**, 165 (1977).

Table 5.5. Surface structures on stepped substrates

Surface	Adsorbed gas	Surface structure	References
Ag[1_2(111) + 1_1(100)]	Xe	hexagonal overlayer	1
Ag[1_1(111) + 2_1(110)]	O_2	disordered	2
		Ag(110)-(2 × 1)-O	2
	Cl_2	(6 × 1)-Cl	2
Au(S)-[5_5(111) + 2_1 (100)]	O_2	oxide	3
Cu[2_2(100) + 1_1 (010)]	O_2	(410), (530) facets	4
Cu[1_2(111) + 1_1 (100)]	Xe	hexagonal overlayer	5
	Kr	hexagonal overlayer	5
Cu[2_1(100) + 1_1 (111)]	Xe	hexagonal overlayer	6
	CO	adsorbed	6
Cu[8_8 (100) + 4_4 (010) + 1_1 (001)]	O_2	(410), (100) facets	4
Cu(S)-[3_3(100) + 1_1 (010)]	CO	not adsorbed	7
	N_2	(1 × 2)-N	7
Cu(S)-[4_4(100) + 1_1 (010)]	O_2	(1 × 1)-O	7
	CO	not adsorbed	7
	N_2	(1 × 3)-N	7
Cu(S)-[6_3 (100) + 1_1 (111)]	H_2S	8($1d$)-S	8
Ir(S)-[5_5(111) + 2_1 (100)]	O_2	(2 × 1)-O	9
	CO	disordered	9
	H_2O	not adsorbed	9
	H_2	adsorbed	10
Ni[2_2(100) + 1_1 (010)]	O_2	facets	11
	N_2	Ni(100)-($6\sqrt{2} \times \sqrt{2}$)R45°−N	11
		Ni(110)-(2 × 3)-N	11
Pd[2_2(100) + 1_1 (010)]	CO	(1 × 1)-CO	12, 13
		(1 × 2)-CO	12, 13
Pd[2_1(100) + 1_1 (111)]	CO	(2 × 1)-CO	12
		3($1d$)-CO	12
Pd(S)-[7_7(111) + 2_1 (110)]	CO	($\sqrt{3} \times \sqrt{3}$)R30°-CO	12
		hexagonal overlayer	12
Pt(S)-[3_3(111) + 2_1 (100)]	H_2	facets	13
Pt(S)-[5_5(111) + 2_1 (100)]	O_2	2($1d$)-O	14
		Pt(111)-(2 × 2)-O	15
		Pt(111)-($\sqrt{3} \times \sqrt{3}$)R30°-O	15
		Pt(111)-($\sqrt{79} \times \sqrt{79}$)R18°7'-O	15
		Pt(111)-(4 × $2\sqrt{3}$)R30°-O	15
		Pt(111)-3($1d$)-O	15
	CO	disordered	14
	H_2	2($1d$)-H	14, 16
		adsorbed	17
		Pt(S)-[11(111) × 2(100)]	17
Pt(S)-[4_8(111) + 2_1 (100)]	H_2	2($1d$)-H	16
Pt(S)-[7_7(111) + 2_1 (110)]	O_2	(2 × 2)-O	18–20
		not adsorbed	14

(continued)

Table 5.5.—Continued

Surface	Adsorbed gas	Surface structure	References
	CO	disordered	14
	H_2	(2×2)-H	14
		adsorbed	21
Pt(S)-[11_{11} (111) + 1_1 ($1\bar{1}1$)]	NO	(2×2)-NO	22
	NH_3	disordered	22
Re(S)-[$14(0001) \times (10\bar{1}1)$]	CO	(2×2)-CO	23
		(2×1)-C	23
Rh(S)[$1_1(111) + 2_1$ (110)]	O_2	$2(1d)$-O	24
		$\begin{pmatrix} 1 & 2 \\ 2 & 0 \end{pmatrix}$-O	24
		$\begin{pmatrix} 1 & 2 \\ 7 & -1 \end{pmatrix}$-O	24
		facets	24
	CO	$\begin{pmatrix} 1 & 2 \\ 5 & -1 \end{pmatrix}$-CO	24
		$\begin{pmatrix} 1 & 2 \\ 2 & 0 \end{pmatrix}$-CO	24
		hexagonal overlayer	24
	CO_2	$\begin{pmatrix} 1 & 2 \\ 5 & -1 \end{pmatrix}$-CO	24
		$\begin{pmatrix} 1 & 2 \\ 2 & 0 \end{pmatrix}$-CO	24
		hexagonal overlayer	24
	H_2	adsorbed	24
	NO	disordered	24
		$\begin{pmatrix} -1 & 1 \\ 3 & 0 \end{pmatrix}$	24
Rh(S)-[5 (111) + 2_1 (100)]	O_2	(2×2)-O	24
		Rh(S)-[$12(111) \times 2(100)$]-(2×2)-O	
		Rh(111)-(2×2)-O	24
	CO	$(\sqrt{3} \times \sqrt{3})$-R30°-CO	24
		(2×2)-CO	24
	CO_2	$(\sqrt{3} \times \sqrt{3})$-R30°-CO	24
		(2×2)-CO	24
	H_2	adsorbed	24
	NO	(2×2)-NO	24
W(S)-[$6(110) \times (1\bar{1}0)$]	O_2	(2×1)-O	25
W(S)-[$8(110) \times (112)$]	O_2	(2×1)-O	25
W(S)-[$10(110) \times (011)$]	O_2	(2×1)-O	26
W(S)-[$12(110) \times (1\bar{1}0)$]	O_2	(2×1)-O	25
W(S)-[$16(110) \times (112)$]	O_2	(2×1)-O	25
W(S)-[$24(110) \times (011)$]	O_2	(2×1)-O	26

1. M.A. Chesters, M. Hussain, and J. Pritchard, Surf. Sci. **35**, 161 (1973).
2. R.A. Marbrow and R.M. Lambert, Surf. Sci. **71**, 107 (1978).
3. M.A. Chesters and G.A. Somorjai, Surf. Sci. **52**, 21 (1975).
4. E. Legrand-Bonnyns and A. Ponslet, Surf. Sci. **53**, 675 (1975).
5. R.H. Roberts and J. Pritchard, Surf. Sci. **54**, 687 (1976).
6. H. Papp and J. Pritchard, Surf. Sci. **53**, 371 (1975).
7. J. Perdereau and G.E. Rhead, Surf. Sci. **24**, 555 (1971).

Table 5.5.—Continued

8. J.L. Domange and J. Oudar, Surf. Sci. **11**, 124 (1968).
9. D.I. Hagen, B.E. Nieuwenhuys, G. Rovida, and G.A. Somorjai, Surf. Sci. **57**, 632 (1976).
10. B.E. Nieuwenhuys, D.I. Hagen, G. Rovida, and G.A. Somorjai, Surf. Sci. **59**, 155 (1976).
11. R.E. Kirby, C.S. McKee, and M.W. Roberts, Surf. Sci. **55**, 725 (1976).
12. H. Conrad, G. Ertl, J. Koch, and E.E. Latta, Surf. Sci. **43**, 462 (1974).
13. A.M. Bradshaw and F.M. Hoffman, Surf. Sci. **72**, 513 (1978).
14. B. Lang, R.W. Joyner, and G.A. Somorjai, Surf. Sci. **30**, 454 (1972).
15. P. Légaré, G. Maire, B. Carière, and J. P. Deville, Surf. Sci. **68**, 348 (1977).
16. K. Baron, D.W. Blakely, and G.A. Somorjai, Surf. Sci. **41**, 45 (1974).
17. G. Maire, P. Bernhardt, P. Légaré, and G. Lindauer, Proc. 7th Int. Vac. Congr. and 3rd Int. Conf. Solid Surf., Vienna, 861 (1977).
18. K. Schwaha and E. Bechtold, Surf. Sci. **65**, 277 (1977).
19. F.P. Netzer and R.A. Wille, J. Catal. **51**, 18 (1978).
20. F.P. Netzer and R.A. Wille, Proc. 7th Int. Vac. Congr. and 3rd Int. Conf. Solid Surf., Vienna, 927 (1977).
21. K. Christmann and G. Ertl, Surf. Sci. **60**, 365 (1976).
22. J. Gland, Surf. Sci. **71**, 327 (1978).
23. M. Housley, R. Ducros, G. Piquard, and A. Cassuto, Surf. Sci. **68**, 277 (1977).
24. D.G. Castner and G.A. Somorjai, Surf. Sci. **83**, 60 (1979).
25. K. Besocke and S. Berger, Proc. 7th Int. Vac. Congr. and 3rd Int. Conf. Solid Surf., Vienna, 893 (1978).
26. T. Engel, T. von dem Hagen, and E. Bauer, Surf. Sci. **62**, 361 (1977).

Table 5.6.   Miller indices, stepped surface designations, and angles between the macroscopic surface and terrace planes for fcc crystals

Alternative microfacet notation	Miller index	Macrofacet notation	Stepped surface designation	Angle between the macroscopic surface and terrace (degrees)
	(544)	$4_8(111) + 1_1 (100)$	(S)-[9(111) × (100)]	6.2
	(755)	$5 (111) + 2_1 (100)$	(S)-[6(111) × (100)]	9.5
	(533)	$3_3(111) + 2_1 (100)$	(S)-[4(111) × (100)]	14.4
	(211)	$1_2(111) + 1_1 (100)$	(S)-[3(111) × (100)]	19.5
	(311)	$2_1(100) + 1_1 (111)$	(S)-[2(111) × (100)]	29.5
	(311)	$2_1(100) + 1_1 (111)$	(S)-[2(100) × (111)]	25.2
	(511)	$3_2(100) + 1_1 (111)$	(S)-[3(100) × (111)]	15.8
	(711)		(S)-[4(100) × (111)]	11.4
	(665)		(S)-[12(111) × (111)]	4.8
$7_7(111) + 2_1 (110)$	(997)	$8_8(111) + 1_1 (11\bar{1})$	(S)-[9(111) × (111)]	6.5
$2_4(111) + 1_1 (110)$	(332)	$\frac{5}{2}_5(111) + \frac{1}{2}_1 (11\bar{1})$	(S)-[6(111) × (111)]	10.0
$1_2(111) + 1_1 (110)$	(221)	$\frac{3}{2}_3(111) + \frac{1}{2}_1 (11\bar{1})$	(S)-[4(111) × (111)]	15.8
$1_1(111) + 2_1 (110)$	(331)	$2_2(111) + 1_1 (11\bar{1})$	(S)-[3(111) × (111)]	22.0
	(331)		(S)-[2(110) × (111)]	13.3
	(771)		(S)-[4(110) × (111)]	5.8
	(610)		(S)-[6(100) × (100)]	9.5
	(410)		(S)-[4(100) × (100)]	14.0
	(310)		(S)-[3(100) × (100)]	18.4
$2_2(100) + 1_1 (010)$	(210)	$1_1 (110) + 1_1 (100)$	(S)-[2(100) × (100)]	26.6
	(210)		(S)-[2(110) × (100)]	18.4
	(430)		(S)-[4(110) × (100)]	8.1
	(10, 8, 7)		(S)-[7(111) × (310)]	8.5

Table 5.7. Surface structures formed by adsorption of organic compounds

Surface	Adsorbed gas	Surface structure	References
Ag(110)	HCN	disordered	1
	C_2N_2	disordered	1
Au(111)	C_2H_4	not adsorbed	2
	n-heptane	not adsorbed	2
	cyclohexene	not adsorbed	2
	benzene	not adsorbed	2
	naphthalene	disordered	2
Au(S)-[5_5(111) + 2_1 (100)]	C_2H_4	not adsorbed	2
	n-heptane	not adsorbed	2
	cyclohexene	not adsorbed	2
	benzene	not adsorbed	2
	naphthalene	disordered	2
Cu(111)	C_2H_4	not adsorbed	4
	Fe-phtalocyanine	adsorbed	4
	Cu-phtalocyanine	adsorbed	4
	H-phtalocyanine	adsorbed	4
	glycine	(8×8)	5
	L-alanine	$(2\sqrt{13} \times 2\sqrt{13})R13°40'$	5
	L-tryptophan	$\begin{pmatrix} 7 & 1 \\ -2 & 4 \end{pmatrix}$	5
	D-tryptophan	$\begin{pmatrix} -8 & 1 \\ -2 & 4 \end{pmatrix}$	5
(100)	C_2H_4	(2×2)	3
	Fe-phtalocyanine	$\begin{pmatrix} 5 & -2 \\ 2 & 5 \end{pmatrix}$	4
	Cu-phtalocyanine	$\begin{pmatrix} 5 & -2 \\ 2 & 5 \end{pmatrix}$	4
	H-phtalocyanine	$\begin{pmatrix} 5 & -2 \\ 2 & 5 \end{pmatrix}$	4
	glycine	(4×2) $\begin{pmatrix} 8 & -4 \\ \frac{4}{5} & \frac{8}{15} \end{pmatrix}$	5
	L-alanine	$\begin{pmatrix} 2 & 1 \\ 2 & -1 \end{pmatrix}$	5
	L-tryptophan	(4×4)	5
	D-tryptophan	(4×4)	5
(110)	C_2H_4	ord. 1*d*	3
Cu(S)-[3_3(100) + 1_1 (010)]	CH_4	not adsorbed	6
	C_2H_4	not adsorbed	6
Cu(S)-[4_4(100) + 1_1 (010)]	CH_4	not adsorbed	6
	C_2H_4	not adsorbed	6
Fe(100)	C_2H_4	$c(2 \times 2)$-C	7
Ir(111)	C_2H_2	$(\sqrt{3} \times \sqrt{3})R30°$	8
		(9×9)-C	8

236

Table 5.7.—Continued

Surface	Adsorbed gas	Surface structure	References
	C_2H_4	$(\sqrt{3} \times \sqrt{3})R30°$	8
		(9×9)-C	8
	cyclohexane	disordered	8
		(9×9)-C	8
	benzene	(3×3)	8
		(9×9)-C	8
(100)	C_2H_2	disordered	9, 10
		$c(2 \times 2)$-C	9, 10
	C_2H_4	disordered	10
		$c(2 \times 2)$-C	10
	benzene	disordered	10
(110)	C_2H_4	disordered	11
		(1×1)-C	11
	benzene	disordered	11
		(1×1)-C	11
Ir(S)-[$5_5(111) + 2_1 (100)$]	C_2H_2	(2×2)	8
	C_2H_4	(2×2)	8
	cyclohexane	(2×2)	8
	benzene	disordered	8
Mo(100)	CH_4	$c(4 \times 4)$-C	12
		$c(2 \times 2)$-C	12
		$c(6\sqrt{2} \times 2\sqrt{2})R45°$-C	12
		(1×1)-C	12
Ni(111)	CH_4	(2×2)	13
		$(2 \times \sqrt{3})$	13
	C_2H_2	(2×2)	14, 15
	C_2H_4	(2×2)	15–17
	C_2H_6	(2×2)	13, 17
		$(2 \times \sqrt{3})$	13
		$(\sqrt{7} \times \sqrt{7})R19°$-C	16
	cyclohexane	$(2\sqrt{3} \times 2\sqrt{3})R30°$	17
	benzene	$(2\sqrt{3} \times 2\sqrt{3})R30°$	17, 18
(100)	CH_4	$c(2 \times 2)$	13
		(2×2)	13
	C_2H_2	$c(2 \times 2)$	19
		(2×2)	19
		$c(4 \times 2)$	20
		(2×2)-C	20
	C_2H_4	$c(2 \times 2)$	19, 21
		(2×2)	19
		$c(4 \times 2)$	20
		(2×2)-C	20
		$(\sqrt{7} \times \sqrt{7})R19°$-C	21
	C_2H_6	$c(2 \times 2)$	13
		(2×2)	13

(continued)

Table 5.7.—Continued

Surface	Adsorbed gas	Surface structure	References
	benzene	$c(4 \times 4)$	18
(110)	CH_4	(2×2)	13
		(4×3)	13
		(4×5)-C	13, 22
		(2×3)-C	22
	C_2H_4	(2×1)-C	23–25
		(4×5)-C	22, 23
		graphite overlayer	24
	C_2H_6	(2×2)	13
	C_5H_{12}	(4×3)	26
		(4×5)	26
Pt(111)	C_2H_2	(2×1)	27
		(2×2)	28–30
	C_2H_4	(2×2)	29–31
		(2×1)	27
		$2(1d)$-C	31
		graphite overlayer	31, 32
	n-butane	$\begin{pmatrix} 2 & 1 \\ -1 & 2 \end{pmatrix}$	33
		$\begin{pmatrix} 2 & 2 \\ -5 & 5 \end{pmatrix}$	33
		$\begin{pmatrix} 3 & -2 \\ 2 & 5 \end{pmatrix}$	33
	n-pentane	$\begin{pmatrix} 2 & 1 \\ 0 & 6 \end{pmatrix}$	33
	n-hexane	$\begin{pmatrix} 2 & 1 \\ -1 & 3 \end{pmatrix}$	33
	n-heptane	$\begin{pmatrix} 2 & 1 \\ 0 & 8 \end{pmatrix}$	33
		(2×2)	31
	n-octane	$\begin{pmatrix} 2 & 1 \\ -1 & 4 \end{pmatrix}$	33
	cyclohexane	$\begin{pmatrix} 4 & -1 \\ 1 & 5 \end{pmatrix}$	33
		disordered	31
		(2×2)	31
		graphite overlayer	31
	benzene	$\begin{pmatrix} -2 & 2 \\ 5 & 5 \end{pmatrix}$	31, 35
		$\begin{pmatrix} 4 & -2 \\ 0 & 4 \end{pmatrix}$	34
		$\begin{pmatrix} 4 & -2 \\ 0 & 5 \end{pmatrix}$	34
		$\begin{pmatrix} -2 & 2 \\ 4 & 4 \end{pmatrix}$	35
	toluene	$3(1d)$	31, 36
		(4×2)	36
		graphite overlayer	36
	naphthalene	(6×6)	35, 37
		napthalene (001)	37

Table 5.7.—Continued

Surface	Adsorbed gas	Surface structure	References
	pyridine	(2×2)	35
	m-xylene	2.6(1*d*)	36
	mesitylene	3.4(1*d*)	36
	T-butylbenzene	disordered	36
	N-butylbenzene	disordered	36
	aniline	3(1*d*)	36
	nitrobenzene	3(1*d*)	36
	cyanobenzene	3(1*d*)	36
(100)	C_2H_2	$c(2 \times 2)$	27, 37–40
	C_2H_4	$c(2 \times 2)$	27, 37–39, 41
		graphite overlayer	32, 41
		(511), (311) facets	32
	benzene	disordered	32
		2(1*d*)	35
	naphthalene	(6×6)	35
	pyridine	(1×1)	35
		$c(2 \times 2)$	35
	toluene	3(1*d*)	36
	M-xylene	3(1*d*)	36
	mesitylene	3(1*d*)	36
	T-butylbenzene	disordered	36
	N-butylbenzene	disordered	36
	aniline	disordered	36
	nitrobenzene	disordered	36
	cyanobenzene	disordered	36
	C_2N_2	(1×1)	42
(110)	HCN	$\begin{pmatrix} 1 & \frac{2}{3} \\ -1 & \frac{2}{3} \end{pmatrix}$	43
		$c(2 \times 4)$	43
		(1×1)	43
	C_2N_2	(1×1)	1, 44
Pt(S)-[3$_3$(111) + 2$_1$ (100)]	C_2H_4	disordered	31
		graphite overlayer	31
		facets	31
	cyclohexane	disordered	31
		(4×2)-C	31
	n-heptane	(4×2)	31
		(4×2)-C	31
	benzene	disordered	31
		graphite overlayer	31
		facets	31
	toluene	disordered	31
		2(1*d*)-C	31
Pt(S)-[5$_5$(111) + 2$_1$ (100)]	C_2H_4	(2×2)	31, 45

(continued)

239

Table 5.7.—Continued

Surface	Adsorbed gas	Surface structure	References
		$\begin{pmatrix} 3 & 2 \\ -2 & 5 \end{pmatrix}$-C	31
		$\begin{pmatrix} 6 & 1 \\ -1 & 7 \end{pmatrix}$-C	31
		$(\sqrt{19} \times \sqrt{19})R23.4°$-C	32
		graphite overlayer	32
	cyclohexane	$2(1d)$	31
	n-heptane	(2×2)	31
		$\begin{pmatrix} 1 & 1 \\ -1 & 2 \end{pmatrix}$	31
		(9×9)-C	31
	benzene	$3(1d)$	31
		(9×9)-C	31
	toluene	disordered	31
		(9×9)-C	31
Pt(S)-[6$_{12}$(111) + 1$_1$ (100) + 2$_2$(100)]	C_2H_4	disordered	31
		graphite overlayer	31
	cyclohexane	disordered	31
	n-heptane	disordered	31
	benzene	disordered	31
	toluene	disordered	31
		graphite overlayer	31
Pt(S)-[4$_8$(111) + 1$_1$ (100)]	C_2H_4	adsorbed	31
	cyclohexane	disordered	31
	n-heptane	(2×2)	31
		$\begin{pmatrix} 1 & 1 \\ -1 & 2 \end{pmatrix}$	31
		(5×5)-C	31
		(2×2)-C	31
		$\begin{pmatrix} 1 & 1 \\ -1 & 2 \end{pmatrix}$-C	31
		$2(1d)$-C	31
	benzene	disordered	31
		$\begin{pmatrix} 1 & 1 \\ -1 & 2 \end{pmatrix}$-C	31
		graphite overlayer	31
	toluene	$3(1d)$	31
		graphite overlayer	31
Pt(S)-7$_7$(111) + 2$_1$ (110)	C_2H_4	disordered	45
		graphite overlayer	46, 47
	N	disordered	48
Pt(S)-[5$_4$(100) + 1$_1$ (111)]	C_2H_4	graphite overlayer	32
		(511), (311), and (731) facets	32
Re(0001)	C_2H_2	disordered	32
		$(2 \times \sqrt{3})R30°$-C	32
	C_2H_4	disordered	32
		$(2 \times \sqrt{3})R30°$-C	32

Table 5.7.—Continued

Surface	Adsorbed gas	Surface structure	References
Rh(111)	C_2H_2	$c(4 \times 2)$	49
	C_2H_4	$c(4 \times 2)$	49
		(8×8)-C	49
		$(2 \times 2)R30°$-C	49
		$(\sqrt{19} \times \sqrt{19})R23.4°$-C	49
		$(2\sqrt{3} \times 2\sqrt{3})R30°$-C	49
		(12×12)-C	49
(100)	C_2H_2	$c(2 \times 2)$	49
	C_2H_4	$c(2 \times 2)$	49
		$c(2 \times 2)$-C	49
		graphite overlayer	49
(331)	C_2H_2	$\begin{pmatrix} -1 & 1 \\ 3 & 0 \end{pmatrix}$	50
	C_2H_4	$\begin{pmatrix} -1 & 1 \\ 3 & 0 \end{pmatrix}$	50
		graphite overlayer	50
Rh(S)-[5$_5$(111) + 2$_1$ (100)]	C_2H_2	disordered	50
	C_2H_4	disordered	50
		(111), (100) facets	50
Si(111)	C_2H_2	disordered	51
(311)	C_2H_2	$c(1 \times 1)$	52
		(2×1)	52
		(3×1)	52
	C_2H_4	$c(1 \times 1)$	52
		(2×1)	52
		(3×1)	52
Ta(100)	C_2H_4	adsorbed	53
W(111)	CH_4	(6×6)-C	54
(100)	CH_4	(5×1)-C	54
(110)	C_2H_4	$(15 \times 3)R\alpha$-C	54
		$(15 \times 12)R\alpha$-C	54

1. M.E. Bridge, R.A. Marbrow, and R.M. Lambert, Surf. Sci. **57**, 415 (1976).
2. M.A. Chesters and G.A. Somorjai, Surf. Sci. **52**, 21 (1975).
3. G. Ertl, Surf. Sci. **7**, 309 (1967).
4. J.C. Buchholz and G.A. Somorjai, J. Chem. Phys. **66**, 573 (1977).
5. L.L. Atanasoska, J.C. Buchholz, and G.A. Somorjai, Surf. Sci. **72**, 189 (1978).
6. J. Perdereau and G.E. Rhead, Surf. Sci. **24**, 555 (1971).
7. C. Bruckner and T. Rhodin, J. Catal. **47**, 214 (1977).
8. B.E. Nieuwenhuys, D.I. Hagen, G. Rovida, and G.A. Somorjai, Surf. Sci. **59**, 155 (1976).
9. T.N. Rhodin and G. Brodén, Surf. Sci. **60**, 466 (1976).
10. G. Brodén, T. Rhodin, and W. Capehart, Surf. Sci. **61**, 143 (1976).
11. B.E. Nieuwenhuys and G.A. Somorjai, Surf. Sci. **72**, 8 (1978).
12. C. Guillot, R. Riwan, and J. Lecante, Surf. Sci. **59**, 581 (1976).
13. G. Marie, J.R. Anderson, and B.B. Johnson, Proc. Roy. Soc. Lond. **A320**, 227 (1970).
14. D.E. Eastman and J.E. Denuth, Proc. 2nd Int. Conf. Solid Surf., 365 (1974).
15. J.E. Demuth, Surf. Sci. **69**, 365 (1977).
16. J.C. Bertolini and G. Dalmai-Imelik, Coll. Int. CNRS, Paris, July 1969.
17. G. Dalmai-Imelik, J.C. Bertolini, J. Massardier, J. Rousseau, and B. Imelik, Proc. 7th Int. Vac. Congr. and 3rd Int. Conf. Solid Surf., Vienna, 1179 (1977).

(continued)

Table 5.7.—Continued

18. J.C. Bertolini, G. Dalmai-Imelik, and J. Rousseau, Surf. Sci. **67**, 478 (1977).
19. C. Casalone, M.G. Cattania, M. Simonetta, and M. Tescari, Surf. Sci. **62**, 321 (1978).
20. K. Horn, A.M. Bradshaw, and K. Jacobi, J. Vac. Sci. Tech. **15**, 575 (1978).
21. G. Dalmai-Imelik and J.C. Bertolini, C.R. Acad. Sci. **270**, 1079 (1970).
22. F.C. Schouter, E.W. Kaleveld, and G.A. Bootsma, Surf. Sci. **63**, 460 (1977).
23. J. McCarty and R.J. Madix, J. Catal. **38**, 402 (1975).
24. J.G. McCarty and R.J. Madix, J. Catal. **48**, 422 (1977).
25. N.M. Abbas and R.J. Madix, Surf. Sci. **62**, 739 (1977).
26. G. Maire, J.R. Anderson, and B.B. Johnson, Proc. Roy. Soc. Lond. **A320**, 227 (1970).
27. A.E. Morgan and G.A. Somorjai, J. Chem. Phys. **51**, 3309 (1969).
28. L.L. Kesmodel, R.C. Baetzold, and G.A. Somorjai, Surf. Sci. **66**, 299 (1977).
29. P.C. Stair and G.A. Somorjai, J. Chem. Phys. **66**, 573 (1977).
30. W.H. Weinberg, H.A. Deaqs, and R.P. Merrill, Surf. Sci. **41**, 312 (1974).
31. K. Baron, D.W. Blakeley, and G.A. Somorjai, Surf. Sci. **41**, 45 (1974).
32. B. Lang, Surf. Sci. **53**, 317 (1975).
33. L.E. Firment and G.A. Somorjai, J. Chem. Phys. **66**, 2901 (1977).
34. P.C. Stair and G.A. Somorjai, J. Chem. Phys. **67**, 4361 (1977).
35. J.L. Gland and G.A. Somorjai, Surf. Sci. **38**, 157 (1973).
36. J.L. Gland and G.A. Somorjai, Surf. Sci. **41**, 387 (1974).
37. L.E. Firment and G.A. Somorjai, Surf. Sci. **55**, 413 (1976).
38. T.E. Fischer and S.R. Kelemen, J. Vac. Sci. Technol. **15**, 607 (1978).
39. T.E. Fischer and S.R. Kelemen, Surf. Sci. **69**, 485 (1977).
40. T.E. Fischer, S.R. Kelemen, and H.P. Bonzel, Surf. Sci. **64**, 85 (1977).
41. B. Lang, P. Légaré, and G. Maire, Surf. Sci. **47**, 89 (1975).
42. F.P. Netzer, Surf. Sci. **52**, 709 (1975).
43. M.E. Bridge and R.M. Lambert, J. Catal. **46**, 143 (1977).
44. M.E. Bridge and R.M. Lambert, Surf. Sci. **63**, 315 (1977).
45. B. Lang, R.W. Joyner, and G.A. Somorjai, Surf. Sci. **30**, 454 (1972).
46. F.P. Netzer and R.A. Wille, J. Catal. **51**, 18 (1978).
47. F.P. Netzer and R.A. Wille, Proc. 7th Int. Vac. Congr. and 3rd Int. Conf. Solid Surf., Vienna, 927 (1977).
48. K. Schwaka and E. Bechtold, Surf. Sci. **66**, 383 (1977).
49. D.G. Castner, B.A. Sexton, and G.A. Somorjai, Surf. Sci. **71**, 519 (1978).
50. D.G. Castner and G.A. Somorjai, Surf. Sci. **83**, 60 (1979).
51. Y.W.Chung, W. Siekhaus, and G.A. Somorjai, Surf. Sci. **58**, 341 (1976).
52. R. Heckingbottom and P.R. Wood, Surf. Sci. **23**, 437 (1970).
53. M.A. Chesters, B.J. Hopkins, and M.R. Leggett, Surf. Sci. **43**, 1 (1974).
54. M. Boudard and D.F. Ollis, in *The Structure and Chemistry of Solid Surfaces*, ed. G.A. Somorjai, Wiley, New York, 1969.

tures formed with organic adsorbates are brought together in Table 5.7. Most of the substrates in Tables 5.2 to 5.4 are low-index faces, and the gases absorbed are, for the most part, small inorganic molecules, such as H_2, O_2, N_2, CO, and NO. Inspection of the tables permits one to propose two general rules[11] that are usually obeyed during the adsorption of these small molecules: (1) the observed surface structures have the same rotational symmetry as the substrate; and (2) the unit cell of the surface structure is the smallest allowed by the molecular dimensions and adsorbate–adsorbate interactions.

The frequent occurrence of ordered fractional-coverage adsorption indicates that adsorbate–adsorbate interactions at close range ($\lesssim 5$ Å) are often repulsive. Island formation can occur simultaneously, showing that at larger separations these interactions can become attractive.

Also, adsorbates of any type have a general tendency to form identical superstructures on different substrates of a given symmetry, showing the effect of adsorbate–adsorbate interactions. For example, oxygen forms a (2 × 2) superstructure on the hexagonal faces of Ag, Cu, Ir, Nb, Ni, Pd, Pt, Re, Rh, and Ru. This is most obvious for the physisorption of rare gases, where the adsorbate–substrate interactions parallel to the surface are so small that a hexagonal close-packed layer is formed even on substrates of different surface symmetry and greater roughness, such as with Xe on Cu(100), Cu(110), Cu(211), and Cu(311). This hexagonal overlayer has been analyzed for Xe on Ag(111) and found to correspond to the (111) plane of the fcc inert-gas solid.

In the last few years, LEED studies of high-Miller-index or stepped surfaces have become more frequent. Almost all of these studies have been on fcc metals, where the atomic structure of these surfaces consists of periodic arrays of terraces and steps. A nomenclature that is more descriptive of the actual surface configuration has been developed for these surfaces, as described in Chapter 4. In Figure 4.14 the locations of these high-Miller-index surfaces are shown on the unit stereographic triangle. As can be seen from that figure, all the stepped surfaces that have low-Miller-index-type steps lie on the (100), (110), and (111) zone lines. For surfaces that lie inside the unit stereographic triangle, the steps themselves have steps, and this type of surface is classified as a kinked surface. The only kinked surface for which surface structures have been reported is the Pt(10,8,7) surface. The real space drawings and LEED patterns of the platinum (111), (755), and (10, 8, 7) surfaces are shown in Figure 5.8.

In calculating the stepped-surface designations, it was assumed that the surfaces were stable in a monatomic step configuration, which is generally the case for clean surfaces. This can readily be verified by LEED. In LEED patterns of stepped surfaces, the step periodicity is superimposed on the terrace periodicity, resulting in the splitting of the terrace diffraction spots into doublets or triplets at certain beam voltages. The direction of the splitting is perpendicular to the step edge, and the magnitude of the splitting is inversely proportional to the terrace width, so that the terrace width can be obtained by measuring the splitting observed in the LEED pattern. By combining the terrace width and step height with the angle between the terrace and step planes, the macroscopic surface plane can be determined.

The stability of stepped surfaces is an important consideration in LEED studies. Although these surfaces have higher surface free energies than do the low-index faces, most of the clean stepped surfaces are stable in a single-step-height configuration from room temperature to

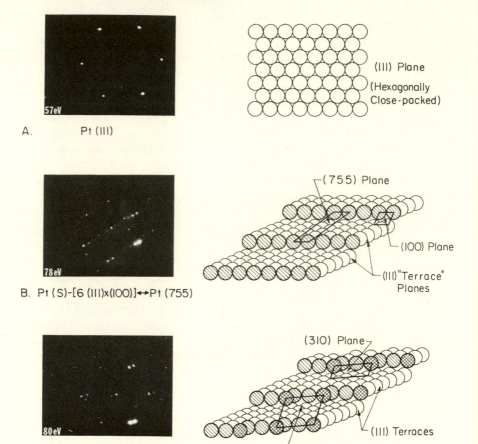

A.　　　Pt (III)

B.　Pt (S)-[6 (III)x(I00)] ↔ Pt (755)

C. Pt (S)-[7 (III)x(3I0)] ↔ Pt (I0,8,7)

Figure 5.8. Electron diffraction patterns and atomic structures of a step-free (A), a stepped (B), and a kinked (C) surface of platinum.

near the melting point of the metals. When gases are adsorbed on these surfaces, however, their stability can noticeably change. Some surfaces reconstruct, forming multiple-height steps and large terraces. Other high-index surfaces form large low-index facets, while some retain the single-step-height configuration.

The surface structures observed for gas adsorption on stepped surfaces are listed in Table 5.5. In this table the stepped surfaces are denoted by either their Miller index label or their stepped-surface designation, depending on which system was used by the original author. By using Table 5.6, one may convert back and forth between these two systems. It is interesting to compare the surface structures formed on stepped surfaces with those formed on the low-index faces given in Tables 5.2 to 5.4. For stepped surfaces with fairly large terrace widths (~six to eight atoms or larger) the surface structure that forms on the terrace is generally the same as the one that forms on the low-index face. The surface structures on the low-index surfaces tend to be better ordered than those on the stepped surfaces. An example of this is the existence of several one-dimensional structures on stepped surfaces. The one-dimensional structures cause streaks to occur in the LEED patterns and are denoted as $n(1d)$ structures in the tables, where n is the number of streaks between rows of the substrate diffraction spots. Also, the adsorption of gases may cause faceting of the substrate due to the high surface free energy of stepped surfaces.

Ordered Organic Monolayers

The adsorption characteristics of organic molecules on solid surfaces are important in several areas of surface science. The nature of the chemical bonds between the substrate and the adsorbate, and the ordering and orientation of the adsorbed organic molecules, play important roles in adhesion, lubrication, and hydrocarbon catalysis. Several studies have been undertaken to determine the molecular structure, ordering, and interaction of monolayers for different groups of organic compounds under well-characterized conditions on low-Miller-index metal crystal surfaces. However, the structures of only two of the small organic molecules, acetylene (C_2H_2) and ethylene (C_2H_4), adsorbed on the (111) crystal face of platinum have been determined using a combination of surface crystallography with the diffraction beam intensities measured by LEED, high-resolution electron energy loss spectroscopy (HREELS), and ultraviolet photoelectron spectroscopy (UPS). These techniques were discussed in Chapter 2.

Over 50 other organic monolayers have been studied by LEED and a combination of other techniques when adsorbed on single-crystal surfaces. Although structure analysis has not been carried out for these systems, their ordering characteristics and the size and orientation of their unit cell have been determined. By studying the systematic variation of their shape and bonding characteristics, correlations can be made between these properties and their interactions with the metal surfaces. Analysis of the changes of surface structure with the shape or size of these molecules often permits unambiguous determination of their location and orientation on the surface, even in the absence of surface crystallography.

Next, we review the surface structures of monolayers of various homologues of organic compounds, the paraffins, the phthalocyanines, a few aromatic systems, and amino acids that have been determined during recent investigations.

Normal Paraffin Monolayers on Platinum and Silver. Straight-chain saturated hydrocarbon molecules from propane (C_3H_8) to octane (C_8H_{18}) were deposited from the vapor phase on platinum and silver (111) crystal surfaces in the temperature range 100 to 200 K. The ordered monolayer was produced first and then, with decreasing temperature, a thick crystalline film was condensed and the surface structures of these organic crystals studied by LEED.[12]

At the highest temperature at which the organic molecule condenses, T_1, at a given vapor flux, a surface structure is formed that exhibits only one-dimensional order. As the temperature is lowered, these monolayer structures become more ordered and form a two-dimensional ordered surface structure at T_2 (Fig. 5.3). Upon further lowering of the temperature to T_3, the rate of condensation of the organic vapor on the surface becomes greater than the rate of evaporation. At this temperature or below, the growth of organic multilayers commences. In Figure 5.3 the phase transition temperatures for the various adsorbed paraffins are plotted as a function of chain length for adsorption and growth on the Pt(111) crystal face. The transition temperatures fall on a smooth curve and increase with increasing chain length. Similar results have been obtained for deposition on the Ag(111) surface.

The paraffins adsorb with their chain axis parallel to the platinum substrate. Thus their surface unit cell increases smoothly with increasing chain length, as shown in Figure 5.9. The *n*-butane molecules, unlike the larger molecules, form several monolayer surface structures as the experimental conditions are varied. It appears that the smaller the paraffin, the more densely packed it is on the surface. Evidently, as the packing becomes too dense for *n*-butane in one surface structure, it forms a different one.

The monolayer adsorption characteristics of the C_4 to C_8 paraffins are very similar on Ag(111) to that on Pt(111). The monolayers are less strongly held on the silver surface, as manifested by the lower temperatures necessary to produce ordered surface structures on silver.

Multilayers condensed upon the ordered monolayers maintain the

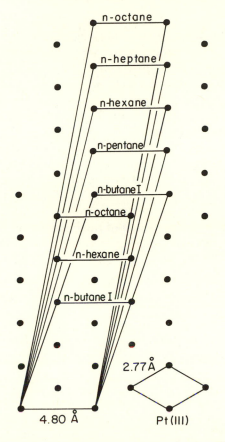

Figure 5.9. Observed surface unit cells for *n*-paraffins on Pt(111).

same orientation and packing as are found in the monolayers. Thus the monolayer structure determines the growth orientation and the surface structure of the growing organic crystal. This phenomenon is called pseudomorphism and, as a result, the surface structures of the growing organic crystals do not correspond to planes in the reported bulk crystal structures. The exception appears to be *n*-octane on the Ag(111) surface, deposited with a (10$\bar{1}$) orientation of its bulk crystal structure.

The saturated hydrocarbons are very susceptible to electron beam damage, both in the monolayer and multilayer forms. Whereas aromatic hydrocarbons and other conjugated systems exhibit minimal or no beam-damage effects during the times necessary to carry out the LEED experiments, the ordered structures of paraffins disappear after ~5 sec of electron beam exposure as a result of desorption or partial dissociation of the organic adsorbates.

Benzene, Cyclohexadiene, Cyclohexene, Cyclohexane, and Naphthalene Monolayers on Platinum and Silver. Benzene adsorbs on the platinum (111) crystal face into a well-ordered metastable $\left(\begin{smallmatrix} 4 & -2 \\ 0 & 4 \end{smallmatrix}\right)$ structure at 300 K under ultrahigh-vacuum conditions.[13] This initial structure transforms into a stable $\left(\begin{smallmatrix} 4 & -2 \\ 0 & 5 \end{smallmatrix}\right)$ structure at a rate that is sensitive to both the sample temperature and the flux of benzene vapor to the surface. I–V curves were taken from the diffraction beams of both surface structures, and these indicate very little change in the carbon–platinum layer spacing during the structural transformation. However, the work-function changes with respect to the clean (111) platinum surface are -1.4 eV for the $\left(\begin{smallmatrix} 4 & -2 \\ 0 & 4 \end{smallmatrix}\right)$ and -0.7 eV for the $\left(\begin{smallmatrix} 4 & -2 \\ 0 & 5 \end{smallmatrix}\right)$ surface structures, a very large variation. Both the surface unit-cell size and the calibrated Auger determination of the surface carbon content indicate that the adsorbate structure is likely to have some of the benzene molecules inclined at an angle to the surface. In the absence of surface-structure analysis, the precise location of the benzene molecules with respect to each other or relative to the surface platinum atoms cannot be identified. However, a complete set of I–V curves is available and should constitute a sufficient data base for structure analysis.

Benzene forms a rotationally disordered structure on the reconstructed (100) platinum surface. However, the work-function changes with increasing surface coverage are similar to those of benzene on the (111) crystal face.

The adsorption of cyclohexadiene on the Pt(111) surface produces the same two surface structures that were found during the adsorption of benzene on this crystal face.[13] Thus on the platinum surface, this molecule readily dehydrogenates to benzene at 300 K.

Cyclohexene adsorbed[13] on the Pt(111) surface produces a $\left(\begin{smallmatrix} 2 & -2 \\ 4 & 2 \end{smallmatrix}\right)$ surface structure at 300 K. The work-function change upon adsorption is -1.7 eV. As the temperature is increased to 450 K, a new $\left(\begin{smallmatrix} 2 & 0 \\ 0 & 2 \end{smallmatrix}\right)$ surface structure appears.

Cyclohexane[14] forms a (9×9) surface structure on the Ag(111) crystal face and a $\left(\begin{smallmatrix} 4 & -1 \\ 1 & 5 \end{smallmatrix}\right)$ surface structure on the Pt(111) crystal face at approximately 200 K. This latter surface structure corresponds to the (001) surface orientation of the monoclinic bulk crystal structure of the molecule. On heating of the platinum crystal face to 450 K, a $\left(\begin{smallmatrix} 2 & 0 \\ 0 & 2 \end{smallmatrix}\right)$ surface structure forms that is identical to the surface structure formed by cyclohexene monolayers at the same temperature. It appears that cyclohexane dehydrogenates at elevated temperatures on platinum to form the same species as that of cyclohexene.

Naphthalene forms[13] a glassy, poorly ordered monolayer on Pt(111) at

300 K. However, upon heating to 450 K, the monolayer orders to form a (6 × 6) surface structure (see Figure 5.4). Adsorbed naphthalene also forms a disordered layer on the Ag(111) crystal face at 300 K. However, below 200 K, an ordered structure appears with a unit cell $\left(\begin{smallmatrix} 2.8 & 0.8 \\ -2.0 & 3.8 \end{smallmatrix}\right)$, and sometimes another, less stable monolayer structure is also detectable.

It is interesting that benzene and naphthalene form monolayer surface structures on the Pt(111) crystal face at 300 K and higher temperatures, whereas monolayer surface structures form only at low temperatures (~200 K) on the Ag(111) crystal face.[14] While these aromatic molecules are held by strong chemical bonds to the platinum, their heats of adsorption must not be greater than the heats of sublimation (10.7 and 17.3 kcal/mol for benzene and naphthalene, respectively) on the silver crystal plane. Thus adsorbate–adsorbate and adsorbate–substrate interaction are of the same magnitude for silver.

Other Organic Adsorbates on Platinum. The adsorption and ordering characteristics of a large group of organic compounds have been studied on the platinum (100) and (111) single-crystal surfaces.[13] Low-energy electron diffraction has been used to determine surface structures. Work-function-change measurements have been made to determine the charge redistribution that occurs on adsorption. The molecules that have been studied are aniline, benzene, biphenyl, *n*-butylbenzene, *t*-butylbenzene, cyanobenzene, cyclopentane, mesitylene, 2-methylnaphthalene, nitrobenzene, propylene, pyridine, toluene, and *m*-xylene. All the molecules studied adsorb on both the Pt(111) and Pt(100)-(5 × 1) surfaces and are electron donors to the metal surface. The adsorbed layers are more ordered on the hexagonally symmetric Pt(111) surface than on the square-symmetric Pt(100) surface. Unsaturated molecules generally adsorb on these crystal faces of platinum by forming π-bonds with the metal surfaces, indicated by work-function-change studies as well as by the large heats of adsorption of these molecules when compared to the heats of adsorption of saturated hydrocarbons.

Phthalocyanine Monolayers and Films on Copper. Monolayer structures and epitaxial growth of vapor-deposited crystalline phthalocyanine films on single-crystal copper substrates were studied using low-energy electron diffraction.[15] Ordered monolayers of three different phthalocyanines—copper, iron, and metal-free—were seen on two different faces of copper, the (111) and (100). The monolayer structures formed were different on the two crystal faces, and the several phthalocyanines yield nonidentical monolayer structures.

Ordered multilayer deposits were grown on both the Cu(111) and Cu(100) substrates. Electron beam damage to the phthalocyanine molecules was not observed. Space-charge effects due to electron bombardment were not apparent below an incident electron energy of 25 eV.

The surface structures observed for the multilayer deposits of the phthalocyanines on both substrate faces, Cu(111) and Cu(100), were not those of any plane in the bulk crystal structure of the phthalocyanines.

The monolayer surface structures observed for the various phthalocyanines on the two copper substrates are summarized in Table 5.7. In all cases the size of the surface unit mesh is consistent with a surface-structure unit cell containing a single planar phthalocyanine molecule oriented parallel to the substrate. The bonding to the copper substrate is largely through the phthalocyanine ligand rather than through the central metal atom, since the metal-free phthalocyanines are found, from thermal desorption experiments, to be bound as strongly to the Cu(100) and Cu(111) surfaces as the Cu- and Fe-phthalocyanines. Although the central metal atom in the phthalocyanines has no effect on the surface structures formed on Cu(100), it does play a major role in determining the surface structure on Cu(111). Not only are the monolayer structures different for the three phthalocyanines, but the epitaxy of the multilayer deposit indicates a fundamental difference in the interaction of the metal and metal-free phthalocyanine with the Cu(111) surface. The metal-free phthalocyanine film grows, in the multilayer deposits, as a number of individual domains or crystallites, each yielding its own diffraction beams, including its own specular reflection, since the surface planes are not parallel to one another. The metal-free phthalocyanine film exhibits sixfold symmetry in crystallite orientation. The Cu(111) surface, although sixfold symmetric in the atomic positions in the top layer, is only threefold symmetric when the positions of the second- and third-layer atoms are included (... ABC ... stacking of an fcc crystal). Thus the metal-free phthalocyanine interacts with the substrate surface either through nonlocalized interactions such as van der Waals forces or bonds with only electrons in the copper which exhibit sixfold symmetry (the metallic s electrons of the top copper layer). Copper- and iron-phthalocyanines exhibit threefold symmetry in crystallite growth. The addition of a metal atom in the phthalocyanine reduces the apparent symmetry of the substrate. The central metal atom is thus involved in bonding to the second-layer copper atoms or to threefold symmetric electron orbitals (for example, d orbitals) of the surface copper atoms.

Deposition of Cu-phthalocyanine on a Pt(111) surface resulted in only poor ordering of monolayer structures and no ordering of multilayer structures. This demonstrates the importance of the details of the adsorbate–substrate interaction even for very large adsorbates that overlap tens of surface atoms. The absence of an ordered multilayer structure on this substrate indicates the role of an initially ordered monolayer in controlling epitaxial growth.

Ordered multilayer deposits of phthalocyanine molecules could be observed by low-energy electron diffraction with no apparent electron-beam-induced chemical effects. This appears to be consistent with the general trend for molecules with highly conjugated electron systems to be more stable under electron bombardment than are other organic molecules.

The surface structures observed for the multilayer phthalocyanine films are summarized in Table 5.7. These structures do not correspond to planes of either of the previously reported crystal structures of vapor-deposited phthalocyanine films, because the unit mesh constants reported in this work are considerably smaller than those previously reported. The unit-cell dimensions correspond much more closely to a unit cell containing one molecule rather than, for example, the four molecules per cell reported for α-phthalocyanine.

Amino Acid Monolayers and Films on Copper. Monolayer structures and ordered multilayer films of several amino acids on single-crystal substrates were studied using low-energy electron diffraction.[16] At monolayer coverage, ordered layers of glycine, alanine, and D- and L-tryptophan were observed on both Cu(100) and Cu(111). For both glycine and alanine on Cu(111), the unit-cell size suggests several molecules per unit cell, considering the dimensions of the nearly close-packed ac plane in bulk glycine. The alanine unit cell on Cu(100) is consistent with a single molecule per unit cell. The unit cell for glycine on Cu(100) requires at least two molecules per unit cell.

The monolayer structures for tryptophan are consistent with one and two molecules per unit cell, respectively, for Cu(100) and Cu(111). The structures observed for D- and L-tryptophan are related by mirror inversion, which is consistent with the symmetry relationship between the two molecules. A mixture of the optical isomers, DL-tryptophan, does not form an ordered monolayer; thus there is no segregation or cooperative interaction between the various isomers.

In addition to forming ordered monolayer structures, ordered multilayer films several hundred angstroms in thickness were also grown for tryptophan. Ordered multilayers could be grown for DL-tryptophan even though the DL-tryptophan monolayer was disordered.

Electron-beam-damage effects followed the general rule that molecular groups in intimate contact with the metal substrate and aromatic groups appear to be relatively stable. Thus in the monolayer, alanine, with a methyl group probably sticking out from the surface, was the only molecule found to be unstable. In multilayer films, only tryptophan, with the aromatic indole group to stabilize the molecule, was found to yield multilayers stable under electron beam radiation.

Surface Crystallography of Ordered Monolayers of Atoms

By the use of LEED and, lately, ion scattering techniques, primarily, the location of many atomic adsorbates, their bond distances, and bond angles from their nearest neighbor atoms have been determined. The substrates utilized in these investigations were low-Miller-index surfaces of fcc, hcp, and bcc metals, in most cases, and a few low-Miller-index surfaces of semiconductors that crystallize in the diamond, zincblende, and wurtzite structures and could be cleaned and ordered with good reproducibility.

Since the substrate on which adsorbates are deposited greatly influences the behavior of those adsorbates, it is important to first examine the substrates themselves. We must distinguish between the clean surface and the same surface when covered with adsorbates, because adsorbates are capable of modifying the geometric (and electronic) structure of the substrate. To allow a convenient comparison, Table 5.8 combines the structures for both clean and adsorbate-covered surfaces, as far as they have been determined with a reasonable degree of precision and reliability by the various surface crystallographic techniques mentioned in Chapter 2.

Effect of the Adsorbate on the Clean Substrate Surface Structure

In the clean surfaces the bond lengths between the atoms in the first and second layer are somewhat contracted with respect to the bond lengths in the bulk (see Chapter 4). The more open the surface (lower coordination number) the larger is this bond contraction. Adsorbed atoms have the effect of removing this bond contraction and restoring the bulk bond length between the surface and the second layer in most

Table 5.8. The structures of clean surfaces and adsorbed atoms obtained by surface crystallography (classified by structural type of substrate surface)

Surface type and surface	Topmost substrate bond length [and layer spacing] relaxation (%)	Adsorption site	Adsorption layer spacing (Å)	Adsorption bond length (Å)	Equivalent bond lengths for non-surfaces	Method	References	Comments
fcc(111)								
Ag(111)	0 [0]					LEED	2	
						LEED and averaging	3	
Ag(111) + $p(1 \times 1)$-Au	0 [0]	cABC...	2.36	2.88	2.88	LEED	2b	
Ag(111) + $(\sqrt{3} \times \sqrt{3})R30°$-I	0 [0]	cABC...	2.25	2.80	2.54–2.85	LEED	4	good results using LEED geometry; cABC... better than bABC...
	0 [0]	cABC...			2.54–2.85	LEED	5	
Ag(111) + incommensurate Xe	0 [0]	b or cABC...	2.34	2.87 ± 0.02	2.54–2.85	SEXAFS	6	
	0 [0]	variable	3.5	variable	3.53	LEED and Fourier transf.	7	
		variable	3.55	variable	3.53	LEED	8	
Al(111)	−1 to +1.5 [−3 to +5]					LEED	1	
	−1 [−3]							
Au(111)	0 [0]					LEED and Fourier transf.	1e	reconstructions often observed
						LEED	2b	
Co(111)	0 [0]					LEED	9	high-temperature phase
Cu(111)	−1.3 to 0 [−4 to 0]					LEED	10	
Ir(111)	−0.8 ± 1.6 [−2.5 ± 5]					LEED and Fourier transf.	11	
Ni(111)	0 [0]					LEED	12	
						LEED and averaging	13	
Ni(111) + $p(2 \times 2)$-2H	0 [0]	b and cABC...	1.15 ± 0.10	1.84 ± 0.06	1.47–1.87	LEED	14	graphitic overlayer geometry
		a, b, and cABC...	1.42–2.02	2.02–2.48		model calc.	15	
Ni(111) + $p(2 \times 2)$-O	0 [0]	b or cABC...	1.20 ± 0.10	1.88 ± 0.06	1.84–2.18	LEED	16	
Ni(111) + $p(2 \times 2)$-S	0 [0]	cABC...	1.40 ± 0.10	2.02 ± 0.06	2.10–2.23	LEED	17	
	0 [0]	cABC...	1.57	2.13	2.10–2.23	LEED	18	

(continued)

Table 5.8.—Continued

Surface type and surface	Topmost substrate bond length [and layer spacing] relaxation (%)	Adsorption site	Adsorption layer spacing (Å)	Adsorption bond length (Å)	Equivalent bond lengths for non-surfaces	Method	References	Comments
Pt(111)	0 [0]					LEED	19	
Rh(111)	−0.3 ± 0.6 [−1 ± 2]					MEIS and HEIS	20	
						LEED	21	
hcp(0001)								
Be(0001)	0 [0]					LEED	22	
Cd(0001)	0 [0]					LEED	23	
Co(0001)	0 [0]					LEED	9	low-temperature phase
Ti(0001)	−0.5 [−2]					LEED	24	
Ti(0001) + p(1 × 1)-Cd	0 [0]	cABA...	2.57 ± 0.05	3.08 ± 0.03	3.01	LEED	23, 25	
Ti(0001) + p(1 × 1)-N	+1 [5]	AcBAB...	1.22 ± 0.05	2.095 ± 0.03	2.12	LEED	26	underlayer in octahedral holes
Zn(0001)	−0.5 [−2]					LEED and averaging	27	
bcc(110)								
Fe(110)	0 [0]					LEED	28	
Na(110)	0 [0]					LEED	29	
W(110)	0 [0]					LEED and averaging	30	
	0 [0]					LEED	31	
W(110) + p(2 × 1)-O	0 [0]	3-fold	1.25 ± 0.10	2.08 ± 0.07	1.75–2.12	LEED	32	central bridge site not excluded
fcc(100)								
Ag(100)	0 [0]					LEED	36	
Ag(100) + c(2 × 2)-Cl	0 [0]	4-fold	1.72 ± 0.10	2.68 ± 0.06	2.36–2.77	LEED	37	
Ag(100) + c(2 × 2)-Se	0 [0]	4-fold	1.91 ± 0.10	2.80 ± 0.07	2.46–2.86	LEED	38	
Al(100)	0 [0]					LEED	1b, 1c, 33	
	0 [0]					MEED	34	
	0 [0]					LEED and Fourier transf.		
Al(100) + c(2 × 2)-Na	0 [0]	4-fold	2.06 ± 0.11	2.88 ± 0.08	2.82–3.00	LEED	35	
Au(100)	0 [0]					LEED	39	metastable surface
Co(100)	−1.5 [−4]					LEED	40	
Co(100) + c(2 × 2)-O	0 [0]	4-fold	0.80	1.94	2.12	LEED	41	

Surface	Coverage	Site	Spacing	Bond length	Range	Method	Ref.	Comments
Cu(100)	0 [0]					LEED	10a, 42, 43	
	0 [0]					LEED and averaging	44	
	0 [0]					LEED and Fourier transf.	1e	
Cu(100) + c(2 × 2)-N	0 [0]	4-fold	1.45 ± 0.04	2.32 ± 0.03	1.993–2.11	LEED and averaging	45	layer spacing of 0.90 is second choice
	0 [0]	4-fold	0.90 ± 0.10	2.02 ± 0.05	1.993–2.11	LEED	46	poor agreement between theory and experiment
Cu(100) + p(2 × 2)-Te	0 [0]	4-fold	1.70 ± 0.15	2.48 ± 0.10	2.51–2.76	LEED	47	
Ni(100)	0 [0]					LEED	12a, 12c, 48	
	0 [0]					LEED and Fourier transf.	1e	
Ni(100) + p(2 × 2)-C	+4 [+8.5]					LEED	49	C position unknown; expansion confirmed by HEIS
Ni(100) + c(2 × 2)-Na	0 [0]	4-fold	2.23 ± 0.11	2.84 ± 0.08	2.80–3.10	LEED	50, 51	
Ni(100) + c(2 × 2)-O	0 [0]	4-fold	0.90 ± 0.10	1.98 ± 0.05	1.84–2.18	LEED	52	
	0 [0]	4-fold	0.90 ± 0.20	1.98 ± 0.10	1.84–2.18	LEIS	53	0 embedded in top Ni layer
						INS	54	this geometry good in ARUPS theory vs. experiment
Ni(100) + p(2 × 2)-O	0 [0]	4-fold	0.90 ± 0.10	1.98 ± 0.05	1.84–2.18	LEED	55	
	0 [0]	4-fold	0.96	2.01	1.84–2.18		56	
	0 [0]	bridge	1.20 ± 0.10	2.13 ± 0.05	1.84–2.18		18	4-fold site expected with larger cluster
	0 [0]	4-fold	0.75	1.91	1.84–2.18		57	
Ni(100) + c(2 × 2)-S	0 [0]	4-fold	1.30 ± 0.10	2.19 ± 0.06	2.10–2.23	LEED	52, 58	this geometry good in ARUPS theory vs. experiment
		4-fold	1.30	2.19	2.10–2.23	LEIS	59	
		4-fold				INS	60	substrate hollow distorted (diamond shape)
Ni(100) + p(2 × 2)-S	0 [0]	4-fold	1.30 ± 0.10	2.19 ± 0.06	2.10–2.23	LEED	55	
	0 [0]	4-fold	1.33	2.21	2.10–2.23		56	
	0 [0]	4-fold	1.31	2.20	2.10–2.23		18	
Ni(100) + c(2 × 2)-Se	0 [0]	4-fold	1.45 ± 0.10	2.28 ± 0.06	2.31–2.53	LEED	52	
	0 [0]	4-fold	1.47	2.29	2.31–2.53		18	
Ni(100) + p(2 × 2)-Se	0 [0]	4-fold	1.55 ± 0.10	2.34 ± 0.07	2.31–2.53	LEED	55	
Ni(100) + c(2 × 2)-Te	0 [0]	4-fold	1.90 ± 0.10	2.58 ± 0.07	2.54–2.85	LEED	52, 61	
Ni(100) + p(2 × 2)-Te	0 [0]	4-fold	1.80 ± 0.10	2.52 ± 0.07	2.54–2.85	LEED	55	
Pt(100)	0 [0]					LEED	62	metastable surface
Rh(100)	0 [0]					LEED	63	
Xe(100)	0 [0]					LEED	64	

(continued)

Table 5.8.—Continued

Surface type and surface	Topmost substrate bond length [and layer spacing] relaxation (%)	Adsorption site	Adsorption layer spacing (Å)	Adsorption bond length (Å)	Equivalent bond lengths for non-surfaces	Method	References	Comments
bcc(100)								
Fe(100)	−0.7 to −1.5					LEED	65	less contraction when less clean
Fe(100) + p(1 × 1)-O	+3 [+7.5]	4-fold	0.48	2.02 and 2.03	2.09–2.15	LEED	66	O closest to second layer Fe atoms
Fe(100) + c(2 × 2)-S	0 [0]	4-fold	0.48	2.02 and 2.03	2.09–2.15	LEED	67	
Mo(100)	−4 [−12]	4-fold	1.15 ± 0.05	2.33	1.99–2.44	LEED	68	
						LEED	69a–c	[−13%] found by surface resonance analysis, see ref. 69d
Mo(100) + c(2 × 2)-N	0 [0]	4-fold	1.02 ± 0.10	2.45 ± 0.05	2.11–2.33	LEED	70	
Mo(100) + p(2 × 1)-O	0 [0]	4-fold	0.70	2.28 and 2.33	1.66–2.07	LEED	71	
Mo(100) + p(1 × 1)-S	0 [0]	4-fold	1.16 ± 0.10	2.51 ± 0.05	2.53	LEED	72	
W(100)	−2 to −4 [−4.4 to −11]					LEED	31, 73	
	0 to −2.5 [0 to −6]					HEIS	74	
W(100)c(2 × 2)	[−6]					LEED	75	low-temperature reconstruction: zigzag rows of touching top-layer W atoms
W(100) + c(2 × 2)-H	[−6]					LEED	75a, 76	probably substrate identical to W(100)c(2 × 2)
W(100) + p(1 × 1)-2H	0 [0]					LEED	76	H probably in bridge sites
fcc(110)								
Ag(110)	−2 to −3 [−6 to −10]					LEED	36b, 79	
						LEED and Fourier transf.	77	
Ag(110) + p(2 × 1)-O						LEED	80	probably long-bridge site
						LEIS	81	probably long-bridge site
Al(110)	−3 to −4.5 [−9 to −15]					LEED	1a–d, 33	
						LEED and Fourier transf.	77	
						MEED	34	
						model calc.	78	
Cu(110)	−3 to −4 [−10 to −12]					LEED	82	second-layer spacing may be also contracted

Surface type and surface	Surface structure, including bond length [and layer spacing] relaxations, relative to bulk	Method	References	Comments
Ir(110)(2 × 1)	−2 [−10]	LEED	83	missing row reconstruction; second-layer atoms pressed sideways somewhat
Ir(110)(1 × 1)	−2.5 [−7.5]	LEED	84	one-quarter monolayer of randomly positioned O prevents reconstruction
Ir(110) + c(2 × 2)-O	short bridge 1.37 ± 0.05 1.93 ± 0.05 0 [0]	LEED	85	one-quarter monolayer of randomly positioned O present
Ni(110)	−1.5 [5] −1.2 [−4] −1.2 [4] +0.3 [+1]	LEED MEIS MEIS	12a, 48 77 86 86	one-third monolayer of randomly positioned O present
Ni(110) + p(2 × 1)-H	−2.5 [−8]	LEED	87	(2 × 1) structure probable due to substrate reconstruction (row pairing); H position unknown
Ni(110) + p(2 × 1)-O	short bridge 1.46 ± 0.05 1.92 ± 0.04 1.84–2.18 0 [0] short bridge 1.84–2.18	LEED 88 LEIS	LEED 89, 90 17	
Ni(110) + c(2 × 2)-S	center 0.93 ± 0.10 2.17 ± 0.10 2.10–2.23 center 0.87 ± 0.03 2.11 ± 0.03 2.10–2.23 +6 ± 3 [+1.5 to 0.75] long bridge 1.04 2.04 2.10–2.23 0 [0]	LEED MEIS	91	S closest to second layer Ni atoms
Rh(110)	−0.9 [−2.7]	LEED	92	
bcc(111)				
Fe(111)	−1.5 [−15]	LEED	93	
fcc(311)				
Cu(311)	−1 [−5]	LEED	94	
			95	

Surface type and surface	Surface structure, including bond length [and layer spacing] relaxations, relative to bulk	Method	References	Comments
Other surfaces				
CoO(111)	O-terminated polar face of NaCl structure; top-layer contraction −5% [−15%]	LEED	96	
GaAs(100)	bulk structure with As termination; no relaxation	LEED	97	As termination forced by molecular beam epitaxy
GaAs(110)	zincblende structure with top Ga and As atoms rotated, respectively, into and out of surface (keeping about constant mutual bond length, rotated by projected angle of 27°); Ga and As back bonds contracted by −2.5% and −3.6%, respectively	LEED	98	
		model calc.	99	agree qualitatively with LEED result

(continued)

Table 5.8.—Continued

Surface type and surface	Surface structure, including bond length [and layer spacing] relaxations, relative to bulk	Method	References	Comments
GaAs(110) + (1 × 1)-AS	substrate has unrelaxed bulk structure: As bonded to surface Ga as in bulk (possible small bond-angle change)	LEED	97a	
MgO(100)	unrelaxed bulk NaCl structure within ±5%	LEED	100	
MoS₂(0001)	layer compound cleaved between two three-plane layers; top contraction by −1.6% [−4.7%]; first van der Waals spacing contracted [−3%]	LEED	32b, 101	
Na₂O(111)	fluorite structure terminated between two Na layers; no relaxation	LEED	102	
NbSe₂(0001)	as MoS₂(0001), but top contraction by −0.2% [−0.6%]; first van der Waals spacing contracted [−1.4%]	LEED	32b, 101	
NiO(100)	unrelaxed bulk NaCl structure within [±5%]	LEED	103	
Si(111)"(1 × 1)"	bulk structure with −2% [−14%] top contraction	LEED	104	impurity-stabilized
	bulk structure with +0.8% [+5%] and +1% [+1%] expansions in two topmost layer spacings, respectively	LEED	105a	annealed at conversion temperature
	bulk structure with −4% [−30%] top contraction	cluster	106a	
	bulk structure with −1% [−6%] top contraction	model calc.	105b	
Si(111)p(2 × 1)	top layer contracted −1% [−8%], buckled 18% [±22%]; second layer spacing contraction −10% [−10%]	LEED	105a	
	qualitatively, as above	cluster	106	
	as LEED, but buckled ±10% [±50%]: no second-layer spacing construction	model calc.	106b	
Si(111)(7 × 7)	top double layer may tend toward one planar (graphitic) layer; buckled and coincident with substrate with period (7 × 7)	LEED	107	
Si(100)p(2 × 1)	top atom pairing (Schlier–Farnsworth model) with elastic relaxations down several layers	LEED	108	
		LEED	109	
		LEED	110	
		model calc.	111	
Si(100) + p(2 × 1)-H	probably: substrate as Si(100)p(2 × 1)	model calc.	111	
Si(100) + p(1 × 1)-2H	unrelaxed unreconstructed substrate	LEED	112	
TiS₂(0001)	as MoS₂, but top contraction by −1.7% [−5%]; first van der Waals spacing contracted [−5%]	LEED	113	
TiSe₂(0001)	as MoS₂, but top expansion by +1.7% [+5%]; first van der Waals spacing contracted [−5%]	LEED	113	
ZnO(0001)	unreconstructed Zn-terminated wurtzite structure with top contraction by −3% [−25%]	LEED	114	
ZnO(1010)	unreconstructed wurtzite structure, top Zn and O pulled into surface somewhat	LEED	115–117	
ZnSe(110)	zincblende structure reconstructed as GaAs (110)	LEED	118	
	qualitatively as above	model calc.	99d, 106b	

Note: Alphabetical order is used within classes, considering only the letters of the chemical species. First the cases with metal substrates are listed, in order of decreasing close packings of the metal surfaces; these are then followed by the other materials. For metal substrates the second column indicates relative bond-length changes between first and second layers of the clean surface or the substrate and the corresponding layer spacing changes [in square brackets], referred to the bulk values: expansions and contractions have positive and negative signs, respectively (values close to 0% are mostly quoted as 0%, mainly when no variation away from 0% was tried in LEED calculations). In the adsorption site description, the stacking sequence for fcc(111) and hcp (0001) is indicated by the familiar ABCABC... or ABABAB... notation, lowercase letters being used for adsorbates; for other substrates, the adsorption site is often characterized by the coordination number ("n-fold" meaning n nearest neighbors). Adsorption bond lengths are compared with bond lengths known from other sources for molecules and bulk compounds that are made up of the combinations of adsorbate and substrate atoms (to be found in the standard structure and crystallographic tables). Analytical methods are abbreviated.

1. (a) D.S. Boudreaux and V. Hoffstein, Phys. Rev. **B3**, 2447 (1971); (b) D.W. Jepsen, P.M. Marcus, and F. Jona, Phys. Rev. **B6**, 3684 (1972); (c) G.E. Laramore and C.B. Duke, Phys. Rev. **B5**, 267 (1972); (d) M.R. Martin and G.A. Somorjai, Phys. Rev. **B7**, 3607 (1973); (e) D.L. Adams and U. Landman, Phys. Rev. **B15**, 3775 (1977).

2. (a) F. Forstmann, Jap. J. Appl. Phys., Suppl. **2**, Pt. 2, 657 (1974). (b) F. Soria, J.L. Sacédon, P.M. Echenique, and D. Titterington, Surf. Sci. **68**, 448 (1977).

3. T.C. Ngoc, M.G. Lagally, and M.B. Webb, Surf. Sci. **35**, 117 (1973).

4. F. Forstmann, W. Berndt, and P. Büttner, Phys. Rev. Lett. **30**, 17 (1973).

5. J.D. Head, K.A.R. Mitchell, and L. Noodleman, Surf. Sci. **61**, 661 (1977).

6. P.H. Citrin, P. Eisenberger, and R.C. Hewitt, J. Vac. Sci. Technol. **15**, 449 (1978).

7. (a) P.I. Cohen, J. Unguris, and M.B. Webb, Surf. Sci. **58**, 429 (1976); (b) M.B. Webb and P.I. Cohen, CRC Solid State Sci. **6**, 253 (1976).

8. N. Stoner, M.A. Van Hove, S.Y. Tong, and M.B. Webb, Phys. Rev. Lett. **40**, 243 (1978).

9. B.W. Lee, R. Alsenz, A. Ignatiev, and M.A. Van Hove, Phys. Rev. **B17**, 1510 (1978).

10. (a) G.E. Laramore, Phys. Rev. **B9**, 1204 (1974); (b) P.R. Watson, F.R. Shepherd, D.C. Frost, and K.A.R. Mitchell, Surf. Sci. **72**, 562 (1978).

11. C.-M. Chan, S.L. Cunningham, M.A. Van Hove, W.H. Weinberg, and S.P. Withrow, Surf. Sci. **66**, 394 (1977).

12. (a) J.E. Demuth, P.M. Marcus, and D.W. Jepsen, Phys. Rev. **B11**, 1460 (1975); (b) R. Feder, Phys. Rev. **B15**, 1751 (1977); (c) G.E. Laramore, Phys. Rev. **B8**, 515 (1973).

13. T.C. Ngoc, M.G. Lagally, and M.B. Webb, Surf. Sci. **35**, 117 (1973).

14. M.A. Van Hove, G. Ertl, K. Christmann, R.J. Behm, and W.H. Weinberg, J. Chem. Phys. **70**, 4168 (1979).

15. S.W. Wang, and W.H. Weinberg, Surf. Sci. **77**, 14 (1978).

16. P.M. Marcus, J.E. Demuth, and D.W. Jepsen, Surf. Sci. **53**, 501 (1975).

17. (a) J.E. Demuth, D.W. Jepsen, and P.M. Marcus, Phys. Rev. Lett. **32**, 1182 (1974); (b) P.M. Marcus, J.E. Demuth, and D.W. Jepsen, Surf. Sci. **53**, 501 (1975).

18. A.B. Anderson, J. Chem. Phys. **66**, 2178 (1977).

19. (a) L.L. Kesmodel and G.A. Somorjai, Phys. Rev. **B11**, 630 (1975); (b) L.L. Kesmodel, P.C. Stair, and G.A. Somorjai, Surf. Sci. **64**, 342 (1977).

20. (a) E. Bogh and I. Stensgaard, Proc. 7th Int. Vac. Congr. and 3rd Int. Conf. Solid Surf., Vienna, A-2757 (1977); (b) J.F. Van der Veen, R.G. Smeenk, and F.W. Saris, Proc. 7th Int. Vac. Congr. and 3rd Int. Conf. Solid Surf., 2515 (1977).

21. D.C. Frost, K.A.R. Mitchell, F.R. Shepherd, and P.R. Watson, Proc. 7th Int. Vac. Congr. and 3rd Int. Conf. Solid Surf., Vienna, A-2725 (1977).

22. J.A. Strozier and R.O. Jones, Phys. Rev. **B3**, 3228 (1971).

23. H.D. Shih, F. Jona, D.W. Jepsen, and P.M. Marcus, Commun. Phys. **1**, 25 (1976).

24. H.D. Shih, F. Jona, D.W. Jepsen, and P.M. Marcus, J. Phys. **C9**, 1405 (1976).

25. H.D. Shih, F. Jona, D.W. Jepsen, and P.M. Marcus, Phys. Rev. **B15**, 5550 and 5561 (1977).

26. (a) H.D. Shih, F. Jona, D.W. Jepsen, and P.M. Marcus, Phys. Rev. Lett. **36**, 798 (1976); (b) H.D. Shih, F. Jona, D.W. Jepsen, and P.M. Marcus, Surf. Sci. **60**, 445 (1976).

(continued)

Table 5.8.—Continued

27. W.N. Unertl and H.V. Thapliyal, J. Vac. Sci. Technol. **12**, 263 (1975).

28. R. Feder and G. Gafner, Surf. Sci. **57**, 45 (1976).

29. (a) S. Andersson, J.B. Pendry, and P.M. Echenique, Surf. Sci. **65**, 539 (1977); (b) P.M. Echenique, J. Phys. **C9**, 3193 (1976).

30. M.G. Lagally, J.C. Buchholz, and G.C. Wang, J. Vac. Sci. Technol. **12**, 213 (1975).

31. (a) R. Feder, Phys. Status Solidi **B62**, 135 (1974); (b) M.A. Van Hove and S.Y. Tong, Surf. Sci. **54**, 91 (1976).

32. (a) M.A. Van Hove and S.Y. Tong, Phys. Rev. Lett. **35**, 1092 (1975); (b) M.A. Van Hove, S.Y. Tong, and M.H. Elconin, Surf. Sci. **64**, 85 (1977).

33. R.H. Tait, S.Y. Tong, and T.N. Rhodin, Phys. Rev. Lett. **28**, 553 (1972).

34. N. Masud, C.G. Kinniburgh, and J.B. Pendry, J. Phys. **C10**, 1 (1977).

35. (a) B.A. Hutchins, T.N. Rhodin, and J.E. Demuth, Surf. Sci. **54**, 419 (1976); (b) M.A. Van Hove, S.Y. Tong, and N. Stoner, Surf. Sci. **54**, 259 (1976).

36. (a) D.W. Jepsen, P.M. Marcus, and F. Jona, Phys. Rev. **B8**, 5523 (1973); (b) W. Moritz, Doctoral thesis (University of Munich, 1976).

37. E. Zanazzi, F. Jona, D.W. Jepsen, and P.M. Marcus, Phys. Rev. **B14**, 432 (1976).

38. A. Ignatiev, F. Jona, D.W. Jepsen, and P.M. Marcus, Surf. Sci. **40**, 439 (1973).

39. R. Feder, Surf. Sci. **68**, 229 (1977).

40. M. Maglietta, E. Zanazzi, and F. Jona, Bull. Am. Phys. Soc. **22**, 355 (1977).

41. M. Maglietta, E. Zanazzi, U. Bardi, F. Jona, D.W. Jepsen, and P.M. Marcus, Surf. Sci. **77**, 101 (1978).

42. (a) G. Capart, Surf. Sci. **26**, 429 (1971); (b) P.M. Marcus, D.W. Jepsen, and F. Jona, Surf. Sci. **31**, 180 (1972); (c) J.B. Pendry, J. Phys. **C4**, 2514 (1971).

43. J.B. Pendry, *Low Energy Electron Diffraction* (Academic Press, London, 1974).

44. G.G. Kleiman and J.M. Burkstrand, Surf. Sci. **50**, 493 (1975).

45. J.M. Burkstrand, G.G. Kleiman, G.G. Tibbetts, and J.C. Tracy, J. Vac. Sci. Technol. **13**, 291 (1976).

46. J.M. Burkstrand, S.Y. Tong, and M.A. Van Hove, unpublished.

47. A. Salwén and J. Rundgren, Surf. Sci. **53**, 523 (1975).

48. R.H. Tait, S.Y. Tong, and T.N. Rhodin, Phys. Rev. Lett. **28**, 553 (1972).

49. M.A. Van Hove and S.Y. Tong, Surf. Sci. **52**, 673 (1975).

50. (a) S. Andersson and J.B. Pendry, J. Phys. **C6**, 601 (1973); (b) S. Andersson and J.B. Pendry, Solid State Commun. **16**, 563 (1975).

51. J.E. Demuth, D.W. Jepsen, and P.M. Marcus, J. Phys. **C8**, L25 (1975).

52. J.E. Demuth, D.W. Jepsen, and P.M. Marcus, Phys. Rev. Lett. **31**, 540 (1973).

53. H.H. Brongersma and J.B. Theeten, Surf. Sci. **54**, 519 (1976).

54. H.D. Hagstrum and G.E. Becker, J. Chem. Phys. **54**, 1015 (1971).

55. M. Van Hove and S.Y. Tong, J. Vac. Sci. Technol. **12**, 230 (1975).

56. S.P. Walch and W.A. Goddard III, Solid State Commun. **23**, 907 (1977); Surf. Sci. **72**, 645 (1978); Surf. Sci. **75**, 609 (1978).

57. C.H. Li and J.W.D. Connolly, Surf. Sci. **65**, 700 (1977).

58. Groupe d'Étude des Surfaces, Surf. Sci. **48**, 577 (1975).

59. J.B. Theeten and H.H. Brongersma, Rev. Phys. Appl. **11**, 57 (1976).

60. H.D. Hagstrum and G.E. Becker, J. Vac. Sci. Technol. **14**, 369 (1977).

61. (a) J.E. Demuth, D.W. Jepsen, and P.M. Marcus, J. Phys. **C6**, L307 (1973); (b) J.E. Demuth, P.M. Marcus, and D.W. Jepsen, Phys. Rev. Lett. **32**, 1182 (1974).

62. R. Feder, Surf. Sci. **68**, 229 (1977).

63. K.A.R. Mitchell, F.R. Shepherd, P.R. Watson, and D.C. Frost, Surf. Sci. **72**, 562 (1978).

64. A. Ignatiev, J.B. Pendry, and T.N. Rhodin, Phys. Rev. Lett. **26**, 189 (1971).

65. (a) R. Feder, Phys. Status Solidi **58**, K137 (1973); (b) K.O. Legg, F. Jona, D.W. Jepsen, and P.M. Marcus, J. Phys. **C10**, 937 (1977).

66. (a) K.O. Legg, F. Jona, D.W. Jepsen, and P.M. Marcus, J. Phys. **C8**, L492 (1975); (b) K.O. Legg, F. Jona, D.W. Jepsen, and P.M. Marcus, Phys. Rev. **B16**, 5271 (1977).

67. A.B. Anderson, Phys. Rev. **B16**, 900 (1977).

68. (a) R. Feder and H. Viefhaus, to be published; (b) K.O. Legg, F. Jona, D.W. Jepsen, and P.M. Marcus, Surf. Sci. **66**, 25 (1977).

69. (a) L.J. Clarke, Proc. 7th Int. Vac. Congr. and 3rd Int. Conf. Solid Surf., Vienna, A-2725 (1977); (b) T.E. Felter, R.A. Barker, and P.J. Estrup, Phys. Rev. Lett. **38**, 1138 (1977); (c) A. Ignatiev, F. Jona, H.D. Shih, D.W. Jepsen, and P.M. Marcus, Phys. Rev. **B11**, 4787 (1975); (d) C. Noguera, D. Spanjaard, D. Jepsen, Y. Ballu, C. Guillot, J. Lecante, J. Paigne, Y. Petroff, R. Pinchaux, P. Thiry, and R. Cinti, Phys. Rev. Lett. **38**, 1171 (1977).

70. A. Ignatiev, F. Jona, D.W. Jepsen, and P.M. Marcus, Surf. Sci. **49**, 189 (1975).

71. L.J. Clarke, Proc. 7th Int. Vac. Congr. and 3rd Int. Conf. Solid Surf., Vienna, A-2725 (1977).

72. A. Ignatiev, F. Jona, D.W. Jepsen, and P.M. Marcus, Phys. Rev. **B11**, 4780 (1975).

73. (a) M.K. Debe, D.A. King, and F.S. March, Surf. Sci. **68**, 437 (1977); (b) R. Feder, Phys. Rev. Lett. **36**, 598 (1976); (c) R. Feder, Surf. Sci. **63**, 283 (1977); (d) J. Kirschner and R. Feder, Verh. Deutsch. Phys. Ges. **2**, 557 (1978); (e) B.W. Lee, A. Ignatiev, S.Y. Tong, and M. Van Hove, J. Vac. Sci. Technol. **14**, 291 (1977).

74. L.C. Feldman, R.L. Kauffman, P.J. Silverman, R.A. Zuhr, and J.H. Barrett, Phys. Rev. Lett. **39**, 38 (1977).

75. (a) R.A. Barker, P.J. Estrup, F. Jona, and P.M. Marcus, Solid State Commun. **25**, 375 (1978); (b) M.K. Debe and D.A. King, Phys. Rev. Lett. **39**, 708 (1977); (c) M.K. Debe and D.A. King, J. Phys. **C10**, L303 (1977); (d) T.E. Felter, R.A. Barker, and P.J. Estrup, Phys. Rev. Lett. **38**, 1138 (1977).

76. A. Ignatiev, B.W. Lee, and M.A. Van Hove, to be published.

77. C.-M. Chan, S.L. Cunningham, M.A. Van Hove, and W.H. Weinberg, Surf. Sci. **67**, 1 (1977).

78. M.W. Finnis and V. Heine, J. Phys. **F4**, L37 (1974).

79. M. Maglietta, E. Zanazzi, F. Jona, D.W. Jepsen, and P.M. Marcus, J. Phys. **C10**, 3287 (1977).

80. E. Zanazzi, M. Maglietta, U. Bardi, F. Jona, D.W. Jepsen, and P.M. Marcus, Proc. 7th Int. Vac. Congr. and 3rd Int. Conf. Solid Surf., Vienna, 2447 (1977).

81. W. Heiland, F. Iberl, E. Taglauer, and D. Menzel, Surf. Sci. **53**, 383 (1975).

82. H.L. Davis, J.R. Noonan, and L.H. Jenkins, Surf. Sci. **83**, 559 (1979).

83. C.-M. Chan, M.A. Van Hove, W.H. Weinberg, and E.D. Williams, to be published.

84. C.-M. Chan, S.L. Cunningham, K.L. Luke, W.H. Weinberg, and S.P. Withrow, J. Vac. Sci. Technol. **16**, 642 (1979).

85. C.-M. Chan, K.L. Luke, M.A. Van Hove, W.H. Weinberg, and S.P. Withrow, Surf. Sci. **78**, 386 (1978).

86. J.F. van der Veen, R.G. Smeenk, R.M. Tromp, and F.W. Saris, Ned. Tijdschr. Vacuumtech. **2/3/4**, 284 (1978).

87. J.E. Demuth, J. Colloid Interface Sci. **58**, 184 (1977).

88. P.M. Marcus, J.E. Demuth, and D.W. Jepsen, Surf. Sci. **53**, 501 (1975).

89. W. Heiland, H.G. Schäffler, and E. Taglauer, Surf. Sci. **35**, 381 (1973).

90. H.H. Brongersma and J.B. Theeten, Surf. Sci. **54**, 519 (1976).

(continued)

Table 5.8.—Continued

91. J.F. van der Veen, R.M. Tromp, R.G. Smeenk, and F.W. Saris, Surf. Sci. **82**, 468 (1979).

92. S.P. Walch and W.A. Goddard III, Surf. Sci. **72**, 645 (1979).

93. D.C. Frost, S. Hengrasmee, K.A.R. Mitchell, F.R. Shepherd, and P.R. Watson, Surf. Sci. **76**, L585 (1978).

94. H.D. Shih, F. Jona, D.W. Jepsen, and P.M. Marcus, Bull. Am. Phys. Soc. **22**, 357 (1977).

95. R.W. Streater, W.T. Moore, P.R. Watson, D.C. Frost, and K.A.R. Mitchell, Surf. Sci. **72**, 744 (1978).

96. A. Ignatiev, B.W. Lee, and M.A. Van Hove, Proc. 7th Int. Vac. Congr. and 3rd Int. Conf. Solid Surf., Vienna, 1733 (1977).

97. (a) B.J. Mrstik, M.A. Van Hove, and S.Y. Tong, Bull. Am. Phys. Soc. **23**, 391 (1978); (b) S.Y. Tong, M.A. Van Hove, and B.J. Mrstik, Proc. 7th Int. Vac. Congr. and 3rd Int. Conf. Solid Surf., Vienna, 2407 (1977).

98. (a) C.B. Duke, A.R. Lubinsky, B.W. Lee, and P. Mark, J. Vac. Sci. Technol. **13**, 761 (1976); (b) A.R. Lubinsky, C.B. Duke, B.W. Lee, and P. Mark, Phys. Rev. Lett. **36**, 1058 (1976); (c) P. Mark, G. Cisneros, M. Bonn, A. Kahn, C.B. Duke, A. Paton, and A.R. Lubinsky, J. Vac. Sci. Technol. **14**, 910 (1977); (d) S.Y. Tong, A.R. Lubinsky, B.J. Mrstik, and M.A. Van Hove, Phys. Rev. **B17**, 3303 (1978).

99. (a) J.E. Rowe, S.B. Christman, and G. Margaritondo, Phys. Rev. Lett. **35**, 1471 (1975); (b) W.E. Spicer, P.W. Chye, P.E. Gregory, T. Sukegawa, and I.A. Babaloba, J. Vac. Sci. Technol. **13**, 233 (1976); (c) J.R. Chelikowsky, S.G. Louie, and M.L. Cohen, Phys. Rev. **B14**, 4724 (1976); (d) C. Calandra, F. Manghi, and C.M. Bertoni, J. Phys. **C10**, 1911 (1977); (e) E.J. Mele and J.D. Joannopoulos, Phys. Rev. **B17**, 1816 (1978); (f) W.A. Goddard III, J.J. Barton, A. Redondo, and T.C. McGill, to be published; (g) D.J. Chadi, Phys. Rev. Lett. **41**, 1062 (1978).

100. (a) C.G. Kinniburgh, J. Phys. **C8**, 2382 (1975); (b) C.G. Kinniburgh, J. Phys. **C9**, 2695 (1976).

101. (a) B.J. Mrstik, R. Kaplan, T.L. Reinecke, M. Van Hove, and S.Y. Tong, Phys. Rev. **B15**, 897 (1977); (b) B.J. Mrstik, R. Kaplan, T.L. Reinecke, M. Van Hove, and S.Y. Tong, Nuovo Cimento, **38B**, 387 (1977).

102. S. Andersson, J.B. Pendry, and P.M. Echenique, Surf. Sci. **65**, 539 (1977).

103. C.G. Kinniburgh and J.A. Walker, Surf. Sci. **63**, 274 (1977).

104. H.D. Shih, F. Jona, D.W. Jepsen, and P.M. Marcus, Phys. Rev. Lett. **37**, 1622 (1976).

105. (a) P.P. Aver and W. Mönch, to be published; (b) L.C. Snyder and Z. Wasserman, Surf. Sci. **77**, 52 (1978).

106. (a) W.S. Verwoerd and F.J. Kok, Ned. Tijdschr, Vacuumtech. **2/3/4**, 303 (1978); (b) D.J. Chadi, Phys. Rev. Lett. **41**, 1062 (1978).

107. (a) J.D. Levine, S.H. McFarlane, and P. Mark, Phys. Rev. **B16**, 5415 (1977); (b) P. Mark, J.D. Levine, and S.H. McFarlane, Phys. Rev. Lett. **38**, 1408 (1977).

108. J.A. Appelbaum and D.R. Hamann, Surf. Sci. **74**, 21 (1978).

109. K.A.R. Mitchell and M.A. Van Hove, Surf. Sci. **75**, 147L (1978); (b) S.Y. Tong and A.L. Maldonado, Surf. Sci. **78**, 459 (1978).

110. S.Y. Tong, private communication.

111. J.A. Appelbaum, G.A. Baraff, and D.R. Hamann, Phys. Rev. Lett. **35**, 729 (1975); Phys. Rev. **B14**, 588 (1976).

112. S.J. White, D.P. Woodruff, B.W. Holland, and R.S. Zimmer, Surf. Sci. **68**, 457 (1977).

113. B. Lau, B.J. Mrstik, S.Y. Tong, and M.A. Van Hove, to be published.

114. C.B. Duke and A.R. Lubinsky, Surf. Sci. **50**, 605 (1975).

115. C.B. Duke, A.R. Lubinsky, S.C. Chang, B.W. Lee, and P. Mark, Phys. Rev. **B15**, 4865 (1977).

116. C.B. Duke, A.R. Lubinsky, B.W. Lee, and P. Mark, J. Vac. Sci. Technol. **13**, 761 (1976).

117. A.R. Lubinsky, C.B. Duke, S.C. Chang, B.W. Lee, and P. Mark, J. Vac. Sci. Technol. **13**, 189 (1976).

118. (a) C.B. Duke, A.R. Lubinsky, M. Bonn, G. Cisneros, and P. Mark, J. Vac. Sci. Technol. **14**, 294 (1977); (b) P. Mark, G. Cisneros, M. Bonn, A. Kahn, C.B. Duke, A. Paton, and A.R. Lubinsky, J. Vac. Sci. Technol. **14**, 910 (1977).

cases (exceptions will be discussed below), presumably by providing a more bulklike environment for the substrate atoms.

Here we shall distinguish between surfaces that, in the clean state, have reconstructed and those that have unreconstructed structures. In the case of reconstructed structures, the surface atoms have moved sufficiently far away from their ideal bulk positions to generate either superlattices (that is, larger two-dimensional structural unit cells) or, if no superlattice is present, at least substantially modified bond lengths or bond angles.

The general rule governing the small atomic displacements on clean unreconstructed surfaces seems to be that bond lengths increase slightly with the number of nearest neighbors (called the coordination number); this is in accordance with long-established knowledge.[17] Thus a surface atom, having lost some nearest neighbors as compared to the bulk, tends to have a reduced bond length to its neighbors. Since the lattice constant parallel to the surface in the top layer is usually forced upon the top layer by the substrate, only the bond lengths to the second-layer atoms should, in general, decrease. This effect is small (at most about 4 percent contraction of bond lengths) and is sometimes presumably drowned in the experimental and theoretical uncertainty (1 to 2 percent of these bond lengths) of the LEED analyses that have produced most of these results. In general, however, the effect is clearly discernible. A closer look at Table 5.8 shows that on clean close-packed faces—such as the fcc(111), hcp(0001), bcc(110), and fcc(100) faces of metals—usually almost no contraction is detected (rare cases of very small expansions have been reported, however). On less closely packed faces—such as the bcc(100), fcc(110), bcc(111), and fcc(311) faces of metals (see Fig. 4.6)—small contractions are systematically detected in LEED analyses. Such results find independent confirmation in ion scattering experiments and theoretical calculations (see the references in Table 5.8). They are also in qualitative agreement with very small (~ 1 percent) bond-length contractions observed, such as in electron diffraction studies of metal clusters of 12 to 92 Å radius.[18]

The physical or chemical origin of these contractions can be explained in different terms. First-principles descriptions are too involved for inclusion here. Instead, we indicate some phenomenological descriptions. First, one can imagine the electron cloud attempting to smooth its surface (as if there were a surface tension), thereby producing electrostatic forces that draw the surface atoms toward the substrate. This effect should be the stronger the less closely packed the surface is. Second, with fewer neighbors, the two-body repulsion energy is smaller, allowing greater atomic orbital overlap and therefore more favorable bonding at shorter bond lengths. Third, one may say that for surface atoms the bonding electrons are partly shifted from the cut bonds to the remaining noncut bonds, thereby increasing the charge content of the latter and so reducing the bond length. On ionic crystal surfaces, the asymmetries in the ionic electrostatic forces at surfaces may explain the contrast between the similar bond-length contractions observed on CoO(111) and the lack of observable contractions on MgO(100) and NiO(100).[19]

The foregoing descriptions of the origin of bond-length contractions at surfaces are consistent with the observations made when adsorbates are deposited on these surfaces: the shortened bond lengths are systematically lengthened again (sometimes to more than their bulk values) by the presence of adsorbates, as is visible in Table 5.8.

Surprisingly, only a half monolayer of adsorbates is often sufficient to restore the bulk bond length between the substrate atoms. This behavior is observed both by LEED and by ion scattering experiments. With Fe(100) + $p(1 \times 1)$-O, the underlying metal bond lengths are expanded to beyond their bulk value, and in that process, the FeO bulk oxide geometry is approached, presumably exhibiting a first stage of the oxidation process at a surface. Ion scattering experiments indicate a similar behavior (not searched for in an earlier LEED analysis) for Ni(110) − $c(2 \times 2)$-S. In this connection it is interesting to note the case of Ti(0001) − (1×1)-N, where a surface slab of three layers essentially identical to the bulk compound TiN is formed by a slight expansion of the topmost Ti–Ti layer spacing and intercalation of nitrogen (Fig. 5.11f). Not properly understood yet is the case of Ni(100) − $p(2 \times 2)$-C, in which an expansion of the topmost Ni–Ni layer spacing also occurs, and possibly some kind of nickel carbide is formed. With O on Al(100) (not studied with LEED), there is considerable evidence for adatom penetration (metal oxide formation).[20] Oxides can, of course, be formed on many surfaces, but details of their geometry and behavior remain to be elucidated.

A variety of different reconstruction geometries are thought to occur on surfaces. On fcc metals large superlattices, such as (5×1) or (5×20), are observed on Pt(100),[21] Ir(100),[22a] and Au(100).[23] Structure analysis points to a hexagonally close-packed restructuring of the topmost atomic layer. A weakly bound adsorbate layer, such as that formed during the physisorption of Xe on Ir(100),[22b] appears not to affect the basic geometry of tbe reconstructed substrate. However, these reconstructions are usually destroyed in favor of the unreconstructed geometry as a result of chemisorption, with its stronger substrate–overlayer bonding. This can even happen with rather small coverage, such as a few percent of a monolayer. In essence, the adatoms simulate the "missing half" of the substrate. These "impurity-stabilized" unreconstructed surfaces [for example, Pt(100) and Au(100)] have the structure known for the other stable, clean, unreconstructed metal surfaces.

The (110) face of these materials (Pt, Ir, and Au) often exhibits a (2 × 1) reconstruction, or more generally (n × 1) reconstructions [such as on Ir(110) with n = 2, 3, or 4],[24] sometimes with a statistical distribution of the values of the integer n [as on Au(110), where n = 2 dominates].[25] Several models for these reconstructions have been suggested, but the "missing-row" model seems to be the most promising in studies of Ir(110) (2 × 1) (Fig. 4.9) and Au(110) (n × 1)[25] with random n. In this model small facets of the hexagonal close-packed (111) face are built [note the analogy with the fcc(100) reconstructions], which is consistent with the knowledge that the (111) face of fcc metals is energetically the most favorable one. These reconstructions also seem to be destroyed by adsorbates in favor of the unreconstructed structure. This has been established in particular with the Ir(110) surface. There the (2 × 1) reconstruction disappears as a result of adsorption of a disordered quarter monolayer of oxygen; in fact, the resulting substrate shows contracted bond lengths, just as with the unreconstructed clean fcc(110) surfaces. Adsorption of an additional half monolayer of oxygen, which orders in a c(2 × 2) arrangement, then removes that bond-length contraction. This two-step process supports the notion that the effect of adsorbates on the substrate becomes more pronounced with increasing coverage. But apparently the effect is sometimes strong (a small fraction of a monolayer can destroy a reconstruction) and sometimes weak [a full film of xenon on Ir(100) seems not to destroy the reconstruction]. A special behavior is found for hydrogen on Ni(110), which produces a (2 × 1) superlattice believed to be due to a pairwise attraction and approach of adjoining rows of surface nickel atoms (the row-pairing model).

Although the (111) face of fcc metals is of the lowest surface free energy—a fact that may explain the reconstructions of the (100) and (110) faces—the (111) face itself may also reconstruct. Au(111) is normally reconstructed with a structure that may nevertheless still involve the hexagonal close-packed layer geometry (since extra sets of hexagonally arranged spots appear in LEED), but with a lattice constant different from that of the bulk.[26]

For bcc metals, only one reconstruction has been thoroughly analyzed, namely that of W(100)c(2 × 2), which occurs at low temperatures (Fig. 5.10). The mechanism responsible for this could be a charge-density wave,[27] which induces a structural wave that can have a wavelength

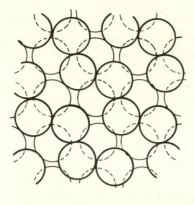

bcc (100) c(2×2)

Figure 5.10. Top view of the W(100)c(2 × 2) surface. Thin-lined atoms are behind the plane of thick-lined atoms. Dotted lines represent atoms in unreconstructed (ideal bulk) positions.

related to the lattice constant [as with W(100)c(2 × 2)] or not related to it [as with Mo(100)].[28]

Hydrogen chemisorption at less than full coverage appears not to change noticeably the structure of the W(100)c(2 × 2) surface.[29] Interestingly, chemisorption of hydrogen at room temperature on an unreconstructed W(100) surface seems to generate the same c(2 × 2) reconstruction obtained by simple cooling.[29,30] At full hydrogen coverage the reconstruction disappears, and W—W bond-length contractions seem to disappear as well.[30]

A type of reconstruction that one might expect to occur but that has not been observed is related to the relatively easy phase transition between hcp and fcc metals. This involves only the shifting of hexagonal close-packed layers of atoms from the ... ABABAB ... to the ... AB-CABC ... stacking arrangement. Such a shift could easily occur for the topmost atomic layer of hcp (0001) or fcc(111) surfaces. Interestingly, it does not seem to take place in reality on the five hcp(0001) and nine fcc(111) surfaces analyzed so far; this includes the case of Co on both sides of its hcp–fcc phase transition.

Reconstructions are particularly frequent on semiconductor surfaces. In three cases the structure and the underlying mechanism have most probably been identified. For GaAs(110) [and ZnSe(110), which behaves like GaAs(110) but whose properties have been less extensively studied], rehybridization of the orbitals around the surface Ga and As atoms occurs, producing new optimum bond angles (different from the tetrahedral angles of the bulk) that force substantial movements of the surface atoms (bond lengths remaining almost unchanged). In this case the atomic movements can be accommodated without enlarging the surface unit cell, so that no superlattice is generated (Figure 4.12). This type of reconstruction is also predicted by several model calculations. The adsorption of a monolayer of arsenic on this reconstructed surface restores the bulk lattice geometry in the topmost substrate layer, as a LEED analysis indicates (the adsorbed As atoms bond to the surface Ga atoms). Oxygen adsorption has the same effect on GaAs(110). Oxygen (which appears to bond to the surface As atoms) approximately restores the bulk geometry of the surface. This is indicated both by UPS data and by theoretical cluster calculations.

For Si(100)p(2 × 1) a long search has produced a structure derived from the Schlier–Farnsworth model, in which adjoining surface atoms (each with two unsatisfied "dangling bonds") simply bond together by bending toward each other and pairing up dangling bonds (one dangling bond per surface atom remains unsatisfied). A substantial bond bending occurs, and this distortion propagates elastically through the lattice down to a few layers in depth (Figure 4.11). Adsorption of hydrogen up to a certain coverage onto this surface seems not to change the nature of this reconstruction. However, a higher coverage of hydrogen destroys the reconstruction and restores the bulk geometry at the silicon substrate surface (apparently the surface Si—Si bonds have been broken and possibly replaced by bonds between the surface silicon atoms and the additional hydrogen atoms).

The special influence of hydrogen on the substrate should be stressed. Whereas other adsorbates leave the substrate essentially unchanged or remove a reconstruction, hydrogen at low coverages can induce a substrate reconstruction (at high coverages, hydrogen behaves as other adsorbates).

Adsorption Geometry of Atoms

Table 5.8 lists the adsorption site, the adsorption layer spacing, and the adsorption bond length for the atomic adsorbates that have been studied. A close scrutiny of the data permits us to draw several important conclusions. The few exceptions to these conclusions will be discussed below. (1) The adsorbed atoms tend to occupy sites where they are surrounded by the largest number of substrate atoms (largest coordination number). This site is usually the one that the bulk atoms would occupy in order to continue the bulk lattice into the overlayer [bABAB . . . and cABCABC . . ., respectively, where the lower case letters represent the atom in the overlayer]. (2) The adsorbed atom-substrate atom bond lengths are similar to the bond lengths in compounds that contain the atom pairs under consideration.

The various adsorption geometries that are found most commonly are displayed in Figure 5.11. The threefold hollow sites on the fcc(111) and hcp(0001) and bcc(110) are shown both in top and side views. Similarly, the fourfold hollow sites on the fcc(100) and bcc(100) crystal faces are shown. Finally, the center, long-bridge, and short-bridge sites on the fcc(100) crystal face and the location of atoms in an underlayer in the hcp(0001) crystal face are also displayed. These are the sites that are most commonly occupied by the adsorbed atoms that were investigated so far.

With atomic adsorption on semiconductor surfaces, one expects adsorbates to choose positions on the substrate that are relatively more predictable than with metal substrates: semiconductor surfaces often have relatively well-defined unsatisfied bonds (dangling bonds) ready to serve as adsorption sites, whereas such a simple argument does not seem to apply to metal surfaces. Such behavior would·also be expected from the geometry of bulk compounds, which tend to be unique and predictable for semiconductors (such as the very common zincblende and wurtzite structures based on tetrahedral arrangements) but much more varied and complicated for compounds containing metal atoms (thus many different crystallographic phases of such compounds exist, with metal oxides, for example). The stronger tendency toward surface reconstructions for semiconductors compared to metals (possibly due to the lack of close packing and ensuing freedom of movement in semiconductors associated with a strong desire to satisfy stoichiometry) adds a different dimension to the structural possibilities. Thus the one established geometry for adsorption on a semiconductor, that for GaAs(110) $- p(1 \times 1)$-As, has the simple structure expected from the bulk geometry of GaAs; As bonds to the surface Ga atoms (Figure 4.12). With oxygen on GaAs(110), it appears from UPS[31] and cluster calculations[32] that the reconstruction is also removed. For hydrogen on Si(100), although the hydrogen positions have not been determined directly, the hypothesis of adsorption to the Si dangling bonds is consistent with the observations of the disappearance of the (2×1) reconstruction of the clean surface and with model calculations (see Figure 4.11).

Turning to metal substrates, in most cases of atomic adsorption on metal surfaces where the adsorption geometry has been determined (see Table 5.8), only one adsorption site is involved; that is, all adatoms have identical surroundings [the exceptions are Ni(111) $- p(2 \times 2)$-2H and

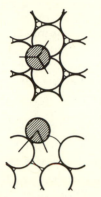

(a) fcc(111), hcp(0001): hollow site   (b) bcc(110): 3-fold site   (c) fcc(100): hollow site

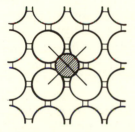
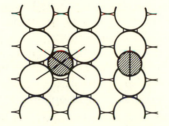
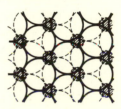

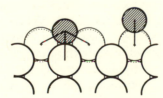

(d) bcc(100): hollow site   (e) fcc(110): center long-and-short bridge sites   (f) hcp (0001): underlayer

Figure 5.11. Top and side views (in top and bottom sketches of each panel) of adsorption geometries on various metal surfaces. Adsorbates are drawn shaded. Dotted lines represent clean surface atomic positions; arrows show atomic displacements due to adsorption.

Ag(111) + Xe, discussed below]. The adsorption can thus be conveniently characterized by the adsorption site and the metal–adsorbate bond lengths.

Adsorption sites (or "registries") on metals differ mainly in the number of nearest metal neighbors (the coordination number) and the two-dimensional symmetry. One might expect the adatom valence to influence the number of nearest metal neighbors and therefore the adsorption site, but there is little evidence for such behavior. The divalent oxygen chooses the twofold coordinated short-bridge site on Ni(110) and Ir(110), while on Ag(110) the long-bridge site with an uncertain coordination number (depending on the unknown bond lengths) may be chosen. Oxygen on other surfaces, and the divalent S, Se, and Te as well as all other adsorbates on various surfaces, do not show this behavior. Instead, one finds the strong tendency for adatoms to occupy sites with the largest available number of nearest metal neighbors (see Table 5.8 and Figure 5.10). Even W(110) $- p(2 \times 1)$-O seems to involve the threefold adsorption site rather than the higher-symmetry central twofold site that one might predict from the oxygen valency or by using sites obtained by extending the substrate lattice out beyond the surface.

It is interesting to note that this tendency toward occupying the site with the largest coordination number during adsorption on metals holds [except with oxygen on fcc(110)] independently of the crystallographic face for a given metal, the metal for a given crystallographic face, and the adsorbate for a given substrate.

It will be observed that adsorption sites with many nearest neighbors are also usually sites of high symmetry. Therefore, one may say that adsorbate atoms appear to favor sites of high symmetry. There is only one exception to this preference: in W(110) $- p(2 \times 1)$-O the oxygen seems to choose a site that by itself (ignoring other adsorbed atoms) has one mirror plane instead of a site that has two orthogonal mirror planes. This may be related to the fact that the overlayer as a whole already has low symmetry (only a twofold axis of rotation). Even for oxygen on fcc(110) surfaces, a site with the highest possible symmetry is chosen; no other site has higher symmetry (although several have the same symmetry —two orthogonal mirror planes).

If one now also takes into consideration the second and deeper substrate layers, one may wonder in particular whether the adsorbate atoms choose an adsorption site consistent with a continuation of the substrate lattice. It appears from the available results that the bulk lattice is in fact usually continued into the overlayer, as if a substrate atom rather than a foreign atom had adsorbed, despite differing bonding characteristics.

This bulk lattice continuation is satisfied by nearly all the cases listed in Table 5.8; O on fcc(110) again is an exception. Ni(111) − $p(2 \times 2)$-2H and Ti(0001) − $p(1 \times 1)$-Cd are also exceptions, belonging to an interesting class of surfaces. These are the hcp(0001) and fcc(111) surfaces, which we describe by the registry sequences ABABAB . . . and ABCAB-C . . ., respectively (surface at left). The continuation of the bulk lattice into the overlayer would imply the sequences bABABAB . . . and cAB-CABC . . ., respectively, with lowercase letters representing overlayers. These sequences are indeed found for fcc Ag(111) − $p(1 \times 1)$-Au, fcc Ni(111) − $p(2 \times 2)$-S, and fcc Ag(111) − $(\sqrt{3} \times \sqrt{3})R30°$-I. For fcc Ni(111) − $p(2 \times 2)$-O, it could not be determined whether the sequence is cABCABC . . . or bABCABC . . . With fcc Ni(111) − $p(2 \times 2)$-2H and its two adatoms per unit cell (see Figure 5.12), LEED studies indicate that both threefold coordinated sites are used; both cABCABC . . . and bABCABC . . . occur simultaneously (with Ni—H bond lengths identical to within 0.1 Å). Some recent model calculations agree with this insofar as they predict that the two threefold sites and the top site have higher binding energies than do other adsorption sites, with the top site possibly less favorable from the point of view of diffusion. On the other hand, hcp Ti(0001) − $p(1 \times 1)$-Cd was found to have the deviating sequence cABABAB . . ., meaning that the cadmium atom is repelled by second-nearest Ti neighbors.

Multilayers of Cd of Ti(0001) have been studied as well, indicating a Cd crystal growth according to the sequence . . . $acac$ABAB . . .; the Cd film has the expected hcp structure known for the bulk material. In this case the Ti and Cd lattice constants are sufficiently close to allow growth of the film in registry with the substrate mesh.

A further question regarding the adsorption registry is whether it depends on adsorption coverage (density of adatoms); this is relevant to

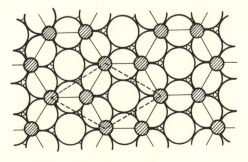

Figure 5.12. Top view of the Ni(111) + $p(2 \times 2)$−2H surface. The dashed lines indicate the unit cell. Adsorbate atoms are drawn shaded.

Ni (III) + p(2×2) 2H

the effects of adatom–adatom interactions. The situation is illustrated by a limited set of results, namely, those for quarter-monolayer and half-monolayer adsorption of O, S, Se, and Te on Ni(100) in $p(2 \times 2)$ and $c(2 \times 2)$ periodicities; the adsorption site is found not to depend on coverage in these cases (the nearest adatom–adatom distances are 4.90 and 3.46 Å for the two coverages, respectively, compared with the largest adatom diameter of about 2.7 Å for Te).

Adsorption in many different adsorption sites simultaneously is expected for overlayers with an incommensurate lattice. This has been confirmed by LEED intensity analyses for the case of an incommensurate overlayer of Xe on Ag(111), where both the substrate and the overlayer consist of hexagonal close-packed layers (with unrelated unit cells) parallel to the surface.

Concerning adsorption bond lengths, it is necessary first to recall the uncertainty in the determination of these. Depending on the case (such as the orientation of the bonds, among other factors), the uncertainty in bond lengths in atomic adsorption determined by LEED varies from about 0.04 to 0.09 Å, corresponding to relative uncertainties of 2 to 4 percent of the bond length. Medium- and high-energy ion scattering and SEXAFS may have uncertainties of about 0.02 Å, or 1 percent, in the few cases examined so far.

Figure 5.13 reproduces the adsorption bond-length information contained in Table 5.8. The first observation is that the bond lengths found at surfaces agree well (with a few exceptions) with those found in other environments, molecules and solid compounds containing the atom pairs under consideration. As is well known,[17] bond lengths in molecules tend to be smaller than those in bulk compounds because of the difference in coordination number (number of nearest neighbors). On the whole, it seems that bond lengths at surfaces lie closer to the bulk values than to the molecular values, which, again on the basis of coordination numbers, seems reasonable.

The uncertainty in the surface bond lengths is sufficiently small that the principal bonding mechanisms can probably be investigated. For example, a partial study[33] of systematics in these results, in the spirit of reference 17, suggests that the long-established concepts of bond order, valency saturation, and resonating bonds are applicable. Model calculations are also beginning to shed light on the particulars of the chemisorption mechanisms.[34]

A nonstructural quantity useful in the understanding of chemisorption is the charge transfer between adsorbates and substrates. This charge transfer (giving rise to dipole moments that influence the work

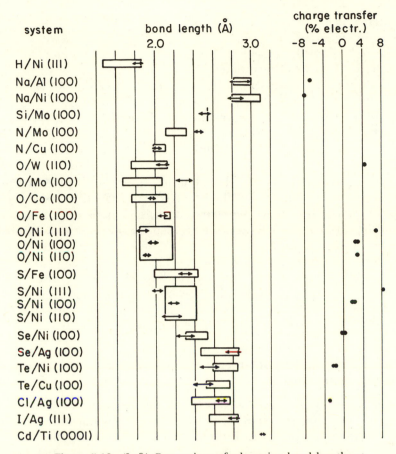

Figure 5.13. (Left) Comparison of adsorption bond lengths at surfaces (arrows show uncertainty) with equivalent bond lengths in molecules and bulk compounds (blocks extending over range of value found in standard tables). (Right) Induced charge transfers (obtained as discussed in the text) for adsorption.

function) can be roughly estimated from the observed work-function change $\Delta\phi$ during adsorption and the relative positions of the surface atoms, using the relation $\Delta\phi = 4\pi e\sigma d$ for the potential change through a dipole layer, where e is the electronic charge and σ the dipolar charge density; d represents the length of the dipoles and may, for example, be taken to be the component of the adsorbate–substrate bond length perpendicular to the surface. Within this simple model, one may now ask what fraction $\Delta e/e$ of an electron transferred at each adsorbate site through a distance equal to this component of the bond length produces the measured work-function change. This fraction, $\Delta e/e$, is plotted in Figure 5.13 for those cases in which work-function changes have been measured. The first observation is that $|\Delta e/e|$ is relatively small—at most about 11 percent—even for alkali adsorbates (which produce the largest work-function changes; the larger bond lengths for alkali adsorption probably explain the larger work-function changes). A number of model calculations for single adsorbates (the low-coverage limit) also predict such small charge transfers. Additionally, at higher coverages (those applicable to the results shown in Figure 5.13) one may invoke the effect of dipole–dipole interactions: these tend to reduce the dipole strengths. This last point is confirmed by the observed coverage dependence of the implied charge transfers; the charge transfer per adatom is reduced when the adsorption coverage is increased. However, the behavior of charge transfers is seen to be even more complicated when one notices that the work-function change and the charge transfer can switch their signs on a variation of the coverage, as happens with Se adsorbed on Ni(100),[35] even though the bonding geometry is not noticeably affected. Furthermore, Te on Ni(100) produces charge transfers opposite in sign to O and S on Ni(100), despite their having the same valency.[36]

Surface Crystallography of Ordered Multiatomic and Molecular Monolayers

Molecules deposited on surfaces may retain their basic molecular character, bonding as a whole to the substrate. They may dissociate into their constituent atoms, which then bond individually to the substrate. Instead, molecules may break up into smaller fragments which become largely independent or recombine into other configurations. Intermediate cases can also occur, such as with relatively strong bonding of molecules with resulting strong distortions. In addition, coadsorption of different atoms or different molecules can occur. The exact form that such multiatomic or molecular adsorption takes among those mentioned above is known to depend strongly on the temperature, as will be discussed below; this provides a way to determine the bond strengths.

The study of the structure of multiatomic and molecular adsorbates is in its early stages, but more and more effort is devoted to it because of its obvious importance in catalysis and other fields of surface science. The techniques of investigation used are primarily LEED, photoemission, and high-resolution electron energy loss spectroscopy. The detailed adsorption geometry has been analyzed so far for a few cases of coadsorption of atoms and a few adsorbed molecules. The ordering characteristics of molecular monolayers have been investigated for a number of small molecules and a sizable family of larger organic molecules.

Coadsorption of Atoms

One small family of surface structures created by coadsorption of two different atomic species has been structurally investigated. The substrate is fcc Ni(100), which is not structurally affected by the adsorption. The adatoms are S and Na, deposited sequentially in that order, each in either half-coverage $c(2 \times 2)$ or quarter-coverage $p(2 \times 2)$ ordered overlayers. With a half monolayer of each species (see Figure 5.14a), the position of the S atoms in hollow sites is not affected by the addition of the Na atoms; the Na atoms choose the unoccupied hollow sites on the substrate, where they have four nearest S neighbors with a Na—S bond length of 2.76 ± 0.1 Å (compared with 2.735 to 3.38 Å in a number of bulk compounds). The Na atoms are 0.2 Å farther away from the substrate than in the absence of S, an increase by 0.15 Å of the Ni—Na bond length. Halving the Na coverage while leaving that of S unchanged does not affect these results (see Figure 5.14b), indicating little charge effect in the bonding. This last impression is confirmed by work-function measurements. For half-monolayer coverage of both species, the work-function change relative to the bare substrate is -2.65 eV (compared with -2.55 eV in the absence of S), while halving the Na coverage yields -2.85 eV; this halving, therefore, induces a charge transfer of the order of 1 percent of an electron between two atoms.

With a quarter monolayer of both S and Na, the position of the S

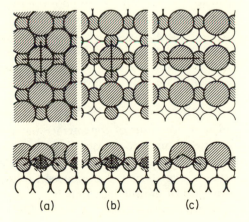

Figure 5.14. Coadsorption geometry of S (small shaded circles) and Na (large shaded circles) on Ni(100) (open circles), in top and side views: (a) half-monolayer of S and half-monolayer of Na; (b) half-monolayer of S and quarter-monolayer of Na; (c) quarter-monolayer of S and quarter-monolayer of Na.

(a)   (b)   (c)

atoms is again insensitive to the addition of Na atoms (see Figure 5.14c), and again the Na atoms choose unoccupied hollow sites on the substrate, but only those sites that provide the closest contact with S atoms, rather than the sites that allow closer contact with the substrate. So an attractive force acts between the coadsorbed species. Again the Na—S bond length is 2.76 ± 0.1 Å, even though the number of nearest S atoms is now reduced from 4 to 2. The work-function change (relative to the bare substrate) is now -3.10 eV, so that again little charge effect is seen in the bonding, despite the fact that Na and S could be expected to have a strong ionic character. The mutual destruction of parallel dipoles seems to play a significant role here, as in all cases of high-coverage adsorption discussed previously.

Dissociative Adsorption of Carbon Monoxide

On many metal substrates, CO dissociates into individual atoms that in some cases still produce an ordered monolayer. The resulting geometry has been investigated for a titanium substrate and for an iron substrate.

With $Ti(0001) - p(2 \times 2)$-CO, preliminary results[37] suggest that the C and O atoms occupy threefold hollow sites, the C atoms forming a $p(2 \times 2)$ array and the O atoms forming a similar but shifted $p(2 \times 2)$ array. Both C and O probably choose the same type of hollow site, but which of the two inequivalent threefold sites (bABC... or cABC...) is not known.

In the case of $Fe(100) - c(2 \times 2)$-CO, the LEED analysis[38] finds that the C and O atoms individually and randomly occupy fourfold hollow sites in a $c(2 \times 2)$ array; that is, a $c(2 \times 2)$ array of unoccupied sites exists, all other sites being occupied at random by either C or O atoms. The average Fe—C and Fe—O bond length is 1.93 Å (C and O usually have very similar radii), somewhat smaller than for $Fe(100) - p(1 \times 1)$-O (where it is about 2.08 Å); however, an expansion of the topmost substrate interlayer spacing has not been considered in this dissociative case (the bulk spacing was assumed), resulting in some uncertainty in the Fe–adsorbate bond length as well.

Adsorption Geometry of Molecules

Determinations of the surface by computing the diffraction beam intensities from low-energy electron diffraction are concentrated in two "frontier" areas at present. One is the determination of the surface structures of ever-larger size adsorbed molecules, and the other is the determination of atomic locations in reconstructed clean solid surfaces.

So far, only a very few adsorbed molecular structures have been analyzed by surface crystallography. The first system studied in detail was acetylene adsorbed on the (111) crystal face of platinum. We shall discuss the complex adsorption and structural characteristics of this small organic molecule in some detail, as it reveals the unique surface bonding arrangements that are possible and points to the importance of the use of additional techniques to complement the diffraction information.

Acetylene forms spontaneously an ordered (2 × 2) surface structure on the Pt(111) surface at 300 K, at low exposure under ultrahigh-vacuum conditions. The intensity profiles reveal that this structure is metastable, and upon heating to 350 to 400 K for 1 hr, it undergoes a transformation to a stable structure with the same (2 × 2) unit cell. Ethylene adsorbs on the Pt(111) surface, and at 300 K it forms an ordered (2 × 2) surface structure that is identical to the stable acetylene structure, as shown by the intensity profiles.

Data were obtained in the electron energy range 10 to 200 eV, but little sensitivity to the organic adsorbate is found above ~100 eV. The observed diffraction pattern arises from three equivalent 120°-rotated domains of (2 × 2) unit cells. The optimum agreement between calculated and experimental intensity data for the metastable acetylene structure is achieved for an atop-site coordination.[39] The molecule is located at a z-distance of 2.5 Å from the underlying surface platinum atom. However, the best agreement is obtained if the molecule is moved toward a triangular site by 0.25 Å, where there is a platinum atom in the second layer, as shown in Figure 5.15. The C—C bond distance is similar to that in ethylene as indicated by HREELS studies.

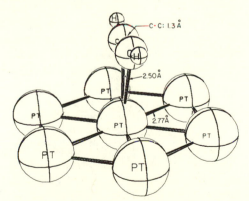

Pt (III) + metastable C_2H_2

Figure 5.15. Perspective view of metastable C_2H_2 on Pt(111) (hydrogen-atom positions are uncertain).

The same system, C_2H_2 on Pt(111), has also been studied by ultraviolet photoelectron spectroscopy (UPS), high-resolution electron energy loss spectroscopy (HREELS), and thermal desorption spectroscopy (TDS). The authors have all reported the presence of at least two states of binding and conversion from one state to the other as a function of temperature.

To solve the stable acetylene surface structure, it was necessary to combine the experimental information that came from LEED and HREELS.[40] The vibrational spectrum indicated that a methyl group was present and that the molecule must be lined up at some angle to the crystal surface. LEED structure analysis determined that the species is coordinated to a threefold site where there are no metal atoms underneath, in the second layer. The C—C axis is normal to the surface within the uncertainty of 15°. The C—C bond length was found to be 1.50 ± 0.05 Å. This value is nearly identical to the carbon–carbon single-bond distance in most saturated organic molecules. There are also three equivalent Pt—C bond lengths of 2.00 ± 0.05 Å. The surface species most consistent with all the studies is ethylidyne ($\geqslant C-CH_3$); its structure is shown in Figure 5.16. This structure is also found in many organometallic acetylene clusters. The ethylidyne group forms readily upon exposure of ethylene (C_2H_4) to the Pt(111) surface, with the transfer of one hydrogen atom to the surface per ethylene. The complete conversion of C_2H_2 to ethylidyne requires the presence of surface hydrogen atoms and proceeds rapidly only at ~350 K. By comparison with reported reaction mechanisms on related transition metal clusters, it seems likely that vinylidene ($> C = CH_2$) is an intermediate during the conversion from the metastable to the stable acetylene structure. An

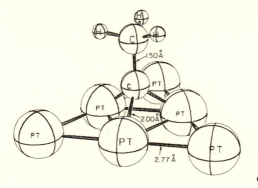

Pt (III) + ethylidyne

Figure 5.16. Perspective view of ethylidyne on Pt(111), the stable structure reached after acetylene adsorption with hydrogen addition.

interesting question concerns the source of hydrogen that must be attached to the molecule to form ethylidyne from acetylene. It appears that there is enough hydrogen on the metal surface from the residual background of the ultrahigh-vacuum diffraction chamber to provide the hydrogen necessary for the one-quarter monolayer of adsorbate. The disproportionation reaction of C_2H_2 can be ruled out, as neither LEED nor HREELS shows evidence for the presence of more than one surface species. Indeed, increased hydrogen partial pressures during the adsorption studies facilitate the formation of the stable surface structure of acetylene from the metastable structure.

Both the long C—C bond distance (1.50 Å) and the very short Pt—C distance (2.0 Å) indicate the strong interaction between the adsorbed molecule and the three platinum surface atoms. The covalent Pt—C distance would be 2.2 Å. The shorter metal–carbon distances indicate multiple metal–carbon bonding, which may be carbene or carbynelike. Compounds with these types of bonds exhibit high reactivity in metathesis and in other addition reactions.[41] The carbon–carbon single-bond distance indicates that the molecule is stretched as much as possible without breaking this chemical bond.

It is likely that the unique surface and catalytic chemistry of platinum are associated with the formation and decomposition of hydrocarbon molecular intermediates of the type produced by the adsorption of C_2H_2 or C_2H_4. Metals to the left of platinum in the periodic table would form stronger metal–carbon bonds. As a result, the carbon–carbon bond would snap, and molecular fragments would form instead of the ethylidyne species even at 300K. LEED and HREELS studies of the structure and bonding of C_2H_2 and C_2H_4 on Ni and Pd(111) surfaces are in progress. Preliminary electron spectroscopy evidence indicates that the molecules remain oriented parallel to these metal surfaces.

The second molecular system that has recently been studied is CO in a $c(2 \times 2)$ arrangement on the Ni(100) crystal face.[42] It appears from LEED that this molecule is bound by its carbon end to one nickel atom with a Ni—C bond length of 1.8 ± 0.1 Å (see Figure 5.17). The carbon-end bonding configuration has long been expected from UPS and IR evidence, and HREELS confirms bonding to a single nickel atom. Photoemission results do favor a perpendicular position for the molecule.

A LEED analysis of CO adsorbed in a $(2\sqrt{2} \times \sqrt{2})$ arrangement on Pd(100) also indicates molecular adsorption,[43] as expected from previous studies. However, there is bridge bonding of the carbon ends to pairs of metal atoms with a Pd—C bond length of 1.90 ± 0.06 Å and no

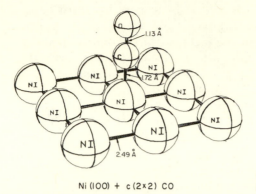

Ni (100) + c(2×2) CO

Figure 5.17. Proposed structure of CO on Ni(100) from LEED and electron spectroscopy studies.

noticeable tilting of the CO axis from the perpendicular to the surface, the CO bond length being 1.15 ± 0.1 Å (see Figure 5.18). There is some indication of relatively large vibration amplitudes for the O atoms, the nature of these vibrations remaining unresolved.

Comparing these results for CO bonded to Ni(100) and Pd(100) with the structure of metal–carbonyl clusters, one finds that the multiply coordinated CO on palladium has relatively smaller metal–carbon bond lengths than in terminal bonding to nickel, suggesting a stronger bonding to palladium. However, the heats of adsorption are rather similar.

High-resolution electron loss spectroscopy studies indicate that on the Ni(111), Pd(111), Pt(111) and Rh(111) crystal planes, CO prefers the lower symmetry on-top site, together with the twofold bridge sites at low

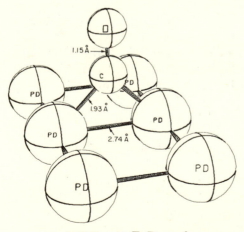

Pd (100) + (2√2×√2) R 45° 2 CO

Figure 5.18. Adsorption structure of CO on Pd(100) at a half-monolayer coverage.

CO partial pressures ($\leq 10^{-4}$ torr).[44,45] It is not clear at present why the molecule seeks out these bonding sites instead of the threefold hollow site on the various transition metals. In addition, there is no evidence so far for the presence of the gem-dicarbonyl species (two CO molecules per metal atom) on metal surfaces that is frequently observed on small metal cluster carbonyls. In the near future, the reasons for these differences will be addressed by both experiment and theory.

References

1. See, for example, J. Bardeen, Phys. Rev. **58,** 727 (1940); J.E. Lennard-Jones, Trans. Faraday Soc. **28,** 28 (1932).
2. I. Langmuir, J. Am. Chem. Soc. **40,** 1361 (1918).
3. E.W. Müller and T.T. Tsany, *Field Ion Microscopy,* Elsevier, New York, 1969; E. W. Müller, Science **149,** 591 (1965).
4. H.P. Bonzel and N.A. Gjostein, Appl. Phys. Lett. **10,** 258 (1967).
5. G.A. Somorjai and H.H. Farrell, Adv. Chem. Phys. **20,** 215 (1971).
6. C.-C. Wang, T.-M. Lu, and M.G. Lagally, J. Chem. Phys. **69,** 479 (1978).
7. G. Ehrlich, Surf. Sci. **63,** 422 (1977).
8. (a) D.G. Castner and G.A. Somorjai, Surf. Sci. **83,** 60 (1979). (b) J. Perdereau and G.E. Rhead, Surf. Sci. **24,** 555 (1971); (c) R.H. Roberts and J. Pritchard, Surf. Sci. **54,** 687 (1976).
9. J.L. Gland, Surf. Sci. **71,** 327 (1978).
10. (a) B. Lang, R.W. Joyner, and G.A. Somorjai, Surf. Sci. **30,** 454 (1972); (b) W. Erley, H. Wagner, and H. Ibach, Surf. Sci. **80,** 612 (1979).
11. F.J. Szalkowski and G.A. Somorjai, J. Chem. Phys. **54,** 389 (1971).
12. (a) L.E. Firment, Ph.D. dissertation, University of California, Berkeley, 1976; (b) L.E. Firment and G.A. Somorjai, J. Chem. Phys. **66,** 2901 (1977).
13. (a) J.L. Gland and G.A. Somorjai, Adv. Colloid Interface Sci. **5,** 203 (1976); (b) P.C. Stair and G.A. Somorjai, J. Chem. Phys. **67,** 4361 (1977).
14. L.E. Firment and G.A. Somorjai, J. Chem. Phys. **69,** 573 (1978).
15. J.C. Buchholz and G.A. Somorjai, J. Chem. Phys. **66,** 573 (1977).
16. L.L. Atanasoska, J.C. Buchholz, and G.A. Somorjai, Surf. Sci. **72,** 189 (1978).
17. L. Pauling, *The Nature of the Chemical Bond,* 3rd ed., Cornell University Press, Ithaca, N.Y., 1960.
18. H.J. Wasserman and J.S. Vermaak, Surf. Sci. **32,** 168 (1972).
19. M.A. Van Hove and P.M. Echenique, Surf. Sci. **82,** L298 (1978).
20. See, for example, R.P. Messmer, and D.R. Salahub, Phys. Rev. **B16,** 3415 (1977), and references therein.
21. S. Hagstrom, H.B. Lyon, and G.A. Somorjai, Phys. Rev. Lett. **15,** 491 (1965).
22. (a) J.T. Grant, Surf. Sci. **18,** 228 (1969); (b) A. Ignatiev, A.V. Jones, and T.N. Rhodin, Surf. Sci. **30,** 573 (1972).
23. (a) D.G. Fedak and N.A. Gjostein, Surf. Sci. **8,** 77 (1967); (b) P.W. Palmberg and T.N. Rhodin, Phys. Rev. **161,** 586 (1967).
24. W.H. Weinberg, private communication.
25. W. Moritz and D. Wolf, private communication.

26. (a) D.M. Zehner, B.R. Appleton, T.S. Noggle, J.W. Miller, J.H. Barrett, L.H. Jenkins, and O.E. Schorr III, J. Vac. Sci. Technol. **12,** 454 (1975); (b) D. Wolf and A. Zimerm, private communication.

27. (a) J.E. Inglesfield, J. Phys. **C11,** L69 (1978); (b) E. Tosatti, Solid State Commun. **25,** 637 (1978).

28. T.E. Felter, R.A. Barker, and P.J. Eastrup, Phys. Rev. Lett. **38,** 1138 (1977).

29. M.K. Debe and D.A. King, J. Phys. **C10,** L303 (1977).

30. A. Ignatiev, B.W. Lee, and M.A. Van Hove, to be published.

31. W.E. Spicer, P.W. Chye, P.E. Gregory, T. Sukegawa, and I.A. Babaloba, J. Vac. Sci. Technol. **13,** 233 (1976).

32. W.A. Goddard III, private communication.

33. A. Madhukar, Solid State Commun. **16,** 461 (1975).

34. (a) Overviews of theoretical methods in surface model calculations are given in chapters by T.B. Grimley and R.P. Messmer, in *The Nature of the Surface Chemical Bond,* ed. T.N. Rhodin and G. Ertl, North-Holland, Amsterdam, 1978; see also: (b) T.L. Einstein and J.R. Schrieffer, Phys. Rev. **B7,** 3629 (1973); (c) T.B. Grimley and C. Pisani, J. Phys. **C7,** 2831 (1974); (d) N.D. Lang and A.R. Williams, Phys. Rev. Lett. **34,** 531 (1975); (e) O. Gunnarson, H. Hjelmberg, and B.I. Lundqvist, Phys. Rev. Lett. **37,** 292 (1976); (f) S.W. Wang and W.H. Weinberg, Surf. Sci., **77,** 14 (1978); (g) D.J.M. Fassaert and A. van der Avoird, Surf. Sci. **55,** 291 and 313 (1976); (h) S.G. Louie, K.M. Ho, J.R. Chelikowsky, and M.L. Cohen, Phys. Rev. Lett. **37,** 1289 (1976); (i) R.V. Kasowski, Phys. Rev. Lett. **37,** 219 (1976); (j) J.A. Appelbaum, G.A. Baraff, and D.R. Hamann, Phys. Rev. **B14,** 588 (1976); (k) W.A. Harrison, Surf. Sci. **55,** 1 (1976); (l) D.W. Bullett and M.L. Cohen, J. Phys. Chem. **10,** 2083 and 2101 (1977); (m) R.P. Messmer, S.K. Knudsen, K.H. Johnson, J.B. Diamond, and C.Y. Yang, Phys. Rev. **B13,** 1396 (1976); (n) I.P. Batra and O. Robaux, Surf. Sci. **49,** 653 (1975); (o) R.F. Marshall, R.J. Blint, and A.B. Kunz, Phys. Rev. **B13,** 3333 (1976); (p) S.P. Walch and W.A. Goddard III, J. Am. Chem. Soc. **98,** 7908 (1976); (q) A.B. Anderson, J. Chem. Phys. **66,** 2173 (1977).

35. J.E. Demuth and T.N. Rhodin, Surf. Sci. **45,** 249 (1974).

36. S. Andersson and J.B. Pendry, J. Phys. **C9,** 2721 (1976).

37. H.D. Shih, F. Jona, D.W. Jepsen, and P.M. Marcus, J. Vac. Sci. Technol. **15,** 596 (1978).

38. F. Jona, K.O. Legg, H.D. Shih, D.W. Jepsen, and P.M. Marcus, Phys. Rev. Lett. **40,** 1466 (1978).

39. L.L. Kesmodel, R. Baetzold, and G.A. Somorjai, Surf. Sci. **66,** 299 (1977).

40. L.L. Kesmodel, L.H. Dubois, and G.A. Somorjai, Chem. Phys. Lett. **56,** 267 (1978).

41. E.O. Fischer and A. Massböl, Angew. Chem. **76,** 345 (1964). R.R. Schrock, J. Am. Chem. Soc. **98,** 5399 (1976).

42. S. Andersson and J.B. Pendry, Surf. Sci. **71,** 75 (1978).

43. H.J. Behm, K. Christmann, G. Ertl, and M.A. Van Hove, Surf. Sci. **88,** L59 (1979).

44. W. Erley, H. Wagner, and H. Ibach, Surf. Sci. **80,** 612 (1979).

45. L.H. Dubois and G.A. Somorjai, Surf. Sci. **91,** 514 (1980).

6. The Surface Chemical Bond

Introduction

This chapter reviews what has been learned about the nature of the surface chemical bonds of adsorbed atoms and molecules from our combined studies of solid surfaces and adsorbates. The clean, solid surface is our reference state, together with the gas-phase atom or molecule that reflects the state of the system before the surface bond is formed.[1] Ideally, we would like to keep the atomic and electronic structures of the surface and gas atoms unchanged as the chemical bond forms, with the exception of those electronic orbitals participating in the bonding. In practice, however, the atoms in the solid surface are likely to relocate and experience charge-density redistribution as the surface bond forms. This is vividly demonstrated by the behavior of the reconstructed (100) crystal faces of platinum[2] and iridium[3] in the presence of carbon monoxide. As the molecules adsorb, the surface atoms snap back into their unreconstructed fourfold symmetry positions. This transformation is readily detectable by LEED. Another example is the relocation of iron surface atoms in the Fe(100) crystal face[4] upon oxygen chemisorption. The metal atoms and oxygen atoms move to reside in the same surface plane when the (2 × 2) surface structure is formed.

Similarly, the adsorbed atom or molecule distorts in the strong electric field created by the surface dipole, and atomic rearrangements in the molecule occur that affect many more electronic orbitals than those that participate directly in forming the surface bond. The reader is reminded of the rehybridization of acetylene on the Pt(111) crystal face, which involves hydrogen shift from one carbon atom to the other and stretching of the 1.2-Å $C \equiv C$ triple bond of the gas-phase molecule to a 1.5-Å $C - C$ single bond in the adsorbed state.[5] Especially in the case of strong chemisorption, major surface reconstruction and molecular rearrangement are likely to take place, as demonstrated by numerous examples in Chapter 5. These changes, in addition to the bonding geometry and electronic structure of the bonding orbitals, all have to be scrutinized to obtain a complete physical picture of the adsorbate–surface system.

Nevertheless, in spite of the complex character of the surface bond, a number of major features have become apparent from recent experimental studies:

1. Several binding sites, even on the more atomically homogeneous low-Miller-index surfaces, are distinguishable by their structure and binding strength for different adsorbates. As a result, a sequential filling of binding sites is frequently observed.
2. The structure of the binding sites and the strength of binding change from crystal face to crystal face.
3. The character of the surface bond may change markedly with temperature. This is perhaps the most novel feature of the surface chemical bond as compared with other types of chemical bonds.
4. The presence of surface irregularities (steps and kinks) and of other adsorbates can markedly influence the surface chemical bond.
5. The unique electronic and structural environment of the surface may cause the formation of distinct surface phases for multicomponent systems, which has no analogue in the bulk-phase diagrams.[6]
6. A model that assumes the surface bond is localized and is similar to that in multinuclear cluster compounds represents fairly well many of the surface bonds.

Next, we discuss the available experimental data that aid us in arriving at these various conclusions about the nature of the surface chemical bond. First, we review what is known about bonding on the clean and relatively smooth low-Miller-index surfaces. Then we discuss the effects of irregularities and other adsorbates on surface bonding and discuss what is known about "surface compounds" and the unique surface phase behavior of multicomponent systems. Finally, we present evidence indicating correlations between bonding at surfaces and in multinuclear clusters.

Heats of Adsorption of Small Molecules on Solid Surfaces

One of the important physical-chemical properties that characterize the interaction of solid surfaces with gases is the bond energy of the adsorbed species. The determination of the bond energy is usually performed indirectly by measuring the heat of adsorption (or heat of desorption) of the gas. The various methods of measuring the heat of adsorption have been reviewed in Chapter 2. Briefly, the techniques are of two types; equilibrium and kinetic. Commonly, the adsorption isotherms (coverage, θ, as a function of equilibrium pressure, p) are measured at various temperatures and ΔH_{ads} is determined by the use of the Clausius–Clapeyron equation,

$$\left(\frac{\partial \ln p}{\partial T}\right)_{\theta = \text{const}} = -\frac{\Delta H_{ads}}{RT^2} \tag{1}$$

The kinetic measurement involves temperature-programmed desorption that was first developed by Ehrlich and Hudda.[7] From the temperature at which the desorption peak appears and from the knowledge of the kinetic order of the desorption process, the heat of desorption is obtained. In the absence of an activation energy for desorption, it is then equated with the heat of adsorption, ΔH_{ads}.

To define the heat of adsorption, let us consider the chemisorption of a diatomic molecule, X_2, onto a site on a uniform solid surface, M. The molecule may adsorb without dissociation to form MX_2. M represents the adsorption site where bonding occurs to a cluster of atoms or to a single atom. In this circumstance, the heat of adsorption, ΔH_{ads}, is defined as the energy needed to break the MX_2 bond.

$$MX_{2(ads)} \xrightarrow{\Delta H_{ads}} M + X_{2(gas)} \tag{2}$$

If the diatomic molecule adsorbs dissociatively, the heat of adsorption for this purpose is defined as

$$2MX_{(ads)} \xrightarrow{\Delta H_{ads}} 2M + X_{2(gas)} \tag{3}$$

From the knowledge of ΔH_{ads} (which is always positive) and the binding energy of the gas-phase molecule, the energy of the surface chemical bond is given by

$$\Delta H_{bond} (MX_2) = \Delta H_{ads} \tag{4a}$$

or

$$\Delta H_{bond} (MX) = \frac{(\Delta H_{ads} + D_{X_2})}{2} \tag{4b}$$

for the cases of molecular and atomic adsorptions, respectively, where D is the dissociation energy of the X_2 gas molecule.

There are several difficulties in determing ΔH_{ads} reliably:

1. The heat of chemisorption can change markedly with the adsorbed

gas coverage, θ. An example of this is shown in Figure 6.1 for CO on the Pd(111) surface. Decreasing ΔH_{ads} with increasing adsorbate concentration is commonly observed due to repulsive adsorbate–adsorbate interactions at higher coverages. In some cases this change in the magnitude of ΔH_{ads} can lead to major changes in chemisorption characteristics. For example, CO_2 was found to dissociatively chemisorb on the clean Rh(111) face to CO_{ads} and O_{ads} as a consequence of the strong metal–oxygen and metal–CO bonds.[8] As the oxygen surface coverage is increased, however, CO_2 no longer dissociates. Because of the decreasing heat of oxygen chemisorption with increasing oxygen coverage, the thermodynamics of dissociative CO_2 chemisorption is no longer favorable. Frequently, the amount adsorbed is not known (during a desorption experiment, for example), because precise measurement of the pressure or the coverage is difficult.

2. The surface is heterogeneous. There are many sites where the adsorbed species have different binding energies.[9] If a polycrystalline surface is utilized for the chemisorption studies instead of a structurally well-characterized single-crystal surface, in this circumstance the measured ΔH_{ads} is an average of the heats of adsorption at the various binding sites.

3. The adsorbate may change bonding as a function of temperature as well as adsorbate concentration.[10] For example, oxygen may be molecularly adsorbed at low temperatures, while it dissociates to atoms at higher surface temperatures. Change of bonding with temperature as well as with coverage is frequently observed.

Thus ΔH_{ads} changes as experimental conditions are varied. Because of

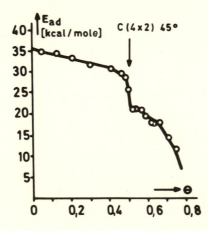

Figure 6.1. Isosteric heat of adsorption for CO on Pd(111) crystal face as a function of coverage. (After H. Conrad, G. Ertl, J. Koch, and E. E. Latta, Surf. Sci. **43**, 462 [1974].)

these and other experimental difficulties, the heat of adsorption measurements should be repeated using different techniques of measurements to verify reproducibility. In order to understand the atomic details of the interaction of the solid surface with gases, single-crystal surfaces should be used whenever possible with well-defined atomic surface structure and composition.

Nevertheless, a review of the heats of chemisorption for small diatomic molecules—H_2, O_2, CO, CO_2, and N_2—provides strong evidence that the bonding of the adsorbed molecules is localized; much of the strong bonding interaction is between the adsorbate and the nearest-neighbor metal atoms. It appears that the molecules may occupy several different sites with different coordination numbers and rotational symmetries on the same surface. For example, on a hexagonal surface structure such as the (111) crystal face of fcc metals, the adsorbates may occupy a threefold site, a twofold bridge site, or an on-top site. In each site the molecule may have different binding energies and thus different heats of chemisorption. Therefore, each "uniform" crystal face may exhibit a variety of bonding interactions and several adsorption sites. When the surface structure of the substrate metal is changed—for example, to a (100) face of an fcc metal or to a stepped or kinked surface—new binding sites appear with the changing atomic surface structure. Thus the nature of the chemical bonding, which is already varied on a given crystal surface, shows even more variation from crystal face to crystal face. This is clearly indicated from the heat of adsorption values obtained for these small molecules on the different crystal faces of transition metals that will be shown below.[11]

A polycrystalline metal surface exhibits all the adsorption sites of the crystal faces from which it is composed. Since these sites are present simultaneously, heats of chemisorption for these surfaces represent an average of the binding energies of the different surface sites weighted according to the relative concentrations of these sites. As a result, the measured heats of adsorption of molecules do not show the large structural variations that are readily detectable for single-crystal surfaces.

Tables 6.A1 to 6.E2 in the Appendix to this chapter list the heats of chemisorption for H_2, O_2, CO, CO_2, and N_2 on the different transition metal surfaces. In Tables 6.A1, 6.B1, 6.C1, and 6.E1, the ΔH_{ads} values obtained for single-crystal surfaces are given. In Tables 6.A2, 6.B2, 6.C2, 6.D, and 6.E2, the average heats of adsorption obtained for polycrystalline metal surfaces are listed. The methods of measurement (technique, temperature range, coverage, and type of sample) are indicated in the tables together with the references.

The heats of chemisorption on single-crystal planes indicate the presence of binding states for an adsorbed small molecule on a given surface with ΔH_{ads} that may differ by as much as 20 kcal/mol. Thus it is not possible to identify *one* value for the heat of chemisorption of an adsorbate on a given transition metal unless the binding state is specified or it is certain that only one binding state exists. The large changes in ΔH_{ads} are particularly pronounced for the high-Miller-index surfaces, which have large concentrations of surface irregularities, steps, and kinks. Several studies report higher binding energies for H_2 or CO at these low-coordination-member sites. In Figure 6.2 the thermal desorption spectra of hydrogen from a flat Pt(111) and a stepped and kinked platinum surface are displayed. The higher-temperature thermal desorption peaks are clearly discernible on the high-Miller-index surfaces. While the stepped platinum surface has two desorption peaks, the kinked surface has three, which can be associated with three different binding sites on this surface: terrace, step, and kink sites.

There have been several attempts to correlate the heats of chemisorption of small molecules on various transition-metal surfaces across the periodic table. We also present the data in this manner. In Figures 6.3a, 6.4a, 6.5a, and 6.6a, the heats of chemisorption that were obtained for

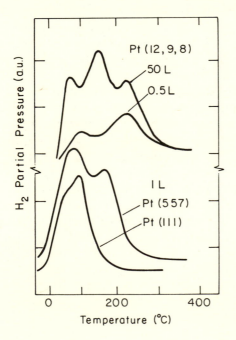

Figure 6.2. Thermal desorption spectra for hydrogen chemisorbed on flat Pt(111), stepped Pt(557), and kinked Pt(12, 9, 8) crystal surfaces.

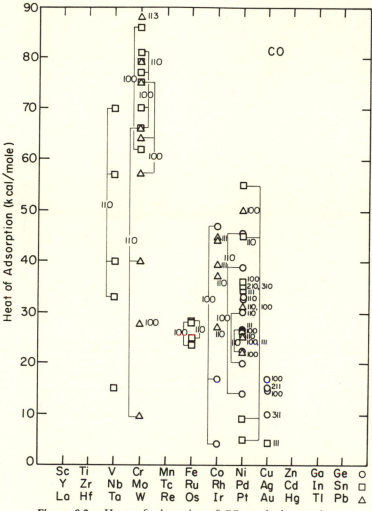

Figure 6.3a. Heats of adsorption of CO on single-crystal surfaces of transition metals.

single-crystal surfaces are plotted as a function of groups of transition metals along the periodic table. Those metals having similar d-electron occupancy but belonging to the $3d$, $4d$, and $5d$ rows of transition metals, respectively, are grouped together. The data are widely scattered because of the large variation of ΔH_{ads} resulting from multiple binding sites for a given crystal plane and changes of ΔH_{ads} from crystal face to

289

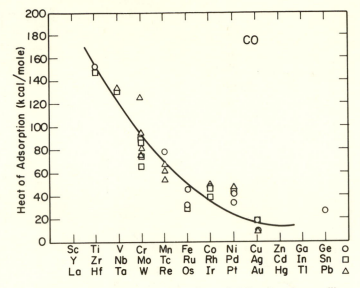

Figure 6.3b. Heats of adsorption of CO on polycrystalline transition-metal surfaces.

crystal face. In general, there is no such thing as a single-valued heat of adsorption for a given metal–adsorbate system. The binding sites of different strengths fill up with adsorbates as the coverage is increased or other experimental variables are changed (temperature, surface structure). The sequential filling of bonding sites by various adsorbates will be amply demonstrated in the next section.

It would be of value to determine trends in ΔH_{ads} for a given surface structure–adsorbate system in the same column in the periodic table. In Figure 6.7 the ΔH_{ads} values for hydrogen are listed for Ni, Pd, and Pt surfaces. Regrettably, no particular trends are apparent from the available data within the accuracy of the experiments.

The heats of chemisorption of these small molecules obtained for polycrystalline surfaces are plotted in Figures 6.3b, 6.4b, 6.5b, 6.6b, and 6.8 as a function of groups of transition metals along the periodic table. Over polycrystalline surfaces one obtains an averaged heat of chemisorption. It appears that this average value decreases from left to right across the periodic table. Similar correlations have been suggested previously by several investigators.

There have been several attempts to correlate the average heats of chemisorption to various thermodynamic parameters, such as the heats

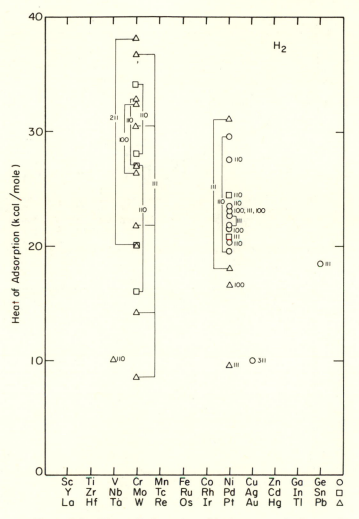

Figure 6.4a. Heats of adsorption of hydrogen on single-crystal surfaces of transition metals.

of formation of oxides, carbides, and hydrides, and to the bond energies of gaseous carbonyls. With the exception of the correlation of ΔH_{ads} for oxygen with the binding energies of the highest-valency oxides, the correlations are rather poor and do not seem to be very useful for predictive purposes.[11]

The structure sensitivity of the heats of chemisorption of small

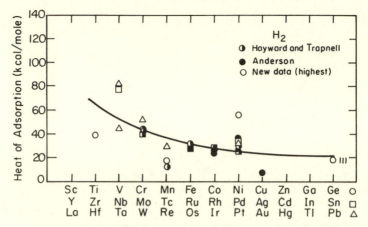

Figure 6.4b. Heats of adsorption of hydrogen on polycrystalline transition-metal surfaces.

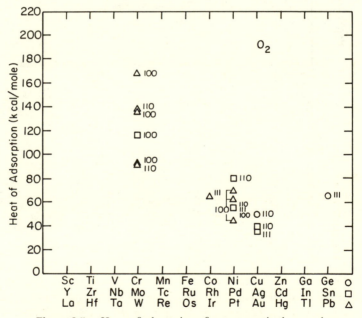

Figure 6.5a. Heats of adsorption of oxygen on single-crystal surfaces of transition metals.

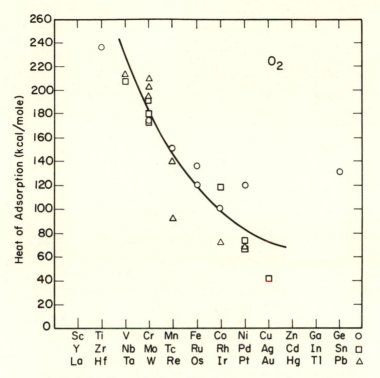

Figure 6.5b. Heats of adsorption of oxygen on polycrystalline transition-metal surfaces.

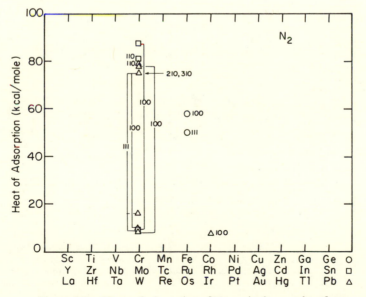

Figure 6.6a. Heats of adsorption of N₂ on single-crystal surfaces of transition metals.

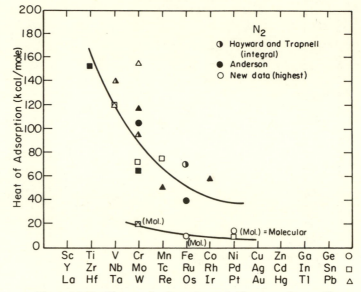

Figure 6.6b. Heats of adsorption of N_2 on polycrystalline transition-metal surfaces.

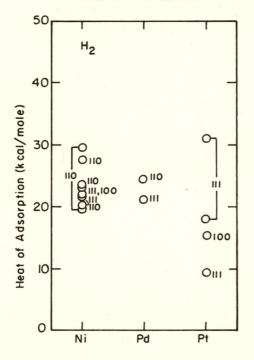

Figure 6.7. Heats of adsorption of H_2 on single-crystal planes of nickel, palladium, and platinum metals in the same group in the periodic table.

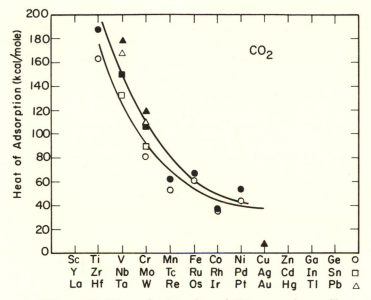

Figure 6.8. Heats of adsorption of CO$_2$ on polycrystalline transition-metal surfaces. Filled symbols indicate the differential heats of adsorption.

molecules on transition-metal surfaces indicates the control of the local surface environment in forming the surface chemical bonds.

Sequential Filling of Surface Binding Sites by Adsorbates

Studies by electron spectroscopy [especially high-resolution electron energy loss (HREELS)] and by low-energy electron diffraction (LEED) have revealed the sequential filling of the various binding sites on surfaces with increasing coverage of the adsorbates. In some cases there is evidence for exchange between atoms in the different binding sites. Next, we present a few examples demonstrating these effects.

The adsorption of hydrogen on the W(100) crystal face was studied as a function of coverage.[12] There is a $c(2 \times 2)$ surface structure which forms at low coverages ($\theta \sim 0.1$ to 0.2) followed by the appearance of a more complex sequence of surface structures as the coverage is increased.[13] The vibrational spectra obtained with increasing amounts of hydrogen on the surface are shown in Figure 2.23. At low coverages ($\theta = 0.4$), only one electron loss peak, at 155 meV, is observed. At higher coverages, another loss peak appears, at 130 meV. The 155-meV peak is associated with the one-coordinated top site of hydrogen atoms; the 130-meV peak, with the twofold bridge sites. There appears to be a reversible exchange of hydrogen between these surface sites. On the other two low-index crystal faces of tungsten [W(111) and W(110)] the on-top sites fill up first, followed by the twofold bridge sites.[14] These conclusions are in good agreement with electron-stimulated desorption[15] as well as photoemission studies.[16]

When oxygen is chemisorbed on the W(100) surface, a series of different LEED patterns are observed with increasing oxygen coverage.[17] HREELS studies reveal a 75-meV loss peak at very low coverages, which

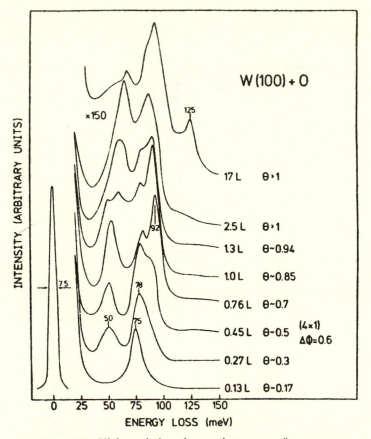

Figure 6.9. High-resolution electron loss spectra from oxygen adsorbed on the W(100) crystal face as a function of coverage. (From H. Froitzheim et al., ref. 18.)

indicates adsorption in the fourfold coordination site.[18] This is shown in Figure 6.9. In the coverage range 0.25 to 0.5, a second loss peak at 50 meV appears, while the LEED patterns yield two domains of a (4 × 1) surface structure. In this circumstance, structure analysis indicates the presence of rows of oxygen atoms arranged in such a way as to correspond to a WO_2 rutile-type compound.[18] At higher coverages, ordered multiple rows form, and finally oxygen occupies even the one-coordinated on-top sites as well.

Similar results are obtained by monitoring the adsorption of carbon monoxide on various crystal surfaces. On the Rh(111) crystal face, CO (forms a ($\sqrt{3} \times \sqrt{3}$)R30° structure at a coverage of about 0.5.[19] At somewhat higher coverage, a "split" (2 × 2) pattern is visible, and near $\theta = 1.0$, a (2 × 2) surface structure forms.[19]

In Figure 6.10, the vibrational spectra of CO adsorbed on the Rh(111) crystal face are displayed as a function of coverage.[20] First, the 1980-cm^{-1} peak appears, which is associated with the C=O stretching frequency of adsorbed CO molecules occupying an on-top site, linearly bound primarily to one metal surface atom (Rh—C=O). As the coverage is increased, a new energy loss peak at 1840 cm^{-1} appears, which is due to the stretching vibration of another type of CO occupying a bridge surface site, bound mostly to two surface atoms

$$
\begin{array}{c}
O \\
\parallel \\
C \\
\diagup \quad \diagdown \\
Rh \quad\quad Rh
\end{array}
$$

With increasing CO coverage these loss peaks shift to 2070cm^{-1} and 1870cm^{-1}, respectively. Thermal desorption studies indicate that the CO molecules at the bridge sites have a 4-kcal lower binding energy than CO molecules at the on-top sites.[20] When the surface coverage is decreased, the bridge-bonded species are removed first from the Rh(111) crystal face.

It is interesting to note that CO prefers to occupy the bridge and on-top sites on many (111) and (100) crystal faces of face-centered cubic metals and not the highest-symmetry threefold or fourfold surface sites. Chemisorbed oxygen or sulfur atoms, however, almost always occupy the highest symmetry sites on these crystal faces, as determined by low-energy electron diffraction. The reasons for these marked differences in bonding-site occupation by the different adsorbates will be subjected to intense theoretical and experimental investigations in the near future.

It should also be noted that while CO adsorbs strongly on the on-top

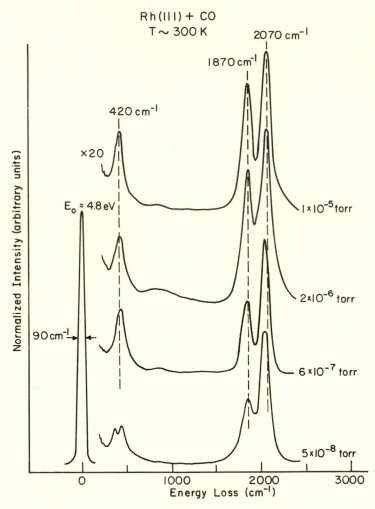

Figure 6.10. High-resolution electron loss spectra of CO chemisorbed on Rh(111) at 300 K as a function of pressure.

sites on rhodium[20] and platinum[21] crystal faces and less strongly on the bridge sites, it behaves in just the opposite manner during chemisorption on nickel[22] and palladium[23] crystal faces. On these surfaces CO occupies the bridge sites first. Thus subtle differences in electronic and atomic surface structures can modify the relative bonding strengths of the same surface sites.

HREELS studies have detected molecular CO in a one-coordinated on-top location on all three tungsten crystal faces, (111), (100), and

(110).[24] Upon CO dissociation the carbon and the oxygen atoms are located on fourfold sites on the W(100) face.

Sequential filling of adsorption sites by molecular NO was also reported by Pirug et al. in their HREELS study on the Pt (100) surface[24b].

Temperature-Dependent Changes in the Character of the Surface Chemical Bond

The importance of temperature in determining the nature of adsorbed species has been known for a long time. Langmuir[25] has called attention to the fact that there were apparently two types of adsorption. One type occurred at low temperatures near the boiling points of the adsorbate atoms or molecules. This type of adsorption would decrease with increasing temperature at a given pressure. For some molecules, however, such as carbon monoxide or oxygen on metals, the amount that could be adsorbed started to increase again as the temperature was increased further. This second type of adsorption was called "activated" by Taylor.[26] The low-temperature type of adsorption has gradually become known as "physical adsorption." It is characterized by low heats of adsorption (less than twice the heat of liquefaction) and a lack of chemical specificity. The second type of adsorption has become known as chemical adsorption or "chemisorption," since usually an energy of activation was involved such as observed in most chemical reactions and, unlike physical adsorption, the activated adsorption was characterized by a high degree of chemical specificity and occurred only for certain substrate–adsorbate combinations.[27]

The conversion from physisorbed to chemisorbed state by a thermally activated process has been commonly displayed by using a one-dimensional potential-energy diagram as shown in Figure 6.11a. The reference states are the gas-phase molecule, the adsorbed molecule, and the dissociated atom, respectively, and the curve crossing simulates the activation energy barrier the molecule has to surmount to atomize and occupy its stronger bound state at a smaller interatomic distance. If the activation energy barrier the molecules have to surmount to dissociate is lower than the heat of desorption, atomization will occur with increasing temperature (Figure 6.11a). However, if the activation energy for dissociation is greater than the heat of desorption, the molecule will be removed from the surface before dissociation can occur, as shown in Figure 6.11b. In reality the height of the activation energy barrier can be modified by a variety of experimental factors, including the surface coverage.[28] In many instances there are several activation energy barriers that separate molecules in their different binding states, and there

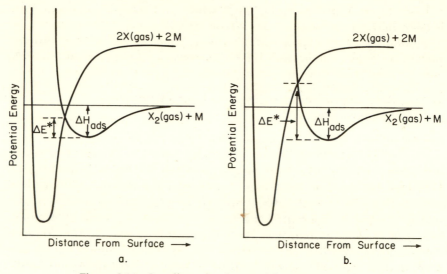

Figure 6.11. One-dimensional potential-energy curves for dissociative adsorption through a precursor or physisorbed state: (a) represents adsorption into the stable state with low activation energy as compared to ΔH_{ads}; (b) represents adsorption in the stable state with higher activation energy than ΔH_{ads}.

may be several reaction channels for bond breaking, all leading to the formation of the chemisorbed state.

Since this physical picture of activated adsorption was developed relatively early in the history of surface chemistry, the experimental techniques were just not available at that time to scrutinize the activated adsorption process in its atomic details. Presently, various electron spectroscopies and LEED hold promise in elucidating this phenomenon.

The adsorption of hydrocarbons has been investigated as a function of temperature in some detail. A striking phenomenon is the sequential bond breaking in these molecules as the temperature is increased. Molecular ethylene adsorbs intact on the W(110) surface at 200 K, for example.[29] However, upon heating to 300 K it converts to acetylene (C_2H_2) as it breaks two C—H bonds, as indicated by electron spectroscopy studies. Subsequent heating to 500 K causes two more C—H bonds to break, and there is evidence from UPS studies for the presence of C_2 units on the tungsten surface. As the temperature is increased even further, to 1100 K, the carbon dimer dissociates and only carbon atoms are left on the surface.[29] Similar studies were carried out on the Fe(100) crystal face.[30] Again, C_2H_4 adsorbs as an intact molecule at 142 K and dehydrogenates to C_2H_2 as the temperature is increased to 200 K. Upon further heating, however, the $C\equiv C$ bond of the molecule is broken at 400 K, and UPS evidence shows the presence of CH and CH_2 molecular fragments on the metal surface. At 800 K, only carbon atoms are detectable. On the (110) iron surface the C_2H_4 molecule does not break up to fragments at 300 K, in contrast to its behavior on the Fe(100) face.[31] This is again a striking confirmation of the importance of the local atomic surface structure of the substrate in controlling the chemical bonding of the adsorbates.

The temperature-dependent changes in the surface chemical bonds of adsorbates can be monitored by photoemission, in addition to LEED and HREELS studies. The bonding of C_2H_4 on the Ni(111) crystal face has been studied by Eastman and Demuth.[32] Upon heating to 230 K the molecule dehydrogenates to C_2H_2. The dominant bonding between the unsaturated hydrocarbon and the transition metal surface occurs via the π-d bonding interaction, similar to that which occurs in organometallic compounds. Bonding involves a mixture of occupied metallic d-states with unoccupied molecular π^* states and occupied molecular π-states with unoccupied metalic d^* states to produce both active and back-bonding components, respectively.

Similar behavior of ethylene molecules has been revealed by more detailed high-resolution electron energy loss spectroscopy studies of nickel surfaces.[33] The molecule adsorbs without decomposition at 150 K. However, it rehybridizes to assume close to a single C—C bond sp^3-type configuration, di-σ-bonded to the Ni(111) surface. At higher coverages there is also evidence of additional hydrogen bonding. At 230 K ethylene dehydrogenates to C_2H_2 and hydrogen. Acetylene adsorbs without decomposition on the same Ni(111) surface up to 400 K.[33] It is rehybridized, however, and has a bond order of 1.5 (between sp^2 and sp^3). Above 400 K the molecules break the C—C bond and CH units are detectable on the metal surface. It appears that the carbon atoms bond to three Ni atoms. These molecular fragments should not be viewed as free radicals, since they exhibit high thermal stability. Since the ELS studies indicate that the CH units occupy the threefold sites on the Ni(111) crystal face,[33] they should be viewed as surface compounds with the formula Ni_3CH.

The platinum and rhodium (111) crystal faces adsorb and rearrange C_2H_4 molecules sequentially, with increasing temperature, but without causing the breakup of the molecules until 450 K is reached. C_2H_4 adsorbs on top of a Pt atom at 300 K, forming a well-ordered (2 × 2) surface structure.[34] As the temperature is increased to 375 K, the molecule rotates into a threefold site; it rearranges to form a C—CH$_3$, ethylidine species while maintaining the same (2 × 2) unit cell.[5] The stronger chemical interaction that occurs as the carbon atom in the molecule interacts with three metal atoms instead of one results in a very strong and short metal–carbon bond (2.0 Å instead of the covalent 2.2 Å Pt—C bond), and an elongation of the C—C bond to a single bond (1.57 Å). The ethylidine molecule has a C—C internuclear axis that is perpendicular to the surface, and thus the two carbon atoms in the molecule are in different chemical environments. This type of stable molecular structure for the rearranged C$_2$H$_4$ molecule on the Pt(111)[35] and Rh(111)[36] crystal faces should be contrasted with the presence of CH molecular fragments on the Ni(111) crystal face in identical experimental conditions.[33] As the temperature is increased further, these molecular surface species dissociate to CH fragments and around 800 K form a graphite overlayer. The basal plane of graphite that is produced is parallel to the surface and readily discernible by LEED studies. The metal surfaces that remain catalytically active in the presence of the ethylidine molecule or in the presence of stable C$_2$ or CH fragments become catalytically inactive in the presence of the graphitic carbon monolayer that forms at higher temperatures.

The free energy of the reaction C$_2$H$_4$ + 6M → 2M—C + 4M—H has large negative values. Nevertheless, the process occurs in sequence. It appears that each bond-breaking step requires an activation energy, and the surface intermediates that form are well protected by these potential-energy barriers in a finite temperature range. From the few experiments that have been carried out so far, there is evidence that the molecules relocate to a different surface site as part of the bond scission and molecular rearrangement process. Similar experimental studies exploring the temperature-dependent changes of bonding of diatomic and polyatomic molecules will help to elucidate the nature of the surface chemical bond and the activated bond-breaking processes.

HREELS studies have recently reported the temperature-dependent dissociation of NO on the Ni(111) face[36b] and that of CO on a stepped nickel surface.[36c] Molecular NO dissociates upon warming the (111) crystal surface of nickel in the 250–350 K range. Molecular CO dissociates on the stepped surface upon warming to 430 K.

Pressure Dependence of Chemical Bond Breaking on Surfaces

When CO is adsorbed on the Ni(111) or Rh(111) crystal face, the bonding geometries of the adsorbed molecules can readily be determined by a variety of experimental techniques, such as LEED and HREELS. As the solid is heated to above 450 to 500 K in $P_{CO} < 10$ torr or at low pressures in vacuum, the adsorbed molecules are desorbed into the gas phase without showing any sign of dissociation prior to desorption.[37] Nevertheless, nickel and rhodium catalyst surfaces exhibit catalytic reaction chemistry that indicates dissociation of CO molecules at 500 K and above. The reason for the differences in chemical behavior when the surface is heated at low pressures or in vacuum or in the high pressures of the reactants lies in the differences of surface coverages in the two experimental circumstances. At low pressures molecules desorb from their molecular binding states without being able to surmount the potential-energy barrier needed to relocate in the dissociated states (Figure 6.11b.). At the higher temperatures (> 500 K in these cases), where the molecules could dissociate, they are no longer in the surface environment, as the heat of desorption is smaller than the activation energy for dissociative chemisorption. However, at higher reactant pressures the surface coverage of CO can be significant, so that the molecules that are forced to remain on the surface can undergo dissociation.

Using pressure, therefore, new reaction channels may open up for the various adsorbates, leading to rearrangements and other chemical changes that compete with the desorption of the intact molecules. Thus low-pressure studies carried out at low coverages may often fail to detect many of the chemical pathways that would be accessible for the molecules. These reaction paths are readily detectable for the high surface concentration of adsorbates that are present at high reactant pressures.

Similar observations have been made in studies of the silver–oxygen system.[38] When the metal was heated in oxygen at higher pressures (> 1 torr), several oxygen surface structures and new oxygen species could be detected at the surface that did not form at low oxygen pressures under otherwise similar conditions. The oxygen species that appears only after higher-pressure oxygen heat treatments appears to be among the active surface species for the epoxidation of ethylene to produce ethylene oxide.

Effects of Surface Irregularities on Bonding

Perhaps the most striking effects of atomic steps and kinks on bonding so far are reported for nickel[33] and platinum.[39] On these metals, several molecules dissociate in the presence of surface irregularities while they remain intact on the smooth low-Miller-index surfaces. Most hydrocarbons show this behavior on platinum, as is well documented in many publications.[39] While ethylene remains molecularly adsorbed on Ni(111) at 200 K, it dissociates readily on a stepped 6(111) × (100) nickel surface at as low as 150 K.[33] Acetylene is completely dehydrogenated on the same stepped nickel surface and forms C_2 units at temperatures as low as 150 K. The C_2 species decompose to carbon atoms below 180 K, as indicated by HREELS studies. In contrast, acetylene remains intact on the Ni(111) surface up to 400 K. At that temperature, however, the C—C bond is broken on the flat surface prior to breaking the C—H bonds.[33] Thus the stepped surface not only facilitates the decomposition of C_2H_2 at much lower temperatures than the (111) face, but it exhibits a different bond-breaking activity (i.e., dehydrogenation instead of hydrogenolysis). Thermal desorption and HREELS studies indicate the presence of both molecular and dissociated carbon monoxide on stepped nickel surfaces and only molecular CO on Ni(111) under identical experimental conditions.[36,40] Whereas CO molecules adsorb intact on the Pt(111) or Pt(100) surfaces, electron spectroscopy studies indicate dissociation of the molecule at kink sites.[41] Once the kinks on the platinum surface are blocked by carbon from the dissociated species, only molecular adsorption is detectable. Steps on Pt were found to affect the dissociative adsorption of NO as well.[42]

Although the effects of steps on hydrocarbon bonding were also reported in studies of iridium[43] and rhodium,[44] they were less pronounced. The reason appears to be the ability of even the smooth low-Miller-index surfaces of these metals to dissociatively adsorb hydrocarbons and other molecules. Thus, for metals left of platinum and nickel in the periodic table, surface irregularities are likely to influence the degree of dissociation and perhaps the kinetics of the various bond-breaking processes rather than exhibiting a unique type of bond-breaking activity that atoms in the smooth surfaces do not.

Molecules with higher dissociation energies than hydrocarbons, such as CO and N_2, are likely to exhibit marked structure sensitivity on these metals. There is experimental evidence for N_2 dissociation at vacant step sites on tungsten. While molecular CO desorbs from the clean Rh(111) surface at low pressures ($< 10^{-4}$ torr) before dissociation may occur, it dissociates on the stepped Rh(775) crystal face under identical conditions.[37,45]

There is evidence for different modes of bonding at steps. For example, hydrogen adsorption at platinum steps causes an increase in the work function, while adsorption of hydrogen on terraces of platinum decreases the work function.[46] The heat of adsorption of hydrogen was reported to be about 3 kcal higher at stepped Pt surfaces and even higher at kinked Pt surfaces than on the Pt(111).[47] Higher heats of adsorption of oxygen at iridium steps[48] and carbon at stepped nickel surfaces[49] have also been reported.

There are several reports of greatly increased adsorption probabilities of gases in the presence of steps. These were found for dissociative adsorption of N_2, O_2 on W,[50] H_2S and O_2 on Cu,[51] and H_2 and O_2 on Pt.[52] These effects are likely to be due to the reduction of a thermal activation energy barrier at the step sites for strong chemical binding, in addition to the increased bonding interaction between the adsorbates and atoms at the step site or at other surface irregularities. Surface irregularities of magnesium oxide[53] and silicon[54] crystal faces also exhibit stronger bonding and increased adsorption probabilities for oxygen and carbon monoxide.

Effects of Other Adsorbates on the Bonding of Adsorbed Gases

Carbon monoxide adsorbs predominantly as a molecule on the clean Rh(111),[44] Rh(100),[44] and Ir(110) crystal faces.[48] However, if these crystal faces are covered with a partial monolayer of carbon, a large fraction of the molecules dissociate, as indicated by thermal desorption studies.[55,56] Recently, HREELS studies showed that the stretching vibrations of CO and of several hydrocarbons on the Ni(111) crystal face shift to lower frequencies in the presence of a partial monolayer of carbon on the metal surface.[57] This shift indicates a weakening of the $C{=}O$ and $C{-}H$ bonds. Thus it appears that the adsorbed carbon strongly interacts with the coadsorbed molecules. Surface crystallography and electron spectroscopy studies show evidence for the carbene or carbynelike character of the surface carbon on platinum[10] and on nickel[33e] surfaces. These active carbon species form multiple metal bonds and are distinctly different in chemical character from the graphitic form that is produced upon heating the metal above 700 to 800 K in the presence of hydrocarbons. These active carbonaceous fragments can readily rehydrogenate and produce methane and other hydrocarbons and can also interact strongly with the various other molecular adsorbates.

LEED studies have shown the existence of coadsorbed surface structures of N_2 and H_2 on W(110)[58] and CO and H_2 on Pt(111).[59] In the presence of both gases, ordered surface structures form that are different from the structures that are produced when only one or the other gas is present. Thus both species participate in forming the surface unit cell. This is indicative of a strong, attractive interaction between the two different molecules to produce the mixed adsorbed monolayer.

As opposed to these strong, attractive interactions with other adsorbates, adsorbed atoms may also block important bonding sites, thereby preventing bond formation or bond breaking. Gold on polycrystalline platinum surfaces prevented the hydrogenolysis of hydrocarbons.[60] The presence of copper on nickel surfaces reduces the rate of adsorption of oxygen,[61] and sulfur prevents the chemisorption of CO.[62]

Potassium, when present on the Fe(100) crystal face, eliminates the activation energy for the dissociated chemisorption of N_2.[63] As a result, the rate of dissociation of N_2 is increased 300-fold over that on the clean iron surface under identical experimental conditions. Similarly, alkali metals increase the rate of dissociative chemisorption of CO on iron.[64] In these circumstances the alkali metal acts as an electron donor that markedly influences the bonding between the adsorbate and the substrate. In fact, electron acceptors such as halogens, or oxygen or electron donors (alkali metals), are frequently added to catalyst surfaces in order to obtain optimum rates and reaction selectivity. These are then called "promoters."

There are then several effects that occur upon the addition of other adsorbates that influence the nature of the surface chemical bond. These additives can (1) block important binding sites (steps or kinks, for example), and (2) they are either electron donors or acceptors, which influence the nature and magnitude of charge transfer between the adsorbate and the substrate. In some cases (3) they can also restructure the surface and thus alter the geometry of the binding sites (sulfur on Ni or Pt).

Surface Compounds

Investigations of the segregation of carbon to nickel surfaces and of sulfur to nickel surfaces determined that these processes are exothermic by 4 to 12 kcal, respectively.[65] Correlations of the heat of adsorption of sulfur on transition metals with the heats of formation of transition metal sulfides clearly indicate that the metal-sulfur surface bonds are markedly stronger than the sulfur bonds in the bulk sulfides.[65b] As a result of their stronger chemical bonding to atoms at the surface as compared to atoms in the bulk, carbon and sulfur will occupy surface sites preferentially on nickel surfaces and many other transition metals. Surface segregation is just one of many experimental indications that surfaces provide unique chemical bonding environments, which may also result in the formation of surface compounds, that have no chemical analogue in the bulk phase diagrams.

The presence of strongly bound oxygen that appears at the surfaces of many transition metals is a case in point. When platinum[66] or rhodium[67] are heated gently (<450 K) in the presence of low pressures of oxygen ($\leq 10^{-5}$ torr), a chemisorbed oxygen layer forms that can be removed readily by hydrogen or carbon monoxide, which react with the chemisorbed species to form H_2O or CO_2, respectively. Upon heating in oxygen to higher temperatures and/or pressures, oxygen partly diffuses into the bulk in the near-surface region, and simultaneously a strongly bound surface oxide forms.[66] This oxide does not easily react with hydrogen or with carbon monoxide but remains intact in the presence of hydrocarbons and other reducing gases. While it decomposes when heated in vacuum, the decomposition temperature is in excess of 1200 K, indicating very high surface binding energies ($\Delta H_{ads} \geq 60$ kcal/mol). For the Pt–O system, there is no known bulk oxide with similar properties of high thermal stability as the strongly bound surface oxide. For rhodium, further heating of the strongly bound surface oxide results in the formation of an epitaxial layer of Rh_2O_3 on the Rh(111) crystal surface. Thus, for the Rh–O system, the surface oxide appears prior to the formation of the bulk oxide phase, and this strongly bound surface oxide is without analogue in the bulk Rh–O phase diagram.[67]

The formation of surface compounds in the Cu(110)-S and Cu(111)-S systems has been reported together with a rather complete surface phase diagram.[68] Two-dimensional surface phases have been identified in the Ag–S[69] and Ni–C[70] systems as well.

Surfaces provide unique atomic and electronic environments, since there is a large change in the number of nearest neighbors, site symmetry, and bonding anizotropy as compared to bonding sites in the bulk of the solid. It is, therefore, not altogether surprising that surface compounds may form that could not form in the bulk of the solid.

The behavior of binary metal alloys changes markedly with increasing surface-to-volume ratio. For very small particles, as their dispersion approaches unity, most of the atoms will be located on the surface, where they have a reduced number of nearest neighbors and different rotational symmetries. Under these conditions the electronic interactions that govern the formation of the bulk phases are altered. New phases or new compounds may form or compounds that are stable in the bulklike environment become unstable. This is strikingly confirmed by the formation of the so-called "bimetallic clusters," which exhibit a very different phase behavior as small particles.[71] For example, Ru and Cu or Ir and Au show very small solubilities in each other's phase. These metals, Ru–Cu and Ir–Au, are largely immiscible, as indicated by their bulk phase diagrams. When codeposited in a large-surface-area dispersed particle form, they exhibit complete miscibility,[72] as shown by their unique chemical and chemisorption behavior. Sinfelt has demonstrated how these bimetallic clusters form at high dispersions, as well as their unique chemical activity.

A theoretical model that explains the changes in phase behavior in multicomponent systems as the surface-to-volume ratio is altered has been developed by Moran-Lopez and Falicov.[73]

Particles made up from two or more components with high surface-to-volume ratios are a fertile area for scrutiny for the existence of surface compounds or surface phases. Such systems should be of unique importance in powder metallurgy. In this way it should be possible to produce materials with new and unusual mechanical properties that could not be produced by working with the same constituents in their bulk form.

Cluster Model of the Surface Chemical Bond

The structure sensitivity of the heats of adsorption, the molecular orientation at the binding site, and bond scission, and the importance of surface irregularities, steps, and kinks in the adsorption and bond-breaking processes bring into focus the importance of localized interactions in forming the surface chemical bonds. The formation of such a bond often involves more than one metal atom for each adsorbate. The adsorbate also tends to occupy sites of high symmetry where equivalent bonding to several substrate atoms is possible.

Recent studies indicate several important similarities between cluster compounds where there are several metal atoms in the cluster and certain substrate–adsorbate systems.[74]

A representative group of metal clusters that contain 3 to 15 metal atoms, all forming metal–metal bonds, are shown in Table 6.1. These clusters exist mostly in solution; thus exploration of their chemistry is usually carried out in the liquid phase.[75] This makes it somewhat difficult to readily correlate their properties with those of adsorbates deposited on the surface of the same metal. Since CO is one of the most frequently utilized ligands, it is a natural choice to explore correlations between its surface and cluster chemistries.

Figure 6.12 shows the correlation between the heats of chemisorp-

Table 6.1. Metal clusters: stable clusters containing three to fifteen metal atoms

Triangles (3)	Tetrahedra (4)	Five-atom
$Os_3(CO)_{12}$	$Ni_4(C_5H_5)_4H_3$	$Pt_3Sn_2(1,5—C_8H_{12})_3Cl_6$
$Pd_3(CNR)_6$	$Ir_4(CO)_{12}$	Bi_5^{3+}
$Re_3Cl_{12}^{2-}$	$Li_4(CH_3)_4$	$Fe_5\underline{C}(CO)_{15}{}^a$
$Fe_3(CO)_9(RC{\equiv}CR)$	$Ni_4[CNR]_7$	$Ni_5(CO)_{12}^{2-}$
Octahedra (6)	**Seven-atom**	**Eight-atom**
$Co_6(CO)_{14}^{2-}$	$Rh_7(CO)_{16}^{3-}$	$Os_8(CO)_{23}$
$Mo_8Cl_{14}^{2-}$	$Rh_7(CO)_{16}I^{2-}$	$Au_8Au[Pr_3]_6^{3+a}$
$Ru_6\underline{C}(C_6H_6)(CO)_{14}{}^a$		Bi_8^{2+}
$Au_6[PR_3]_6^{2+}$		$Co_4B_4H_4(C_5H_5)_4$
Nine-atom	**Ten-atom**	**Twelve-atom**
Bi_9^{5+}	$Au_{10}Au[SCN]_3[PR_3]_7$	$Pt_{12}(CO)_{24}^{2-}$
Sn_9^{4-}		$Rh_{12}\underline{Rh}(CO)_{24}H_3^{2-a}$
$Pt_9(CO)_{18}^{2-}$		
	Fifteen-atom	
	$Pt_{15}(CO)_{30}^{2-}$	
	$Rh_{15}\underline{C}_2(CO)_{28}{}^a$	

[a] Metal cages. An atom, underscored, is at the center of the polyhedron.

tion of CO on transition-metal surfaces and the binding energies for M—CO in metal carbonyls.[11] The correlation is certainly not unreasonable. More important, electron spectroscopy studies revealed that the photoelectron spectra of three and four metal–carbonyl clusters is almost identical to the spectra of CO adsorbed on the surface of the same transition metals[75] [$Ir_3(CO)_4$ and $Ru_4(CO)_4$ versus CO on Ir(100) and Ru(100)]. The cluster had to contain no fewer than three metal atoms to obtain good agreement between the electronic structures of the cluster and adsorbate systems.

The ethylidine or ethyledene molecules that form by the rearrangement of adsorbed acetylene and are located at a triangular site on the Pt(111) and Rh(111) surfaces have their metallorganic cluster compound equivalents. Osmium and cobalt form almost identical trinuclear ethylidine clusters, as determined by X-ray crystallography.[5] The vibrational spectra of these metallorganic systems are very similar to the vibrational spectra of the adsorbed C_2H_2 molecule on the Pt(111) and Rh(111) crystal faces as determined by HREELS. Thus it appears that, in

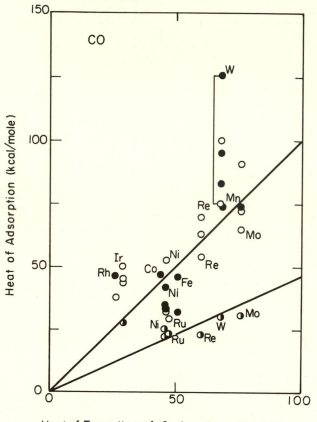

Figure 6.12. Heats of adsorption of CO on transition-metal surfaces as a function of the heats of formation of gaseous carbonyls. Filled circles indicate the results of caloriemetric studies; open circles, thermal desorption; half-filled circles, thermal desorption studies displaying the lowest energy-binding states.

at least a limited number of systems, the cluster model of the surface chemical bond is appropriate. Many more studies are needed to probe the validity of the cluster model for other adsorbate–substrate systems.

While structural studies that determine the location of the molecules (ligands), their site symmetry, bond distances and bond angles, and electronic and vibrational structures would greatly aid in establishing chemical correlations between surfaces and clusters, there are other studies that could also be performed. For example, the exchange rates of ligands such as carbonyls among the on-top (one-coordinated), bridge (twofold), and triangular (threefold) sites in the cluster can be determined by nuclear magnetic resonance (NMR). From the temperature dependence of the NMR spectra, the activation energies for the exchange can also be obtained.[76] A great deal of information is already available about the rates of exchange of CO among these sites and activation energies for the exchange from NMR studies of clusters in solution. The same bonding sites are also available on many solid surfaces. There is, however, little or no information of similar type available for adsorbates on surfaces. It is hoped that through the use of HREELS or other surface-site-sensitive techniques, these data will become available in the near future, so that the surface–adsorbate and cluster systems could be better compared.

There are reasons, of course, why a simple cluster model will not describe the surface bond in many circumstances. The surface environment is asymmetric and there is a surface dipole that may modify the bonding. Several different sites on the surface in the proximity of the bonding site may facilitate the transition from one bonding configuration to another. These sites would not all be available for a four-atom cluster compound of high structural symmetry. The high charge density in the bulk metal can modify the charge density of the surface metal atoms to prevent the stabilization of higher oxidation states at the surface. Nevertheless, investigation of the similarities and differences in bonding and reactivity of adsorbates on surfaces and ligands in multinuclear clusters is a promising area of research to explore the nature of the surface chemical bond.

References

1. G.A. Somorjai, Angew. Chem. **16,** 92 (1977).
2. A. Morgan and G.A. Somorjai, J. Chem. Phys. **51,** 3309 (1969).
3. J.T. Grant, Surf. Sci. **18,** 228 (1969).
4. K. Legg, F. Jona, D. Jepsen, and P. Marcus, J. Phys. **C8,** 492 (1975).
5. (a) P. Stair and G.A. Somorjai, Chem. Phys. Lett. **41,** 391 (1976); (b) L.L. Kesmodel, L.H. Dubois, and G.A. Somorjai, Chem. Phys. **70,** 2180 (1979).
6. J. Oudar, Int. Met. Rev., 57, 1978 (2).
7. G. Ehrlich and F. Hudda, J. Chem. Phys. **35,** 1421 (1962).
8. L.H. Dubois and G.A. Somorjai, Surf. Sci. **88,** L13 (1979).
9. G.A. Somorjai and J. Buckholz, Acc. Chem. Res. **9,** 3331 (1976).
10. G.A. Somorjai, Surf. Sci. **89,** 496 (1979).
11. I. Toyoshima and G.A. Somorjai, Catal. Rev.-Sci. Eng. **19,** 105 (1979).

12. H. Froitzheim, H. Ibach, and S. Lehwald, Phys. Rev. Lett. **36**, 1549 (1976).
13. R. Barker and P. Estrup, Phys. Rev. Lett. **41**, 1307 (1978).
14. C. Backx, R. Willis, B. Feuerbacher, and B. Fitton, Surf. Sci. **68**, 516 (1977); **163**, 193 (1977).
15. D. Menzel, in *Topics and Applied Physics*. vol. 4, ed. R. Gomer, Springer-Verlag, New York, 1975, pp. 12, 101.
16. J. Anderson and G. Lapayre, Phys. Rev. Lett. **36**, 376 (1976).
17. G. Ertl and J. Küppers, *Low Energy Electrons and Surface Chemistry*, Verlag Chemie, Weinheim, 1974.
18. H. Froitzheim, H. Ibach, and S. Lehwald, Phys. Rev. **B14**, 1362 (1976).
19. D.G. Castner, B. Sexton, and G.A. Somorjai, Surf. Sci. **71**, 519 (1978).
20. L.H. Dubois and G.A. Somorjai, Surf. Sci. **91**, 514 (1980).
21. (a) H. Froitzheim, H. Ibach, and S. Lehwald, App. Phys. **13**, 147, 1977; (b) A. M. Baro and H. Ibach, J. Chem. Phys. **71**, 4812 (1979).
22. (a) S. Andersson, Solid State Commun. **20**, 229 (1976); (b) W. Erley, H. Wanger, and H. Ibach, Surf. Sci. **80**, 612 (1979).
23. A. Bradshaw and F. Hoffman, Surf. Sci, **72**, 513 (1978).
24. (a) H. Froitzheim, H. Ibach, and S. Lehwald, Surf. Sci. **63**, 56 (1977); (b) G. Pirug, H.P. Bonzel, H. Hopster, and H. Ibach, J. Chem. Phys. **71**, 593 (1979).
25. I. Langmuir, J. Am. Chem. Soc. **40**, 1361 (1918).
26. H. Taylor, J. Am. Chem. Soc. **53**, 578 (1931).
27. P. Emmett, in *Critical Reviews in Solid State Science*, vol. 4, CRC Press, Cleveland, Ohio, 1974, pp. 127–150.
28. (a) H. Bonzel, Surf. Sci. **58**, 236 (1977); (b) T. Engel and G. Ertl, J. Chem. Phys. **69**, 1267 (1978).
29. See G.A. Somorjai, Acc. Chem. Res. **9**, 248 (1976), and references.
30. T. Rhodin, C. Bruckner, and A. Anderson, J. Phys. Chem., to be published.
31. G. Brodén, G. Gafner, and H. Bonzel, J. Appl. Phys. **13**, 333 (1977).
32. D. Eastman and J. Demuth, Phys. Rev. Lett. **32**, 1123 (1974).
33. (a) H. Ibach, H. Hopster, and B. Sexton, Appl. Surf. Sci. **1**, 1 (1977); (b) S. Lehwald, W. Erley, H. Ibach, and H. Wagner, Chem. Phys. Lett. **62**, 360 (1979); (c) S. Lehwald and H. Ibach, Surf. Sci. **89**, 425 (1979); (d) J. E. Demuth and H. Ibach, Surf. Sci. **85**, 365 (1979); (e) J. E. Demuth and H. Ibach, Surf. Sci. **78**, L238 (1978).
34. P. Stair and G.A. Somorjai, J. Chem. Phys. **66**, 2036 (1977).
35. (a) H. Ibach and S. Lehwald, J. Vac. Sci. Technol. **15**, 407 (1978); (b) L.H. Dubois, L. Kesmodel, and G. A. Somorjai, Chem. Phys. Lett. **56**, 267 (1978).
36. (a) L.H. Dubois, D.G. Castner, and G.A. Somorjai, J. Chem. Phys., **72**, 5234 (1980); (b) S. Lehwald, J.T. Yates, and H. Ibach, ECOSS3 Proceedings, to be published, 1980; (c) W. Erley, H. Ibach, S. Lehwald, and H. Wagner, Surf. Sci. **83**, 585 (1979).
37. L.H. Dubois and G.A. Somorjai, Adv. Chem. series, to be published, 1980.
38. (a) K. Muller, Z. Naturforsch **20A**, 153 (1965); (b) H. Albers, W. Van der Wal, and G. Bootsma, Surf. Sci. **68**, 47 (1977); (c) G. Rovida, F. Fratesi, M. Maglietta, and E. Ferroni, Surf. Sci. **43**, 230 (1974).
39. (a) B. Lang. B. Joyner, and G.A. Somorjai, Proc. Roy. Soc. (Lond.) **A331**, 335 (1972); (b) S. Bernasek, W. Siekhaus, and G.A. Somorjai, Phys. Rev. Lett. **30**, 1202 (1973); (c) S. Ceyer, R. Gale, S. Bernasek, and G.A. Somorjai,

J. Chem. Phys. **64,** 1934 (1976); (d) D. Blakely and G.A. Somorjai, J. Catal. **42,** 141 (1976).

40. W. Erley, H. Wagner, and H. Ibach, Surf. Sci. **80,** 612 (1979).
41. R. Mason and G. A. Somorjai, Chem. Phys. Lett. **44,** 468 (1976).
42. H. Bonzel, G. Brodén, and G. Pirug, J. Catal. **53,** 96 (1978).
43. B. Niewenhuys and G.A. Somorjai, Surf. Sci. **72,** 8 (1978).
44. D.G. Castner, B. Sexton, and G.A. Somorjai, Surf. Sci. **71,** 519 (1978).
45. K. Besocke and H. Wagner, Surf. Sci. **87,** 457 (1979).
46. K. Christman and G. Ertl, Surf. Sci. **60,** 365 (1976).
47. S.M. Davis and G.A. Somorjai, Surf. Sci. **91,** 73 (1980).
48. D. Hagen, B. Niewenhuys, G. Rovida, and G.A. Somorjai, Surf. Sci. **57,** 632 (1976).
49. L. Isett and J. Blakely, J. Vac. Sci. Technol. **12,** 237 (1976).
50. (a) K. Besocke and S. Berger, Proc. 7th Int. Vac. Congr. and 3rd Int. Conf. Solid Surf., Vienna, **2,** 893 (1977); (b) D. King, ibid., **1,** 769, and references therein.
51. J. Perdereau and G. Rhead, Surf. Sci. **24,** 555 (1971).
52. (a) K. Christman and G. Ertl, Surf. Sci. **60,** 365 (1976); (b) G. Pirug. G. Brodén, and H. Bonzel, Proc. 7th Int. Vac. Congr. and 3rd Int. Conf. Solid Surf., Vienna, **2,** 907 (1977).
53. C. Harkins, W. Shang, and T. Leland, J. Phys. Chem. **73,** 130 (1969).
54. H. Ibach, Surf. Sci. **53,** 444 (1975).
55. B. Sexton and G.A. Somorjai, Surf. Sci. **71,** 519 (1978).
56. B. Nieuwenhuys and G.A. Somorjai, Surf. Sci. **72,** 8 (1978).
57. H. Ibach and G.A. Somorjai, App. Surf. Sci. **3,** 293 (1979).
58. J. May, L. Germer, and C. Chang, J. Chem. Phys. **45,** 2383 (1966).
59. (a) A. Morgan and G.A. Somorjai, Trans. Am. Crystallogr. Assoc. **4,** 59 (1958); (b) A. Morgan and G.A. Somorjai, Surf. Sci. **12,** 405 (1968).
60. D. Hagen and G.A. Somorjai, J. Catal. **41,** 466 (1976).
61. J. B. Sinfelt, Science **195,** 641 (1977).
62. J. Blakely, to be published.
63. G. Ertl, M. Weiss, and S. Lee, Chem. Phys. Lett. **60,** 391 (1979).
64. G. Broden, G. Gafner, and H. Bonzel, Surf. Sci. **84,** 295 (1979).
65. (a) L. Isett and J. Blakely, Surf. Sci. **58,** 397 (1976): (b) J. Oudar.
66. C. Smith, J.P. Biberian, and G.A. Somorjai, J. Catal. **57,** 426 (1979).
67. D.G. Castner and G.A. Somorjai, App. Surf. Sci., to be published (1980).
68. (a) J. Domange, J. Vac. Sci. Technol. **9,** 682 (1972); (b) J.M. Moison and J. Domange, Surf. Sci. **67,** 336 (1977).
69. M. Lagues and J. Domange, Surf. Sci. **47,** 77 (1975).
70. J. Blakely and J. Shelton, *Surface Physics of Materials*, vol. 1, New York, Academic Press, 1975, p. 189.
71. J. Sinfelt, Science **195,** 641 (1977).
72. J. Clarke, Chem. Rev. **75,** 291 (1975).
73. (a) J. Moran-Lopez and L. Falicov, Surf. Sci. **79,** 109 (1979); (b) J. Moran-Lopez and L. Falicov, Phys. Rev. **B18,** 2542 (1978).
74. T. Rhodin, E. Band, C. Brucker, and W. Pretzer, Chem. Rev., April 1979.
75. E.W. Plummer, in *Photoemission and the Electronic Properties of Surfaces*, eds. B. Feuerbacher, B. Fitton, and R. Willis, Wiley, London, 1979.
76. E. Muetterties, Science **196,** 839 (1977).

Appendix*

Table 6.A1. Heat of adsorption of O_2 on single-crystal surfaces[a]

Metal	ΔH_{ads} (kcal/mol)	Technique	References
Ag(110)	~40	TD isothermal	27
(111)	35–45	TD	28–30
Cu(110)	50	TD	31
Ge(111)	64.6 ± 11.5	mass desorbed GeO	32
Ir(111)	65	TD	33
Mo(100)	118	TD	34
Pd(110)	80~48	isothermal	35
(100)	55	LEED	36
(111)	50–60	TD	36
Pt(100)	44.7 (β_1)	TD	37
	62.1 (β_2)		
	69.3 (β_3)		
W(100)	120–173	isosteric TD	38
(100)	136	TD	34
(100)	93.2	TD atom	39
(110)	139	TD	34
(110)	92.0	FEM, TD	40
(111)	23.1	TD	41

[a] RT, room temperature; cal., calorimetry; TD, thermal desorption; FEM, field-emission microscopy.

Table 6.A2. Heat of adsorption of O_2 on polycrystalline surfaces[a]

Metal	Form	Temperature	ΔH_{ads} (kcal/mol)	Technique	References
Ag	plate	216~226 K	8.3	cal.	1
	powder	383 K	130 → 10 ($\Theta 0 \to 1$)	cal.	2
	powder	773 K	41.8 ± 3.5	isotherm	3
Al	film	RT	211–159	cal.	4
Au	powder	140 K	103→	cal. (wide-range Θ)	5
Co	film	RT	100→	cal.	4
	film	RT	98→	cal.	6
	film	77 K	84→	cal.	6
Cr	film	RT	174	cal.	4
Cu	plate	207~223 K	5.2	cal.	1
	film	273 K	109.8–20	cal. (2-type adsorption state)	7

*References for the tables in this appendix follow Table 6.E2.

Table 6.A2.—Continued

Metal	Form	Temperature	ΔH_{ads} (kcal/mol)	Technique	References
Fe	film	RT	136→	cal.	4
	film	RT	71→	cal.	8
	film	273 K	99–120→	cal.	9
Ge	film	RT	131	cal.	10
Ir	ribbon	RT	38.1 ± 3	TD (second order)	11
			71.5 ± 2		
Mn	film	RT	152→	cal.	4
Mo	film	RT	172→	cal.	4
	film	RT	192→	cal.	6
	film	77 K	170→	cal.	6
	filament	1800–2200 K	132	TD	12
	ribbon	RT	60	TD (two states)	13
			110		
Ni	film	RT	120~130	cal.	14
	film	RT	115→	cal.	4
	film	RT	107→	cal.	6
	film	RT	140~150→	cal.	15
	film	273 K	76~89→	cal.	16
	film	90 K	71→	cal.	6
	film	77 K	70→	cal.	6
Pd	film	RT	67	cal. mean value	4
Pt	film	RT	67~72	cal.	4
	filament	298 K	58	TD atomization	17
	filament	RT	54 (β_4)	TD (first order)	18
			39 (β_2)	TD (first order)	18
Re	film	RT	175	cal.	10
	filament	RT	≤92	TD	19
	filament	1600–2200 K	127	TD	12
	filament	RT	127	TD	20
Rh	film	RT	120	cal.	4
Si	film	RT	218	cal.	10
Ta	film	RT	212→	cal.	4
Ti	film	RT	236→	cal.	4
W	film	RT	183	cal.	6
	film	RT	194	cal.	4
	filament	303 K	139	cal.	21
	film	77 K	184	cal.	6
	film	90 K	184	cal.	6
	ribbon	RT	53 }	TD	22
			106 }	TD	
	tip	RT	92~115	FEM	23
	ribbon	RT	139 }	TD (atomic desorption)	24
			97 }	oxide evaporation	
	filament	RT	76.1 }	two states	19
			110.7 }		
	filament	2000–2150 K	135	TD, FEM	25
	filament	RT	161	TD	12
	ribbon	200–2700 K	145 }	$\theta = 0$ TD	
			83 }	$\theta = 1$ TD	26

[a] See Table 6.A1 for abbreviation key.

Table 6.B1. Heat of adsorption of H_2 on single-crystal surfaces[a]

Metal	Temperature	ΔH_{ads} (kcal/mol)	Technique	References
Cu(311)	251–297 K	$9.3 \to 9.2 \to 9.9$	IR isosteric	86
Ge(111)	?	18.5 ± 2.3	TD	32
Mo(100)	78 K	16 (β_1) 20 (β_2) 27 (β_3)	TD	87
(110)	78 K	$28 \pm$ (β_1) $34 \pm$ (β_2)	TD	88
Ni(100)	RT	23.1	TD	89
(110)	298–315 K 299–345 K	19.5 (α) 29.5 (β)	isosteric TD isothermal TD	90
(110)	282 K	20.3 (low θ)	isothermal TD	91
(110)	333–400 K	27.7	isosteric TD	92
(111)	RT	21.4 ± 1 calc. 22.7 ± 1 exp.	isothermal TD	93, 47
(100) (110) (111)	RT	23.0 (β_1, β_2) 21.5 (β_1, β_2) 23.0 (β_1, β_2)	isosteric	94
(110)	RT	23.5	TD (two states)	95
Pd(110) (110) (111) (111)	RT [Pd(S)9(111) × (111)]	$\left.\begin{array}{l}24.4\\22.8\end{array}\right\}$ D_2 $\left.\begin{array}{l}20.8\\23.8\end{array}\right\}$ H_2	isosteric ($\Theta = 0$)	96
Pt(100)	RT	15~16	TD	97
(111)	RT	18 31	molecular dissociative TD	98
(111)	150 K	9.5 (β_2)	isothermal	99
(111)	RT	15~20 30~32	TD	100
Ta(110)	RT	10	est.	101
W(100)	78 K	26.3 (β_1) 26.6 (β_1) 32.3 (β_2) 32.6 (β_2)	H_2, TD D_2, TD H_2, TD D_2, TD	102
(110)	78 K	27.0 ± 1 (β_1) 32.7 (β_2)	TD	103
(111)		14.1 (β_1) 21.7 (β_2) 30.4 (β_3) 36.6 (β_4) 6~11 (γ)		
(211)	110 K	16 (β_1) 35 (β_2)	TD	104
(211)	78, 300 K	46	FEM	72
(100)	78 K	26 (β_1) 32 (β_2)	TD	87
(110)	180–220 K	65–82 (boundary-free) 60 (boundary)	FEM	60

[a] See Table 6.A1 for abbreviation key.

Table 6.B2. Heat of adsorption of H_2 on polycrystalline surfaces[a]

Metal	Form	Temperature	ΔH_{ads} (kcal/mol)	Technique	References
Co		RT	24.1 ($\theta \to 0$)	cal.	42
Cr	film	RT	45.0	cal.	43
Cu	filament	250–400 K	43.5 (calculated)	TD	44
	film	242–337 K	9.6–11.9	work function	45
	film		8~12	work function	46
Fe	film	RT	32	cal.	43
	film	90 K	32	cal.	113
	film	77 or 90 K	27.5→	cal.	48, 14
	film	306–583 K	~16 (9.3–21.7)	isosteric ($\theta = 0.8$)	49
	filament	RT	20	TD	50
Ir	filament	100, 300 K	24	TD	51
	filament	RT	26.1 ($\theta \to 0$)		42
Mn	film	78, 193, 295 K	12.0	TD	52
	film	298 K	17.1	isothermal	53
Mo	film	293 K	40	cal. ($\theta = 0$)	54
	film	77–373	34	isosteric ($\theta = 0.45$)	55
	filament	300–1000 K	~7 (estd., $\theta = 0.6$)	TD	81
Ni	film	RT	25.5	cal.	56
	film	294 K	28	cal.	54, 57
	film	RT	11–38.4	cal.	15
	film	RT	30.0	cal.	14
	film	RT	31.0	cal.	43
	disk	~273 K	30	cal., isosteric	134
	film	77, 273 K	18.0→	cal. ($\theta = 0 \to ~0.8$)	58, 59
		77 K	18		
		273 K	18→		
	tip	4.2 K	68–72	FEM	60
	tip	345–357 K	46 ± 4	FEM (low θ)	61
	film	RT	31.5 (extrapolated)	isosteric ($\theta = 0$)	62
	film	273 K	20	isosteric ($\theta = 0.9$)	63
	film	RT	25–10	isosteric ($\theta = 0.8$)	64
	film	RT	16.0 ($\theta = 0.1, 0.75$) 15.0 ($\theta = 0.5$) }	2 states	65
	filament	86 K 279 K	38.2 52.6 }	cal.	66
	film	90 K	30	cal.	113
		296 K	31		
Pd	filament	200 K	22 (β_1) 25 (β_2) 35 (β_3) 13~14 (α) }	TD	67
	film	RT	28	cal.	43

(continued)

Table 6.B2.—Continued

Metal	Form	Temperature	ΔH_{ads} (kcal/mol)	Technique	References
Pt	film	278–295 K	8 (γ) 12 (β_1) 21 (β_2)	TD	68
	tip	4.2–320 K	16	FEM	69
	foil	140–600 K	16~<5	TD (isosteric)	70
	filament	300 K	26.0 (β)	TD	71
	tip	83–293 K	15	work function	194
Re	ribbon	RT	27.5 (first order) 67 (second order)	TD	73
Rh	film	RT	28	cal.	43
	filament	100, 300 K	18	TD	51
Ta	film	300 K	30	cal.	113
	film	RT	45	cal.	43
	filament	RT	82	TD	74
Ti	film	273 K	39 (39~22)	cal.	75
	film	77 K	27.5 (27.5~<10)	cal.	76
W	film	RT	45	cal.	43, 14
	film	RT	46.5	cal.	56
	film	294 K	52	cal.	57
	filament	79 K	45	cal.	77
	filament	RT	28	cal.	78
	film	90 K	35→	cal.	56
	film	195	42	cal.	56
	filament	77 K	31	TD isothermal	79
	filament		36	TD	80
	filament	300–1000 K	~7 (estd., $\theta = 0.6$)	TD	81
	filament		35	TD	82
	filament	RT	25.5, 29.0	TD	83
	tip	78, 300 K	20–46 ⎫ 15–20 ⎬ 8–14 ⎪ 6–10 ⎭	FEM (4 states)	72
	filament	100, 300 K	35	TD	51
	tip	4.2 K	64	FEM (low θ)	60, 84
	tip	4.2 K	6	TD (110 boundary)	84
	film	90–273 K	42	isosteric	85
	filament	300 K	15	TD isosteric ($\theta = 0.7$)	79
	filament	77 K	31	isosteric ($\theta = 0\uparrow$)	79
	—	RT	45.9 (cal.) ⎱ $\theta \to 0$ 44.0 (ob.) ⎰	—	42

a See Table 6.A1 for abbreviation key.

Table 6.C1. Heat of adsorption of CO on single-crystal surfaces[a]

Metal	Temperature	ΔH_{ads} (kcal/mol)	Technique	References
Ag(111)	66–123 K	6.5, 4.3	isothermal, isosteric	132
Co(0001)	300 K	4.2	TD	133
		16.8		
		47.0		
Cu(100)	77 K	14.5	isosteric	135
(100)	77–300 K	16.6, $\theta = 0$	isosteric	136
		11.5, $\theta = 0.5$	isosteric	136
(311)	77–300 K	14.6–10.8	isobar	137
Ir(110)	RT	37 (mean)	TD	138
(111)	RT and below	35 ± 1	isothermal $\theta = 0.33$	139
(111)	RT	45	$\theta < 0.33$ TD	140
(111)	RT	39 ± 3	$\theta = 0.33$ TD	140
(111)	RT	43.9	$\theta < 0.3$ TD	141
Ni(100)	350 K	30.0	isothermal	136
(110)	300 K	3.9, 13.8, 45.7	TD	133
(100)	200, 298 K	26.1	tracer isothermal	142
(110)	RT	3.9, E_3 (α), 45.7, E_1 (β)	TD	125
		13.8, E_2 (intermediate)		
(110)	182, 298, 287 K	25.3	tracer isothermal	142
(110)	RT	25	$\theta > 0.7$ isothermal	143
		~30	$\theta < 0.7$ isothermal	
(110)	128–300 K	25.4	TD isosteric	144
		30.0		
(110)	RT	22.5	TD	145
		26.4		
		33 ± 1	relaxation	146
		32.7	TD	147
(111)	RT	23.5 ⎱	$\theta = 0.33$ isothermal TD	148
		26.5 ⎰	$\theta = 0$	
Mo(100)	RT	62 ± 4	TD	149
		79 ± 4		
		86 ± 5		
(100)	RT	65.7	TD	150
		70.3		
		74.7		
		77.3		
		81.6		
Nb(110)	300 K	69 ⎫		
		15		
		33.7 ⎬	TD	133
		39.9		
		57.0		
		6.9 ⎭		

(continued)

Table 6.C1.—Continued

Metal	Temperature	ΔH_{ads} (kcal/mol)	Technique	References
Pd(100)	77 K	35.7 ($\theta = 0$), 27.7 ($\theta = 0.5$)	isothermal isosteric	151
(100)	RT	36	LEED	152, 154
(110)	300 K	4.8 9.0 45.2 55.3	TD	133
(100)		30	isothermal	36
(111)	RT	34 → 32	isothermal	153
(110)		30	isothermal	36
(110)	RT	41 (5 × 2), 27 (2 × 1)	isothermal	155
(111)		30	isothermal	36
Pt(100)	RT	32		156
(110)	RT	25.1 (α)(satd.)	TD	157
		30.8 (β)(satd.)	TD	
		31.8 ± 0.9 (→satd.)	isosteric	
(110)	333 K	25.1 28.7 31.4	TD	158
Ru(100)	RT	29.6	$\theta \le 0.6$	160
		22.2	$\theta > 0.6$	
		23.5 28	TD	
(101)	300 K	25.1, 28.2	TD	161
W(100)	300 K	93 ± 5 (β_3) 74 ± 4 (β_3) 62 ± 4 (β_2) 57 ± 4 (β_1); 21 ± 2 (α)	TD	162
(100)	RT	27 ± 4 57 ± 4 64 ± 4 82 ± 5	TD	149
(110)	320 K	6.8, 10, 10–25	isothermal D	163
	1000–1100 K	55 (β_1) Arrhenius		
	—	69.5 ± 3 (β_1)		
	320 K	9.5	stepwise D	163
	>1000 K	40 (β_1)		
	20–400 K	66.0 (β_2)		
(110)	300–1100 K	80.7–23.1	TD ($\theta\uparrow$)	164
(113)	300–1100 K	87.6–27.7	TD ($\theta\uparrow$)	164
(110)	4.2 K	52 (boundary) 90 (boundary-free)	FEM	60

[a] See Table 6.A1 for abbreviation key.

Table 6.C2. Heat of adsorption of CO on polycrystalline surfaces[a]

Metal	Form	Temperature	ΔH^{ads} (kcal/mol)	Technique	References
Ag	supported on Al_2O_3	303 K	18~20	IR	105
	film supported on Al_2O_3	194.5, 209.5 K	8.7	isothermal	106
		303 K	18–20	IR	105
Co	film	273 K	47→	cal.	107
		303 K	18–20	IR	105
Cu	film supported on Al_2O_3	194.5, 209.5 K	9.3	isothermal	106
		303 K	18–20	IR	105
Fe	film	273 K	46	cal.	107
	film	RT	32	cal. isothermal	108
	film	77 K	10	cal. isothermal	14
	film	90–306 K	4–20	$\theta = 0.8$	49
Ge		RT	>26	TD	109
Ir	supported on Al_2O_3	303 K	>50	IR	105
Mn	film	273 K	78→	cal.	107
Mo	film	273 K	74→	cal.	107
	filament	RT	4.6	TD	110
			20.3, 65.7	physical	
		1073 K	30.4		
		1123 K	74.7	tracer TD	112
		1073, 1123 K	78.6		
		1073, 1123, 1173 K	83.2		
			86.9		
		1123, 1173 K	91.1		
	ribbon	RT	35 (α)	TD, ESD	111
	filament	300 K	20.3, 30.4, 65.7	TD	133
Nb	film	273 K	125	cal.	107
Ni	film	273 K	42→	cal.	107
	film	RT	35→	cal.	14
	film	RT	33	cal.	15

(continued)

Table 6.C2.—Continued

Metal	Form	Temperature	ΔH_{ads} (kcal/mol)	Technique	References
	film	273 K	25.0–30.1	cal.	114
	film	90 K	8.4–25	(diff.) two states	115
	film	RT	35 ($\theta = 0.74$) → 4.5 ($\theta = 1$)	TD	116
			4.4 (α) 32 (β)		
	plate	RT	25	TD tracer	117
	filament	300 K	6.5, 52.6	TD	133
	film	RT	24–17	isosteric	118
Pd	film supported on Al_2O_3	273 K	43→	cal.	107
		303 K	≥38	IR	105
Pt	supported on Al_2O_3				
	tip	303 K	≥44	IR	105
	film	RT	32 (low θ)	FEM	72
	foil	273 K	48→	cal.	107
		RT	$\theta = 0$–0.8 34 ± 2 (28 ± 3) $\theta = 0.8\sim6$ 28 ± 2 (19 ± 3) $\theta = 6\sim9$ 22 ± 2 (9 ± 4)	TD (first order)	119
Re	ribbon	300 K	32.27	TD	159
	filament	RT	36 (α) 54 (β)	TD	120
	filament	RT	50 (α) 69.6 (β)	TD four states	121
	filament	RT	23.1 (α_1) 27.7 (α_2) 46.2 (β_1) 55.3 (β_2) 62.3 (β_3)	TD exchange	122
Rh	film supported on Al_2O_3	273 K	46→	cal.	107
		303 K	≥44	IR	105
Ta	film	273 K	134→	cal.	107

Ti	film	273 K	153→	cal.	107
W	filament	RT	53 75 100	TD	123
	filament	RT	20 75 100	TD	124
	filament	RT	18.7 51.7 67.6 77.5	TD	125
	filament	RT	29 59 69.5 75.5	TD	126
	filament	RT	23 63 77 90	TD	127
	filament	300 K	18.7, 51.7 67.6, 77.5	TD	133
	tip	20 K	16 44 76	FEM heat of adsorption	128
	tip	78 K	20 52 70 100	FEM	72
	tip	4.2 K	36 52	FEM	129
	film	RT	14.8	cal. ($\theta = 0.72$) isothermal	130
	film	RT	8.0 ($\theta = 0.84$), 3.6 ($\theta = 1.02$)	cal. isothermal	130
	filament	RT	13 30 89	TD	131
Zr	film	273 K	150→	cal.	107

[a]See Table 6.A1 for abbreviation key.

Table 6.D. Heat of adsorption of CO_2 on polycrystalline surfaces (form, film; RT) [165]

Metal	ΔH_{ads} (kcal/mol)	Integral	Calc.
Co	37↓	35	29
Cr	115↓	81	—
Fe	67↓	61.5	46.5
Mn	63↓	53	85
Mo	107↓	89	92
Nb	162↓	132	168
Ni	54↓	44	28
Ta	182→	168	172
Ti	187↓	163	203
W	120→	109	155

Table 6.E1. Heat of adsorption of N_2 on single-crystal surfaces[a]

Metal	Temperature	ΔH_{ads} (kcal/mol)	Technique	References
Fe(100)	RT above	50–60 (58)	isothermal TD	189
(111)	RT	50	isothermal TD	190
Ir(100)	80 K	13–14 (γ_3), 7–8 (γ_1), 10–11 (γ_2)	FEM	191
Mo(100)	78	9.7 ± 1 (γ) 87 ± 3 (β)	TD	87
(110)	78	81 ± 3 (β)	TD	93
W(100)	78	10.5 (γ^-), 9.2 (γ^+) 78.0 (β_2)	TD	192
	~1200–~1500 K			
(110)	100 K	<9 (γ)	TD	179
(100)	110 K	~10, ~11 (γ^+, γ^-) 75 (β)	TD	
	300 K			
(111)	110 K	~9 (γ)	TD	179
	220 K	16 (α)		
	300 K	~75 (β)		
(110)	195, 300 K	79 (β)	TD	193
(110)	77, 300 K	117.6	FEM	178
(100)	78	9.2 (γ) 10.5 (γ) 80 (β_2)	TD	87
(100)	78–300 K	9–75	TD	179

[a] See Table 6.A1 for abbreviation key.

Table 6.E2. Heat of adsorption of N_2 on polycrystalline surfaces[a]

Metal	Form	Temperature	ΔH_{ads} (kcal/mol)	Technique	References
Fe	film	77 K	10	cal.	113
	film	RT	70	cal.	108
	film	90 K	5	cal.	166
Ni	film	78 K	<7	TD	167
			6~10		
			9–14		
	film	77 K	10	cal.	113
Nb	filament	RT	>120	TD	168
Ir	filament	RT	58	TD	169
Mo	filament	RT	62.3 (ζ), 71.5 (ϵ)	TD	170
	filament	RT	69.2 (ζ), 83.0 (ϵ)	TD	195
	filament	RT	60.5 ± 2	TD	171
	filament	RT	~80		168
	film	78–300 K	6–20	TD	172
Pd	film	78 K	<7 6~10 } two states	TD	167
Re	filament	RT	67.9–74.9 (β)	TD	122
	filament	RT	66.9–73.8 (β)	TD exchange	122
Ta	film	300 K	140→	cal.	113
	filament	RT	>120	TD	168

(continued)

Table 6.E2.—Continued

Metal	Form	Temperature	ΔH^{ads} (kcal/mol)	Technique	References
W	filament	RT	92.2 → 80.7 → 115	cal.	173
	film	RT	95→	cal.	14
	ribbon	196, 300, 473 K	116	($\theta \to 0$) isosteric	174
	film	195, 290 K	14–24	(molecular)	176
	filament	RT	~80	TD	168
	tip	20, 77, 290 K	153 (E_{des})	FEM	177
		243 K	81	TD	180
		77, 300 K	154.5	FEM	178
	filament	~115 K	~9 (γ)	TD	175
		<400 K	~20 (α)		
		~100 K	58–81		
	ribbon/filament	RT	110/75	TD	181
	filament	<373 K	diff. 95, 150	TD	182
	ribbon		92 (des.)	TD	183
	ribbon	>RT	0.1 ($\theta = 0 - \sim1.0$), 5.0 ($\theta = 2.0$)	TD	23
	filament	RT	1.84 (ϵ), 1.61 (ζ), 18 (α)	TD	184
			93	TD	185
	filament	RT	80, 82, 89	TD	186
					187
	filament	RT	~45–60	TD	188

[a] See Table 6.A1 for abbreviation key.

1. F. Kelemen and A. Neda, Stud. Univ. Babes-Bolyai, Ser. Math.-Phys. **17,** 11 (1972).
2. V.E. Ostrovskii and M.F. Temkin, Kinet. Katal. **7,** 529 (1966).
3. W. Kollen and A.W. Czanderna, J. Colloid Interface Sci. **38,** 152 (1972).
4. D. Brennan, D.O. Hayward, and B.M.W. Trapnell, Proc. Roy. Soc. (Lond.) **A256,** 81 (1960).
5. N.N. Dobrovol'skii and V.E. Ostrovskii, Kinet. Catal. **12,** 1324 (1972).
6. D. Brennan and M.J. Graham, Disc. Faraday Soc. **41,** 95 (1966).
7. G.I. Murgullescu, G. Ilie, and M.I. Vass, Rev. Roum. Chem. **14,** 1201 (1969).
8. J. Bagg and F.C. Tompkins, Trans. Faraday Soc. **51,** 1071 (1955).
9. G. Wedler, Z. Phys. Chem. (N.F.) **27,** 388 (1961).
10. D. Brennan, D.O. Hayward, and B.M.W. Trapnell, J. Phys. Chem. Solids **14,** 117 (1960).
11. V.N. Ageev and N.I. Ionov, Sov. Phys.-Tech. Phys. **16**(10), 1742 (1972).
12. W. Greaves and R.E. Stickney, Surf. Sci. **11,** 395 (1968).
13. P.A. Redhead, Can. J. Phys. **42,** 886 (1964).
14. O. Beeck, Adv. Catal. **2,** 151 (1951).
15. D.F. Klemperer and F.S. Stone, Proc. Roy. Soc. (Lond.) **A243,** 375 (1957).
16. G. Wedler, Z. Phys. Chem. (N.F.) **24,** 73 (1960).
17. B. Weber, J. Fusy, and A. Cassuto, J. Chim. Phys. **66,** 708 (1969).
18. Y.K. Peng and P.T. Dawson, Can. J. Chem. **52,** 3507 (1974).
19. W. Weiershausen, Ann. Phys. (Folge 7) **15,** 150 (1965).
20. B. Weber, J. Fusy, and A. Cassuto, J. Chim. Phys. Phys.-Chim. Biol. **73,** 455 (1976).
21. J.K. Roberts, Proc. Roy. Soc. (Lond.) **A152,** 464 (1935).
22. J.A. Becker, E.J. Becker, and R.G. Brandes, J. Appl. Phys. **32,** 411 (1961).
23. J.A. Becker, Adv. Catal. **7,** 135 (1955).
24. Yu.G. Ptushinskii and B.A. Chuikhov, Surf. Sci. **6,** 42 (1967).
25. W. Engelmaier and R.E. Stickney, Surf. Sci. **11,** 370 (1968).
26. H.-W. Wassmuth, H. Werner, and A.K. Mazumdar, J. Vac. Sci. Technol. **9,** 835 (1972).
27. H.A. Engelhardt and D. Menzel, Surf. Sci. **57,** 591 (1976).
28. G. Rovida, F. Ferroni, F. Maglietta, and F. Pratesi, in *Adsorption–Desorption Phenomena* (Proc. 2nd Int. Conf., April 1971), ed. F. Ricca, Academic Press, New York, 1972, p. 417.
29. G. Rovida, F. Pratesi, M. Maglietta, and E. Ferroni, J. Vac. Sci. Technol. **9,** 796 (1972).
30. G. Rovida, F. Pratesi, M. Maglietta, and E. Ferroni, Surf. Sci. **43,** 230 (1972).
31. G. Ertl, Surf. Sci. **7,** 309 (1967).
32. Yu.I. Belyakov and T.N. Kompaniets, Sov. Phys.-Tech. Phys. **17,** 674 (1972).
33. V.P. Ivanov et al. Surf. Sci. **61,** 207 (1976).
34. N.P. Vas'ko, Yu.G. Ptushinskii, and B.A. Chuikhov, Surf. Sci. **14,** 448 (1968).
35. G. Ertl and P. Rau, Surf. Sci. **15,** 443 (1969).
36. G. Ertl and J. Koch, in Proc. 5th Int. Congr. Catal. 1972 (publ. 1973), ed. J.W. Hightower, p. 969.
37. G. Kneringer and F.P. Netzer, Surf. Sci. **49,** 125 (1975).
38. M. Bacal, J.L. Desplat, and T. Alleau, J. Vac. Sci. Technol. **9,** 851 (1972).
39. M.G. Wells and D.A. King, J. Phys. **C7,** 4053 (1974).
40. C. Kohrt and R. Gomer, J. Chem. Phys. **52,** 3283 (1970).
41. Yu.P. Zingerman and V.A. Ishchuk, Sov. Phys.-Solid State **9,** 623 (1967).
42. J.R. Anderson, *Structure of Metallic Catalysts,* Academic Press, New York, 1975.
43. O. Beeck, Rev. Mod. Phys. **17,** 61 (1945).
44. M.D. Malev, Sov. Phys.-Tech. Phys. **17,** 2009 (1973).
45. C.S. Alexander and J. Pritchard, J. Chem. Soc., Faraday Trans. I, **68,** 202 (1972).
46. J. Pritchard and F.C. Tompkins, Trans. Faraday Soc. **56,** 540 (1960).
47. J. Lapujoulade and K.S. Neil, J. Chem. Phys. **57,** 3535 (1962).
48. O. Beeck, W.A. Cole, and A. Wheeler, Disc. Faraday Soc. **8,** 314 (1950).
49. A.S. Porter and F.C. Tompkins, Proc. Roy. Soc. (Lond.) **A217,** 544 (1953).
50. E. Chornet and R.W. Coughlin, J. Catal. **27,** 246 (1972); **28,** 414 (1973).
51. V.J. Mimeault and R.S. Hansen, J. Chem. Phys. **45,** 2240 (1965).
52. R.I. Bickley, M.W. Roberts, and W.C. Storey, J. Chem. Soc. **A1971,** 2774 (1971).
53. O. Beeck, Disc. Faraday Soc., **8,** 118 (1950).
54. S. Černý, V. Ponec, and L. Hládek, J. Catal. **5,** 27 (1966).
55. R.A. Pasternak and N. Endow, J. Phys. Chem. **70,** 4044 (1966).
56. D. Brennan and F.H. Hayes, Trans. Faraday Soc. **60,** 589 (1964).

57. M. Wahba and C. Kemball, Trans. Faraday Soc. **49**, 1351 (1953).
58. F.M. Bröcker and G. Wedler, Disc. Faraday Soc. **41**, 87 (1966).
59. G. Wedler and F.M. Bröcker, Surf. Sci. **26**, 454 (1971). California, Berkeley, 1971.
60. R. Gomer, Disc. J. Chem. Phys. **27**, 1099 (1957).
61. R. Wortman, R. Gomer, and R. Lundy, J. Chem. Phys. **27**, 1099 (1957).
62. F. Sweet and E. Rideal, 2nd Proc. Int. Congr. Catal., Paris, 1960 (publ. 1961).
63. P.M. Grundy and F.C. Tompkins, Trans. Faraday Soc. **53**, 218 (1957).
64. C.M. Quinn and M.W. Roberts, Trans. Faraday Soc. **58**, 569 (1962).
65. G. Wedler and G. Fisch, Ber. Kernforschungsanlage Juelich **1**, 215 (1972).
66. D.D. Eley and P.R. Norton, Proc. Roy. Soc. (Lond.) **A314**, 319 (1969).
67. A.W. Aldag and L.D. Schmidt, J. Catal. **22**, 260 (1971).
68. J.J. Stephan, V. Ponec, and W.M.H. Sachtler, J. Catal. **37**, 81 (1975).
69. R. Lewis and R. Gomer, Surf. Sci. **17**, 333 (1969).
70. M. Procop and J. Völter, Surf. Sci. **33**, 69 (1972).
71. H.U.D. Wiesendanger, J. Catal. **2**, 538 (1963).
72. W.J.M. Rootsaert, L.L. va Reijen, and W.M.H. Sachtler, J. Catal. **1**, 416 (1962).
73. R. Ducros, J.J. Ehrhardt, M. Alnot, and A. Cassuto, Surf. Sci. **55**, 509 (1976).
74. S.M. Ko and L.D. Schmidt, Surf. Sci. **42**, 508 (1974).
75. G. Wedler and H. Strothenk, Z. Phys. Chem. (N.F.) **48**, 86 (1966).
76. G. Wedler and H. Strothenk, Ber. Bunsenges. Phys. Chem. **70**, 214 (1966).
77. J.K. Roberts, Proc. Roy. Soc. (Lond.) **A152**, 445 (1935).
78. J.K. Roberts and B. Whipp, Proc. Camb. Phil. Soc. **30**, 376 (1936).
79. T.W. Hickmott, J. Chem. Phys. **32**, 810 (1960).
80. Yu.G. Ptushinskii and B.A. Chuikhov, Kinet. Catal. **5**, 444 (1964).
81. G.E. Moore and F.C. Unterwald, J. Chem. Phys. **40**(9), 2626 (1964).
82. R. Hansen, Exptl. Method of Cat. React., ed. R.B. Anderson. Academic Press, New York, 1968.
83. L.J. Rigby, Can. J. Phys. **43**, 1020 (1965).
84. R. Gomer, R. Wortman, and R. Lundy, J. Chem. Phys. **26**, 1147 (1957).
85. E.K. Rideal and B.M.W. Trapnell, J. Chem. Phys. **47**, 126 (1950).
86. J. Pritchard, T. Catterick, and R.K. Gupta, Surf. Sci. **53**, 1 (1975).
87. H.R. Han and L.D. Schmidt, J. Phys. Chem. **75**, 227 (1971).
88. M. Mahning and L.D. Schmidt, Z Phys. Chem. (N.F.) **80**, 71 (1971).
89. J. Lapujoulade and K.S. Neil, Surf. Sci. **35**, 288 (1973).
90. G. Ertl and D. Küppers, Ber. Bunsenges. Phys. Chem. **75**, 1017 (1971).
91. J. Lapujoulade and K.S. Neil, J. Chim Phys. Phys.-Chim. Biol. **70**, 798 (1973); C.R. Acad. Sci. Paris, **C274**, 2125 (1972).
92. L.H. Germer and A.U. MacRae, J. Chem. Phys. **37**, 1382 (1962).
93. J. Lapujoulade and K.S. Neil, C.R. Acad. Sci. Paris **C274**, 2125 (1972).
94. K. Christmann, O. Schober, G. Ertl, and M. Neumann, J. Chem. Phys. **60**, 4528 (1974).
95. J. McCarty, J. Falcona, and R.J. Madix, J. Catal. **30**, 235 (1973).
96. H. Conrad, G. Ertl, and E.E. Latta, Surf. Sci. **41**, 435 (1974).
97. F.P. Netzer and G. Kneringer, Surf. Sci. **51**, 526 (1975).
98. W.H. Weinberg, J. Vac. Sci. Technol. **10**, 89 (1973); V.A. Lampton, Ph.D. thesis, University of California, Berkeley, 1971.
99. K. Christmann, G. Ertl, and T. Pignet, Surf. Sci. **54**, 365 (1976).
100. W.H. Weinberg, D.R. Monroe, V. Lampton, and R.P. Merrill, J. Vac. Sci. Technol. **14**, 444 (1977).
101. D.L. Fehrs and R.E. Stickney, Surf. Sci. **8**, 267 (1967).
102. P.W. Tamm and L.D. Schmidt, J. Chem. Phys. **51**, 5352 (1969); P.J. Estrup and J. Anderson, J. Chem, Phys. **45**, 2254 (1966).
103. P.W. Tamm and L.D. Schmidt, J. Chem. Phys. **54**, 4775 (1971); K.J. Matysik, Surf. Sci. **29**, 324 (1972).
104. R.R. Rye, B.D. Barford, and P.G. Cartier, J. Chem. Phys. **59**, 1693 (1973).
105. N.N. Kartaradage and N.P. Sokolova, Dokl. Phys. Chem. (Proc. Acad. Sci. USSR, Phys. Chem. Sect.) **172**, 39 (1967).
106. B.M.W. Trapnell, Proc. Roy. Soc. (Lond.) **A218**, 566 (1953).
107. D. Brennan and F.M. Hayes, Phil. Trans. Roy. Soc. (Lond.) **A258**, 347 (1965).
108. J. Bagg and F.C. Tompkins, Trans. Faraday Soc. **51**, 107 (1955).
109. Ch. Kleint and W. Moldenhauer, Trans. 3rd Int. Vac. Congr., Stuttgart, 1965 (publ. 1966).
110. D.A. Degras and J. Lecante, J. Chim. Phys. Phys.-Chim. Biol. **64**, 405 (1967).
111. P.A. Redhead, Appl. Phys. Lett. **4**, 166 (1964).
112. A.D. Crowell and L.D. Matthews, Surf. Sci. **7**, 79 (1967).

113. O. Beeck, W.A. Cole, and W. Wheeler, Disc. Faraday Soc. **8**, 314 (1950).
114. G. Wedler and G. Schroll, Z. Phys. Chem. (N.F.) **85**, 216 (1973).
115. M. McD. Baker and E.K. Rideal, Trans. Faraday Soc. **51**, 1597 (1955).
116. N. Nayasaki, J. Watanabe, and K. Kawasaki, Denki. Gijutsu Sogo. Kenkyujo, Ih. **36**, 1576 (1960).
117. A.D. Crowell, J. Chem. Phys. **32**, 1576 (1960).
118. A.M. Hogan and D.A. King, in *The Structure and Chemistry of Solids*, ed. G.A. Somorjai, **57**-I, Wiley, New York, 1969.
119. Y. Nishiyama and H. Wise, J. Catal. **32**, 50 (1974); P.A. Redhead, Vacuum **12**, 203 (1962).
120. R.P.H. Gasser, R. Thwaites, and J. Willkinson, Trans. Faraday Soc. **63**, 195 (1967).
121. M. Alnot, J.J. Ehrhardt, J. Fusy, and A. Cassuto, Surf. Sci. **46**, 81 (1974).
122. J.T. Yates, Jr., and T.E. Madey, J. Chem. Phys. **51**, 334 (1969).
123. G. Ehrlich, J. Chem. Phys. **36**, 1171 (1962).
124. G. Ehrlich, J. Chem. Phys. **34**, 39 (1961).
125. D.A. Degras, Nuovo Cimento, Suppl. **5**, 408 (1967).
126. P.A. Redhead, Trans. Faraday Soc. **57**, 641 (1961).
127. L.J. Rigby, Can. J. Phys. **42**, 1256 (1964); **43**, 532 (1964).
128. L.W. Swanson and R. Gomer, J. Chem. Phys. **39**, 2813 (1963).
129. R. Klein, J. Chem. Phys. **31**, 1306 (1959).
130. E.K. Rideal and B.M.W. Trapnell, Proc. Roy. Soc. (Lond.) **A205**, 409 (1951).
131. Yu.K. Ustinov, V.N. Ageev, and N.I. Ionov, Sov. Phys.-Tech. Phys. **10**, 851 (1965).
132. G. McElhiney, H. Papp, and J. Pritchard, Surf. Sci. **54**, 617 (1976).
133. D.A. Degras, Nuovo Cimento, Suppl. **5**, 420 (1967).
134. J. Lapujoulade, Nuovo Cimento, Suppl. **5**, 433 (1967).
135. M.A. Chesters, J. Pritchard, and M.L. Sims, in *Adsorption–Desorption Phenomena* (Proc. 2nd Int. Conf., April 1971), ed. F. Ricca, Academic Press, New York, 1972.
136. J.C. Tracy, J. Chem. Phys. **56**, 2736, 2748 (1972).
137. H. Papp and J. Pritchard, Surf. Sci. **53**, 371 (1975).
138. J.E. Demuth and T.N. Rhodin, Surf. Sci. **45**, 249 (1974); K. Christmann and G. Ertl, Z. Naturforsch. **28A**, 1144 (1973).
139. C.M. Comrie and W.H. Weinberg, J. Chem. Phys. **64**, 250 (1976).
140. D.I. Hagen, B.E. Nieuwenhuys, G. Rovida, and G.A. Somorjai, Surf. Sci. **57**, 632 (1976); **59**, 177 (1976).
141. V.N. Ageev and N.I. Ionov, Kinet. Catal. **14**, 592 (1973).
142. K. Kleier, A.C. Zettlemoyer, and H. Leidheiser, Jr., J. Chem. Phys. **52**, 589 (1970).
143. H.H. Madden, J. Kuppers, and G. Ertl, J. Chem. Phys. **58**, 3401 (1973).
144. T.N. Taylor and P.J. Estrup, J. Vac. Sci. Technol. **10**, 26 (1973).
145. J.M. McCarty and R.J. Madix, Surf. Sci. **54**, 121 (1976).
146. C.R. Helms and R.J. Madix, Surf. Sci. **52**, 677 (1975).
147. J.L. Falconer and R.J. Madix, Surf. Sci. **48**, 393 (1975).
148. K. Christmann, O. Schober, and G. Ertl, J. Chem. Phys. **60**, 4719 (1974).
149. K. Christmann, O. Schober, and G. Ertl, J. Chem. Phys. **60**, 184 (1974).
150. L.D. Mattew, Surf. Sci. **24**, 248 (1971).
151. J.C. Tracy and P.W. Palmberg, J. Chem. Phys. **51**, 4852 (1969).
152. H. Conrad, G. Ertl, and E.E. Latta, Surf. Sci. **43**, 462 (1974).
153. G. Ertl and J. Koch, Z. Naturforsch. **25A**, 1906 (1970); G. Ertl and J. Koch, in *Adsorption and Desorption Phenomena* (Proc. 2nd Int. Conf., April 1971), ed. F. Ricca, Academic Press, New York, 1972, p. 345.
154. G. Ertl and J. Koch, Z. Phys. Chem. (N.F.) **69**, 323 (1970).
155. G. Ertl and P. Rau, Surf. Sci. **15**, 443 (1969).
156. R. Lewis and R. Gomer, Nuovo Cimento **5**, 506 (1967).
157. C.M. Comrie and R.M. Lambert, J. Chem. Soc. Faraday I **72**, 1659 (1976).
158. H.P. Bonzel and R. Ku, J. Chem. Phys. **58**, 4617 (1973).
159. R.A. Shigeishi and D.A. King, Surf. Sci. **58**, 379 (1976).
160. T.E. Madey and D. Menzel, 2nd Int. Conf. Solid Surf., 1974 (publ. 1974), eds. H. Kumagai and T. Toya, pp. 229-235.
161. P.D. Reed, C.M. Comrie, and R.M. Lambert, Surf. Sci. **59**, 33 (1976).
162. L.R. Clavenna and L.D. Schmidt, Surf. Sci. **33**, 11 (1972).
163. C. Kohrt and R. Gomer, Surf. Sci. **24**, 77 (1971); **40**, 71 (1973).
164. V.S. Ageikin and Yu.G. Ptushinskii, Sov. Phys.-Solid State **10**, 1698 (1969).
165. D. Brennan and D.O. Hayward, Phil. Trans. Roy. Soc. (Lond.) **A258**, 375 (1965).
166. G. Wedler, D. Borgmann, and K.-P. Geuss, Surf. Sci. **47**, 592 (1975).
167. D.A. King, Surf. Sci. **9**, 375 (1968).
168. S.M. Ko and L.D. Schmidt, Surf. Sci. **42**, 508 (1974).

169. V.J. Mimeault and R.S. Hanson, J. Phys. Chem. **70**, 3001 (1966).
170. T. Oguri, J. Phys. Soc. Jap. **19**, 77 (1964).
171. A.A. Parry and J.A. Pryde, Br. J. Appl. Phys. **18**, 329 (1967).
172. D.A. King and F.C. Tompkins, Trans. Faraday Soc. **64**, 496 (1968).
173. P. Kisliuk, J. Chem. Phys. **31**, 1605 (1959).
174. P. Kisliuk, J. Chem. Phys. **30**, 174 (1959).
175. G. Ehrilich, J. Chem. Phys. **34**, 29 (1961).
176. D.O. Hayward, D.A. King, and F.C. Tompkins, Proc. Roy. Soc. (Lond.) **A297**, 305 (1967).
177. G. Ehrlich and F.G. Hudda, J. Chem. Phys. **32**, 942 (1960).
178. G. Ehrlich and F.G. Hudda, J. Chem. Phys. **36**, 3233 (1962).
179. T.A. Delchar and G. Ehrlich, J. Chem. Phys. **42**, 2686 (1968).
180. G. Ehrlich, Adv. Catal. **14**, 255 (1963); J. Chem. Phys. **36**, 1171 (1962).
181. T.W. Hickmott and G. Ehrlich, J. Phys. Chem. Solids **5**, 47 (1958).
182. G. Ehrlich and T.W. Hickmott, J. Chem. Phys. **26**, 219 (1957).
183. G. Ehrlich, J. Phys. Chem. **60**, 1388 (1956); O. Beeck, Adv. Catal. **2**, 151 (1950).
184. J.A. Becker and C.D. Hartman, J. Phys. Chem. **153**, 159 (1953).
185. T. Oguri, J. Phys. Soc. Jap. **18**, 1280 (1963).
186. M.P. Hill, S.M.A. Lecchini, and B.A. Pethica, Trans. Faraday Soc. **62**, 229 (1966).
187. T.E. Madey and J.T. Yates, Jr., J. Chem. Phys. **44**, 1675 (1966).
188. H.F. Winters and D.E. Horne, Surf. Sci. **24**, 587 (1971).
189. G. Ertl, M. Grunze, and M. Weiss, J. Vac. Sci. Technol. **13**, 314 (1976).
190. G. Ertl. M. Grunze, and M. Weiss, private communication.
191. B.E. Nieuwenhuys, D.Th. Meijer, and W.M.H. Sachtler, Surf. Sci. **40**, 125 (1973).
192. L.R. Clavenna and L.D. Schmidt, Surf. Sci. **22**, 365 (1970).
193. P.W. Tamm and L.D. Schmidt, Surf. Sci. **26**, 286 (1971).
194. J.C.P. Mignolet, J. Chim. Phys. Phys.-Chim. Biol. **54**, 19 (1957).
195. T. Oguri, J. Phys. Soc. Jap. **18**, 1280 (1963).

7. Energy Transfer In Gas-Surface Interactions

Introduction

Even the simplest gas–surface interaction involves several steps that begin with the collision of the incident atoms or molecules with the surface. As the gas species nears the surface, it experiences an attractive potential whose range depends on the electronic and atomic structures of the collision partners. The interaction may vary in strength in proportion to the reciprocal of the distance between the collision partners[1] (for example, between incident gas ions and surface ions of opposite charge), and therefore it is of long range. It may be much shorter in range and vary in proportion with the inverse third or sixth power of the distance. A certain fraction of the incident gas atoms are trapped in this attractive potential well, and once trapped they can move along the surface by diffusion. The adsorbed species may desorb from the surface if sufficient energy is imparted to it at a given surface site to overcome the attractive surface forces.

During the collision of the adsorbed gas atom with the surface, it exchanges kinetic or translational energy, T, with the vibrational modes, V_s, of the surface atoms. The type of energy transfer that takes place in this circumstance is often called the T-V_s energy exchange. During the collision of gas molecules with the surface they may exchange internal energy, including rotation, R, or vibration, V, for example, with the vibrating surface atoms. In this circumstance there are also R-V_s and V-V_s energy-transfer processes. The various energy-transfer processes are depicted schematically in Figure 7.1.

To understand the dynamics of the gas–surface interaction, it is essential that we determine how much energy is exchanged between the gas and surface atoms through the various energy-transfer channels. By

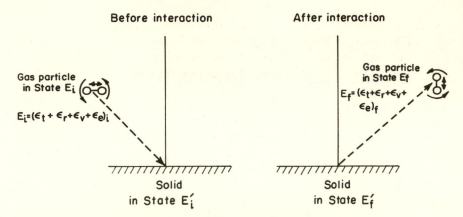

Figure 7.1. Scheme of energy transfer in gas–surface interaction. E_i and E_f indicate initial and final energy states. For the gas molecules these are the sum of translational (E_t), rotational (E_r), vibrational (E_v), and electronic (E_e) states.

determining the residence time of the adsorbed atoms or molecules on the surface at the various surface sites that are available for bonding, the dependence of the energy transfer on residence time and the surface structure can be determined. Finally, we would like to know the kinetic parameters, rate constants, activation energies, and preexponential factors for each of the elementary surface steps of adsorption, surface migration, and desorption to obtain a complete description of the gas–surface energy-transfer process.

Perhaps the most versatile method for studying gas–surface collision dynamics is molecular beam–surface scattering, shown schematically in Figure 2.26. A well-collimated beam of molecules strikes the surface, and the species that are desorbed are detected by a mass spectrometer. By chopping the incident beams with a slotted disk, the beam is interrupted with a frequency of ω. The molecules that arrive at the detector lag behind a reference light signal which is chopped at the same frequency, because of the long time of flight of these molecules and because of the finite residence time on the solid surface. From the phase shift, ϕ, the time of flight can be computed. For example, if $\phi = 36°$ and $\omega = 10^3$ Hz, the time of flight $\tau = \phi/360\omega = 10^{-4}$ sec. If there is reason to believe that the surface residence time is much shorter (for example, surface atom vibration times are approximately 10^{-12} sec), the average beam velocity is also obtained from knowledge of the flight path, d. For $d = 10$ cm, $\bar{u}$ $(d/\tau) = 10^5$ cm/sec. The angular distribution of the scattered molecules can be obtained by a rotation of the mass spectrometer about the sample.[2] Often, we would also like to know the velocity of the molecules after scattering, independent of their properties before the collision. In this circumstance the scattered molecules are interrupted by another rotating disk chopper and their time of flight is measured by the detector.[3] Thus the molecules whose velocities are detected left the surface before chopping. Their time of flight reflects the velocity they have acquired as a result of energy-transfer processes at the surface. By using suitable correlation choppers, not only the average time of flight (or average velocity) but the velocity distribution in the beams can also be determined. By measurement of the velocity distribution of the incident as well as the scattered molecules and the angular distribution of the scattered species, the dynamics of the gas–surface scattering process can be completely described.

The angular distribution of scattered molecules is usually displayed by plotting the intensity (flux or number density) of the arriving molecules per detector solid angle versus the angle of scattering, θ, which is usually measured with respect to the surface normal. This is called the rectilinear plot.[4] The velocity distribution at a given angle of scattering is displayed by plotting the intensity of arriving molecules of given velocity, $I(u)$, as a function of the velocity. Both angular distribution and velocity measurements yield significant information on the energy exchange between the gas and surface atoms. In Figure 7.2 the angular distributions

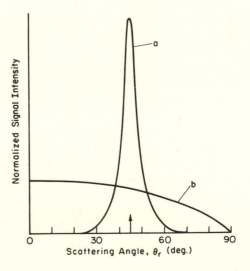

Figure 7.2. Rectilinear plot displaying the (a) specular scattering and (b) cosine angular distribution of scattered beams. The arrow indicates the angle of incidence.

in the two limiting cases of gas–surface interactions are shown: specular and cosine scattering. The arrow on the abscissa denotes the angle of incidence. When the angle of incidence and the scattering angle are identical for most molecules, this is a case of specular scattering; in this circumstance usually little or no energy transfer takes place between the incident molecules and the surface. When the scattered intensity decreases as cos θ with respect to the surface normal, this is a case of the cosine scattering distribution;[5] this type of angular distribution is expected upon complete thermal accommodation (that is, when the molecules desorb with kinetic "temperature" or velocity distribution that is the same as the temperature of the solid and is thought to reflect complete energy accommodation. Of course, direct measurement of the velocity distribution of the scattered molecules helps us to verify the extent of the kinetic energy transfer. Figure 7.3, for example, displays the velocity distribution of oxygen molecules scattered from hot oxidized tungsten surfaces. It is clear from the velocity distribution data that the kinetic energy of the scattered molecules is much lower than that corresponding to the surface temperature of the solid, indicating rather poor $T-V_s$ energy transfer.

Comsa[5] has pointed out that the desorbing molecules are usually only a fraction of all molecules leaving the surface in a molecular beam–

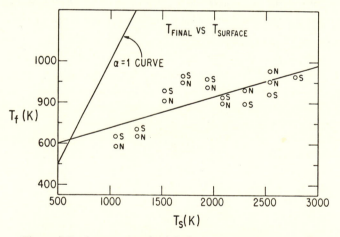

Figure 7.3. Average scattered O_2 beam temperatures, T_f, fitted to the O_2 velocity distribution as a function of the surface temperature of tungsten, T_s, detected at the specular angle [S = specular (45°)] and at the surface normal (N). The curve expected for complete thermal accommodation ($\alpha = 1$) is also shown. (After D. Auerbach et al., ref. 9a.)

surface scattering experiment. Some of the molecules diffract or scatter specularly without energy accommodation. Some will undergo only partial energy transfer on account of a potential energy barrier to adsorption that gives rise to a low "sticking probability." By applying detailed balancing—that is, the condition in which at equilibrium the angular distribution of the adsorbed fraction of the molecules is identical with the angular distribution of the desorbed fraction—it was shown that the desorbing molecules would, in general, not obey the cosine law or exhibit a Maxwellian distribution corresponding to the temperature at the surface. It should be noted that, only in equilibrium, the ensemble of molecules leaving the surface must have cosine angular and Maxwellian velocity distributions.

At present most of our information about the nature of energy transfer in gas–surface interactions comes from angular distribution measurements. This is due to the considerable experimental difficulty associated with the measurement of velocity distributions (poor signal-to-noise ratio). We shall discuss our present understanding of gas–surface energy transfer by referring primarily to angular distribution studies. Nevertheless, work is in progress in several laboratories to measure velocity distributions of scattered molecules. No doubt, velocity distribution measurements will be carried out with increasing frequency in the near future. The experimental parameters that are usually varied in these experiments are the temperatures at the surface, T_s; the kinetic temperature of the incident gas, T_g (where T_g is defined[5] as $T_g = mu^2/4R$, where m is the molecular weight of the molecules in the beam); the angle and intensity of the incident beam, θ and I; and the atomic surface structure and the surface composition. The data that are obtained comprise the intensity of the scattered molecules, I, as a function of angle (the beam intensity is usually normalized to the incident beam intensity, I_0) and the phase shift of the scattered beam, Φ. Both I/I_0 and Φ are usually measured as a function of temperature at the surface and at various chopping frequencies in the range $\omega = 1$ to 10^3 Hz. The higher the chopping frequency, the shorter the residence times of the molecules that are being analyzed as a function of T_s and θ. At very low chopping frequencies ($\omega \simeq 1$ Hz), most of the desorbing molecules, fast or slow, are detected, and the experimental results reflect nearly steady state conditions.[6]

Before the advent of the application of the molecular beam–surface scattering technique, it was customary to obtain information about the efficiency of energy transfer during gas–surface collisions from measurement of the thermal accommodation α_T. It is defined as

$\alpha_T = (T_g - T_{g'},)/(T_g - T_s)$, where T_g and $T_{g'}$, are the kinetic temperatures of the incident and desorbed atoms, respectively, and T_s is the temperature at the surface.[7] T_g is obtained by measuring the temperature of the gas source in thermal equilibrium; this allows the direct determination of the average gas velocity, u. The degree of energy transfer is measured by accurately determining the change of surface temperature as the gas is admitted. The thermal accommodation coefficient is unity if $T_s = T_{g'}$, and very inefficient if $T_{g'}$, approaches T_g.

Let us discuss separately the information available on the nature of the various energy-transfer processes, $T-V_s$, $R-V_s$, and $V-V_s$, before we review what is known about the gas–surface interaction potential and about the kinetic parameters of the elementary surface steps of adsorption, surface diffusion, and desorption.

T–V_s Energy Transfer

The available experimental data indicate poor energy transfer when atoms are scattered from clean ordered crystal surfaces near 300 K.[6] Siekhaus et al.[8] have monitored the average velocity of inert gas atoms that were scattered from graphite surfaces as a function of sample temperature. Their results are shown in Figure 7.4. As the surface temperature increases, one expects a corresponding increase in the kinetic temperature of the reflected beam if there is good $T-V_s$ energy accommodation. This is not found, however. The reflected gas velocity hardly changes with large changes of the temperature at the surface, indicating that the $T-V_s$ energy transfer is very poor indeed. Similar results were obtained during scattering of oxygen molecules from oxidized tungsten surfaces.[9a] The undissociated molecules back-scatter from the surface without adsorbing much of the thermal energy from the hot surface atoms (Figure 7.3).

Recent measurements of the mean kinetic energy of D_2 molecules desorbing from nickel surfaces indicate a strong angular dependence.[9b] The molecules were produced by the recombination of atoms diffusing through the nickel samples. The angular flux distribution was sharply peaked in the forward direction, and at the surface normal, the mean kinetic temperature of the desorbing molecules, $(<E>/2k)$, was 700 K higher than the temperature at the surface, T_s. At glancing angles, however, the mean kinetic temperature was 400 K below T_s, which was held at $T_s = 1143$ K. A simple model that assumes that the molecules encounter a potential energy barrier with holes was proposed to explain the results.[9c]

Most of our present information about $T-V_s$ energy transfer is in-

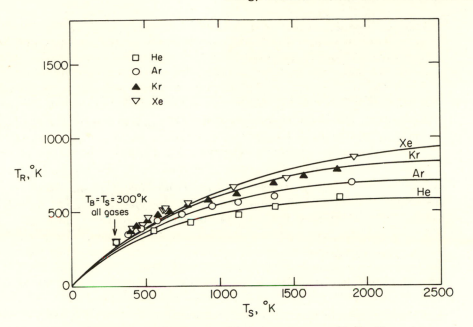

Figure 7.4. Average scattered beam temperatures, T_R, for He, Ar, Kr, and Xe as a function of the graphite surface temperature, T_s. (After W. Seikhaus et al., ref. 8.)

ferred from angular distribution measurements. The shape of the scattering distribution, I versus θ, and its behavior as a function of surface temperature and angle of incidence are used to classify the degree of energy exchange between the gas atoms and the surface. Merrill et al. found that a dimensionless attractive potential-well depth for the scattering of inert gases from Pt, W, and Ag can be used to compute the magnitude of the energy transfer.[10] They distinguish inelastic and trapping interactions. A less intense scattering distribution, not centered at the specular angle, is the result of the inelastic interaction. A trapping-dominated interaction is characterized by a very broad intensity distribution with the maximum intensity near the surface normal, presumably due to the large contribution of trapped particles emitted from the surface in a cosine distribution that has a maximum at the surface normal. In general, the angular distributions of inert gases scattered from clean metal surfaces have features that are in between the purely specular and cosine angular distributions—indicating relatively poor T–V_s energy transfer in all cases that were studied.

Several models have been proposed to permit computation of the scattering distributions and their behavior as a function of incident angle and surface temperature. The soft-cube model[11] and trajectory calculations[12] have been used reasonably successfully to obtain qualitative agreement with the experimental data.

Effect of Diatomic Adsorbates on the $T-V_s$ Energy Transfer

When gas atoms or molecules are scattered from a layer of adsorbed carbon monoxide as compared to a clean metal surface, their angular distribution changes completely.[13] This is shown for H_2 for the Pt(111) surface and carbon monoxide–covered Pt(111) surfaces in Figure 7.5. While the scattering is largely specular from the clean metal, it is cosinelike from the carbon monoxide–covered surface. Figure 7.6 shows the same effect obtained for other monatomic and diatomic molecules from the adsorbate-covered surface. This effect is not due to changes of the mass of the surface species, because when hydrogen or other gases are scattered from a layer of graphite,[13] the same effect is not noticeable and the scattering distribution remains fairly specular. It is clear that the very efficient $T-V_s$ energy transfer in this circumstance is due to some molecular property of the adsorbate. Since carbon monoxide is bound to the surface through the carbon atom and the oxygen end of the molecule is away from the metal surface, the molecule has low-frequency bending modes that can readily be excited on impact by the incident gas atoms or molecules. The adsorbed molecule acts as a pillow by absorbing the momentum of the gas species, perhaps via these low-frequency bending modes.

These results indicate that the cleanliness of the surface can influence markedly the efficiency of the $T-V_s$ energy transfer. It is relatively poor on clean surfaces and excellent on surfaces covered with CO. Therefore, great care has to be taken in experiments to assure the cleanliness of the surface.

$R-V_s$ Energy Transfer

One may study the scattering behavior of diatomic molecules to learn about this type of energy-transfer process. A great deal of evidence indicates that the excitation of molecular rotations by the vibrating surface atom is efficient provided that the rotational energy levels are within the energy range of lattice vibrations. The angular distributions of H_2, D_2, and HD scattered from nickel or platinum[14] surfaces clearly demonstrate this (Figures 7.7a and 7.7b). While H_2 scatters specularly, the intensity of scattered D_2 and HD are greatly reduced in the specular

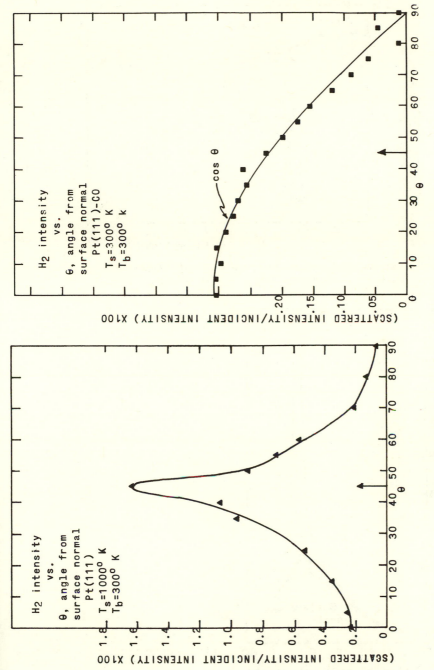

Figure 7.5. Scattering distribution of H_2 from clean Pt(111) (left) and carbon monoxide-covered platinum crystal surface (right). Arrows indicate the angle of incidence.

339

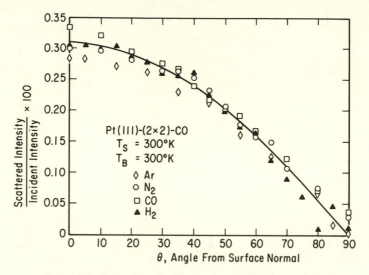

Figure 7.6. Scattering distributions of N_2, H_2, CO, and Ar from carbon monoxide–covered platinum (111) crystal face.

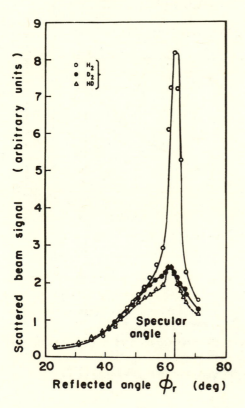

Figure 7.7a. Scattering distributions of hydrogen isotopes, H_2, HD, and D_2 from a (111) orientation nickel film. (After R. Palmer et al., ref. 14a.)

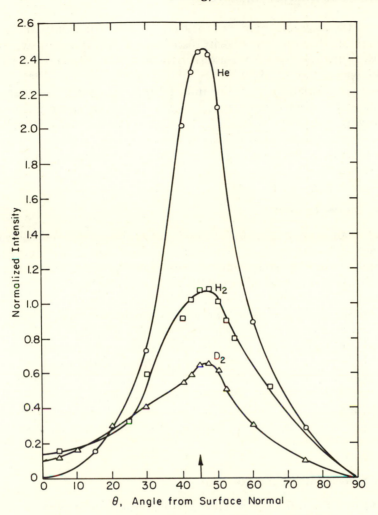

Figure 7.7b. Scattering distributions of H_2, D_2, and He from a Pt(1̄11) crystal face.

direction. For H_2, the rotational energy levels are too far apart (larger than 1000 cal) to be excited by the thermal energy of the surface atoms. For D_2 and HD, however, as well as for all heavier diatomic molecules, the rotational energy levels are spaced close enough so that R-V_s energy transfer by a single phonon interaction is feasible. This energy exchange causes broadening of the angular distribution as demonstrated by both experiments, utilizing other diatomic molecules and theory. The excitation of the rotational modes of the back-scattered molecules should have a marked effect on their velocity distribution. This is demonstrated indirectly by the rotationally inelastic diffraction of H_2, D_2, and HD from the (100) crystal face of MgO.[15]

Recent calculations by Doll[16] indicate that changes of internal energy contents by either rotational or vibrational excitation can influence the trapping probability of incident molecules. The excited internal modes may be coupled with translation during the collision. Thus the momentum of the colliding molecules can be increased in a way that decreases the sticking probability.

V–V_s Energy Transfer

This process might be studied using diatomic and polyatomic molecules by vibrationally exciting the incident species with a laser or an electron beam. Such experiments are feasible but have not been successfully carried out as yet.[17] Another method of study of V–V_s transfer is through isotope effects. Most promising for this purpose are molecules containing hydrogen or its isotopes, as a relatively large change in the mass and hence in the vibrational frequency spectrum can be obtained by isotopic substitution. There is evidence for a large isotope effect during the decomposition of CH_4 and CD_4 on hot tungsten surfaces. CH_4 dissociates at higher rates and with smaller activation energy than CD_4. The mechanism of methane dissociation is thought to involve hydrogen tunneling from the carbon bond to the metal atoms.[18] The thermal dissociation of CD_4 is slower because of the lower tunneling probability of the heavier isotope.

Gas–Surface Interaction Potential

Most of the information concerning the attractive potential experienced by the atoms incident on the surface has come from helium atomic beam diffraction experiments. Two types of measurement facilitate the extraction of the gas–surface potential parameters. First, the scattering distributions need to be measured as a function of the reflected angle, so that the relative intensities of the diffraction peaks are known. The relative intensities of the diffraction peaks depend on the strength of the repulsive part of the potential. In addition to rigorous calculations, a simple relationship between the strength of the repulsive part of a Lennard-Jones potential and the number of diffraction peaks has been established to allow an approximate calculation of the repulsive strength parameter.[19] Second, the intensity of a diffraction peak must be monitored as a function of incident polar or azimuthal angle in order to observe the angular location of the minimum or maximum in intensity that is created when an atom falls into a bound-state resonance. The angular location of the bound states is essential to calculation of the energy levels, well depth, and range parameter of the attractive part of the potential.

Alkali–Halide and Graphite Surfaces

The scattering of He and Ne atomic beams, H_2 and D_2, from LiF(001) NaF(001), NaCl(001), and graphite surfaces was studied.[20,21] The angular locations of the diffraction peaks agree well with the values calculated from knowledge of the de Broglie wavelength and the surface crystallographic structure. The diffraction probabilities for each channel are extracted from the experimental peak intensities by taking the velocity distribution of the supersonic incident beam into consideration. Satisfactory agreement is obtained between the experimental diffraction probabilities and those calculated by an approximate quantum theory of Levi et al.[22-24] The theory employs a simple corrugated hard wall as the interaction potential but takes into account the accelerating effect of an attractive square well on the incident atom. The incident atom is not allowed to be bound at the surface in this well, nor is multiple scattering explicitly allowed. Thus a comparison of the theoretical values of the diffraction probabilities with the experimental probabilities is valid only at particular incident and azimuthal angles that do not correspond to a resonance condition, and, indeed, the theory does not reproduce the observed variation in diffraction probability as a function of azimuthal angle. However, such comparison between experiment and a theory that is not realistic about all aspects of the interaction is valuable to the theorist for an orderly and progressive understanding of the gas–surface interaction. Although not applied to the foregoing set of data, recent approximate quantum theories that have varying degrees of realistic interaction potentials include the semiclassical theory of Doll[25-27] and Masel et al.,[28,29] the quasiclassical theory of Ray and Bowman,[30] and the CCGM theory.[31-33] For the most part, these theories await rigorous comparison to good scattering data produced by monoenergetic beam scattering from clean crystal surfaces at low temperatures. However, a comparison of these approximate theories to exact quantum results has been presented.[34,35]

Several bound states of He on LiF, NaF, and graphite have been found through investigation of the intensity of the specular beam as a function of azimuthal angle.[20,21] These bound-state resonances are detected by sharp drops in intensity of the specularly scattered beam. A knowledge of the incident and azimuthal angles at which these minima occur, as well as the incident beam energy, enables calculation of the bound-state energies. They range in value from 0.1 to 6 meV for He bound to LiF. Tsuchida[36] predicted the energies of the bound states of the He/LiF system by numerically solving the Schrödinger equation, which employed a realistic potential obtained from a pairwise summation of a Lennard-Jones 12-6 potential between a He atom and each crystal atom. His calculated values agree well with the observed bound-state energies.[37,38]

H and D have also been scattered from the alkali halides.[39] The three observed bound-state energies fit a Morse potential's energy-level spacings well. Surprisingly, however, the well depth of H on LiF and H on NaF is identical. The strength of the exponentially repulsive part of the potential is calculated.[40,41] The results are in fair agreement with the experimental diffracted intensities.

It becomes apparent from comparison with existing data that gas-surface scattering, even for these systems, cannot yet be fully described theoretically. Both the quantum-mechanical theory and the experimental data are incomplete. Progress in theory development, however, is being made with the use of simple corrugated hard-wall potentials for optimization of the scattering model, the extension to more realistic interaction potentials,[42,43] and the critical evaluations of the various assumptions in the calculations.[44,45] The exact quantum-mechanical caluclations of Wolken[46] and Tsuchida[38] indicate that a reliable and realistic gas-surface interaction potential can be extracted from scattering data.

Metal Surfaces

The discussion so far has been concerned with the determination of interaction potentials between an atom and an alkali-halide crystal. Corresponding early diffraction studies of single-crystal metal surfaces have not been very successful, owing to the absence of well-defined diffraction peaks. Beeby has discussed the absence of diffraction from metal surfaces as the result of a smaller value of Θ_D, the surface Debye temperature, in contrast to the alkali halides, thus leading to a larger fraction of inelastic transitions.[47-49] It has become apparent from quantum[50] and semiclassical calculations[29] that the periodic part of the gas–metal surface interaction potential on low-Miller-Index, smooth, closely packed surfaces is too weak to produce strong nonspecular diffraction.[51] This idea is supported by the experimental evidence from the first pure metal single crystal to exhibit diffraction peaks. Helium diffraction was observed from the W(112) crystal only in the direction in which the periodic surface structure is more pronounced.[52,53] A semiclassical calculation employing a Lennard-Jones potential fits the scattering distribution qualitatively. Regrettably, an effusive thermal He beam was used in the experiments that makes the theoretical distribution insensitive to variation of the well depth of the potential employed in the calculation. Helium diffraction was readily observed from the atomically rougher crystal faces of copper and nickel with the use of more monoenergetic nozzle-beams.[54,55]

He and H_2 diffraction has been observed from a Ag(111) surface in high-resolution apparatuses.[56,57] In this study, the surface was cooled to below or near the surface Debye temperature in order to minimize inelastic transitions. The first-order diffraction peaks are of the order of 1000 times less intense than the specular peak. Interestingly, the intensity of H_2 diffraction is significantly greater than that for He diffraction.

Diffraction of gas atoms from metal surfaces will certainly yield information about the interaction potential of chemically important systems in the future. Already, diffraction of He from stepped platinum surfaces identified a different interaction potential at a step as compared to that at a terrace.[58]

Rainbow Peaks

Bilobular scattering or rainbow scattering has been observed from LiF[59] and stepped metal surfaces.[60,61] Observation of this structure in the scattering distribution also allows determination of the gas–surface interaction potential. Helium atoms incident along the step edges of a Pt(553) single crystal scatter specularly, while rainbow peaks are evident if the He atoms are incident perpendicular to the step edges.[61] Thus, in agreement with semiclassical calculations,[29] it appears that a strong periodic structure of the surface is as essential for the occurrence of the rainbow phenomenon as it is for diffraction. The classical trajectory calculations of McClure,[62] and to a lesser extent those of Steele,[63] who uses a different interaction potential, fit well the experimental scattering distributions of Ne from LiF.[58,59] The potential parameters of McClure are in close agreement with those found by Finzel et al.[39] Rainbow peaks have been described as an accumulation of trajectories at two angles that correspond to the maximum and minimum scattering angles introduced by the inflection points of a periodic potential[64] Jewsbury[65] has recently reinterpreted the bilobate structure as the result of single and double scattering events.

Chemistry in Two Dimensions: Surfaces

Elementary Surface Processes: Introduction

A chemical reaction that takes place on a surface has several elementary steps that should be distinguished. These are (1) adsorption on the surface (the incident molecules are trapped by the attractive surface potential); (2) diffusion of the adsorbates from one binding site along the surface to other sites; (3) surface reactions that involve bond breaking, insertion of atoms, or molecular rearrangments; and (4) desorption of the product molecules.

These processes are schematically displayed in Figure 7.8. We would like to study the kinetics of each of these elementary surface steps separately and determine their rates and various rate parameters (activation energy, preexponential factor) and their dependence on the atomic surface structure and surface composition. To achieve this, the experimental conditions must be carefully selected.

For example, most surface reactions are carried out at high pressures (1 to several atmospheres) to establish conditions for optimum rate or for similarity to the chemical environment of our planet. In this circumstance the surface is covered by at least a monolayer of adsorbates. Since the activation energies for adsorption and surface diffusion are generally small (a few kT) in most cases, appreciably lower than the activation energy for desorption, equilibrium among the different surface species, reactants, reaction intermediates, and products is readily established. For this reason we have proposed the "two-dimensional phase" approximation discussed in Chapter 1. This model has been successful in describing the kinetics of crystal growth,[66] evaporation,[67] and many high-pressure surface reactions. Under these conditions either a slow surface reaction step or desorption control the overall rate of the surface process. In this circumstance it is difficult to obtain information about the various elementary surface steps since they are near equilibrium.

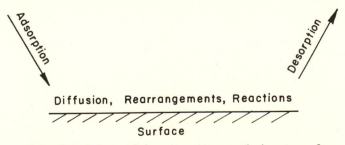

Figure 7.8. Scheme of elementary processes during gas–surface interactions.

However, if we start reducing the pressure, the concentration of adsorbates is also reduced. Molecules no longer occupy all the sites, and the different adsorbates may no longer equilibrate. Only those surface sites are now occupied where the adsorbates are held more tightly. The pressure may be reduced to such an extent that the rate of arrival of the reactants begins to control the surface reaction. In this situation the incident reactants have a choice of surface sites, some of them more preferred than others. Therefore, the effects of the surface structure or surface composition become detectable. As the surface coverage is changed, the relative rates of the various elementary reaction steps may also change. Thus, as a function of pressure and temperature, we may distinguish several or all of the reaction steps and obtain a more complete picture of the reaction mechanism.

We first briefly discuss the two most commonly used techniques for studies of the dynamics of elementary surface processes. Then we discuss our present state of understanding of each elementary surface reaction step, adsorption, surface diffusion, desorption, and surface reactions.

Perhaps the most versatile technique for studying the elementary steps of surface processes is molecular beam–surface scattering.[4,6] Although it can be utilized only at low pressures (usually less than 10^{-4} torr beam pressures), its pressure range permits wide variations of surface coverages. The reaction probability upon a single scattering can be determined (if it is greater than 10^{-3} for most instruments) together with the surface residence times of adsorbates. As discussed in the preceding section, the surface kinetic information is obtained by measurements of the intensity and the phase shift of the product molecules with respect to the reactant flux. Residence times in the range 10^{-6} to 1 sec can be monitored with relative ease, and activation energy is determined from the temperature dependences of the intensities and the phase shifts. The phase shift of the product molecules is usually measured at different chopping frequencies of the incident beam. At a given chopping frequency, only those product molecules are detected that are formed in the surface process and desorbed in less time than the chopping period.

While the reactive molecular beam scattering experiment is being carried out, the reaction chamber must be continuously pumped to remove the scattered product efficiently. Otherwise, the pressure would rise in the chamber and the surface coverage would increase. Thus the reaction would take place on a contaminated gas-covered surface instead of on the clean surface. The pressure in the scattering chamber is maintained at 10^{-8} torr or less throughout the scattering experiment, to assure surface cleanliness. The surface structure and composition is characterized by LEED and Auger spectroscopy before and after the experiment (and also during the experiment if desired) or by other surface probes.

Another method of separating the various surface processes from each other involves the preadsorption of reactants. Once the desired surface coverage is reached, the adsorbates are removed by temperature-programmed thermal desorption[68] (see Chapter 2). By measuring the rate of desorption, the product distribution, and the thermal desorption spectrum (amount desorbed versus temperature) at different coverages, one can determine the kinetics of desorption (activation energies, preexponential factors, order of desorption). Only one of the surface processes, the usually slowest desorption, is studied this way and is thereby separated from the other elementary surface processes.[69]

Modes of Signal Detection

Detection of the product signal in a molecular beam–surface scattering experiment can be performed in different modes: differential and integral. At high chopping frequency (100 to 1000 Hz) the experiments are usually carried out in the differential mode. For this measurement the mass spectrometer is positioned at a certain scattering angle from the surface normal. The signal detected is composed of two different contributions for a given mass number. One part is due to the phase shift on account of the residence times of adsorbed molecules that are directly emitted from the surface into the detector; the second contribution is from the modulated partial pressure of gases in the scattering chamber that enter the detector as a result of their finite residence time in the ultrahigh-vacuum chamber before being pumped away. The modulated background must be subtracted from the total measured signal to obtain the true differential signal. At each scattering angle the modulated background can be measured by covering the aperture in the ionizer of the mass spectrometer facing the crystal so that molecules emitted from the surface cannot enter it directly. The true differential signal is normalized to the incident intensity to give the differential reaction probability. To obtain the total reaction probability, these results must be integrated over all angles, or the mass spectrometer must be moved from angle to angle to obtain the angular distribution of the product.

At low modulation frequencies, in the range 1 to 100 Hz, it is advantageous to use the second method of detection that is called the "integral mode." The mass spectrometer is located behind the crystal at a fixed position out of the direct line of sight of the surface such that only the modulated partial pressure of the product is measured. There are reasons for using the integral detection mode. First, it can be shown that in the first approximation the intensity of the modulated background, $I_{background}$, changes with the frequency of modulation, ω, as $I_{background}$ is proportional to $[\omega^2 + (S/V)^2]^{-\frac{1}{2}}$ where S is the pumping speed (liters/ sec) and V the volume (liters) of the chamber. For very low modulation frequencies (less than 2 Hz), the contribution of this modulated background is 10 times larger than the signal due to the species emitted directly from the surface into the solid angle accepted by the mass spectrometer ionizer aperture. However, at 100 Hz, for example, in studies of H_2–D_2 exchange, the differential signals are a factor of three larger than the modulated background. Second, this integral signal is proportional to the total amount of product emitted by the surface and does not need to be integrated as in the differential mode to maintain an integral reaction probability. Furthermore, any possible effect of the angular distribution of the products is eliminated. One no longer need assume that the products are emitted in a cosine distribution.

When the reaction product signal is measured at low modulation frequencies, therefore, integral mode measurement of the incident beam is also required. This is accomplished by rotating the sample holder such that the beams strike the back of the crystal support. Integral-mode measurements of the product signal are normalized to integral-mode measurements of the incident beam to give the total reaction probability. The differential mode of detection must be used in studies of angular scattering distribution. However, using the integral mode, one obtains the integrated reaction probability with greater accuracy. Also, since the chopping frequencies are very low, conditions obtained under steady-state flux reaction conditions are approached.

Analysis of the Molecular Beam Surface Scattering and Thermal Desorption Data

The experimental data that are available are intensities and phase shifts of the product molecules as a function of T_s and ω. Since the incident flux as well as the scattered molecules are detected under these reaction conditions, a mass balance of reactants and products can be set up.[70] A model of the surface reaction is now assumed using the steady-state approximation; the rate constants for the various elementary steps are expressed as a function of I/I_0 and ϕ.[71] Let us consider a simple dissociative adsorption atomic desorption process to illustrate the technique of data analysis. A beam of diatomic molecules, A_2, of intensity I_0, is incident on the surface. The beam is chopped by a gating function, $g(t) = g_1 e^{i\omega t}$. The molecules interact with the surface with an adsorption probability, η, and desorb with a rate constant k_d.

$$A_{2(gas)} + 2S \xrightarrow{\eta} 2[SA_{ads}] \xrightarrow{k_d} 2S + 2A_{(gas)} \tag{1}$$

A surface mass balance for the adsorbed A atoms yields

$$\frac{d[SA_{ads}]}{dt} = 2\eta I_0 g(t) - k_d[SA_{ads}] \tag{2}$$

Substituting $g(t)$ and a trial solution for $[SA_{ads}]$ gives

$$i\omega[SA_{ads}]^* e^{i\omega t} = 2\eta I_0 g_1 e^{i\omega t} - k_d[SA_{ads}]^* e^{i\omega t} \tag{3}$$

This is solved for $[SA_{ads}]^*$ to give

$$[SA_{ads}]^* = \frac{2\eta I_0 g_1}{k_d + i\omega} \tag{4}$$

Writing the complex number in polar form and solving for the rate of desorption ($k_d[SA_{ads}]^*$) yields

$$k_d[SA_{ads}]^* = \frac{2\eta I_0 g_1{}^{-i\tan(\omega/k_d)}}{\sqrt{1 + (\omega/k_d)^2}} \tag{5}$$

In the limit of low reaction probability (I/I_0), the product to the incident flux ratio is equal to the ratio of $k_d[SA_{ads}]^*$ to $I_0 g_1$ modified by a phase factor related to the surface residence time of the product,

$$\frac{I}{I_0} = \left(\frac{k_d[SA_{ads}]^*}{I_0 g_1} \right) e^{-i\phi} = \epsilon e^{-i\phi} \tag{6}$$

By substituting into Equation (5), we have

$$\epsilon e^{-i\phi} = \frac{2\eta e^{-i\tan(\omega/k_d)}}{\sqrt{1 + (\omega/k_d)^2}} \tag{7}$$

Thus I/I_0 and ϕ are given by

$$\frac{I}{I_0} = \frac{2\eta}{\sqrt{1 + (\omega/k_d)^2}} \tag{8}$$

and

$$\phi = \tan^{-1} \frac{\omega}{k_d} \tag{9}$$

By measuring I/I_0 and ϕ as a function of ω, k_d, the rate constant, and η, the adsorption probability, are determined. It is perhaps most advantageous to plot I/I_0 versus $1/T_s$, the reciprocal surface temperature, and to plot log tangent ϕ versus $1/T_s$ for a family of different frequencies, ω. Expressing the rate constant in the Arrhenius form, $k_d = A\exp(-E/RT)$, both the preexponential, A, and the activation energy for the reaction, E, can be obtained.[70–72] Rate equations for various models of surface reactions have been solved. These include models of the first order (linear) and nonlinear surface reactions, reactions with several steps in series that include surface diffusion as well, and also branched (parallel) reactions. The calculated log I/I_0 versus $1/T_s$ and log tangent ϕ versus $1/T_s$ curves look very different for the different models. These are shown in Figure 7.9. Data analysis involves fitting the experimentally obtained intensity-phase data to models of different sequences of reactions; an example of an analysis of this type will be described later in the chapter.

For analysis of the temperature-programmed desorption data (see Chapter 2), desorption rate for a species in a binding state i is written[68]

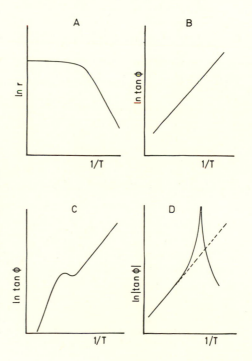

Figure 7.9. Arrhenius plots of the experimental parameters; the amplitude, r; and the tangent of the phase lag ϕ predicted by different reaction models. The variation in the tangent of the phase lag ϕ as a function of $1/T$ is strikingly different for a single step reaction (first or second order) as shown in curve B, for a branched mechanism with two reaction paths operating in parallel (curve C), and for two reaction channels acting in series (curve D). The form of the amplitude r versus $1/T$ is similar for the three reaction mechanisms and is shown in curve A.

$$-\frac{d\eta_i}{dt} = A_i\eta_i^{x_i} \exp\left(-\frac{E_i}{RT}\right) \tag{10}$$

where x_i is the reaction order from state i, A_i the preexponential factor, E_i the desorption activation energy, and n_i the surface concentration in state i. This expression is fairly general for most desorption processes, with a few exceptions. We shall briefly discuss its behavior for the two most common cases, first- and second-order desorption.

For a first-order process the desorption peak maximum temperature T_p and the peak at half-width are independent of the initial coverage. An example of this behavior is shown in Figure 7.10 for xenon on a tungsten crystal surface.

For a second-order desorption process the desorption peak maximum temperature T_p and the half-width both change with increasing initial

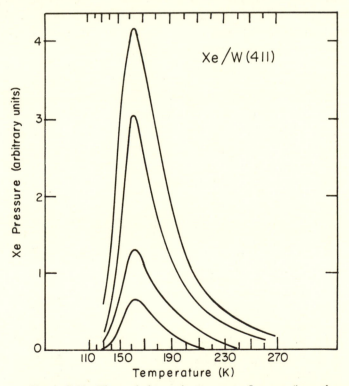

Figure 7.10. Thermal desorption spectra of xenon from the tungsten (411) crystal surface for different surface coverages, illustrating a first-order process. (After D. King, ref. 68.)

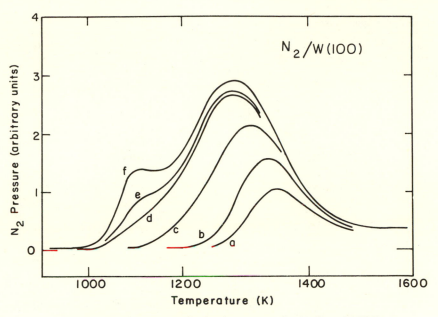

Figure 7.11. Thermal desorption spectra for N_2 from W(100) obtained with increasing surface coverages a to f. The high-temperature peak illustrates the second-order desorption behavior. (After D. King, ref. 68.)

coverage. An example of this type of behavior is the desorption of N_2 from W(100), clearly indicating the dissociative adsorption of nitrogen before desorption under the conditions of the experiment (Figure 7.11).

A variety of procedures for analyzing the desorption spectra have been developed. These are reviewed in detail elsewhere.[73]

Adsorption

When a gas atom or molecule impinges on the surface, we would like to know how long it stays in the attractive potential wells of the surface atoms before desorption. The surface residence times may range from times of a single-atom vibration, approximately 10^{-12} sec, to several seconds or longer. It should be noted that the detectable time range available in most experiments is about 10^{-6} sec to seconds. The residence time depends on the efficiency of $T\text{-}V_s$ energy transfer at the site of impact or at other sites where the atom will migrate within its surface residence time. For a beam of atoms or molecules there may be a range of adsorption probabilities and residence times, depending on the

353

Table 7.1. Experimental values of sticking probabilities of gases on single-crystal surfaces

Surface	Gas	T (K)	η	References
W(110)	N_2	300	0.004	1, 2
(110)	N_2	300	$<10^{-3}$	3
(110)	N_2	300	<0.01	3a
(100)	N_2	300	0.25–0.59	1, 4
(111)	N_2	300	<0.04	1
(111)	N_2	300	0.08	3a
(310)	N_2	300	0.25–0.72	5
(210)	N_2	300	0.28	5
(320)	N_2	300	0.73	5b
(111)	N_2	300	0.08	5b
(110)	H_2	300	0.07	6
(110)	H_2	80	$<10^{-4}$	7
(100)	H_2	300	0.18	6
(111)	H_2	425	0.24	6
Re(0001)	N_2	300	$<10^{-5}$	8
(0001)	N_2	300	0.002	9
Fe(110)	N_2	300	10^{-7}	10
(100)	N_2	300	10^{-7}	11, 12
(111)	N_2	300	10^{-4}	12
Pt(100)				
(5 × 20)	O_2	300	4×10^{-4}	13
(1 × 1)	O_2	300	0.1	14, 15
(110)	O_2	300	0.4	16
(111)	O_2	550	0.02	17, 18
(100)				

atomic structure and the composition of the surface at the place of impact. For example, if the incident atom strikes a site already occupied by an adsorbed atom, they will be different than if it strikes an unoccupied surface site.

Several different types of measurements indicate that the adsorption probability depends intimately on the atomic surface structure and surface composition. One of the frequently determined adsorption parameters is the time-average "sticking coefficient," which detects the fraction of incident gas particles that adsorb. In Table 7.1 some of the sticking coefficient values are listed that were measured in recent years.[74] While these values are reproducible by following a certain procedure of surface preparation, they may vary by orders of magnitude as reported by different laboratories for the same gas–substrate system.

Table 7.1.—Continued

Surface	Gas	T (K)	η	References
(5 × 20)	H_2	300	0.17	19
(5 × 20)	H_2	125	0.07	20
(110)	H_2	125	0.33	20
(111)	H_2	125	0.016	20
(111)	H_2	150	0.1	21
(211)	H_2	125	0.1	20

1. T.A. Delchar and G. Ehrlich, J. Chem. Phys. **42**, 2686 (1965).
2. P.W. Tamm and L.D. Schmidt, Surf. Sci. **26**, 286 (1971).
3. (a) T.E. Madey and J.T. Yates, Nuovo Cimento, Suppl. **5**, 483 (1967); (b) D.A. King and M.G. Wells, Surf. Sci. **29**, 454 (1972); Proc. Roy. Soc. (Lond.) **A339**, 245 (1974).
4. L.R. Clavenna and L.D. Schmidt, Surf. Sci. **22**, 365 (1970).
5. (a) D.L. Adams and L.H. Germer, Surf. Sci. **27**, 21 (1971); (b) S.P. Singh-Boparai, M. Bowker, and D.A. King, Surf. Sci. **53**, 55 (1975).
6. L.D. Schmidt, in *Interactions on Metal Surfaces,* ed. R. Gomer, Springer-Verlag, Berlin, 1975, p. 63.
7. R.S. Polizzotti, Ph.D. thesis, University of Illinois, Urbana, 1974.
8. R. Liu and G. Ehrlich, J. Vac. Sci. Technol. **13**, 310 (1976).
9. A. van Oostrom, in: Proc. 8th Int. Conf. Phenom. Ionized Gases, Int. Atomic Energy Agency, Vienna, 1968.
10. G. Brodén, G. Gafner, and H.P. Bonzel, Appl. Phys. **13**, 333 (1977).
11. G. Ertl, M. Grunze, and W. Weiss, J. Vac. Sci. Technol. **13**, 314 (1976).
12. F. Bozso, G. Ertl, M. Grunze, and M. Weiss, J. Catalysis, **49**, 18 (1977).
13. G. Kneringer and F.P. Netzer, Surf. Sci. **49**, 125 (1975).
14. G. Brodén, G. Pirug, and H.P. Bonzel, Proc. 7th Int. Vac. Congr. and Int. Conf. on Solid Surf., Vienna, 1977, p. 907.
15. C.R. Helms, H.P. Bonzel, and S. Kelemen, J. Chem. Phys. **65**, 1773 (1976).
16. R. Ducros and R.P. Merrill, Surf. Sci. **55**, 227 (1976).
17. H.P. Bonzel and R. Ku, Surf. Sci. **40**, 85 (1973).
18. H. Hopster, H. Ibach, and G. Comsa, J. Catal. **46**, 37 (1977).
19. F.P. Netzer and G. Kneringer, Surf. Sci. **51**, 526 (1975).
20. K.E. Lu and R.R. Rye, Surf. Sci. **45**, 677 (1974).
21. K. Christmann, G. Ertl, and T. Pignet, Surf. Sci. **54**, 365 (1976).

There are two main reasons for this. The sticking probability is strongly structure-sensitive, as indicated by Table 7.1. In general, its value is higher for atomically rough crystal planes. Also, it markedly depends on the coverage of adsorbates. In Figure 7.12 the sticking coefficients of O_2 versus the oxygen coverage for a smoother Pt(111) and a stepped platinum surface are displayed.[75] They decrease by an order of magnitude for an oxygen atom surface coverage of only 8 percent. The stepped surface has a factor-of-2 larger sticking probability at any oxygen surface concentration as compared to the smooth (111) platinum surface.

Another measurement that reflects the surface composition sensitivity of the adsorption is the monitoring of the T–V_s energy transfer by molecular beam–surface scattering studies. As described in the preced-

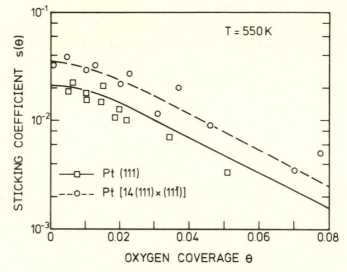

Figure 7.12. Sticking coefficients for oxygen on the flat Pt(111) and on the stepped platinum surfaces as a function of oxygen coverage at 550 K. (After H. Hopster et al., ref. 75.)

ing section, the available data indicate that the $T-V_s$ energy exchange is poor for most atoms and molecules from clean crystal surfaces, but appears to be very good if the surface is covered with a near-monolayer of adsorbed carbon monoxide.[13]

In order to explain the low sticking probability of incident gases and only partial energy transfer upon collision with the surface, a precursor state for adsorption is postulated.[76,77] It is assumed that the incident particles are trapped at the surface and held weakly by a shallow-well-depth attractive potential similar to physical adsorption. Particles in this state have a certain lifetime during which they may reach a site where stronger chemical binding will occur (chemisorption), or they desorb. There is considerable experimental evidence for the existence of this precursor state from atomic beam–surface scattering studies of inert gases from graphite and transition metal surfaces.[8,10] These indicate that inelastic scattering occurs that is accompanied by partial $T-V_s$ energy transfer. However, only a small fraction of the available thermal energy can be transferred to the atoms leaving the surface. Kisliuk[78] has developed a "precursor" model to describe the adsorption kinetics, assuming a certain residence time for gas particles in the second layer where they are weakly held, and has derived an expression for the sticking coefficient as a function of coverage.

It should be noted that CO and NH_3 appear to have high adsorption probabilities, whether probability is measured by time-averaged sticking probability or by molecular-beam-scattering experiments.[79,80] For these molecules, nearly complete energy transfer seems to occur at the point of impact, regardless of surface structure or composition.

Considerable experimental evidence has been gathered which indicates that adsorption leading to dissociation requires an activation energy for many molecules. In molecular beam–surface scattering experiments, this is determined by variation of kinetic temperature (average velocity) of the incident beam by variation of the source temperature. If there is an activation energy for adsorption, the adsorption probability (and the probability of other successive surface processes) increases with increasing beam temperature. An activation energy of 0.5 to 1.5 kcal/mol was determined for the dissociative adsorption of H_2 on Pt(111) crystal faces.[72] A larger 5 kcal/mol activation energy was found for H_2 adsorption over copper crystal surfaces.[81] The activation energy of H_2 dissociation is 1.4 kcal/mol on tantalum surfaces.[82] The rate of NO dissociation on Pt(100) was strongly influenced by the kinetic temperature of the incident gas molecules.[83]

The activation energy for dissociative adsorption of H_2 depends strongly on the atomic surface structure of platinum. On stepped surfaces there is no activation energy.[72] Thus the rate of dissociative adsorption and subsequent surface reactions is about an order of magnitude larger on stepped surfaces, since the presence of an activation energy for adsorption on the (111) crystal face exponentially decreases the rate. Therefore, even a small activation-energy difference from crystal face to crystal face may change the rates of surface processes by an order of magnitude or more. A thorough study of the structure sensitivity of the activation energy for dissociative adsorption of oxygen on silicon crystal faces was reported by Ibach.[84] The activation energy decreases with increasing step density.

Using the absolute-rate theory, the rates of the various elementary surface processes may also be calculated.[85] The preexponential factor A_n is defined for a reaction involving n species as

$$k_n = A_n \exp\left(-\frac{E}{RT}\right) \tag{11}$$

where k_n, E, R, and T have their usual meaning. In the absence of an activation energy for the process, the magnitude of A_n yields the maximum rate. In order to calculate A_n for adsorption, let us consider the change in surface concentration (n_s) of molecules with a gas-phase concentration (n). We take a surface of area (a), and θ is the fraction of surface covered by molecules. The average gas-phase velocity of the molecules ($\bar{c}$) enables us to calculate a volume swept out per unit time of the gas-phase molecules that will strike the surface $\frac{1}{4}\,\bar{c}\cdot a$. The number of collisions per unit time with uncovered surface is given by the preexponential part of Eq. (12):

$$a\,\frac{dn_s}{dt} = \tfrac{1}{4}\bar{c}a(1 - \theta)n \exp\left(- \frac{E}{RT} \right) \tag{12}$$

A Boltzmann factor is multiplied by the right side of Eq. (12),

$$k_n = A_n \exp\left(- \frac{E}{RT} \right) \tag{13}$$

to account for the fraction of molecules having activated adsorption with activation energy E. This situation can apply when adsorption requires an activation energy, as in H_2–D_2 exchange reaction on Cu. This is determined by the rate dependence of surface reaction on beam temperature. The preexponential factor giving the rate of formation of adsorbed surface molecule per unit area, when these molecules are colliding with uncovered surface and are removed rapidly by reaction, is

$$A_1 = \tfrac{1}{4}\,\bar{c}\,(1 - \theta) \tag{14}$$

Note that the units of Eq. (14) are length/time.

Let us consider adsorption as the rate-limiting step of a reaction involving gas-phase and surface-reactant species. We will apply Eq. (14) to the case where product molecules are detected in the gas phase of volume V. First, we replace $1 - \theta$ by θ in Eq. (14), since reaction is considered to take place only when the gas-phase species collides directly with a surface species. Thus $A_1 = \tfrac{1}{4}\,\bar{c}\,\theta$ for this case. Under steady-state conditions for the product species formed on the surface,

$$V\,\frac{d(\text{product})}{dt} = k^{\text{psuedo}}n\cdot a \tag{15}$$

where the preexponential part of k^{pseudo} equals $\tfrac{1}{4}\,\bar{c}\,\theta$. The effective pseudo first-order preexponential for product formation becomes

$$A_1 = \frac{\tfrac{1}{4}\,\bar{c}\,\theta a}{V} \tag{16}$$

which has the units of reciprocal time. This expression applies to the rate of product formation as detected after desorption into the gas phase of volume V after gas-phase species collide directly with the surface species.

Typical values that can be obtained for A for desorption are in the range 10^2 to 10^4 sec^{-1} assuming that $\bar{c} = 4 \times 10^4$ cm/sec.[85]

Surface Diffusion

The migration of atoms or molecules along the surface is one of the most important steps in surface reactions. As we shall see later, the rate-determining step for many reactions that have been studied at low pressures is surface diffusion. Nevertheless, only a few experiments provide information about this surface process. These indicate that the rate of migration of adsorbates along the surface is rapid and the activation energies for surface diffusion are not much larger than kT.[68] Many of the quantitative investigations were carried out with the use of field-emission tips. Gas was adsorbed on one side of the tip near 4 K and the movement of the gas boundary was observed on heating.[86,87] High mobility was observed for oxygen at 27 K, for nitrogen at 50 K, and at less than 20 K for hydrogen on tungsten surfaces. Activation energies for surface migration in the weakly adsorbed precursor states were in the range 1 to 3 kcal/mol. For the migration of chemisorbed species such as O, H, N, and CO, the activation energies are about 10 to 30 percent of the adsorbate–substrate binding energies, in the range 10 to 30 kcal/mol, and they depend on the adsorbate coverage.[88]

There are very few experimental determinations of adsorbate surface diffusion rates, since these are difficult experiments to perform. Butz and Wagner studied the diffusion of Au and Pd on W crystal planes by scanning AES.[89] Radiotracer techniques should be useful for these studies, and it is hoped that they will be used in the near future with increasing frequency.

The preexponential factor for surface diffusion may also be calculated.[85] Surface diffusion may be treated for a surface species of concentration n_s by a random-walk type of analysis. We assume that the molecule adsorbs on a random site, then diffuses to the active site where the reaction occurs. The frequency (f) of jumps of distance d is given by

$$f = \nu e^{-E_D/RT} \tag{17}$$

where ν is the surface vibration frequency, and E_D, the diffusion energy, is averaged over the various surface regions.

The random-walk expression is

$$t = \frac{3\langle x^2 \rangle}{fd^2} \tag{18}$$

where t is the time required to diffuse a distance d. We calculate a surface diffusion velocity ($v = (x^2)^{\frac{1}{2}}/t = fd^2/3(x^2)^{\frac{1}{2}}$). Thus the surface area swept out by diffusing species having a molecular diameter (b) is $v \cdot b$. The collision number of diffusing species with active sites of density N_s is $v \cdot b \cdot N_s$. We evaluate the diffusion distance as the average separation between active sites. This approximation gives $< x^2 >^{\frac{1}{2}} = N_s^{-\frac{1}{2}}$. Combining the collision number with a Boltzmann factor to account for collisions with sufficient energy for reaction (E), we obtain

$$\frac{-dn_s}{dt} = v \cdot b \cdot N_s \cdot n_s \exp\left(-\frac{E}{RT}\right) \tag{19}$$

$$k = \frac{vd^2b}{3} N_s^{3/2} \exp\left[-\frac{E + E_D}{RT}\right]$$

The preexponential factor is

$$A_1 = \frac{vd^2b}{3} N_s^{3/2} \tag{20}$$

As in gas phase collision theory, a steric factor is sometimes multiplied by the right side of Eq. (20) to account for the required orientation of the reactants.

If we assume that the active-surface-site density, N_s, can be in the range 10^{11} to 10^{14} sites/cm^2, $d = 3 \times 10^{-8}$ cm, $b = 3 \times 10^{-8}$ cm, and $v = 10^{13}$ sec^{-1}, the preexponential factors for surface diffusion should be in the range 10^7 to 10^{13} sec^{-1}.

Desorption

The desorption processes have been studied both by temperature-programmed thermal desorption and by molecular beam–surface scattering techniques. The desorption of an adsorbed molecule can be considered as an unimolecular process,[85] that was treated by Ibach et al.[90] in considerable detail. For the desorption of atoms that are mobile on the surface the preexponential becomes $A_1 = kT/h$ and its magnitude is of the order of 10^{13} sec^{-1}. For adsorbed molecules that are immobilized on the surface prior to desorption,[90] the preexponential factor would be in the range of 10^{16} to 10^{17} sec^{-1}. If the molecule is dissociated on the surface, a bimolecular association reaction may take place prior to molecular desorption. This reaction step may be rate-limiting and will be treated in the following section.

Desorption is often the rate-limiting step in surface processes because it is endothermic. For the desorption of CO from the Ni(111) surface the preexponential factor was 6×10^{16} sec^{-1}, in good agreement with calculations.[90]

By measuring the phase shifts as a function of temperature during the scattering of CO molecular beams from the Pd(111) crystal face, a preexponential factor of 3×10^{12} sec^{-1} and a desorption energy of 33 kcal/mol was determined.[79,91] The desorption energy is in good agreement with the isosteric heat of adsorption, indicating that the adsorption process is not activated.

Thermal desorption experiments often indicate second-order kinetics for the process.[68] In these circumstances association reactions that precede the desorption step control the rate of desorption. The influence of a precursor state that the desorbing molecules may pass through during desorption has been discussed by King.[92] The effects of lateral interactions between the desorbing species and the atomic surface structure around the adsorption site or with other adsorbates have been considered by Roberts,[93] Wang,[94] Gaymour and King,[95] and Adams.[96]

For certain surface reactions where desorption appears to be the rate-limiting step, large values for the preexponential factor, in the range 10^{15} to 10^{16} sec^{-1}, have been found.[97] An example of this is the decomposition of formic acid on the Ni(110) crystal face.[98] This large value is thought to be due to a gain in rotational entropy of the desorbing molecule (in this case, CO_2) in the transition state.

Surface Reactions

More detailed kinetic information on surface reactions has only recently become available. The kinetic parameters for these reactions are listed in Table 7.2. Because of the experimental conditions that have fixed the surface concentrations of all but one of the reactants in many instances, the rate constants and the preexponential factors appear pseudo-unimolecular and are also tabulated in this manner.[85] That is, the preexponential factor includes reactants with fixed concentrations throughout the reaction. The preexponential factors that were determined by experiments are in the range 10^2 to 10^{16} sec^{-1}; that is, they vary by 14 orders of magnitude. Most of these values are small as compared to those reported for gas-phase unimolecular ($\sim 10^{13}$ sec^{-1}) or bimolecular

Table 7.2. Preexponential factors, activation energies, and reaction probabilities for several surface reactions studied by molecular beam scattering

Reaction	A^*	E_a (kcal/mol)	Reaction probability	Reference
H + D $\xrightarrow{Pt}$ HD (<600 K)	8×10^4 (sec^{-1})	13.0	$\sim 10^{-1}$	1
H + D $\xrightarrow{Pt}$ HD (>600 K)	2.7×10^5 (sec^{-1})	15.6	$\sim 10^{-1}$	1
D + O$_2$ $\xrightarrow{Pt}$ D$_2$O (700 K)	—	12	—	2
CO + O $\xrightarrow{Pt}$ CO$_2$ (700 K)	—	20	$\sim 10^{-3}$	3
C$_2$H$_4$ + O$_2$ $\xrightarrow{Ag}$ CO$_2$ (800 K)	—	8	$<10^{-2}$	4
2H $\xrightarrow{\text{graphite}}$ H$_2$ (800–1000 K)	1.06×10^{-2} cm^2/atom-sec	15.9	10^{-3}–10^{-2}	5
H$_2$ $\xrightarrow{\text{Ta}}$ 2H (1100–2600 K)	—	75	4×10^{-1}	6
HCOOH $\xrightarrow[\text{decomp}]{\text{Ni}}$ CO$_2$ (<455 K)	10^{12} (sec^{-1})	20.7	—	7
HCOOH $\xrightarrow[\text{decomp}]{\text{Ni}}$ CO$_2$ (>455 K)	5.8×10^8 (sec^{-1})	2.5	~ 0.9	7
C + O$_2$ $\rightarrow$ CO (1000–2000 K)	2.5×10^7 (sec^{-1})	30	10^{-3}–10^{-2}	8
C + O$_2$ $\rightarrow$ CO (1000–2000 K)	3×10^{12} (sec^{-1})	50	10^{-3}–10^{-2}	8
C + 4H $\rightarrow$ CH$_4$ (500–800 K)	1.27×10^{-18} cm^4/atom-sec	3.3	10^{-3}–10^{-2}	5
2C + 2H $\rightarrow$ C$_2$H$_2$ (>1000 K)	1.59 cm^2/atom-sec	32.5	10^{-2}–10^{-2}	5
Ge + O$_2$ $\rightarrow$ GeO (750–1100 K)	10^{16} (sec^{-1})	55	2×10^{-2}	9

Ge + O →	GeO (750–1100 K)	10^{16} (sec^{-1})	3×10^{-1}	55	10
Ge + O$_3$ →	GeO (750–1100 K)	10^{16} (sec^{-1})	5×10^{-1}	55	11
Ge + Cl$_2$ →	GeCl$_2$ (750–1100 K)	10^{7} (sec^{-1})	3×10^{-1}	25	7
Ge + Br$_2$ →	GeBr$_2$ (750–1100 K)	10^{7} (sec^{-1})	3×10^{-1}	20	12
	GeBr$_4$				
Si + Cl$_2$ →	SiCl$_2$ (1100–1500 K)	10^{8} (sec^{-1})	3×10^{-1}	40	7
Ni + Cl$_2$ →	NiCl (900–1400 K)	10^{7} (sec^{-1})	8×10^{-1}	30	13

*For bimolecular reactions, the preexponential factor also includes the surface concentration of one of the reactants that is held constant during the experiments.

1. M. Salmeron, R. Gale, and G.A. Somorjai, J. Chem. Phys. **70**, 2807 (1979).
2. J.N. Smith and R.L. Palmer, J. Chem. Phys. **56**, 13 (1972).
3. H.P. Bonzel and R. Ku, Surf. Sci. **33**, 91 (1972).
4. J.N. Smith, R.L. Palmer, and D.A. Vroom, J. Vac. Sci. Technol. **10**, 373 (1973).
5. M. Balooch and D.R. Olander, J. Chem. Phys. **63**, 4772 (1975).
6. R.A. Krakowski and D.R. Olander, J. Chem. Phys. **49**, 5027 (1968).
7. R.J. Madix, in *Physical Chemistry of Fast Reactions*, ed. D.O. Hayward, vol. 2, Plenum Press, New York, 1976.
8. D.R. Olander, R.H. Jones, J.A. Schwartz, and W.J. Siekhaus, J. Chem. Phys. **57**, 421 (1972).
9. J.B. Anderson and M. Boudart, J. Catal. **3**, 216 (1964); R.J. Madix and M. Boudart, J. Catal. **7**, 240 (1967).
10. R.J. Madix and A.A. Susu, Surf. Sci. **20**, 377 (1970).
11. R.J. Madix, R. Parks, A.A. Susu, and J.A. Schwartz, Surf. Sci. **24**, 288 (1971).
12. R.J. Madix and A.A. Susu, J. Vac. Sci. Technol. **9**, 915 (1972).
13. J.D. McKinley, J. Chem. Phys. **40**, 120 (1964).

363

reactions. The typical preexponential factor for a bimolecular reaction is 10^{-11} cm^3/molecule-sec, which becomes 10^3 sec^{-1} under pseudo-first-order conditions with a reactant pressure of 10^{-4} torr. In fact, frequently when low preexponential factors were obtained in gas-phase studies, the presence of surface reactions was suspected.

We have already computed the preexponential factors that are expected if adsorption, surface diffusion, or desorption steps are rate-limiting in the surface reactions. Using the transition-state theory, let us compute the preexponential factors for surface reactions of different order. Then we shall be able to compare the experimentally obtained values that are shown in Table 7.2 with those A values that are calculated for the different types of surface reactions.

The preexponential factors for the Nth-order reaction of i species is expressed as

$$A_n = \frac{(KT/h)Q^{\ddagger}}{\prod\limits_{i=1}^{N} Q_i} \tag{21}$$

where K is Boltzmann's constant, T the temperature, h is Planck's constant, Q_i the partition function for reactant species i, and $Q^{\ddagger}$ the partition function for transition state.

The total partition function for a mobile surface polyatomic species containing n atoms is

$$Qi = q_T^2 q_R q_v^{3n-3} \tag{22}$$

where q_T, q_R, and q_v are the translational, rotational, and vibrational partition functions, respectively. The total partition function for the transition-state species is calculated by Eq. (22), except that one vibrational component along the bond being broken is removed. The significant component of Q_i is the translational partition function q_T:

$$q_T = \frac{(2\pi MKT)^{1/2}}{h} \tag{23}$$

which typically has the value 10^8 cm^{-1}. Vibrational partition functions have typical values near unity and free rotational partition functions have values near 10. Vibrational partition function components will be neglected for surface species, since their magnitude is near unity.

We have tabulated the expressions for preexponential factors for the various rate-limiting steps in Table 7.3. Note that pseudo-first-order preexponential values are listed that are most useful for comparison

Table 7.3. Values of preexponential factors

Rate-limiting step	Expression for A_n	Pseudo first-order A_n (sec^{-1})	Typical value (sec^{-1})
Adsorption	$\frac{1}{4}\bar{c}\theta a/V$	$\frac{1}{4}\bar{c}\theta a/V$	10^2–10^4
Desorption	ν_0	ν_0	10^{13}
Surface diffusion	$\dfrac{\nu d^2 b N_s^{1/2}}{3}$	$\dfrac{\nu d^2 b N_s^{1/2}}{3}$	10^7–10^{12*}
Surface reaction control Order			
$N = 1$	KT/h	KT/h	$10^{13\dagger}$
$N = 2$	KT/hq_T^2	$\dfrac{KT}{hq_T^2}\,n_s$	$10^{9\dagger}$
$N = 3$	KT/hq_T^4	$\dfrac{KT}{hq_T^4}\,n_s^2$	$10^{5\dagger}$
$N = 4$	KT/hq_T^6	$\dfrac{KT}{hq_T^6}\,n_s^3$	$10^{1\dagger}$

*N_s varies from 10^{11} to 10^{14} sites/cm^2.
†Values are $n_s = 10^{12}$ atoms/cm^2, $KT/h = 10^{13}$ sec^{-1}, $\nu = 10^{13}$ sec^{-1}, $d = 3 \times 10^{-8}$ cm, $b = 3 \times 10^{-8}$ cm, $c = 4 \times 10^4$ cm/sec.

with molecular beam experimental results. Expressions for reactions with various rate-limiting steps are presented in which only translational partition functions are considered. Note that for these reactions the Nth-order preexponential decreases by a factor of 10^{16} cm^{-2} molecule^{-1} for each increase in reaction order by one. We assume that n_s, the surface concentration of the reactant species in excess, is constant as a function of temperature in making the estimations of values in Table 7.3. This assumption, of course, depends upon the experimental conditions and system of interest, and, therefore these relative values are intended only for comparison with each other.

Application of Eq. (23) assumes free-particle motion for the reactant molecules. In those cases where this condition is not met, the calculated preexponential factor for the rate of reaction will be too high. Surface diffusion will be the more likely slow step in that case.

If we use the similarity of the computed and experimental preexponential factors, reaction mechanisms can be suggested for each of the surface reactions where the kinetic parameters have been determined. Although the mechanism may not always be uniquely determined, one can certainly rule out several mechanisms based on the large discrepancies of calculated and experimental A values.

Surface reactions that are controlled by slow adsorption or a slow surface reaction step have preexponential factors in the range 10^2 to 10^5

sec $^{-1}$. For surface-diffusion-controlled reactions, the preexponential factor varies from 10^7 to 10^{12} sec^{-1}, depending upon the density of active sites. Desorption or unimolecular reaction-controlled surface reactions have factors typically greater than 10^{12} sec^{-1}, in the range of 10^{13} to 10^{17} sec^{-1}.

In the gas phase the preexponential factors are typically 10^{13} sec^{-1} for first-order reactions, 10^{-12} to 10^{-10} cm^3/molecule-sec for second-order reactions, and 10^{-37} to 10^{-31} cm^6/molecule^2-sec for third-order reactions. The values measured for surface reactions can deviate significantly from the gas-phase values, depending upon the rate-limiting step, the detailed mechanism, and the reactant concentrations. It appears that the statistical models useful for gas-phase kinetics can also be employed to reproduce experimental preexponential factors for surface reactions.

The surface reactions that were studied and listed in Table 7.2 may be divided into two classes. In the corrosion reactions the surface is one of the reactants, for example, Si(solid) + Cl$_2$(gas) → SiCl$_2$(gas). In catalytic surface reactions the surface atoms act as catalysts and do not appear among the reaction products, for example, H$_2$ + D$_2$ $\xrightarrow{\text{Pt}}$ 2HD. These two types will be reviewed separately.

Corrosion Reactions

Let us now discuss the results of several of the solid-gas corrosion reactions listed in Table 7.2. Olander et al.[99] have studied the oxidation of both the basal plane and the prism plane of graphite. The product of the oxidation reaction is CO, although a small CO$_2$ signal was also detectable during the oxidation of the prism plane. The reaction rate was monitored as a function of temperature. From the chopping-frequency dependence of the reaction probability, they have concluded that there must be two parallel reactions, one slow and one faster, taking place on the graphite surface. For the basal plane the fast step is attributed to the migration of atomic oxygen over the surface to the reaction sites where oxidation occurs. The slow reaction step appears to be the desorption of CO; its rate constant is listed in Table 7.2. There are two types of reaction sites postulated in surface concentrations, of 10^{11} cm^{-2} and 10^8 cm^{-2}, respectively. Grain boundary and possible bulk diffusion of oxygen were found to be important steps in the oxidation of the prism plane. The oxidation of germanium was studied by Anderson and Boudart[100] and Maddix and Boudart;[101] the reaction probability was 0.02 and independent of temperature. However, the oxidation rate on the surface was dependent on the oxygen beam temperature, which indicated 100 to 200 cal/mol activation energy for the adsorption. It appears that the dissociative adsorption of oxygen is the rate-determining step of the reaction. The oxidation reaction using oxygen atoms instead of oxygen molecules was also investigated. The reaction probability in this case is 0.2 to 0.3, much higher than for oxygen molecules at surface temperatures in the range 830 to 1110 K. The difference in the activities appears to be due to the requirement that both atoms in the oxygen molecule interact simultaneously with the surface atom; thus the interaction probability depends on the orientation of the incident oxygen molecule. The interaction between chlorine molecular beams and nickel surfaces has been studied by McKinley et al.[102] The reaction probability is 0.8 at 1000 K and both NiCl and $NiCl_2$ are detectable among the reaction products. The formation of NiCl predominates at low temperatures, while $NiCl_2$ forms almost exclusively above 1400 K. The mechanism that is proposed involves an Ni_2Cl_2 surface intermediate. The formation of this dimer appears to be the rate-limiting step in the overall reaction, while the desorption of NiCl and $NiCl_2$ are rapid reaction steps. The residence time of desorbing NiCl is 916 μsec at 1450 K and 140 μsec at 1700 K, and the $NiCl_2$ residence times are even longer. The reaction probability of oxygen with molybdenum was determined by Ullman and Maddix to be approximately 0.3.[103] Surface diffusion of oxygen atoms to grain boundaries on the polycrystalline molybdenum surface appears to control the formation of the volatile molybdenum oxide. Two important features of elementary surface reactions are exhibited in these studies:

1. It appears that the desorption energy ΔH_{des} is sufficiently high to assure long residence times on the surface of most of the products as compared to vibrational times of surface atoms. As a result, spectroscopic experiments of many types may be successfully performed to study the properties of reaction intermediates.

2. Several of the reactions show rates that increase with increasing reactant beam temperatures. Thus there appears to be a small but detectable activation energy for the adsorption process. Similar activated adsorption processes are also found during many catalyzed surface reactions.

Catalyzed Surface Reactions

In these reactions the surface acts as a catalyst for reactions that either would not likely take place in the gas phase under the same experimental circumstances used for the surface reaction or would take place much more slowly. These reactions may be classified as several types:

1. *Exchange reactions.* With the use of isotope exchange such as the H_2–D_2 exchange, a necessary elementary reaction step is the breaking of the strong chemical bonds, such as the H—H bond. Without bond breaking, the isotope atom cannot be inserted to form the product molecules. By monitoring the rate of formation of the product, one learns about the efficiency of this bond scission and about the participation of the other surface atoms that lead to the formation of the product. Since H—H, C—H, C—C, C—O, and N—N bond-breaking steps are part of the most important catalytic reactions, the isotope exchange gives us a great deal of information on how these processes might occur. The exchange reaction is also likely to be athermic.

2. *Dissociation reactions.* These types of reactions also probe the nature of chemical bond breaking. Since the detected products are the molecular fragments, kinetic information about their formation and their product distribution may be used to learn about the reaction intermediates. These reactions play an important role at high temperatures, an important regime for studies of gas–surface interactions. The decomposition reactions are most often endothermic. However, under some circumstances, such as for formic acid decomposition, this reaction may be exothermic, as will be demonstrated.

3. *Exothermic small-molecule reactions and atom recombination reactions.* In these surface reactions excess chemical energy becomes available that will be distributed among the reaction products and the surface. The molecules that form may desorb with a velocity distribution that is no longer controlled by the surface temperature, T_s. The angular distribution is also likely to deviate from the cosine distribution. The oxidation of CO to CO_2 and of NH_3 to NO and H_2 and O_2 to H_2O are examples of small-molecule reactions of this type. Recombination of atoms on the surface may also liberate a great deal of chemical energy that must be redistributed among the desorbing products and the surface. The recombination of H, O, or Cl, or the reactions of these atoms with each other, are examples of typical recombination processes. Other exothermic surface reactions include the reactions of atoms with adsorbed molecules, such as the reaction of adsorbed ethylene, C_2H_4, with atomic hydrogen to form ethane.

Let us now discuss the results of these different catalytic surface-reaction studies.

H_2–D_2 Exchange

One of the fundamental questions of heterogeneous catalysis is how surfaces lower the activation energy for simple reactions on an atomic scale so that they proceed readily on the surface while the same reactions in the gas phase are improbable. The reaction of hydrogen and deuterium molecules to form hydrogen deuteride, HD, is one of the simple reactions that take place readily on metal surfaces even at temperatures below 100 K. The same reaction is completely inhibited in the gas phase by the large dissociation energy of H_2 (103 kcal/mol). Once the H_2 molecule is dissociated, the successive atom–molecule reaction, $H + D_2 = HD + D$ in the gas phase, still has a potential-energy barrier of roughly 10 kcal/mol.

This reaction was studied on the (111) and stepped surfaces of platinum.[72, 104, 105] The integrated reaction probability was found to be 0.07 on the (111) crystal face and much higher, 0.35, on the stepped surfaces under identical experimental conditions ($T_s = 1100$ K, $\omega = 10$ Hz, $T_g = 300$ K). The reaction probability was also found to depend markedly on the direction of approach of the reactant to the stepped structure. The production of H_2 is highest when the reactant strikes the open side of the step structure, decreasing by approximately a factor of 2 when the inner edge of the step is shadowed.[106] These results are displayed in Figures 7.13 and 7.14, where the intensity of the normalized H_2 signal is monitored as a function of angle of incidence and also as a function of azimuthal angle.

In Figure 7.13, the production of H_2 as a function of the polar angle of incidence is presented. The azimuthal angle, ϕ, is $+ 90°$ in curve a, that is, the plane of beam incidence is perpendicular to the step edges, as shown in the insert. The projection of the reactant beam to the surface is parallel to the step edges in curve b, where ϕ is 0°. These experiments were performed with a beam modulation frequency of 10 Hz with a surface temperature of 1100°C. The dependence of the reaction probability for H_2–D_2 exchange on azimuthal angle is shown in Figure 7.14 for a fixed polar angle of incidence of 45° measured from the macroscopic surface normal. As found in both angle of incidence experiments, the production of HD is highest when the reactants strike the open side of the step structure, increasing by approximately a factor of 2, from $\phi = -90°$ to $\phi = +90°$. To investigate whether the angular dependence observed can be attributed to an activation energy for adsorption, the molecular beam was heated from 300 to 600 K and the production of HD was monitored. No significant variations were observed within the experimental accuracy. The experimental results indicate that the activation energy for adsorption is independent of angle of incidence. The dependence of the H_2–D_2 exchange probability on direction of approach reflects the structural asymmetry of the surface.[106] Although the cross-

369

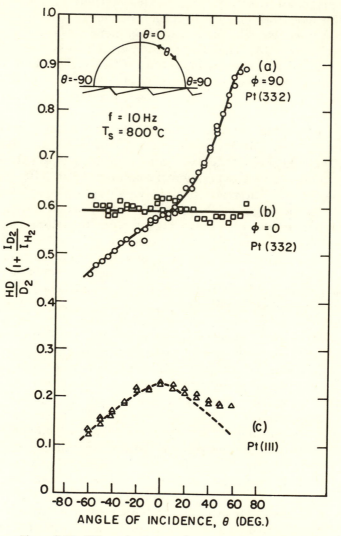

Figure 7.13. HD production as a function of angle of incidence, θ, of the molecular beam, normalized to the incident D_2 intensity. (a) Pt(332) surface with the step edges perpendicular to the incident beam ($\Phi = 90°$); (b) Pt(332) where the projection of the beam on the surface is parallel to the step edges ($\Phi = 0°$); (c) Pt(111).

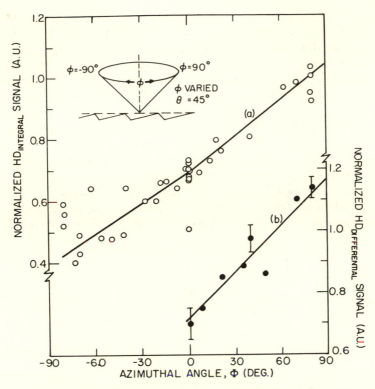

Figure 7.14. HD production on the Pt(332) crystal face as a function of azimuthal angle, ϕ, with the incident angle fixed at $\theta = 45°$: (a) data measured in the integral mode; (b) data obtained in the differential mode.

sectional area of the beam on the surface changes with angle of incidence, the concentration of surface intermediates should not change, since the diffusion time of the adsorbed H and D atoms is much shorter than the reaction time (about 20 ms). This is confirmed by the fact that no change in phase lag occurs for the HD product signals when the angle of incidence is varied. Also, the same trend in HD production is observed with variation in azimuthal angle where the cross-sectional area of the beam remains constant.[106]

As stated before, the reaction probability for H_2–D_2 exchange is found highest when the reactants approach the open side of the step structure. This result clearly is a unique property of step sites for platinum for H—H bond breaking. The site associated with the inner corner atom was found to be the most active. One can estimate the activity of the step in reference to the activity of the (111) terrace by dividing the surface into two parts,[106] the step and the terrace. The incident beam divided between the available area of the step sites and that of the terrace. The reaction probability per unit area at a step site may be used as an adjustable parameter, while the experimental results for the reaction probability per unit area on the Pt(111) surface are used for the flat surface.

The results of these calculations are shown in Figure 7.15. If R is the ratio of the reaction probability per unit area of the step sites to that of the terrace sites of normal incidence, remarkable agreement is found between calculation and the experimental results when the step sites are assumed to be seven times as active as the terrace sites.

The kinetics of the H_2–D_2 exchange on the stepped and (111) platinum surfaces were also studied by measuring the normalized intensities as a function of surface temperature and at a variety of frequencies

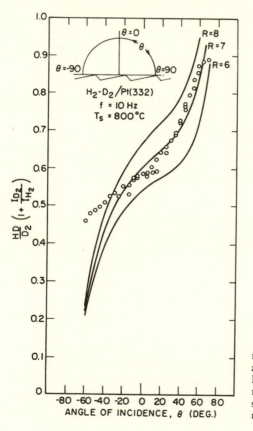

Figure 7.15. Comparison of experimental and calculated HD production as a function of angle of incident, θ, on the Pt(332) surface. R is the ratio of the reaction probability per unit area of the step site to that of a terrace site at normal incidence.

from 1 to 200 Hz, as well as by measuring the phase shifts of the HD as a function of surface temperature at a variety of frequencies. The results yield identical phase and intensity dependence for the step and (111) platinum surfaces, indicating that the reaction mechanisms on these two types of surfaces are identical.[72] The difference lies in the higher reaction probability for dissociative adsorption of the hydrogen and deuterium molecules on the stepped surfaces. After dissociative adsorption has occurred, the subsequent reaction steps are identical on these two types of surfaces. The normalized intensities and the phase shifts as a function of surface temperature are displayed in Figures 7.16 and 7.17. The data may be divided into a high-temperature and a low-temperature regime. At low temperatures in the range 700 to 1100 K, the observed intensity and phase relationships can be fitted to a model that assumes parallel reactions to take place. Data in the high-temperature regime can be fitted to a model which assumes that consecutive reactions take place and in which several steps, including surface diffusion, can be distinguished.[72] The reason for the lower reaction probability for dissociative adsorption of H_2 or D_2 on the Pt(111) crystal face is a small 0.5 to 1.5 kcal/mol activation energy barrier for adsorption.[72] This was revealed by variation of the beam kinetic temperature in molecular beam surface scattering studies. There is no activation energy for adsorption on the stepped platinum surfaces.

The H_2–D_2 reaction was also studied on copper (111) and stepped copper surfaces by Balooch et al.[81] The adsorption of H_2 and D_2 is activated on both copper surfaces. The activation energy of adsorption is

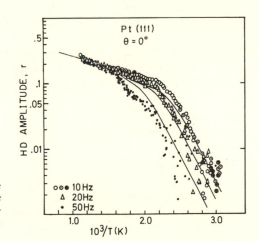

Figure 7.16. Arrhenius plots of the amplitude of the HD product on the Pt(111) surface at several beam modulation frequencies. The mixed H_2–D_2 beam impinges normal to the surface.

373

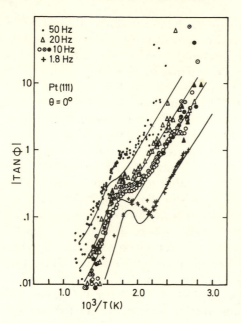

Figure 7.17. Arrhenius plots of the absolute value of the tangent of the phase lag ϕ for the HD product frequencies. The mixed H_2–D_2 beam impinges normal to the surface. The filled triangles and circles correspond to the negative values of tan ϕ ($\phi > 90°$).

about 5 kcal/mol, determined from the beam-temperature dependence of the reaction probability. The H_2–D_2 exchange reaction was also studied by Palmer et al.[107] on platinum and nickel surfaces that were prepared as evaporated thin films. They observed angular distributions that peak near the surface normal following a $\cos^n \theta$ relationship where θ is the angle of desorption from the surface relative to the surface normal. The value of n varies from 2.5 to 4.0 on the different surfaces studied. The peaked angular distribution could be related to the presence of impurities, such as carbon and sulfur contamination, on the metal surfaces. There is evidence from the experiments of Stickney et al.[108] that as the surface is cleaned of sulfur, oxygen, or carbon, n approaches unity. It should be noted that a cosinelike angular distribution for the desorbing HD molecule has been found on both (111) and stepped surfaces of platinum.[72] Copper, on the other hand, seems to show noncosine angular distribution for scattered HD, even from the clean surfaces.

Dissociation Reactions

The atomization of HD on tungsten and tantalum[82] surfaces has been studied. The reaction probability on tungsten surfaces was about 0.3 at 300 K. The angular distribution of the H atom was cosine like, indicating thermal equilibration between the hot surface and the desorbing hydrogen atoms. Long residence times of the order of microseconds for the surface species were measured. On tantalum surfaces, just as on tungsten surfaces, the dissociation probability of hydrogen increases with increasing beam temperature.[82] From the beam-temperature dependence of the reaction probability, it appears that there is an activation energy of 1.4 kcal/mol for the atomization surface reactions. The atoms once formed cannot undergo recombination before desorption from the hot surface. The rate-determining step in the reaction appears to be desorption of the hydrogen atoms with an activation energy of about 75 kcal/mol. The surface diffusion of hydrogen atoms does not seem to require activation energy. The dissociation of N_2O on hot tungsten[109] and platinum[110] surfaces was also studied. On tungsten the decomposition probability approaches unity at high temperature. The product N_2 was emitted with a cosine distribution. It appears that oxygen reacts with hot tungsten surface during the decomposition of N_2O. Both N_2 and NO were found in the scattered beam and the ratio of the two species was approximately 12:1 at 2500 K. On platinum surfaces when N_2O dissociates again, both NO and N_2 species could be detected,[110] the NO concentration being larger than that found upon desorption from tungsten surfaces. The angular distribution of the product NO formed by dissociation on the same platinum surface was of cosine type, indicating accommodation of the molecules on the platinum (100) surface prior to reemission. However, the angular distribution of NO product molecules was different when they were emitted from a carbon-covered platinum surface at 1125 K. The distribution is certainly not cosine; it peaks near the specular angle. Such a peaked distribution reflects a lack of energy accommodation during the surface dissociation reactions on NO on the carbon-covered platinum (100) surface. N_2O may undergo a variety of chemical reactions on the carbon-covered platinum surfaces. These chemical reactions are largely exothermic, to yield CN, CO, and CO_2 reaction products in addition to N_2 and NO. On the clean platinum surface the N_2O molecule can undergo only endothermic reactions.

The decomposition of formic acid was studied by Maddix et al.[98] using molecular beam scattering and thermal desorption. In these studies the clean Ni(100) surfaces and nickel–copper alloy surfaces and carbon-contaminated nickel surfaces were employed. From the clean Ni(100) surface, the decomposition products H_2, CO_2, and CO are all desorbed at the same temperature, 370 K. The desorption of the different products at the same temperature indicates that they must come from a common reaction intermediate. From the product distribution and the reaction-rate parameters it could be concluded that the reaction intermediate is formic anhydride on the surface. On carbon-contaminated and oxygen-contaminated Ni(100) surfaces and on copper–nickel alloys, temperature-programmed desorption of the decomposition products are different as compared to the Ni(100) surface. Also, the product distributions are markedly different, indicating that the majority of the products are CO_2 and H_2. From the results it was concluded that the reaction intermediates have changed as a result of adding impurities or alloying components to the nickel surface from formic anhydride to HCOO. Various reaction sequences show that the decomposition of anhydride is likely to yield H_2, CO_2, and CO in equal quantities, whereas the decomposition of the HCOO should yield only CO_2 and H_2.

These studies demonstrate clearly that by changing the surface composition or the surface structure, or both, the reaction intermediates change, which changes the product distribution as well. Also, the kinetics of the reaction reflect this change of mechanism. In Table 7.4 the activation energy, preexponential factor, and the postulated reaction intermediates for formic acid decomposition are tabulated for clean and contaminated nickel surfaces as well as for the copper–nickel alloys.

Exothermic Small-Molecule Reactions and Atom-Recombination Reactions

The oxidation of carbon monoxide with molecular oxygen was recently studied on platinum surfaces.[111] The velocity distributions of the desorbed CO_2 molecules were measured. The surface temperature of platinum was 800 K; however, the desorbing gas temperature at the surface normal (0°) was measured to be 3500 K, while at 45° it was 2400 K. Thus it appears that the probable velocity decreases monotonically from the surface normal. These large kinetic temperatures indicate that

Table 7.4. Apparent activation energies and preexponential factors for formic acid decomposition from nickel and copper (110) crystal faces of different surface composition[a]. (After I. Wachs and R. Maddix, ref. 98.)

Surface	E_{app} (kcal/gmol)	A_{app} (sec^{-1})	T_p (K)	Intermediate
Ni(110) (clean)	26.6	2×10^{15}	~370	anhydride
Ni(110) (2 × 1)-C	25.5	5×10^{12}	~430	HCOO
Ni(110)-O	N.D.	N.D.	~440	HCOO
Cu/Ni(110) (65/35)	27.6	N.D.	~400	HCOO
Cu(110)	31.9	1×10^{14}	~475	HCOO

[a] N.D., no data.

some of the reaction energy was converted to the translational energy of the product molecule. Although the velocity of the CO_2 molecule was appreciable, the amount of chemical energy that shows up in translation is no more than 10% of the available chemical energy. Nitrogen atom recombination studies reveal that some of the chemical energy that becomes available can be converted to the kinetic energy and internal energy of the desorbing molecules.[112] These studies were carried out on platinum and contaminated platinum.

Study of the elementary reaction steps of chemical surface reactions is in its infancy. Many important catalytic reactions have not been studied this way and the kinetic parameters are available for only a handful of surface reactions. These reactions play important roles in the chemical technology, in astrophysics, and in most high-temperature processes. It is hoped that in the near future molecular beam scattering and other studies will be carried out to determine the rate parameters for many surface reactions, so that we may learn about the elementary surface reaction steps. Once a critical mass of data on surface reactions becomes available, one may see certain classes of surface reactions and certain elementary reactions predominate in well-defined experimental conditions of temperature, pressure, and surface parameters. This way we may also learn to alter the surface structure and surface composition to accelerate or inhibit chemical reactions.

References

1. L. Pauling, *The Nature of the Chemical Bond*, Cornell University Press, Ithaca, N.Y., 1960, p. 7 (and references therein).
2. See, for example, H. Schaeffer III, *The Electronic Structure of Atoms and Molecules*, Addison-Wesley, Menlo Park, Calif., 1972, and references therein.
3. S.T. Ceyer and G.A. Somorjai, Annu. Rev. Phys. Chem. **28** (1977).
4. J. Brumbach and G.A. Somorjai, Crit. Rev. Solid State Sci. **4,** 429 (1974).
5. (a) N. Ramsey, *Molecular Beams*, Oxford University Press, New York, 1969; (b) G. Comsa, Proc. 7th Intern. Congr. and 3rd Intern. Conf. Solid Surfaces, Vienna, 1977, 1317; (c) G. Comsa, Vacuum, **19,** 277 (1969).
6. S.L. Bernasek and G.A. Somorjai, Prog. Surf. Sci. **5,** 377 (1975).
7. L. Thomas, in *Fundamentals of Gas–Surface Interactions*, eds. H. Saltzburg et al., Academic Press, New York, 1967, p. 346.
8. W. Seikhaus, J. Schwarz, and D. Olander, Surf. Sci. **33,** 445 (1972).
9. (a) D. Auerbach, C. Becker, J. Cowin, and L. Wharton, Appl. Phys. **14,** 141 (1977); (b) G. Comsa, R. David, and B.J. Schumacher, Surf. Sci. **85,** 45 (1979); (c) G. Comsa and R. David, Chem. Phys. Lett. **49,** 512 (1977).
10. (a) W. Weinberg and R. Merrill, J. Chem. Phys. **56,** 2881 (1972); (b) R. Sau and R. Merrill, Surf. Sci. **34,** 268 (1973).

11. A. Modak and P. Pagni, J. Chem. Phys. **59**, 2019 (1973).
12. (a) J. McClure, J. Chem. Phys. **57**, 2810 (1972); (b) J. McClure, J. Chem. Phys. **57**, 2823 (1972); (c) W. Steele, Surf. Sci. **38**, 1 (1973).
13. S. Bernasek and G.A. Somorjai, J. Chem. Phys. **50**, 4552 (1974).
14. (a) R. Palmer, H. Saltsburg, and J. Smith, J. Chem. Phys. **50**, 4661 (1969); (b) S. Bernasek and G.A. Somorjai, J. Chem. Phys. **62**, 3149 (1975).
15. R.G. Rowe and G. Ehrlich, J. Chem. Phys. **63**, 4648 (1975).
16. (a) J. Doll, Chem. Phys. **3**, 257 (1974); (b) J. Doll, J. Chem. Phys. **61**, 954 (1974); (c) D. Dion and J. Doll, Surf. Sci. **58**, 415 (1976).
17. S. Brass, D. Reed, and G. Ehrlich, J. Chem. Phys. **70**, 5244 (1979).
18. H. Winters, J. Chem. Phys. **64**, 3495 (1976).
19. J. Doll, Chem. Phys. Lett. **29**, 195 (1974).
20. (a) G. Boato, P. Cantini, and L. Mattera, Surf. Sci. **55**, 141 (1976); (b) G. Derry, D. Wesner, S.W. Krishnaswami, and D.R. Frankl, Surf. Sci. **74**, 245 (1978).
21. (a) G. Boato, P. Cantin, U. Garibaldi, A. Levi, L. Mattera, P. Spadacini, and G. Tommei, J. Phys. **C 6**, L394 (1973); (b) G. Derry, D. Wesner, W. Carlos, and D.R. Frankl, Surf. Sci. **87**, 629 (1979).
22. U. Garibaldi, A. Levi, R. Spadacini, and G. Tommei, Proc. 2nd ICSS (Kyoto) Jap. J. Appl. Phys. Suppl. **2**, Pt. 2, p. 563 (1974).
23. U. Garibaldi, A. Levi, R. Spadacini, and G. Tommei, Surf. Sci. **48**, 649 (1975).
24. C. Chiroli and A. Levi, Surf. Sci. **59**, 325 (1976).
25. J. Doll, Chem. Phys. **3**, 257 (1974).
26. J. Doll, J. Chem. Phys. **61**, 954 (1974).
27. D. Dion and J. Doll, Surf. Sci. **58**, 415 (1976).
28. R. Masel, R. Merrill, and W. Miller, Surf. Sci. **46**, 681 (1974).
29. R. Masel, R. Merrill, and W. Miller, J. Chem. Phys. **64**, 45 (1976).
30. C. Ray and J. Bowman, J. Chem. Phys. **63**, 5231 (1975).
31. F. Goodman, N. Cabrera, V. Celli, and R. Manson, Surf. Sci. **19**, 67 (1970).
32. F. Goodman, Surf. Sci. **19**, 93 (1970).
33. F. Goodman, J. Chem. Phys. **58**, 5530 (1973).
34. R. Masel, R. Merrill, and W. Miller, J. Vac. Sci, Technol. **13**, 355 (1976).
35. R. Masel, R. Merrill, and W. Miller, J. Chem. Phys. **65**, 2690 (1976).
36. A. Tsuchida, Surf. Sci. **46**, 611 (1974).
37. J. Meyers and D. Frankl, Surf. Sci. **51**, 61 (1975).
38. (a) A. Tsuchida, Surf. Sci. **52**, 685 (1975); (b) R. J. LeRoy, Surf. Sci. **59**, 541 (1976).
39. H. Finzel, H. Frank, H. Hoinkes, M. Luschka, H. Nahr, H. Wilscia, and U. Wonka, Surf. Sci. **49**, 577 (1975).
40. H. Chow and E. Thompson, Surf. Sci. **54**, 269 (1976).
41. J. Beeby, J. Phys. **C6**, 1229 (1973).
42. N. Cabrera and F. Goodman, J. Chem. Phys. **56**, 4899 (1972).
43. F. Goodman, W. Liu, and N. Cabrera, J. Chem. Phys. **57**, 2698 (1972).
44. F. Goodman and W. Tan, J. Chem. Phys. **58**, 5527 (1973).
45. G. Wolken, Jr., J. Chem. Phys. **61**, 456 (1974).
46. G. Wolken, Jr., Chem. Phys. **58**, 3047 (1973).
47. J. Beeby, J. Phys. **C4**, L359 (1971).

48. J. Beeby, Proc. 2nd ICSS (Kyoto) Jap., J. Appl. Phys. Suppl. 2, Pt. 2, p. 537 (1974).
49. J. Beeby, Commun. Solid State Phys. **7,** 1 (1975).
50. R. Masel, R. Merrill, and W. Miller, Phys. Rev. **B12,** 5545 (1975).
51. (a) W. Weinberg, J. Phys. **C5,** 2098 (1972); (b) G. Comsa, J. Phys. **C6,** 2648 (1973).
52. D. Tendulkar and R. Stickney, Surf. Sci. **29,** 516 (1971).
53. A. Stoll, J. Ehrhardt, and R. Merrill, J. Chem Phys. **64,** 34 (1976).
54. (a) J. Lapujoulade and Y. Lejay, Surf. Sci. **69,** 354 (1977); (b) J. Lapujoulade, Y. Lejay, and N. Papanicolaou, Surf. Sci. **90,** 133 (1979).
55. K.H. Rieder and T. Engel, Surf. Sci., in press (1980).
56. G. Boato, P. Cantini, and R. Tatarek, J. Phys. **F6,** L237 (1976).
57. J.M. Horne and D.R. Miller, Surf. Sci. **66,** 365 (1977).
58. G. Comsa, G. Mechtersheimer, B. Poelsema, and S. Tomoda, Surf. Sci. **89,** 123 (1979).
59. (a) A. Stoll and R. Merrill, Surf. Sci. **40,** 405 (1973); (b) A. Stoll, R. White, J. Ehrhardt, R. Masel, and R. Merrill, J. Vac. Sci. Technol. **12,** 192 (1975).
60. R. White, J. Ehrhardt, and R. Merrill, J. Chem. Phys. **64,** 41 (1975).
61. S.T. Ceyer, R.J. Gale, S.L. Bernasek, and G.A. Somorjai, J. Chem. Phys. **64,** 1934 (1976).
62. J. McClure, J. Chem. Phys. **57,** 2823 (1972).
63. W. Steele, Surf. Sci. **38,** 1 (1973).
64. (a) W. Weinberg and R. Merrill, J. Chem. Phys. **56,** 2893 (1972); (b) J. Horne and D. Miller, J. Vac. Sci. Technol. **13,** 351 (1975).
65. P. Jewsbury, Surf. Sci. **52,** 325 (1975).
66. See, for instance, G. Goodman, *Crystal Growth,* Plenum Press, New York, 1972.
67. See, for example, P. Hirth et al., *Condensation and Evaporation,* Pergamon Press, Oxford, 1963.
68. D. King, to be published in *Surface Science, Recent Progress and Perspectives,* ed. R. Vanselow, CRC Press, Cleveland, Ohio, 1980.
69. D. Menzel, Top. Appl. Phys. **4,** 101 (1975).
70. D. Olander and A. Ullman, Int. J. Chem. Kinet. **8,** 625 (1976).
71. I. Wachs and R. Maddix, Surf. Sci. **58,** 590 (1976).
72. M. Salmerón, R. Gale, and G.A. Somorjai, J. Chem. Phys. , 2807 (1979).
73. L. Schmidt, Catal. Rev. **9,** 115 (1974).
74. H. Bonzel, Surf. Sci. **68,** 236 (1977).
75. H. Hopster, H. Ibach, and G. Comsa, J. Catal. **46,** 32 (1977).
76. (a) J. Morrison and J. Roberts, Proc. Roy. Soc. (Lond.) **A173,** 1 (1939); (b) L. Clavenna and L. Schmidt. Surf. Sci. **22,** 365 (1970).
77. D. King and M. Wells, Proc. Roy. Soc. (Lond.) **A339,** 245 (1974).
78. P. Kisliuk, J. Phys. Chem. Solids **3,** 95 (1957); **5,** 78 (1958).
79. (a) H. Conrad, G. Ertl, B. Koch, and E. Latta, Surf. Sci. **43,** 462 (1974); (b) A.M. Bradshaw and F.M. Hoffmann, Surf. Sci. **72,** 513 (1978).
80. L. West and G.A. Somorjai, J. Chem. Phys. **57,** 5143 (1972).
81. M. Balooch, M. Cardillo, D. Miller, and R. Stickney, Surf. Sci. **46,** 358 (1974).
82. W. Siekhaus, J. Schwarz, and D. Olander, Surf. Sci. **33,** 445 (1972).

83. S. Singh-Boparai, M. Bowker, and D. King, Surf. Sci. **53,** 55 (1975).
84. H. Ibach and S. Lehwald, 4th Rolla Conf. Surf. Prop., University of Missouri, August 1977.
85. R. Baetzold and G.A. Somorjai, Acc. Chem. Res. **9,** 392 (1976).
86. (a) R. Gomer, Disc. Faraday Soc. **28,** 23 (1959); (b) *Field Emission and Field Ionization,* Harvard University Press, Cambridge, Mass., 1961, chap. 4.
87. G. Ehrlich and F. Hudda, J. Chem. Phys. **35,** 1421 (1961).
88. D. King, in *Surface Science, Recent Progress and Perspectives,* ed. R. Vanselow; see Table 3 (1980).
89. R. Butz and H. Wagner, Surf. Sci. **87,** 69 (1979).
90. H. Ibach, W. Erley, and H. Wagner, Surf. Sci. **92,** 29 (1980).
91. T. Engel and G. Ertl, J. Chem. Phys. **69,** 1267 (1978).
92. D. King, Surf. Sci. **64,** 43 (1977).
93. J. Roberts, Proc. Roy. Soc. (Lond.) **A152,** 445 (1935).
94. J. Wang, Proc. Roy. Soc. (Lond.) **A161,** 127 (1937); Proc. Camb. Phil. Soc. **34,** 238 (1938).
95. C. Gaymour and D. King, J. Chem. Soc. Faraday I **69,** 736 (1973).
96. D. Adams. Surf. Sci. **42,** 12 (1974).
97. (a) R. Maddix and M. Boudart, J. Catal. **13,** 216 (1967); (b) R. Maddix, G. Ertl, and K. Christmann, Chem. Phys. Lett. **62,** 38 (1979).
98. I. Wachs and R. Maddix, Surf. Sci. **65,** 287 (1977).
99. (a) D. Olander, W. Siekhaus, R. Jones, and J. Schwarz, J. Chem. Phys. **57** 408 (1972); (b) ibid., pp. 421–433.
100. J. Anderson and M. Boudart, J. Catal. **7,** 216 (1964).
101. R. Maddix and M. Boudart, J. Catal. **18,** 240 (1967).
102. J. McKinley, J. Chem. Phys. **40,** 120 (1964).
103. A. Ullman and R. Maddix, High Temp. Sci. **6,** 342 (1974).
104. S. Bernasek and G.A. Somorjai, J. Chem. Phys. **62,** 3149 (1975).
105. M. Salmerón, R. Gale, and G.A. Somorjai, Phys. Rev. Lett. **38,** 1027 (1977).
106. M. Salmeron, R. Gale, and G.A. Somorjai, J. Chem. Phys. **67,** 5324 (1977).
107. J.N. Smith and R.L. Palmer, J. Chem. Phys. **56,** 13 (1972).
108. T.L. Bradley, A.E. Dabiri, and R.E. Stickney, Surf. Sci. **29,** 590 (1972).
109. R.N. Coltharp, J.T. Scott, and E.E. Murchlitz, J. Chem. Phys. **51,** 5180 (1969).
110. L.A. West and G.A. Somorjai, J. Vac. Sci. Technol. **9,** 668 (1972).
111. C.A. Becker, J.P. Cowin, and L. Whorton, J. Chem. Phys. **67,** 3395 (1977).
112. R.P. Thorman, D. Anderson, and S.L. Bernasek, Phys. Rev. Lett. **44,** 743 (1980).

8. Catalyzed Surface Reactions: Principles

Catalytic Action

Virtually all chemical technologies and many technologies in other fields use catalysis as an essential part of the process. The role of the catalyst was first defined in 1836 by Berzelius,[1] who identified substances that accelerate chemical reactions without visibly undergoing chemical change as possessing a surface "catalytic force." Indeed, one of the major functions of a catalyst is to aid in rapidly achieving chemical equilibrium for certain chemical reactions.

Two of the simpler, although important reactions that demonstrate this type of catalytic action are the formation of water from oxygen and hydrogen ($1/2 O_2 + H_2 \rightarrow H_2O$) and the formation of ammonia from hydrogen and nitrogen ($3H_2 + N_2 \rightarrow 2NH_3$). Water has a standard free energy of formation of $\Delta G_{298} = -58$ kcal/mol. Yet O_2 and H_2 gas mixtures may be stored indefinitely in a glass bulb without showing signs of any chemical reaction. Just by dropping a high-surface-area platinum gauze into the mixture, the reaction occurs instantaneously and explosively—as demonstrated to the delight of freshman chemistry students in the introductory chemistry courses. The reason for this striking effect can be explained as follows. H_2 and O_2 have large activation energies for the reaction in the gas phase. First, one of the diatomic molecules must be dissociated. Dissociation energies are very large compared with thermal energies, RT (103 kcal/mol for H_2 and 117 kcal/mol for oxygen).[2] The subsequent atom—molecule reactions ($H + O_2$ or $H_2 + O$) still require an activation energy of about 10 kcal/mol.[3] Thus the gas-phase reaction is very improbable under any circumstances. In the presence of a properly structured platinum surface, however, both molecules dissociate to atoms with zero activation energies ($2Pt + H_2 \rightarrow$

2Pt—H, or 2Pt + $O_2 \rightarrow$ 2Pt—O),[4,5] as shown by low-pressure surface studies. In addition, the atom–atom or atom–molecule reactions that take place subsequently have lower or require no activation energies as compared to that in the gas phase.[4] Thus the surface catalytic action involves its ability to atomize large binding-energy diatomic molecules by forming chemisorbed atomic intermediates and to lower the activation energy for the reaction on the surface that follows.

Similarly, the synthesis of ammonia from dinitrogen and hydrogen ($N_2 + 3H_2 \rightarrow 2NH_3$) required the "activation" of the N≡N bond to dissociate the molecule. The nitrogen atoms that form then must react with hydrogen atoms or molecules to produce NH_3. The very large dissociation energy of N_2 (ΔE = 280 kcal/mol) makes it virtually impossible for this reaction to occur in the gas phase. On an iron surface, however, N_2 has a fairly high heat of adsorption[6] ($\Delta H_{ads} \sim$ 9.7 kcal/mol), which assures a fairly long residence time. The molecule then dissociates on a properly structured surface [the (111) crystal face, for example] with a small activation energy (~3 kcal). This is the key initiation step for the catalytic reaction. Iron also readily atomizes the hydrogen molecules. The chemisorbed nitrogen atoms then react with hydrogen atoms on the surface to produce NH, NH_2, and finally NH_3 molecules that desorb into the gas phase.

Selective Catalysis

Another major function of a catalyst is to provide reaction *selectivity*. Under the conditions in which the reaction is to be carried out, there may be many reaction channels, each thermodynamically feasible, that lead to the formation of different products. The selective catalyst will accelerate the rate of only one of these reactions so that only the desired product molecules form with near-theoretical or 100 percent efficiency. One example of this is the dehydrocyclization of *n*-heptane to toluene,[7]

$$CH_3-(CH_2)_5-CH_3 \rightarrow \underset{}{\overset{CH_3}{\langle O \rangle}} + 4H_2$$

This is a highly desirable reaction that converts aliphatic molecules to aromatic compounds. The larger concentration aromatic component in the gasoline, for example, greatly improves its octane number. However, *n*-heptane may participate in several competing simpler reactions. These include hydrogenolysis, which involves C—C bond scission to form smaller-molecular-weight fragments (methane, ethane, propane); partial dehydrogenation, which produces various olefins; and isomerization, which yields branched chains. All of these reactions are thermodynamically feasible, and since they appear to be less complex than dehydrocyclization, they compete effectively with it. A properly prepared platinum catalyst surface, however, catalyzes the selective conversion of n-heptane to toluene without permitting the formation of other products. The catalyst selectivity is equally important for the reactions of small molecules (such as the hydrogenation of CO to produce a desired hydrocarbon) or very large molecules of biological importance, where enzyme catalysts provide the desired selectivity.[8]

Kinetic Expressions

Catalysis is a kinetic phenomenon; we would like to carry out the same reaction at an optimum rate over and over again using the same catalyst. In most cases such a steady-state operation is desirable and aimed for. Therefore, in the sequence of elementary reactions that include adsorption, surface migration, chemical rearrangements, and reactions in the adsorbed state, and desorption of the products, the rate of each step must be of steady state. The rate of the overall catalytic reaction per unit area of catalyst surface can be expressed as (moles of product/catalyst area × time). The rate can often be expressed as the product of the apparent rate constant k and a reactant pressure (or concentration)-dependent term,

$$J_{cat} = k_{cat} \times f(P_i) \tag{1}$$

where P_i is the partial pressure of the reactants. The rate constant for the overall catalytic reaction may contain the rate constants of many of the elementary reaction steps that precede the rate-determining step. Since the slowest rate-reaction step may change as the reaction conditions vary (temperature, pressure, relative surface concentrations of reactants, catalyst structure), k may also change to reflect the changing reaction mechanism. Nevertheless, k can be defined using the Arrhenius expression

$$k_{cat} = A \exp (-E^*_{cat}/RT) \tag{2}$$

where A is the temperature-independent preexponential factor and E^*_{cat} is the apparent activation energy measured under the catalytic reactions conditions.

Since it is often desirable to know how many product molecules are produced over a catalyst site in unit time, it is customary to express the rate as (molecules of product/number of catalyst sites × time),[9] which is frequently called the "turnover frequency" or "turnover number," TN. For most heterogeneously catalyzed small molecule reactions under the usual reaction conditions of 400 to 800 K and a few atmospheres of total reactant pressures, TN usually varies between 10^{-2} and 10^{2} $\mathrm{sec^{-1}}$.

While the turnover number provides us with a figure of merit for the catalyst activity (the reaction rate under certain chosen conditions), the reaction probability, RP, reveals the overall efficiency of the catalyst under the same conditions. The reaction probability is defined as

$$\text{reaction probability} = \frac{\text{rate of formation of product molecules}}{\text{rate of incidence of reactant molecules}}$$

RP can readily be obtained by dividing the turnover number by the rate of molecular incidence, J_{inc}, which is obtained from the pressure using the kinetic theory expression $J_{inc} = P/(2\pi MRT)^{\frac{1}{2}}$. The turnover number TN and the reaction probability RP so obtained ($RP = TN/J_{inc}$) are displayed in Figure 8.1 for the platinum-catalyzed dehydrogenation and hydrogenation of cyclohexene to benzene and cyclohexane, respectively.[10] These reaction parameters are plotted as a function of reactant pressure

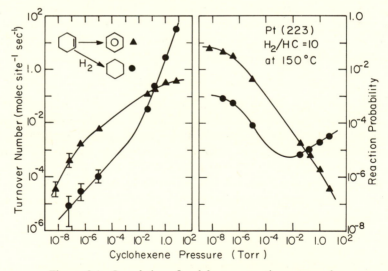

Figure 8.1. Correlation of cyclohexene reaction rates and reaction probabilities over a pressure range of 10 orders of magnitude. The reactions were performed at 150°C over the stepped Pt(223) crystal surface with an H/HC ratio of 10.

over a 10-orders-of-magnitude pressure range, 10^{-8} to 10^2 torr. For thermodynamic reasons dehydrogenation dominates the product distribution at low pressures. Overall, the turnover numbers for hydrogenation and dehydrogenation vary by factors of 10^7 and 10^4, respectively, for a 10^9-fold increase in the cyclohexene pressure. The reaction probability for dehydrogeneation—that is, the fraction of incident cyclohexene molecules converted to benzene—declines steadily from ~0.05 at 10^{-7} torr to less than 10^{-6} at 77 torr. Thus even though the reaction rate increases with increasing reactant pressure, the "efficiency" of the catalyst in converting the reactant to the products decreases markedly with increasing pressure. This is also the case, to a lesser extent, for the hydrogenation reaction. For this reaction the probability decreases by only 2 orders of magnitude ($\sim10^{-3}$ to 10^{-5}), with increasing pressure over the entire pressure range.

The enormous decline in the dehydrogenation probability with increasing pressure is associated with the lengthy mean reaction time that is required for the surface reaction and desorption processes. It is on the order of 10 sec at 450 K. At higher pressures the reaction times are much longer than the period between collisions of the reactant molecules with the surface. Under these conditions most of the surface sites where chemical changes may occur are continuously blocked by adsorbates already there. As a result, most of the incident reactant molecules desorb from the surface before the surface reaction can take place.

The determination of the rates of the net catalytic reactions and how the rates change with temperature and pressure is of great practical importance. Although there are many excellent catalysts that permit the achievement of chemical equilibria (for example, Pt for oxidation of CO and hydrocarbons to CO_2 and H_2O), most catalyzed reactions are still controlled by the kinetics of one of the surface processes. Therefore, from the knowledge of the activation energy and the pressure dependencies of the overall reaction, the catalytic process can be modeled[11] and the optimum reaction conditions can be calculated. Such kinetic analysis,[12] based on the macroscopic rate parameters, is vital for developing chemical technologies based on catalytic reactions.

The rates of reactions are extremely sensitive to small changes of chemical bonding of the surface species that participate in the surface reaction. Since the energy necessary to form or break the surface bonds appears in the exponent of the Arrhenius expression for the rate constant for the overall reaction, it can increase or decrease the rate in an exponential fashion. For example, a change of 3 kcal in the activation energy alters the reaction rate by over an order of magnitude at 500 K. Thus small variations of chemical bonding at different surface irregularities, steps, and kinks, as compared to atomic terraces, can give rise to a very strong structure sensitivity of the reaction rates and the product distribution. Rate measurements exponentially magnify the energetic alterations that occur on the surface and could provide a very sensitive probe of structural and electronic changes at the surface and changes of surface bonding on the molecular scale.

Tabulated Kinetic Parameters for Catalytic Reactions

A great deal of kinetic information has been obtained for different types of catalyzed hydrocarbon reactions carried out over metal catalyst surfaces. These reactions include dehydrogenation, hydrogenation, hydrogenolysis and cracking, ring opening, dehydrocyclization, and isomerization. The kinetic parameters for these reactions are listed in Tables 8.A1 to 8.E4 in the Appendix to this chapter. In these tables the catalyst systems that were used are listed together with the temperature range of the investigation. Since these reactions are always carried out in the presence of hydrogen, both the hydrocarbon and hydrogen concentrations (in molecules/cm^3) are tabulated in a logarithmic form. X and Y are the power-rate-law dependencies of the (HC) and (H$_2$) concentrations, respectively. These exponents are also displayed whenever they are determined. From these data the changes of the reaction rates with reactant concentrations can be determined. The rate of reaction at a given temperature, in the range used in the experimental study, is also calculated and listed, together with the apparent activation energy for the reaction, E_a, and the logarithm of the preexponential factor, log A. From the rate and reactant concentrations a reaction probability (RP) can be calculated. This is also displayed in Tables 8.A1 to 8.E4 for the various catalytic reactions, as -log RP. Fractional selectivities, S, are also supplied when reported. These are defined as the ratio of the rate of the specific reaction to the total reaction rate.

There is a great deal of scatter in the kinetic parameters obtained for a given reaction on different catalyst systems. This is expected, since the structure and bonding characteristics of the different metal catalysts vary widely. Nevertheless, several conclusions may be reached from the inspection of the data. The reaction probabilities are very low under the conditions where these reactions were carried out. They range from 10^{-8} to 10^{-5} for hydrogenation to 10^{-12} to 10^{-8} for most of the other reactions. The apparent activation energies are the lowest for hydrogenation and cyclopropane ring opening, 9 to 15 kcal/mol. For dehydrogenation of cyclohexane and for the hydrogenolysis of C_4 to C_6 alkanes, E_a is in the range 16 to 25 kcal/mol. For most of the other reactions, which include hydrogenolysis (the most frequently studied reaction) of ethane, propane, and other alkanes; cracking of olefins and benzene; dehydrogenation of alkanes; and isomerization of C_5 to C_6 hydrocarbons, the apparent activation energies are in the range 25 to 50 kcal/mol.

The kinetic information displayed in Tables 8.A1 to 8.E4 can be useful in establishing the optimum reaction conditions and catalyst systems. It is hoped that reliable kinetic parameters will become available for many other important catalyzed hydrocarbon reactions in the near future.

Large Surface Area

One of the important considerations in catalysis is the need to provide a large contact area between the reactants and the surface. The total rate (moles of product/time) is proportional to the surface area. As a consequence, a great deal of effort is expended to prepare large surface-area catalysts and to accurately measure the surface area. Several important catalysts are displayed in Figure 8.2. Most technologically important catalysts are used in small particle form (1 to 10 μm in diameter). For example, the iron catalyst utilized in ammonia synthesis or for the hydrogenation of carbon monoxide is employed in this manner. Surface areas of 1 to 10 m^2/g of catalyst can be obtained in this type of configuration. Often, "textural promoters" (for example, CaO or Al_2O_3 in Fe) are added to the catalyst, which slow down sintering and grain growth, which would reduce the surface area. Another group of catalysts, the zeolites, which are aluminosilicates used for the "cracking" of hydrocarbons to lower-molecular-weight fragments, have crystal structures full of pores 8 to 20 Å in size. The structure of one of the many zeolites used for catalysis, mordenite, is shown in Figure 8.3. Since the catalytic reactions occur inside the pores, an enormous inner surface area, of several hundred square meters per gram of catalyst,[13] is available in these catalyst systems. Transition-metal catalysts are generally employed in a small, 10 to 100-Å-diameter particle form dispersed on large-surface-area "supports." The support can be a specially prepared alumina or silica framework (or a zeolite) that can be produced with surface areas in the 10^2-m^2/g range. These "supported" metal catalysts are often available with near-unity dispersion (dispersion is defined as the number of surface atoms per total number of atoms in the particle) of the metal particles and are usually very stable in this configuration during the catalytic

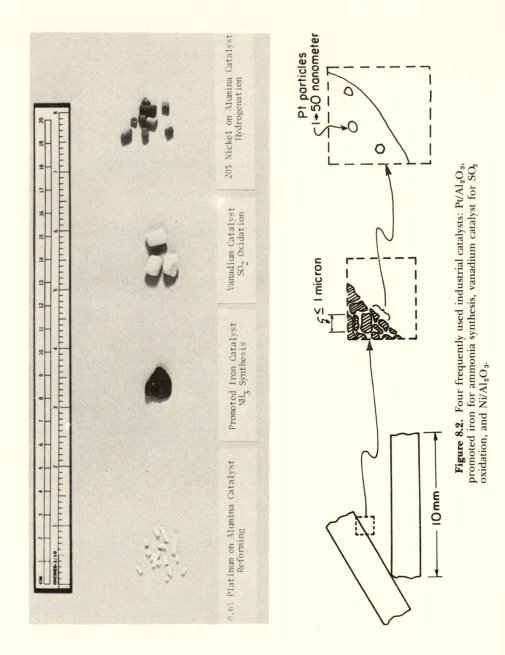

Figure 8.2. Four frequently used industrial catalysts: Pt/Al_2O_3, promoted iron for ammonia synthesis, vanadium catalyst for SO_2 oxidation, and Ni/Al_2O_3.

0.6% Platinum on Alumina Catalyst
Reforming

Promoted Iron Catalyst
NH_3 Synthesis

Vanadium Catalyst
SO_2 Oxidation

20% Nickel on Alumina Catalyst
Hydrogenation

Pt particles
1→50 nanometer

≤ 1 micron

10 mm

388

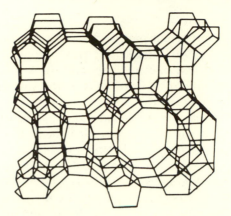

Figure 8.3. One of the important zeolites, mordenite, $Na_8Al_8Si_{40}O_{96} \cdot 24H_2O$, viewed along [001] axis.

reaction. The metal is frequently deposited from solution as a salt. It is then reduced under controlled conditions. Alloy catalysts and other multicomponent catalyst systems can also be prepared in such a way that small alloy clusters are formed on the large-surface-area oxide supports.

The measurement of catalyst surface areas has occupied an important role in the development of surface chemistry and catalysis. The techniques of selective adsorption and the interpretation of the adsorption isotherm (amount adsorbed as a function of equilibrium pressure) had to be developed in order to determine the surface areas and learn about the chemical nature of adsorption.[14] At present, catalyst-surface-area measurements by gas adsorption is a fairly straightforward procedure. From a knowledge of the amount adsorbed (number of molecules) and the area occupied per molecule (16.2 $Å^2$ for N_2 and 25.6 $Å^2$ for Kr, for example), the total surface area that is covered by the adsorbed gas can be calculated. To measure the surface areas of the support and the catalyst or two catalyst components separately, selective chemisorption of well-chosen molecules is frequently employed.

Surface Intermediates

In 1926, Sebatier[15] suggested that most catalytic reactions take place through the formation of intermediate compounds between the reactants or products and the surface. There is little doubt that the catalyst surface atoms form strong chemical bonds, in most cases with the incident molecules. This strong chemical surface–adsorbate interaction provides the driving force for breaking large-binding-energy chemical

389

bonds (C—C, C—H, H—H, N≡N, and C=O bonds), which are often an important part of the catalytic reaction.

A good catalyst will also permit rapid bond breaking between the adsorbed intermediates and the surface, and the speedy release or desorption of the products. If the surface bonds are too strong, the reaction intermediates block the adsorption of new reactant molecules and the reaction stops. For too weak adsorbate–catalyst bonds, the necessary bond-scission processes may be absent. Hence the catalytic reaction will not occur. A good catalyst is thought to be able to form chemical bonds of intermediate strength. These bonds should be strong enough to induce bond scission in the reactant molecules. However, the bond should not be too strong to assure only short residence times for the surface intermediates and rapid desorption of the product molecules, so that the reaction can proceed with a large turnover number.

This is strikingly demonstrated by studies by Sachtler and Fahrenfort[16] of the decomposition of formic acid that is catalyzed by various transition metals. The decomposition produces H_2, CO, H_2O, and CO_2. By plotting the temperatures necessary to obtain the *same rate of reaction* over the different metals as a function of the heat of formation of the bulk metal formate, a volcano-shaped curve is produced, as shown in Figure 8.4. It was concluded that for Au and Ag there is an activation energy for forming the surface intermediates that slows down the reaction. For Ni, Co, Fe, and W the stability of the surface intermediate is too high, so that the decomposition rate is slow. The best catalysts are Pt, Ir, Pd, Rh, and Ru, which form bonds of intermediate strength with the adsorbates.

Another example to demonstrate the importance of intermediate bonding strength for adsorbates to optimize the "activity" (the rate) of the catalyst is the catalyzed hydrogenolysis of ethane, $C_2H_6 + H_2 \rightarrow 2CH_4$. This reaction was studied in detail by Sinfelt[17] over many transition-metal catalysts. His results are shown in Figure 8.5. Os, Ru, Ir, and Rh exhibit the highest activity under identical reaction conditions, while the activity decreases for metals both to the left and right of these groups of transition metals in the periodic table.

The reader should review the heats of chemisorption data obtained for small molecules on polycrystalline transition metal surfaces in Chapter 6. In Figures 6.3 to 6.8, the heats of chemisorption are plotted as a function of atomic numbers of transition metals for the various columns of metals in the periodic table. Using the principle of intermediate binding to exhibit optimum catalytic activity, it is possible to select a group of metals that are likely to have high activity for a given chemical reaction.

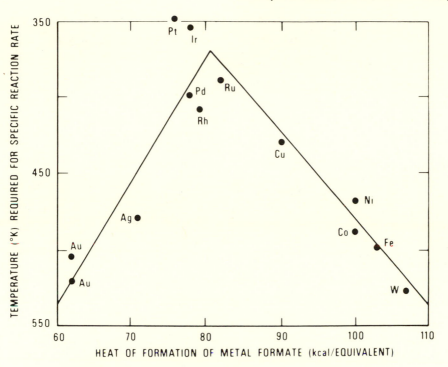

Figure 8.4. Correlation between the catalytic activity of different metals for the decomposition of formic acid (HCOOH) and the corresponding heat of formation for the bulk metal formate. The temperature required for a specific rate of decomposition on a metal is plotted against the heat of formate formation. A low value of temperature implies a low activation energy for the catalytic decomposition. Note the inverted temperature scale. (After W.M.H. Sachtler and J. Fahrenfort, ref. 16.)

Of course, activity is only one of many parameters that are important in catalysis. The selectivity of the catalyst, its thermal and chemical stability and dispersion, are among the other factors that govern our choices. While macroscopic chemical-bonding arguments can explain catalytic activity in some cases, atomic-scale scrutiny of the surface intermediates, catalyst structure, and composition, and an understanding of the elementary rate processes, are necessary to develop the optimum selective catalyst for any chemical reaction.

One of the important directions of research in catalysis is the identification of the reaction intermediates. The surface residence times of many of these species are longer than 10^{-5} sec under most catalytic reaction conditions (as inferred from the turnover frequency). They

391

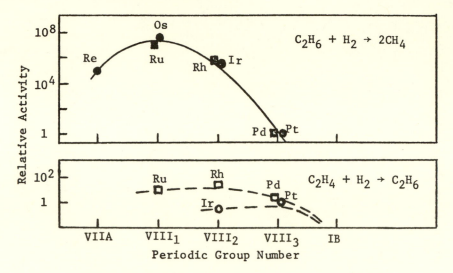

Figure 8.5. Catalytic activity of transition metals for ethane hydrogenolysis. (After J. H. Sinfelt, ref. 17.)

may be detected by suitable spectroscopic techniques either during the steady-state reaction or when isolated by appropriately interrupting the catalytic process.

Active Sites

The idea that the surface posseses active sites in numbers that are smaller than the total number of surface atoms and that these are essential for the catalytic activity was first proposed by Taylor[18, 19] in 1925. He found that while the ability of a catalyst surface to adsorb gases may not be much affected by heat treatments or small concentrations of impurities, the catalytic activity is drastically reduced or "poisoned" during the annealing and poisoning experiments. He suggested that the active sites may be surface irregularities, where surface atoms have a lower degree of coordination.

The presence of unique atomic sites of low coordination and different valency that are very active in chemical reactions has recently been clearly demonstrated by atomic-scale studies of metal and oxide surfaces. Atomic height steps and kinks were discovered to exhibit unique chemical activity on Pt[20] and Ni[21] surfaces. For example, molecular beam–surface scattering studies revealed that the dissociation of H_2 occurs without an activation energy at steps on Pt surfaces, while there is an 0.5

to 2 kcal/mol activation energy on atomic terraces.[4] This causes a seven-fold increase in the rate of H_2–D_2 exchange over stepped platinum surfaces as compared to surfaces without a large concentration of steps [Pt(111)]. Kink sites on Pt surfaces dissociate CO as detected by photo-electron spectroscopy,[22] whereas step and terrace sites adsorb CO molecularly and do not dissociate it. On stepped Ni surfaces C—H bond breaking was reported from HREELS[21] studies of acetylene and ethylene adsorption at temperatures as low as 100 K, while smooth Ni surfaces were much less active for bond breaking and exhibited different chemical behavior.

On oxide surfaces electron spin resonance (ESR) studies revealed the importance of point defects at the surface to enhance chemical activity.[9] Studies of MgO[23,24] revealed that oxygen ion vacancies are active sites for H_2–D_2 exchange. Kokes et al.[25] have identified an active Zn–O pair site in zinc oxide surfaces for the chemisorption and exchange of H_2 and D_2. The importance of W^{3+} ion as an active site in WS_2 for the hydrogenation of benzene was revealed by combined ESR and catalytic studies.[26]

The validity of the active-site concept of Taylor is being demonstrated by atomic-scale studies on an increasing number of systems. Surface irregularities, atomic steps, kinks, and point defects are identified as sites

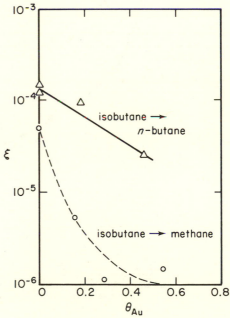

Figure 8.6. Isomerization and hydrogenolysis reaction probabilities of isobutane as a function of the fraction of the surface covered by gold.

of unique bond breaking or molecular rearrangement activity. These irregularities have surface concentrations often very much less than a monolayer. Studies of catalyst activation or deactivation by additives also indicate that below monolayer quantities (5 to 20 percent of a monolayer) change the reaction rates by orders of magnitude. For example, the effect of gold on the hydrogenolysis activity of platinum[27] is shown in

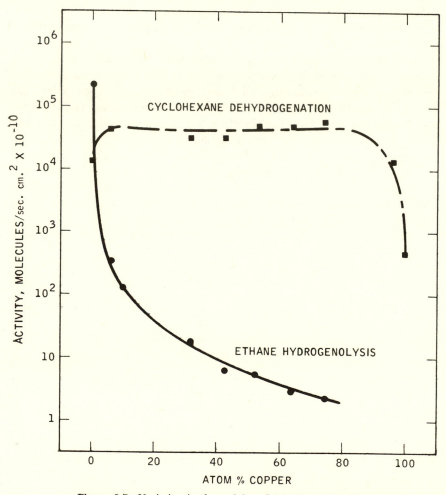

Figure 8.7. Variation in the activity of Cu–Ni catalysts for different hydrocarbon reactions. Activity for cyclohexene dehydrogenation and ethane hydrogenolysis are plotted as a function of the bulk alloy composition.

Figure 8.6. Adding less than 20 percent of a monolayer of Au reduced the rate of this reaction by three orders of magnitude. Thus the active site for C—C bond breaking was removed by the addition of small amounts of gold. Similar effects occur when copper is added to nickel[28] as shown in Figure 8.7. While the rate of hydrogenolysis decreased by three orders of magnitude upon the addition of less than 20 percent of copper, the dehydrogenation rate of cyclohexane has not been affected by this treatment. This observation demonstrates that while the active sites for one reaction may be removed by a small concentration of additives, the sites necessary to carry out other reactions may not be affected or may be influenced to a lesser extent. Controlling the active-site concentrations by additives provides a means to carry out reactions selectively. The increasing importance of multicomponent catalysts, among these alloy catalysts, is due partly to their greater selectivity for chemical reactions.

Acid or Base Character of a Catalyst

Most surface reactions and the formation of surface intermediates involve charge transfer, either an electron transfer or a proton transfer. These processes are often viewed as modified acid–base reactions. It is common to refer to an oxide catalyst as acidic or basic according to its ability to donate or accept electrons or protons.[29]

The electron transfer capability of a catalyst is expressed according to the Lewis definition. A Lewis acid is a surface site capable of receiving a pair of electrons from the adsorbate. A Lewis base is a site having a free pair of electrons that can be transferred to the adsorbate. The acidity of metal ions of equal radius increases with the increasing charge of the metal ions: $Na^+ < Ca^{2+} < Y^{3+} < Th^{4+}$. The strength of the Lewis acidity is measured by determining the binding energies of the charge-transfer complexes that form by this type of electron-transfer process.

The proton-transfer capability of a catalyst is expressed according to the Brønsted definition. A Brønsted acid is a surface site capable of losing a proton to the adsorbate. A Brønsted base is a site that can accept a proton from the adsorbed species. The Brønsted acidity of the catalyst is usually determined by ion exchange from solution (surface proton is substituted by alkali ions Li^+ Na^+, etc.) or by the adsorption of weak acids or bases, such as phenol and pyridine, on the surface. In this way the proton-transfer ability of the surface can be titrated. The Brønsted acidity for oxides has also been related to the metal–oxygen bond energies. In general, the acidity increases with an increase of charge on the metal ion. In the series of oxides Na_2O, CaO, MgO, Ag_2O, BeO, Al_2O_3, CdO, ZnO, SnO, H_2O, B_2O_3, FeO, SiO_2, Ca_2O_3, Fe_2O_3, P_4O_6, SnO_2, GeO_2, TiO_2, SO_2, SO_3, N_2O_5, and Cl_2O_7, those that are to the left of water are bases, and those to the right are acids.[29]

Perhaps the most widely used catalysts, the zeolites, best represent the group of oxides that exhibit acid–base catalysis.[30] Zeolites are alumina-silicates, some of which are among the more common minerals in nature. Modern synthesis techniques permit the preparation of families of zeolite compounds with different Si/Al ratios.[31] Since the Al^{3+} ions lack one positive charge in the tetrahedrally coordinated silica, Si^{4+}, framework, they are sites of proton or alkali-metal affinity. Therefore, variation of the Si/Al ratio gives rise to a series of substances of controlled but different acidity.[32] By using various organic molecules during the preparation of these compounds that build into the structure, subsequent decomposition leaves an open pore structure, where the pore size is controlled by the skeletal structure of the organic deposit. Very high internal surface area catalysts ($10^2 m^2/g$) can be obtained this way with controlled pore sizes of 8 to 20 Å and controlled acidity [(Si/Al) ratio]. These find applications in the cracking and isomerization of hydrocarbons that occur in a shape-selective manner as a result of the uniform pore structure and are the largest volume catalysts utilized in petroleum refining at present.[33] They are also the first of the "high-technology" catalysts where the chemical activity is tailored by atomic-scale study and control of the internal surface structure and composition.

Catalyst Additives

The preparation of a catalyst to provide optimum activity and selectivity is usually achieved by additives that modify the catalytic behavior of the pure catalyst substance. The most frequently used additives are *electron donors,* such as alkali metals (K, Na), or *electron acceptors,* such as oxygen and chlorine.[34] For example, potassium is added to the iron catalyst used for ammonia synthesis or for CO hydrogenation. Various alkali metals are added to the silver catalyst applied for the oxidation of ethylene during the catalyst formulation, while chlorine is added in the form of ethylchloride that is introduced with the reactant mixture of ethylene and oxygen. In the petroleum industry, chlorine and oxygen are often added to commercial platinum catalysts used for re-forming reactions by which aliphatic straight-chain hydrocarbons are converted to aromatic molecules (dehydrocyclization) and to branched isomers (isomerization).

These additives accomplish several tasks during the reaction. By changing the chemical bonding of some of the surface intermediates, the steady-state concentration of these intermediates may be altered, and thus a somewhat higher concentration of the catalytically active species is obtained. In this way the rate of the reaction is increased and the selectivity may be improved. For example, potassium on iron catalyst surfaces eliminates the ~3-kcal activation energy needed to dissociate dinitrogen to produce the atomic nitrogen species.[6] It also increases the heat of adsorption of N_2, thereby shifting the N_2(ads) → 2N(ads) equilibrium toward the formation of atomic nitrogen.

Since the formation of N atoms controls the rate of ammonia production, the addition of K improves the catalytic ammonia synthesis process. When potassium oxide is added to iron oxide to be used for the hydrogenation of CO, it appears to have several beneficial functions.[35,36] It stabilizes the iron oxide in the reducing gas mixture of CO and H_2, which would otherwise be reduced rapidly. It also catalyzes the removal of the carbon that deposits on the catalyst surface as a result of CO dissociation. This seems to occur by the formation of a K_2CO_3 intermediate, which then dissociates to $CO_{2(gas)}$ and $K_2O_{(solid)}$.[37] Thus the surface is effectively cleaned and protected against carbon deposition by the added potassium oxide. Chlorine is thought to shift the equilibrium among oxygen surface intermediates during the epoxidation of ethylene to ethylene oxide. When added to platinum catalysts, it is thought to be a "redispersion" agent; that is, the presence of chlorine stabilizes the small particle structure of the catalyst by retarding particle-size growth, which would reduce the surface area and change the surface structure.[37] The mechanisms of these processes initiated by the halogen additives are yet to be explored in molecular detail. Recent studies indicate that chemisorbed oxygen changes the surface structures of platinum crystal surfaces as compared to their surface structures in the clean state or in the presence of adsorbed carbon.[38] Halogens, which are better electron acceptors than oxygen, may have similar influence on the surface structures of platinum particles. They may stabilize the catalytically more active surface structures which would otherwise be unstable in the hydrocarbon–hydrogen reaction mixture.

Textural promoters are effective additives to stabilize the particle size and surface area of the catalyst. The use of CaO and Al_2O_3 in iron catalysts for this purpose is one example of this activity. Often, the catalyst support plays this stabilizing role. There is evidence for solid-state reactions between Al_2O_3 and Fe, for example, that produce $FeAl_2O_4$,[39] and for compound formation between Ni and TiO_2.[40] A strong chemical *interaction between the catalyst supporting oxide and the metal catalyst* can stabilize not only the structure and size of the catalyst particles but also the oxidation state of the catalyst. Recent reports[41] indicate that the product distribution in the $CO-H_2$ reaction over rhodium catalyst changes from mostly methane to methanol when the support is changed from silica to MgO or La_2O_3. It is likely that the oxidation state of rhodium is altered by the rhodium–support interaction, thereby leading to different reaction intermediates and reaction paths.

Other noncatalytic additives are being utilized to block surface sites of undesirable catalytic activity. For example, gold on platinum or copper on ruthenium blocks the hydrogenolysis reactions (C—C bond breaking) of hydrocarbons without impeding the dehydrogenation or isomerization reactions. Sulfur, a notorious "poison" of catalytic activity for many metal catalysts, can also be a beneficial additive. It is frequently added to the active Pt–Re alloy catalyst in a pretreatment procedure that improves the catalyst selectivity. It is thought that sulfur in this circumstance blocks sites of hydrogenolysis activity.

Often *multicomponent catalyst systems* are utilized to carry out reactions consisting of two or more active metal components or both oxide and metal constituents. For example, a Pt–Rh catalyst facilitates the removal of pollutants from car exhausts. Platinum is very effective for oxidizing unburned hydrocarbons and CO to H_2O and CO_2, and rhodium is very efficient in reducing NO to N_2, even in the same oxidizing environment. Dual functional or multifunctional catalysts are frequently used to carry out complex chemical reactions. In this circumstance the various catalyst components should not be thought of as additives, since they are independently responsible for different catalytic activity. Often there are synergistic effects, however, whereby the various components beneficially influence each other's catalytic activity to provide a combined additive and multifunctional catalytic effects.

It should be clear from the discussion above that the working, active, and selective catalyst is a complex, multicomponent chemical system. This system is finely tuned and buffered to carry out desirable chemical reactions with high turnover frequency and to block the reaction paths for other thermodynamically equally feasible but unwanted reactions. Thus an iron catalyst or a platinum catalyst is composed not only of iron or platinum but of several other constituents as well to assure the necessary surface structure and oxidation state of surface atoms for optimum catalytic behavior. Additives are often used to block sites, prevent side reactions, and alter the reaction paths in a variety of ways.

So far, we have discussed only those additives that are added to the catalyst during preparation. The reaction mixture itself often provides the most important surface modifiers that change the catalytic activity. During hydrocarbon reactions over platinum or during the $CO-H_2$ reaction, a carbonaceous overlayer of monolayer thickness is deposited on the catalyst, which remains catalytically active and is the property of the active catalyst system.[42] The activity of the metal–carbon surface bonds that form controls the reaction paths and rates. During the oxidation of ammonia over platinum, an oxide monolayer forms on the metal surface[43] that is part of the active catalyst system. During ammonia synthesis, however, the iron surface remains probably clean, as indicated by studies by Emmett[44] and his coworkers, and recently by Ertl.[45] An understanding of the catalyst system must include the atomic-scale scrutiny of the bonding and catalytic activity of the monolayer deposits that often form under steady-state reaction conditions on the catalyst surfaces.

Deactivation and Regeneration

Catalysts live long and active lives, but they do not last forever. The type of supported metal catalysts that are used in petroleum refining produces in the range of 200 to 800 barrels of products per pound of catalyst (1 barrel = 36 gallons). Once the catalyst is deactivated, it is either regenerated or replaced. There can be many reasons for the deactivation. At the operating temperatures some of the reactant hydrocarbons may completely decompose and deposit a thick layer of inactive carbon on the catalyst surface. For many catalysts the deactivation is slow enough that they are used in steady-state operation. The liquid or gaseous reactants are passed through the catalyst with a well-defined "space velocity" that is normally measured as the weight hourly space velocity (WHSV), the pound of liquids or gas passed over the catalyst per hour. For other active catalysts, deactivation is so rapid that they are used in a cyclic fashion; the reactors "swing" between running the catalytic reactions and regenerating. Thus understanding the causes of deactivation and developing new catalysts that are more resistant to "poisoning" are constant concerns of the catalytic chemist.

One of the major causes of deactivation is the deposition on the catalyst surface of metallic impurities that are present as compounds in the reactant mixture. Vanadium and titanium containing organometallic compounds decompose and not only deactivate the catalyst surface but often plug the pores of the high-surface-area supports, thereby impeding the reactant–catalyst contact during petroleum refining. Many of the catalyst poisons act by blocking active surface sites. In addition, poisons may change the atomic surface structure in a way that reduces the catalytic activity. Sulfur, for example, is known to change the surface structure of nickel.[46] By forming chemical bonds of different strengths on the different crystal planes, it provides a thermodynamic driving force for the restructuring of the metal particles. Sometimes the rate of deactivation of metal catalysts by small concentrations of sulfur can indeed be dramatic. The catalytic automotive converter necessitated the removal of tetraethyllead from gasoline, one of the best antiknocking agents, because it readily poisoned the Pt–Pd catalyst by depositing on the noble-metal surfaces.

A freshly prepared catalyst may not exhibit optimum catalytic activity upon its first introduction into the reactant stream. There may be efficient but undesirable side reactions that need to be eliminated. For this purpose a small amount of "poison" is often added to the reaction mixture or introduced in the form of pretreatment. Thus deactivating impurities may also be used, in small quantities, to improve the selectivity of the working catalyst.

Some Frequently Used Concepts of Catalysis

During the operation of complex catalyst systems, several macroscopic experimental parameters have been uncovered that provide useful practical information about the nature of the catalyst or the catalyzed surface reaction. A catalytic reaction is defined to be *structure sensitive* if the rate changes markedly as the particle size of the catalyst is changed.[47] Conversely, the reaction is *structure-insensitive* on a given catalyst if its rate is not influenced appreciably by changing the dispersion of the particles under the usual experimental conditions. In Table 8.1 we list several reactions that belong to these two classes. Clearly, variations of particle size give rise to changes of atomic surface structure. The relative concentrations of atoms in steps, kinks, and terraces are altered. Nevertheless, no clear correlation has been made to date between variations of macroscopic particle size and the atomic surface structure.

During the development of mechanistic interpretations of catalytic reactions using the macroscopic rate equations that were determined by experiments, two types of reaction models found general acceptance. In one of them the rate-determining surface reaction step involves interaction between two atoms or molecules, both in the adsorbed state. This reaction model is called the *Langmuir–Hinshelwood mechanism*.[48] In the other the rate-determining reaction step involves a chemical reaction between a molecule from the gas phase and one in the adsorbed state.

Table 8.1. Structure-sensitive and structure-insensitive catalytic reactions

Structure-sensitive	Structure-insensitive

Hydrogenolysis
  Ethane: Ni
  Methylcyclopentane: Pt
	Ring opening
	  Cyclopropane: Pt

Hydrogenation
  Benzene: Ni
	Hydrogenation
	  Benzene: Pt

Isomerization
  Isobutane: Pt
  Hexane: Pt
	Dehydrogenation
	  Cyclohexane: Pt

Cyclization
  Hexane: Pt
  Heptane: Pt	

This is called the *Rideal–Eley mechanism.*[49] Many reactions have rate equations that fit one of these two mechanisms. Recently, the oxidation of CO has been identified by molecular-scale studies as obeying the Langmuir–Hinshelwood reaction mechanism.[50] However, correlation of these reaction mechanisms, suggested by inspection of the macroscopic rate equations, with molecular-level studies of the elementary surface reactions remains one of the future challenges of catalysis.

During studies of a given catalyzed reaction over catalysts that were prepared in different ways, an interesting phenomenon was found, called the *compensation effect.*[51] Using the Arhhenius expression for the rate constant, both the preexponential factor and the activation energy for the reaction were found to have varied greatly from catalyst to catalyst. However, they varied in such a way as to compensate each other, so that the rate constant (or the reaction rate under the same conditions of pressure and temperature) remained almost constant. For example, for the methanation reaction (that is, the hydrogenation of CO), the following empirical relationship was found to hold between A and E^*_{cat}:

$$\ln A = \alpha + \frac{E^*_{cat}}{R\Theta} \qquad (3)$$

where α is a constant and Θ is called the isokinetic temperature, at which the rates on all the catalysts are equal. For the methanation reaction.[52] $\alpha \approx 0$ and $\Theta = 436$ K. Thus $\ln A_{CH4} \approx 1.1 E^*_{cat}$ kcal/mol. Figure 8.8 shows the compensation effect for the methanation reaction for eight different metal catalysts. The $\ln A_{CH4}$ versus E^*_{cat} plots yield a straight-line relationship.

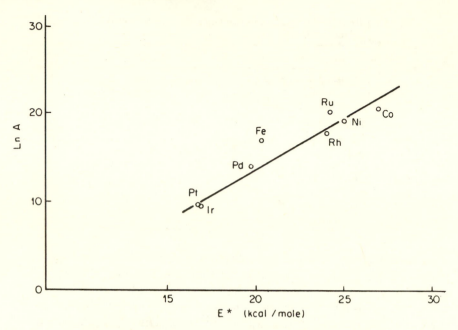

Figure 8.8. Compensation effect for the methanation reaction. The logarithm of the preexponential factor is plotted against the apparent activation energy, E^*_{cat}, for this reaction over several transition metal catalysts.

The compensation effect has been rationalized in a variety of ways. It is thought that one catalyst may have a large concentration of active sites where the reaction requires a high activation energy, while the other catalyst, which is prepared differently, has a small concentration of active sites that have low activation energies for the same surface reaction. An atomic-level explanation of the compensation effect remains the task of scientists in the future.

The Business of Catalysis

It may be instructive to review how widely catalysts are applied in the various technologies and to identify some of the most frequently used materials. There are three major areas of catalyst application at present:[53] automotive, fossil-fuel refining, and production of chemicals. Over $1 billion per year is spent at present to purchase new catalysts and to regenerate old ones. Table 8.2 lists the chemical processes that are the

Table 8.2. Chemical processes that are the largest users of heterogeneous catalysts at present and the catalyst systems they utilize most frequently

Reactions	Catalysts
CO, HC oxidation in car exhaust	Pt, Pd on alumina
NO$_x$ reduction in car exhaust	Rh on alumina
Cracking of crude oil	zeolites
Hydrotreating of crude oil	Co–Mo, Ni–Mo, W–Mo
Re-forming of crude oil	Pt, Pt–Re, and other bimetallics on alumina
Hydrocracking	metals on zeolites or alumina
Alkylation	sulfuric acid, hydrofluoric acid
Steam reforming	Ni on support
Water-gas-shift reaction	Fe–Cr, CuO, ZnO, alumina
Methanation	Ni on support
Ammonia synthesis	Fe
Ethylene oxidation	Ag on support
Nitric acid from ammonia	Pt, Rh, Pd
Sulfuric acid	V-oxide
Acrylonitrile from propylene	Bi, Mo-oxides
Vinyl chloride from ethylene	Cu-chloride
Hydrogenation of oils	Ni
Polyethylene	Cr, Cr-oxide on silica

largest users of heterogeneous catalysts and the catalyst systems that are employed most frequently at present.

The automotive industry uses mostly noble metals, platinum, rhodium, and palladium, for catalytic control of car emissions: unburned hydrocarbons, CO, and NO. These highly dispersed metals are supported on oxide surfaces, and the catalyst system is specially prepared to be active at the high space velocities of the exhaust gases and over a wide temperature range. In petroleum refining, zeolites are most widely used for cracking of hydrocarbon in the presence of hydrogen. The important hydrodesulfurization process uses mostly oxides of molybdenum and cobalt on an alumina support. The "re-forming" reactions to produce cyclic and aromatic molecules and isomers from alkanes to improve the octane number are carried out mostly over platinum or platinum-containg bimetallic catalysts, such as Pt–Re and Pt–Sn. Sulfuric and hydrofluoric acids are the catalysts for alkylation. In the chemical technologies steam re-forming of natural gas (mostly methane) to produce hydrogen, CO, and then methanol from CO and H$_2$, and ammonia from H_2 and N_2, is an important catalytic process. Activated carbon and zinc oxide are used as catalysts in large quantities to remove the sulfur. Then the purified gas is reacted with steam to form CO and H_2, mostly over supported nickel catalyst. The water-gas-shift reaction (CO + H_2O → CO_2 + H_2) is then employed to produce more hydrogen. The most frequently used catalyst for this purpose is iron-based. Copper oxide and zinc oxide are also used for the shift reaction, as well as for the production of methanol from CO and H_2. Nickel is the catalyst for methanation from CO and H_2, and iron is the major catalyst for the ammonia synthesis.

Catalytic hydrogenation processes primarily use nickel as catalyst. Hydrogenation of nitrile groups to amines and various edible and inedible oils for the preparation of margarine, salad oils, and stearine are some of the major applications. Selective hydrogenation of olefins is also an important catalytic process. Among the larger-volume oxidation reactions, the oxidation of ammonia to nitric acid uses noble metals: Pt, Pt–Rh, and Pt–Pd–Rh. The oxidation of SO_2 to SO_3 to produce sulfuric acid uses mostly vanadium oxide as catalyst. Ammoxidation, which makes acrylonitrile from propylene, uses bismuth and molybdenum oxides as catalysts. Oxychlorination to make vinyl chloride from acetylene and HCl uses copper chloride as a catalyst. Polymerization reactions of ethylene and propylene are catalyzed by titanium trichloride, aluminum alkyls, chrome oxide on silica, and peresters. While these are the catalysts that are used in the largest quantity, many other highly selective catalysts serve as the basis of entire chemical technologies. In fact, the value of a very selective catalyst that aids a complex chemical transformation and the production of precious life-saving pharmaceuticals is without compare.

Most of the catalysts employed in the chemical technologies are heterogeneous. The chemical reaction takes place on surfaces, and the reactants are introduced as gases or liquids. Homogeneous catalysts, which are frequently metalloorganic molecules or clusters of molecules, also find wide and important applications in the chemical technologies.[54] Some of the important homogeneously catalyzed processes are listed in Table 8.3. Carbonylation, which involves the addition of CO and H_2 to a C_n olefin to produce a C_{n+1} acid, aldehyde, or alcohol, uses rhodium and cobalt complexes. Cobalt, copper, and palladium ions are used for the oxidation of ethylene to acetaldehyde and to acetic acid. Cobalt(II) acetate is used mostly for alkane oxidation to acids, especially butane. The air oxidation of cyclohexane to cyclohexanone and cyclohexanol is also carried out mostly with cobalt salts. Further oxidation to adipic acid uses copper(II) and vanadium(V) salts as catalysts. The hydrocyanation

Table 8.3. Chemical processes that are the largest users of homogenous catalysts at present and the catalyst systems they utilize most frequently

Reactions	Catalysts
Hydroformylation (aldehydes and alcohols from olefins)	cobalt, rhodium compounds
Carbonylation (acetic acid from methanol)	rhodium complexes and methyl iodide
Oxidation	
Adipic acid from cyclohexane	Cu(II), V(V) salts, nitric acid
Terephtalic acid from *p*-xylene	Co(II), Mn(II) salts, bromide ion
Acetic acid from butane or acetaldehyde	Co(II), Cu(II), Mn(II) salts
Olefin polymerization	
Polyethylene	$TiCl_4$, alkylaluminium, dialkyl magnesium compounds
Ethylene–propylene–diene copolymers	$VOCl_3$, VCl_4, alkylaluminum
Ethylene–butadiene copolymers	$RhCl_3$
Polypropylene, polystyrene	peroxides
Poly(vinyl chloride)	percarbonates
Urethane	amines

of butadiene to adiponitrile uses zero-valent nickel complexes. Polymerization technologies also frequently use homogeneous catalysts. The manufacture of polyethylene terephtalate uses antimony salts, and the copolymerization of ethylene and propylene to produce rubber uses alkylvanadium compounds.

Synthetic Approach to Catalytic Chemistry

The purpose of basic science studies of catalyst systems is to understand how they work on the atomic scale. One aims to identify the active sites where bond breaking and rearrangement take place and to detect surface intermediates that form. Studies are conducted to determine how the atomic surface structure and surface composition determine activity and selectivity. Once such an atomic-scale understanding is obtained, more active and selective catalysts can be designed, or one might find substitutes for precious-metal catalysts that are not readily available. Working catalyst systems have complicated structures, however, that do not lend themselves easily to atomic-scale investigations. The large-surface-area internal pore structure of the support hides the metal particles and makes it difficult to study their structure, oxidation state, and composition, which determine both activity and selectivity. Characterization of these complex but practical catalyst systems are the aims of many laboratories.

There is a different approach to the study of catalyst systems, which I would like to call the synthetic approach.[55] It is similar to the technique used by synthetic organic chemists to prepare complex organic molecules by linking the smaller segments one by one until the final product is obtained. The catalyst particle is viewed as composed of single-crystal surfaces, as shown in Figure 8.9. Each surface has different reactivity and the product distribution reflects the chemistry of the different surface sites. One may start with the simplest single crystal surface [for example, the (111) crystal face of platinum] and examine its reactivity. It is expected that much of the chemistry of the dispersed catalyst system would be absent on such a homogeneous crystal surface. Then high-Miller-index crystal faces are prepared to expose surface irregularities, steps, and kinks of known structure and concentration, and their catalytic behavior is tested and compared with the activity of the dispersed supported catalyst under identical experimental conditions. If there are still differences, the surface composition is changed systematically or other variables are introduced until the chemistries of the model system and the working catalyst become identical. This approach is described by the sequence:

structure of crystal surfaces and adsorbed gases

$\downarrow\uparrow$

surface reactions on crystals at low pressures ($\leq 10^{-4}$ torr)

$\downarrow\uparrow$

surface reactions on crystals at high pressures (10^{+3} to 10^{+5} torr)

$\downarrow\uparrow$

reactions on dispersed catalysts

Investigations in the first step define the surface structure and composition on the atomic scale and the chemical bonding of adsorbates. Studies in the second step, which are carried out at low pressures, reveal many of the elementary surface reaction steps and the dynamics of surface reactions. Studies in the third and fourth steps establish the similarities and differences between the model system and the dispersed catalyst under practical reaction conditions.

The advantage of using small-area catalyst samples is that their surface structure and composition can be prepared with uniformity and can be characterized by the many available surface diagnostic techniques. However, the small catalyst area that must be used in studies of this type necessitated the development of new instrumentation, which will be described next.

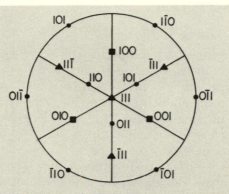

Standard Cubic (III) Projection

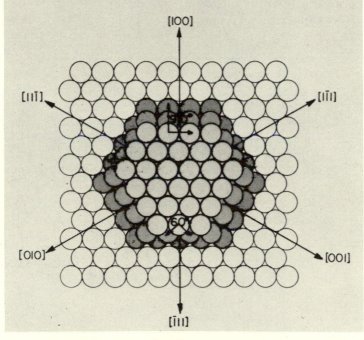

Figure 8.9. Catalyst particle viewed as a crystallite, composed of well-defined atomic planes.

Techniques to Characterize and Study the Reactivity of Small-Area Catalyst Surfaces

High-Pressure Reactors

In our synthetic approach to catalytic reaction studies, it is imperative that we determine the surface composition and surface structure in the same chamber where the reactions are carried out, without exposing the crystal surface to the ambient atmosphere. This necessitates the combined use of an ultrahigh-vacuum enclosure, where the surface characterization is to be carried out, and a high-pressure isolation cell, where the catalytic studies are performed. Such an apparatus is shown in Figure 8.10. The small-surface-area (approximately 1 cm²) catalyst is placed in the middle of the chamber, which can be evacuated to 10^{-9} torr. The surface is characterized by LEED and AES and by other desired surface diagnostic techniques. Then the lower part of the high-pressure isolation cell is lifted to enclose the sample in a 30-cm³ volume that is sealed by a copper gasket (approximately 2000-psi pressure is needed to provide a leak-free seal). The isolation chamber can be pressurized to

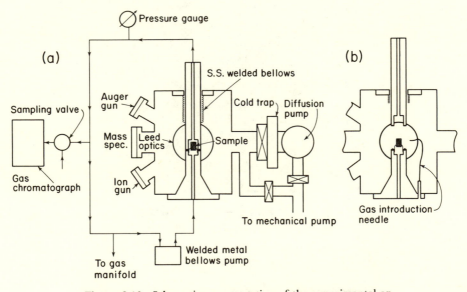

Figure 8.10. Schematic representation of the experimental apparatus to carry out catalytic-reaction-rate studies on single-crystal surfaces of low surface area at low and high pressures in the range 10^{-7} to 10^4 torr.

100 atm if desired and is connected to a gas chromatograph that detects the product distribution as a function of time and surface temperature. The sample may be heated resistively, both at high pressure or in ultrahigh vacuum. After the reaction study the isolation chamber is evacuated, opened, and the catalytic surface is again analyzed by the various surface-diagnostic techniques. Ion-bombardment cleaning of the surface or means to introduce controlled amounts of surface additives by vaporization are also available. The reaction at high pressures may be studied in the batch or the flow mode.

Low-Pressure Reactors

A LEED diffraction chamber can be used that is modified to study the catalyzed surface reaction at low pressures in the range 10^{-8} to 10^{-4} torr. The detector in this case is a quadrupole mass spectrometer that is placed a few centimeters away from the sample surface. By closing a gate valve, the system may be operated as a batch reactor. In this circumstance reaction probabilities as low as 10^{-7} may be adequate to detect products. However, poisoning of the surface by the accumulated product molecules can cause difficulty. By opening the gate valve and by control of the pumping speed between 1 and 100 liters/sec, the chamber is converted to a flow reactor and the reaction may be carried out at a steady state. At the higher pumping speeds, however, the detection sensitivity is somewhat reduced. Still, reactions with probabilities as low as 10^{-4} are detectable. The differences between the low-pressure reactor and the reactive molecular-beam-scattering experiments should be pointed out. Both work in a pressure regime where the surface coverage of the adsorbates is expected to be similar and much lower than in the high-pressure experiments. However, in the low-pressure batch or flow reactor, reactants and product molecules may undergo multiple collisions with the surface. In the directed molecular beam experiment using a differential mode of detection, the products are detected only when produced upon a single scattering.

The low-pressure reactor may be modified to accept an isolation chamber that can be pressurized to 1 atm. In this case the usual sample manipulator that is employed with LEED studies can be utilized as part of the high-pressure enclosure. Detailed description of the various high-pressure apparatuses and their development is given elsewhere.[55]

It is essential to test the high-pressure chamber to make sure that the measured reaction rates using the small-surface-area sample can be readily compared to reaction rates obtained on large-surface-area

Table 8.4. Comparison of initial specific rate data for the cyclopropane ring opening on platinum catalysts

Data source	Type of catalyst	Calculated specific reaction rate at $P°_{CP} = 135$ torr and $T = 75°C$		Comments
		$\dfrac{\text{moles } C_3H_8}{\text{min} \cdot \text{cm}^2 \text{ Pt}}$	$\dfrac{\text{molecules } C_3H_8}{\text{Min} \cdot \text{Pt site}}$	
Kahn et al.[1]	stepped platinum crystal surfaces	2.1×10^{-6} 1.8×10^{-6} 1.8×10^{-6} 2.1×10^{-6}		rate on Pt(s)-[6(111) × (100)] single crystal based on $E^{*a} = 12.2$ kcal/mol
	average	1.95×10^{-6}	812	value of 812 based upon 87% (111) orientation and 13% polycrystalline orientation
Hegedus[2]	0.04 wt% Pt on η-Al$_2$O$_3$	7.7×10^{-7} based on 100% Pt dispersion	410	value of 410 based upon average Pt site density of 1.12×10^{15} atoms/Pt site; this value would be nearly equal to average of values above if dispersion was approximately 50%
Boudart et al.[3]	0.3% and 2.0% Pt on η-Al$_2$O$_3$	8.9×10^{-7}	480	$\eta_{CP} = 0.2$, $E^{*a} = 8.5$ kcal/mol
Dougharty[4]	0.3% and 0.6% Pt on γ-Al$_2$O$_3$	2.5×10^{-6}	1340	$\eta_{CP} = 0.6$, $E^{*a} = 8.5$ kcal/mol

[a] Dougharty reports that $E^* = 8$–9 kcal/mol and $\eta = 0.2$–0.6.

1. D.R. Kahn, E.E. Petersen, and G.A. Somorjai, J. of Catal. **34**, 294 (1974).
2. L.L. Hegedus and E.E. Petersen, J. Catal. **28**, 150 (1973).
3. M. Boudart, A. Aldag, N.A. Dougharty, and C.G. Harkins, J. Catal. **6**, 92 (1966).
4. N.A. Dougharty, Ph.D. thesis, University of California at Berkeley, 1964.

catalysts. This comparison has been made using the ring opening of cyclopropane[56] and the hydrogenation of CO[57] as test reactions. Table 8.4 shows the turnover numbers and the activation energies obtained for the ring opening of cyclopropane to form propane on small-area single-crystal platinum and on dispersed platinum catalysts under identical experimental conditions. The agreement is indeed excellent. This is a structure-insensitive reaction at high pressures that lends itself well to such correlative studies. For structure-sensitive reactions, marked differences are found, with the single-crystal catalyst being much more active in general. Similarly, excellent agreements among rates, activation energies, and product distribution were obtained for the hydrogenation of carbon monoxide over polycrystalline rhodium foils and dispersed, silica-supported rhodium catalyst particles. This is shown in Table 8.5.

Let us use our synthetic approach to investigate catalyst surfaces on the atomic scale and utilize these experimental techniques, which permit us to monitor the activity of small area catalysts over a broad pressure range. As a first step we should find out how some of the best catalysts work on the molecular level. To this end the reactivities of platinum,

Table 8.5. Comparison of polycrystalline Rh foil with a 1 percent Rh/Al_2O_3 catalyst in the $CO-H_2$ reaction at atmospheric pressure

	Polycrystalline Rh[a] foil	Supported 1% Rh/Al_2O_3[b]
Reaction conditions	300°C, 3:1 H_2/CO, 700 torr	300°C[c] 3:1 H_2/CO, 760 torr
Type of reactor	batch	flow
Conversion	<0.1%	<5%
Product distribution	90% $CH_4 \pm 3$ 5% $C_2H_4 \pm 1$ 2% $C_2H_6 \pm 1$ 3% $C_3H_8 \pm 1$ <1% C_4^+	90% CH_4 8% C_2H_6 2% C_3 <1% C_4^+
Absolute methanation rate at 300°C (turnover no.)	0.13 ± 0.03 molecule/site/sec	0.034[c] molecule/site/sec
Activation energy (kcal)	24.0 ± 2	24.0

[a] The values in this column are from B.A. Sexton and G.A. Somorjai, J. Catal. **46**, 167 (1977).
[b] The values in this column are from M.A. Vannice, J. Catal. **37**, 462 (1975).
[c] Data adjusted from 275°C.

rhodium, and iron metal surfaces and that of two oxides, titanium oxide and strontium titanate (TiO_2 and $SrTiO_3$), have been studied in my laboratory. The atomic surface structure and surface composition of platinum were varied to uncover the reasons for its unique selectivity in hydrocarbon conversion reactions. We have investigated the reasons for the reactivity and selectivity of rhodium and iron surfaces for carbon monoxide hydrogenation. Finally, we have investigated the mechanism of photochemical dissociation of water on the TiO_2 and $SrTiO_3$ surfaces. The studies of these three types of catalyzed reactions will be described in Chapters 9 to 11 in some detail. We shall see that atomic-scale scrutiny of the catalyst surfaces reveals many of the important characteristics that explain their reactivities. With this understanding, we can attempt to improve these catalyst systems and the building of new systems.

References

1. (a) J. Berzelius, *Jahres-Bericht über die Fortschritte der Physichen Wissenschaften*, Tübingen, 1836, p. 243; (b) P. Emmett, Crit. Rev. Solid State Sci. **4,** 127 (1974).
2. A. Gaydon, *Dissociation Energies*, Chapman Hall, London, 1953.
3. G.W. Koeppl, J. Chem. Phys. **59,** 3425 (1973).
4. M. Salmerón, R. J. Gale, and G. A. Somorjai, J. Chem. Phys. **70,** 2807 (1979).
5. H. Bonzel, G. Brodén, and G. Pirug, J. Catal. **53,** 96 (1978).
6. G. Ertl, M. Weiss, and S. B. Lee, Chem. Phys. Lett. **60,** 391 (1979).
7. F. Ciapetta and D. N. Wallace, Catal. Rev. **5,** 67 (1971).
8. G.G. Hammes, Adv. Chem. Phys. **39,** 1, (1970).
9. T.E. Madey, J.T. Yates, Jr., D.R. Sandstrom, and R.J.M. Voohoeve, in *Treatise on Solid State Chemistry* **6B,** ed. B. Hannay, Plenum Press, New York, 1976.
10. S.M. Davis and G.A. Somorjai, J. Catal., **65,** 78 (1980).
11. E.E. Petersen, *Chemical Reaction Analysis*, Prentice-Hall, Englewood Cliffs, N.J., 1965.
12. M. Boudart, *Kinetics of Chemical Processes*, Prentice-Hall, Englewood Cliffs, N.J., 1968.
13. J.R. Anderson, *Structure of Metallic Catalysts*, Academic Press, New York, 1975; J.M. Thomas and W.J. Thomas, *Introduction to the Principles of Heterogeneous Catalysis*, Academic Press, New York, 1967.
14. S. Brunauer, *The Adsorption of Gases and Vapors*, Princeton University Press, Princeton, N.J., 1940; A. Clark, *The Theory of Adsorption and Catalysis*, Academic Press, New York, 1970; F.C. Tompkins, *Chemisorption of Gases on Metals*, Academic Press, London, 1978; A. Ozaki, *Isotopic Studies of Heterogeneous Catalysis*, Academic Press, New York, 1976.
15. R.L. Burwell, Jr., in *Catalysis—Progress in Research*, ed. F. Basolo and R.L. Burwell, Jr., Plenum Press, New York, 1973.

16. W.M.H. Sachtler and J. Fahrenfort, in *Actes du Deuxième Congrés International de Catalyse*, 1960, pp. 831–852.
17. J.H. Sinfelt, Adv. Catal. **23**, 91 (1973).
18. H.S. Taylor, Proc. Roy. Soc. (Lond.) **A108**, 105 (1925).
19. M. Boudart, Am. Sci. **57**, 97 (1969).
20. D.W. Blakely and G.A. Somorjai, J. Catal. **42**, 181 (1976).
21. S. Lehwald and H. Ibach, Surf. Sci. **89**, 425 (1979).
22. R. Mason and G. A. Somorjai, Chem. Phys. Lett. **44**, 468 (1976).
23. J.H. Lunsford, J. Phys. Chem. **68**, 2312 (1964).
24. M. Boudart, A. Delbouille, E. G. Derouane, V. Indovina, and H. B. Walters, J. Am. Chem. Soc. **94**, 6622 (1972).
25. R.J. Kokes, A.L. Dent, C.C. Chang, and L.T. Dixon, J. Am. Chem. Soc. **94**, 4429 (1972).
26. R.J.H. Voorhoeve, J. Catal. **23**, 236 (1971).
27. D.I. Hagen and G.A. Somorjai, J. Catal. **41**, 466 (1976).
28. J.H. Sinfelt, Catal. Rev. **9**, 147 (1974).
29. O.V. Krylov, *Catalysis by Non-Metals*, Academic Press, New York, 1970.
30. J.A. Rabo, ed. *Zeolite Chemistry and Catalysis*, A.C.S. Monograph 171, American Chemical Society, Washington, D.C., 1976.
31. J. Turkevich, Catal. Rev. **1**, 1 (1967).
32. J.W. Ward, J. Catal. **11**, 238 (1968).
33. P.B. Venuto and E.T. Habib, Catal. Rev. **18**, 1 (1978).
34. G.A. Somorjai, Surf. Sci. **89**, 496 (1979).
35. G. Brodén, G. Gafner, and H.P. Bonzel, Surf. Sci., **84**, 295 (1979).
36. G.A. Somorjai, Catal. Rev. **18**, 173 (1978).
37. D.W. Blakely, Ph.D. thesis, University of California, Berkeley, 1976.
38. D.W. Blakely and G.A. Somorjai, Surf., Sci. **65**, 419 (1977).
39. H. Topsoe, J.A. Dumesic, and M. Boudart, J. Catal. **28**, 477 (1973).
40. S.J. Tauster and S.G. Fung, J. Catal. **54**, 29 (1978).
41. M. Ichikawa, Bull. Chem. Soc. Jap. **51**, 2268 (1978).
42. D.W. Blakely and G.A. Somorjai, Nature **258**, 580 (1975).
43. R.W. McCabe and L.D. Schmidt, Surf. Sci. **65**, 189 (1977).
44. P.H. Emmett, in *The Physical Basis of Heterogeneous Catalyis*, ed. E. Draugles and R.I. Jaffe, Plenum Press, New York, 1975, p. 1.
45. G. Ertl, private communication.
46. G.A. Somorjai, J. Catal. **27**, 453 (1972).
47. M. Boudart, Adv. Catl. **20**, 153 (1969).
48. H.P. Bonzel and R. Ku, Surf. Sci. **33**, 91 (1972); I. Langmuir, Trans. Faraday Soc. **17**, 621 (1922).
49. D.D. Eley, *Chemi. Ind.*, January 3, 1976.
50. T. Engel and G. Ertl, J. Chem. Phys. **69**, 1267 (1978).
51. A.K. Galwey, Adv. Catal. **26**, 247 (1977).
52. M.A. Vannice, J. Catal. **37**, 462 (1973).
53. D.P. Burke, Chem. Week, March 28, 1979.
54. G.W. Parshall, J. Mol. Catal. **4**, 243 (1978).
55. G.A. Somorjai, Adv. Catal. **26**, 1 (1977).
56. D.R. Kahn, E.E. Petersen, and G.A. Somorjai, J. Catal. **34**, 294 (1974).
57. B.A. Sexton and G.A. Somorjai, J. Catal. **46**, 167 (1977).

Table 8.A1. Kinetic parameters for ethane hydrogenolysis over metal catalysts*

Catalyst	T (°C)	log (HC)	log (H$_2$)	E_a (kcal/mol)	log A	x	y	Rate (molecules/cm^2/sec) at 250°C	$-$log RP	References
SiO$_2$-supported										
5% Co	220–260	17.87	18.69	30	25.48	1.0	−0.8	9×10^{12}	9.0	1
5% Ni	180–220	17.87	18.69	41	31.69	1.0	−2.4	4×10^{14}	7.4	1
5% Cu	290–330	17.87	18.69	21	—	1.0	−0.4	—	—	2
1% Ru	170–270	17.87	18.69	33	29.5	—	—	3×10^{15}	6.5	3
5% Ru	180–210	17.87	18.69	32	28.11	0.8	−1.3	6×10^{14}	7.2	1
0.1% Rh	190–250	17.87	18.69	42	31.4	0.8	−2.1	1×10^{14}	7.9	4
0.3% Rh	190–250	17.87	18.69	42	31.4	0.8	−2.1	1×10^{14}	7.9	4
1% Rh	190–250	17.87	18.69	42	31.9	0.8	−2.1	3×10^{14}	7.5	4
5% Rh	190–225	17.87	18.69	42	31.76	0.8	−2.2	2×10^{14}	7.7	1
10% Rh	190–250	17.87	18.69	42	31.8	0.8	−2.1	2×10^{14}	7.6	1
5% Pd	345–375	17.87	18.69	58	33.57	0.9	−2.5	2×10^{9}	12.6	1
5% Re	230–265	17.87	18.69	31	26.26	0.5	0.3	2×10^{13}	8.6	1
5% Os	125–160	17.87	18.69	35	30.85	0.6	−1.2	2×10^{16}	5.7	1
5% Ir	180–210	17.87	18.69	36	28.72	0.7	−1.6	5×10^{13}	8.2	1
5% Pt	345–385	17.87	18.69	54	31.77	0.9	−2.5	2×10^{9}	12.7	1
1% Rh–0.13% Cu	170–270	17.87	18.69	~33	28.5	—	—	3×10^{14}	7.5	3
1% Ru–0.32% Cu	170–270	17.87	18.69	~33	28.1	—	—	1×10^{14}	7.8	3
1% Ru–0.63% Cu	170–270	17.87	18.69	~33	26.3	—	—	2×10^{12}	9.6	3
1% Os–0.10% Cu	170–270	17.87	18.69	~35	29.1	—	—	1×10^{14}	7.8	3
1% Os–0.17% Cu	170–270	17.87	18.69	~35	28.2	—	—	2×10^{13}	8.6	3
Unsupported								at 250°C		
Ni, powder	225–270	17.87	18.69	43	31.3	1.0	−2.1	2×10^{13}	8.6	5
Ni, powder	165–275	17.51	18.81	40	26.9	—	—	2×10^{10}	11.3	6
Ni, film	254–273	17.10	18.16	58	35.8	—	—	5×10^{11}	9.4	7
Ru, powder	160–200	17.87	18.69	32	28.4	—	—	1×10^{15}	6.9	8

414

Ru, powder	140–300	17.51	18.51	28	25.1	0.9	-0.9	3×10^{13}	8.1	9
Rh, powder	190–250	17.87	18.59	39	29.3	0.9	-1.6	1×10^{13}	8.8	4
Rh, powder	125–175	17.51	18.51	36	30.9	0.8	-2.1	9×10^{15}	5.6	9
Pd, film	270–360	17.10	18.16	50	31.9	1.0	—	1×10^{11}	10.0	10
W, film	170–180	17.10	18.16	27	26.4	—	—	1×10^{15}	6.0	7
Ir, powder	235–300	17.51	18.51	40	28.4	—	—	7×10^{11}	9.7	9
Pt, film	270–340	17.10	18.16	57	34.2	1.0	—	3×10^{10} at 250°C	10.6	10
Unsupported alloys										
Ni–Cu, powder										
($X_{Cu} = 0.062$)	310–340	17.87	18.69	51	31.6	0.9	-1.3	2×10^{10}	11.7	5
($X_{Cu} = 0.103$)	325–355	17.87	18.69	51	31.1	1.0	-1.0	6×10^{9}	12.2	5
($X_{Cu} = 0.315$)	355–395	17.87	18.69	50	29.9	1.0	-1.3	8×10^{8}	13.0	5
($X_{Cu} = 0.633$)	380–440	17.87	18.69	48	28.5	0.8	-1.2	2×10^{8}	13.7	5
Ru–Cu, powder										
($X_{Cu} = 0.018$)	280–350	17.87	18.69	~25	~22	—	—	1×10^{12}	9.9	8
($X_{Cu} = 0.03$)	280–350	17.87	18.69	~25	~22	—	—	6×10^{11}	10.2	8
($X_{Cu} = 0.05$)	350–430	17.87	18.69	~25	21	—	—	6×10^{10}	11.2	8
Other supported and unsupported systems										
0.5% Ru/Al$_2$O$_3$	160–220	19.17	19.60	—	—	1.0	-2.0	3×10^{12} at 180°C	10.7	11
Ir, film	80–205	14.7	14.85	~23	26	—	—	9×10^{14}	4.0	12
								at 300°C		
Ni, powder	315	18.21	0.0	38	25.7	—	—	2×10^{11}	11.0	13
5% Ni/SiO$_2$	300	18.82	19.26	—	—	—	—	5×10^{14}	8.2	14
3.7% Ni/SiO$_2$–Al$_2$O$_3$	300	18.82	19.26	—	—	—	—	2×10^{13}	9.6	11
2.5% Ni/Na–Y	300	18.82	19.26	—	—	—	—	9×10^{12}	9.9	11
2.4% Ni/Li–Y	300	18.82	19.26	—	—	—	—	7×10^{12}	10.0	11
2.3% Ni/Ca–Y	300	18.82	19.26	—	—	—	—	7×10^{12}	10.0	11
1.2% Ni/Mg–Y	300	18.82	19.26	—	—	—	—	8×10^{12}	10.0	11
2% Pt/Al$_2$O$_3$	300	18.39	19.35	—	—	—	—	8×10^{9}	12.6	15

*References for the tables in this Appendix follow Table 8.E4.

(continued)

Table 8.A1.—Continued

Catalyst	T (°C)	log (HC)	log (H_2)	E_a (kcal/mol)	log A	x	y	Rate (molecules/cm²/sec)	$-$log RP	References
								at 350°C		
Co, powder	310–360	17.51	18.81	27	20.3	—	—	8×10^{10}	10.7	6
8% Ni/SiO$_2$–Al$_2$O$_3$	350	18.0 (pulse)	19.39	—	—	—	—	3×10^{14}	—	16
8% Ni/SiO$_2$–Al$_2$O$_3$ + 0.73% Na$^+$	350	18.0 (pulse)	19.39	—	—	—	—	2×10^{14}	—	16
8% Ni/SiO$_2$–Al$_2$O$_3$ + 2.4% Na$^+$	350	18.0 (pulse)	19.39	—	—	—	—	1×10^{14}	—	16
Ni–Cu, films										
(X$_{Cu}$ = 0.0)	310–350	16.56	17.64	28	23.2	—	—	3×10^{13}	7.1	17
(X$_{Cu}$ = 0.04)	310–350	16.56	17.64	22	20.9	—	—	2×10^{13}	7.3	17
(X$_{Cu}$ = 0.18)	310–350	16.56	17.64	22	20.4	—	—	5×10^{12}	7.9	17
(X$_{Cu}$ = 0.54)	310–350	16.56	17.64	18	18.8	—	—	3×10^{12}	8.1	17
(X$_{Cu}$ = 0.82)	310–350	16.56	17.64	21	19.0	—	—	5×10^{11}	8.9	17
(X$_{Cu}$ = 0.96)	310–350	16.56	17.64	16	16.7	—	—	1×10^{11}	9.5	17
3.1% Pt/SiO$_2$ oxidized at 200°C	205–350	18.21	18.21	—	—	—	—	2×10^{14}	8.1	18
3.6% Pt/SiO$_2$ oxidized at 500°C	205–350	18.21	18.21	—	—	—	—	3×10^{13}	8.8	18
1.8% Mo–4.4% Pt/SiO$_2$	205–350	18.21	18.21	33	27.0	1.0	−1.0	3×10^{15}	6.8	18
Pt, powder	370–460	17.51	18.81	53	26.9	0.9	−1.9	3×10^{8}	13.1	19
Pt, film										
clean	300–410	17.02	18.02	53	30.6	—	—	1×10^{12}	9.1	20
steady state	300–410	17.02	18.02	36	24.7	—	—	1×10^{12}	9.1	20
									at 390°C	
3.8% Pt/Na–Y	390	18.03	19.01	—	—	—	—	8×10^{13}	8.2	21
10.8% Pt/NH$_4$–Y	390	18.03	19.01	—	—	—	—	1×10^{14}	8.1	21
2.6% Pt/SiO$_2$	390	18.03	19.01	—	—	—	—	2×10^{13}	8.8	21

Table 8.A2. Kinetic parameters for propane hydrogenolysis over metal catalysts

Catalyst	T (°C)	log (HC)	log (H$_2$)	E_a (kcal/mol)	log A	x	y	Rate (molecules/cm^2/sec)	$-\log RP$	References
								at 150°C		
0.5% Ru/Al$_2$O$_3$	140–170	18.77	19.32	36	31.9	1.0	-1.5	3×10^{13} ($S_{C_2} = 0.89$)	9.3	11
Pd–Rh, powder										
($X_{Rh} = 0.0$)	150	17.81	18.11	—	—	—	—	6×10^{9}	12.0	22
($X_{Rh} = 0.2$)	150	17.81	18.11	—	—	—	—	1×10^{11}	10.6	22
($X_{Rh} = 0.8$)	150	17.81	18.11	—	—	—	—	2×10^{11}	10.4	22
($X_{Rh} = 1.0$)	150	17.81	18.11	—	—	—	—	2×10^{11}	10.4	22
Pd–Pt, powder										
($X_{Pt} = 0.0$)	150	17.81	18.11	—	—	—	—	6×10^{9}	12.0	22
($X_{Pt} = 0.2$)	150	17.81	18.11	—	—	—	—	3×10^{10}	11.3	22
($X_{Pt} = 0.4$)	150	17.81	18.11	—	—	—	—	1×10^{11}	10.8	22
($X_{Pt} = 0.8$)	150	17.81	18.11	—	—	—	—	1×10^{11}	10.7	22
($X_{Pt} = 1.0$)	150	17.81	18.11	—	—	—	—	3×10^{11}	10.3	22
								at 250°C		
6.7% Fe/Al$_2$O$_3$–K$_2$O	270–340	18.79	19.57	31	23.2	1.2	2.0	2×10^{10} ($S_{C_2} = 0.06$)	12.4	23
6.3% Co/SiO$_2$	230–270	18.79	19.57	22	22.8	0.8	-0.7	5×10^{13} ($S_{C_2} = 0.2$)	9.1	23
7% Co/SiO$_2$	245–265	18.79	19.57	34	28.6	0.7	-1.0	3×10^{14} ($S_{C_2} = 0.19$)	8.3	23
4.1% Ni/SiO$_2$	250–300	18.79	19.57	42	30.1	0.7	-1.6	3×10^{12} ($S_{C_2} = 0.53$)	10.1	23

(continued)

Table 8.A2.—Continued

Catalyst	T (°C)	log (HC)	log (H₂)	E_a (kcal/mol)	log A	x	y	Rate (molecules/cm²/sec)	−log RP	References
								at 250°C		
16.7% Ni/SiC	280–305	18.79	19.57	52	34.6	0.9	−2.4	1×10^{13} ($S_{C_2} = 0.56$)	9.8	23
15% Ni/Mg–SiC	250–280	18.79	19.57	46	32.1	0.7	−2.3	1×10^{13} ($S_{C_2} = 0.5$)	9.8	23
Ni, powder	250	18.21	18.55	39	27.4	—	—	2×10^{11}	10.9	13
7.5% Ni/Al₂O₃	210–250	18.19	19.03	49	35.6	1.0	−1 to −2.3	2×10^{15}	6.9	24
Ni, powder clean	200–270	17.51	18.51	33	27.1	0.9	0 to −0.6	9×10^{11} ($S_{C_2} = 0.65$)	9.5	25
steady state	200–270	17.51	18.51	—	—	0.9	0 to −0.6	4×10^{9} ($S_{C_2} = 1.0$)	11.9	25
Ni, film	217–267	17.10	18.16	31	26.4	—	—	3×10^{13}	7.6	7
0.5% Ru/Al₂O₃	150–180	18.79	19.57	37	32.2	0.8	−2.0	7×10^{16} ($S_{C_2} = 0.99$)	5.9	23
W, film	180–190	17.10	18.16	18	21.7	—	—	2×10^{14}	6.8	7
								at 300°C		
Ni, powder	315	18.21	0	36	24.8	—	—	1×10^{11}	11.0	13
2% Pt/Al₂O₃	300	18.39	19.35	—	—	—	—	1×10^{12}	10.3	15
								at 360°C		
Pt, powder	370–440	17.51	18.81	24	19.2	1.0	−1.6	9×10^{10}	10.5	19
Pt, film clean	360	17.02	18.02	—	—	—	—	1×10^{13}	8.0	20
steady state	360	17.02	18.02	—	—	—	—	5×10^{12}	8.3	20

Table 8.A3. Kinetic parameters for cyclopropane ring opening over metal catalysts

Catalyst	T (°C)	log (HC)	log (H_2)	E_a (kcal/mol)	log A	Rate (molecules/cm^2/sec)	$-$log RP	References
Rh, film	-78	17.51	18.51	—	—	8×10^{13}	7.6	26
Supported and unsupported systems						at 25°C		
35% Co/kieselguhr	25	19.12	19.05	$\sim$12	$\sim$21	1×10^{12}	11.0	27
14–40% Ni/Al$_2$O$_3$	25	19.12	19.05	11	22.1	6×10^{13}	9.3	27
Ni, film	-46–0	17.51$^{-0.1}$	18.51$^{0.6}$ (D$_2$)	8	19.3	5×10^{13}	7.8	26
10% Mo/Al$_2$O$_3$	25	19.12	19.05	—	—	2×10^{12}	10.8	27
10% Ru/SiO$_2$	0–80	17.87	18.69	12	21.1	2×10^{12}	9.5	28
0.5–5% Rh/Al$_2$O$_3$	25	19.12	19.05	—	—	4×10^{14}	8.5	27
5% Rh/C	25	19.12	19.05	—	—	3×10^{14}	8.6	27
0.36% Rh/SiO$_2$	-35 to -10	17.87	18.69	11	23.4	2×10^{15}	6.5	28
Pd, powder	25	19.12	19.05	—	—	2×10^{12}	10.8	27
5% Pd/Al$_2$O$_3$	25	19.12	19.05	—	—	1×10^{14}	9.1	27
10% Pd/C	25	19.12	19.05	—	—	4×10^{13}	9.5	27
10% Pd/SiO$_2$	-10 to -25	17.87	18.69	16	25.5	7×10^{13}	8.0	28
Pd, film	-46 to -8	17.51$^{-0.9}$	18.51$^{0.1}$ (D$_2$)	15	25.3	5×10^{14}	6.8	26
10% Os/SiO$_2$	0–60	17.87	18.69	13	22.2	5×10^{12}	9.1	28
10% Ir/SiO$_2$	0–30	17.87	18.69	13	22.4	8×10^{12}	8.9	28
Pt, powder	25	19.12	19.05	—	—	5×10^{13}	9.4	27
0.5–5% Pt/Al$_2$O$_3$	25	19.12	19.05	—	—	7×10^{13}	9.2	27
Pt(557)	75	18.64	19.35	12	25.8	7×10^{16}	5.8	29

(continued)

Table 8.A3.—Continued

Catalyst	T (°C)	log (HC)	log (H_2)	E_a (kcal/mol)	log A	Rate (molecules/cm^2/sec)	$-$log RP	References
						at 25°C		
0.3–2% Pt/η-Al$_2$O$_3$	75	18.64	19.35	8.5	23.1	8×10^{16}	5.7	30
0.3–0.6% Pt/γ-Al$_2$O$_3$	75	18.64	19.35	9	23.9	2×10^{17}	5.3	31
7.1–81% Pt/SiO$_2$	−10 to −21	18.22	19.36	10	22.3	1×10^{15}	7.2	32
0.6% Pt/SiO$_2$	−20 to −30	17.87	18.69	11	22.6	4×10^{14}	7.2	28
Pt, film	−78 to −23	$17.51^{-0.2}$	$18.51^{0.2}$ (D$_2$)	11	24.6	3×10^{16}	4.9	26
3.8% Pt/Na-Y	22	17.51	18.98	—	—	1.4×10^{15}	6.4	21
3.5% Pt/NH$_4$-Y	22	17.51	18.98	—	—	1.8×10^{15}	6.2	21
1.1% Pt/Ce-Y	22	17.51	18.98	—	—	4×10^{15}	5.9	21
1.7% Pt/Al$_2$O$_3$	22	17.51	18.98	—	—	1.3×10^{13}	8.4	21
2.6% Pt/SiO$_2$	22	17.51	18.98	—	—	4×10^{13}	7.9	21
Pt, powder	24	17.09	19.37	—	—	2×10^{13}	7.8	33
0.6–4.8% Pt/Al$_2$O$_3$	24	$17.09^{0.6}$	$19.37^{0.0}$	—	—	3×10^{13}	7.7	33
						at 140°C		
Fe, film	50–150	$17.51^{0.0}$	$18.51^{1.0}$ (D$_2$)	23	24.4	2×10^{12}	9.2	26
8% Ni/SiO$_2$–Al$_2$O$_3$	140	18.8 (pulse)	19.3	—	—	1.5×10^{15}	—	16
8% Ni/SiO$_2$–Al$_2$O$_3$ + 0.44% Na$^+$	140	18.8 (pulse)	19.3	—	—	4×10^{14}	—	16
8% Ni/Si$_2$–Al$_2$O$_3$ + 2.4% Na$^+$	140	18.8 (pulse)	19.3	—	—	5×10^{13}	—	16

Table 8.A4. Kinetic parameters for cyclopropane hydrogenolysis over metal catalysts

Catalyst	T (°C)	log (HC)	log (H_2)	E_a (kcal/mol)	log A	Rate (molecules/cm²/sec)	$-$log RP	References
						at 25°C		
35% Co/kieselguhr	25	19.12	19.05	—	—	5×10^{11}	11.4	27
14–40% Ni/Al_2O_3	25	19.12	19.05	—	—	8×10^{13}	9.2	27
Ni, film	−46–0	17.51	18.51 (D_2)	~8	~19	1.5×10^{13}	8.4	26
10% Mo/Al_2O_3	25	19.12	19.05	—	—	1×10^{12}	11.1	27
10% Ru/SiO_2	0–80	17.87	19.69	12	20.3	4×10^{11}	10.3	28
5% Rh/C	25	19.12	19.05	—	—	8×10^{12}	10.2	27
0.5% Rh/Al_2O_3	25	19.12	19.05	—	—	4×10^{13}	9.5	27
10% Os/SiO_2	0–60	17.87	18.69	13	21.4	8×10^{11}	9.9	28
						at 140°C		
Fe, film	50–150	17.51	18.51 (D_2)	23	24.4	2×10^{12}	9.2	26
8% Ni/SiO_2–Al_2O_3	140	18.7 (pulse)	19.3	—	—	2×10^{14}	—	16
8% Ni/SiO_2–Al_2O_3 + 0.44% Na^+	140	18.7 (pulse)	19.3	—	—	1×10^{14}	—	16

421

Table 8.A5. Kinetic parameters for n-butane hydrogenolysis over metal catalysts

Catalyst	T (°C)	log (HC)	log (H$_2$)	E_a (kcal/mol)	log A	Rate (molecules/cm^2/sec)	$-\log RP$	S_{C_3}	S_{C_2}	References
							150°C			
0.5% Ru/Al$_2$O$_3$	150	18.35$^{0.9}$	19.35$^{-1.35}$	48	39.7	1×10^{15}	7.3	—	—	34
Rh, powder	80–130	17.51$^{0.5}$	18.51$^{-1.3}$	29	28.9	1×10^{14}	7.4	~0.6	~0.3	9
W, film	144–164	17.10	18.16	7	16.4	6×10^{12}	8.2	—	—	7
							at 250°C			
Ni, film	188–209	17.10	18.16	34	28.5	2×10^{14}	6.6	—	—	7
Re/Al$_2$O$_3$	240	18.39	19.35	24	24.0	1×10^{14}	8.2	—	—	35
Os/Al$_2$O$_3$	240	18.39	19.35	34	30.3	1×10^{16}	6.2	—	—	35
Ir/Al$_2$O$_3$	240	18.39	19.35	43	33.8	8×10^{15}	6.4	—	—	35
Ir, powder	180–230	17.51$^{0.4}$	18.51$^{-0.3}$	24	22.9	9×10^{12}	8.5	≤0.05	~0.15	9
Pt/Al$_2$O$_3$	240	18.39	19.35	34	25.3	1.5×10^{11}	11.1	—	—	35
Pt, powder	370–460	17.51$^{1.0}$	18.81$^{-0.6}$	23	19.4	7×10^{9}	11.6	~0.65	~0.33	19
Pt, film	300–400	17.02	18.02	22	20.5	2×10^{11}	9.6	—	—	20
Pt-Re/Al$_2$O$_3$, bimetallic (X_{Re} = 0.25)	240	18.39	19.35	36	27.4	3×10^{12}	9.9	—	—	35
Pt-Re/Al$_2$O$_3$, bimetallic (X_{Re} = 0.55)	240	18.39	19.35	40	30.1	3×10^{13}	9.9	—	—	35
(X_{Re} = 0.75)	240	18.39	19.35	43	31.6	6×10^{13}	8.6	—	—	35
							at 300°C			
Ni, powder	315	18.2	0.0	44	27.8	1.3×10^{11}	11.0	—	—	13
Pd, film	276–310	17.10$^{-0.3}$	18.16$^{0.0}$	38	27.4	1×10^{13}	8.0	~0.85	~0.11	10
Pt/Al$_2$O$_3$	300	18.39	19.35	—	—	1×10^{13}	9.3	—	—	36
2% Pt/Al$_2$O$_3$	300	18.39	19.35	—	—	1×10^{13}	9.3	—	—	15
Pt, film	256–300	17.10$^{0.7}$	18.16$^{1.4}$	21	21.0	1×10^{13}	8.0	~0.6	~0.3	10
Pt(111), film	320	17.10	18.16	~20	—	$\sim6 \times 10^{12}$	8.2	~0.6	~0.3	10
Pt(100), film	300	17.10	18.16	—	—	2×10^{13}	7.7	~0.55	~0.4	10

Table 8.A6. Kinetic parameters for isobutane hydrogenolysis over metal catalysts

Catalyst	T (°C)	$\log$ (HC)	$\log$ (H$_2$)	E_a (kcal/mol)	$\log A$	Rate (molecules/cm^2/sec)	$-\log RP$	S_{C_3}	S_{C_2}	References
0.5% Ru/Al$_2$O$_3$	150	$18.35^{0.7}$	$19.35^{-0.7}$	36	32.8	2×10^{14}	at 150°C 8.0	—	—	34
							at 250°C			
Ni, film	200–220	17.10	18.16	30	26.8	2×10^{14}	6.7	—	—	7
Pd, film	270–310	$17.10^{-0.2}$	$18.16^{0.1}$	21	21.0	2×10^{12}	8.7	~0.94	~0.02	10
Ta, film	200–295	16.51	17.51	~3	~14	~3×10^{12}	7.9	~0.6	~0.2	37
W, film	150–195	16.51	17.51	~12	~19	~4×10^{13}	6.8	~0.7	~0.2	37
Re, film	200–270	16.51	17.51	~5	~15	~2×10^{12}	8.1	~0.4	~0.3	37
Pt, film	265–299	$17.10^{0.5}$	$18.16^{-1.4}$	21	20.7	1×10^{12}	9.0	~0.8	~0.04	10
							at 300°C			
Pd, film	265–310	17.10	18.16	21	20.8	7×10^{12}	8.1	~0.96	~0.02	38
Re–Au, films										
($X_{Re} = 0.71$)	290–390	16.51	17.51	~16	~17	~3×10^{10}	9.9	0.0	~0.8	37
($X_{Re} = 0.51$)	350–410	16.51	17.51	~3	~12	~7×10^{10}	9.5	~0.2	~0.25	37
Pt/Al$_2$O$_3$	300	18.39	19.35	—	—	8×10^{12}	9.4	—	—	36
Pt, powder	370–460	$17.51^{1.0}$	$18.81^{-1.7}$	26	19.9	1×10^{10}	11.4	~0.6	~0.2	19
Pt, film	265–299	$17.10^{0.5}$	$18.16^{-1.4}$	21	20.7	6×10^{12}	8.3	~0.8	~0.04	10
Pt(111), film	294–305	$17.10^{0.5}$	$18.16^{-1.4}$	19	18.7	3×10^{11}	9.5	~1.0	0.0	10
Pt(100), film	299	$17.10^{0.5}$	$18.16^{-1.4}$	—	—	7×10^{12}	8.2	~0.6	~0.6	10
							at indicated T(°C)			
Pt/Na–Y (10 Å)	320	18.19	19.36	—	—	2×10^{13}	8.8	—	—	39
Pt/Na–Y (15–20 Å)	320	18.19	19.36	—	—	4×10^{12}	9.5	—	—	39
Pt, film	360	17.02	18.02	—	—	2×10^{13}	7.8	—	—	20

Table 8.A7. Kinetic parameters for methylcyclopropane ring opening over metal catalysts

Catalyst	T (°C)	log (HC)	log (H$_2$)	E_a (kcal/mol)	log A	Rate (molecules/cm^2/sec)	$-$log RP	S_{nC_4}	S_{iC_4}	References
						at indicated T(°C)				
Ni, film	-46	17.51	18.51	—	—	2×10^{13}	8.3	0.23	0.77	26
Pd, film	-23	17.51	18.51	—	—	2×10^{13}	8.1	0.15	0.85	26
Pt, film	-64	17.51	18.51	—	—	1×10^{13}	8.4	0.02	0.98	26
						at 0°C				
7.1% Pt/SiO$_2$	0	18.21	19.36	9	22	0.3–1.2×10^{15}	7.1	0.05	0.95	40
27% Pt/SiO$_2$	0	18.21	19.36	9	22	3–9×10^{14}	7.2	0.06	0.94	40
81% Pt/SiO$_2$	0	18.21	19.36	9	22	3–9×10^{14}	7.2	0.03	0.97	40
27% Pt/SiO$_2$	0	18.21	19.36	10	22.3	2×10^{14}	7.8	0.06	0.94	32
0.5% Pt/SiO$_2$	0	17.51	18.21	—	—	6×10^{13}	7.7	0.04	0.96	41
0.5% Pt/SiO$_2$–Al$_2$O$_3$ (BF)	0	17.51	18.21	—	—	6×10^{14}	6.6	1.0	0.0	41
2.5% Pt/SiO$_2$	0	17.51	18.21	—	—	2×10^{13}	8.2	0.04	0.96	41
						at 24°C				
Pt, powder	24	17.09	19.37	—	—	4×10^{12}	8.4	0.05	0.95	33
0.3% Pt/Al$_2$O$_3$	24	17.09	19.37	—	—	3×10^{13}	7.5	0.13	0.97	33
4.8% Pt/Al$_2$O$_3$	24	17.09	19.37	—	—	2×10^{13}	7.7	0.13	0.87	33
3% Pt/SiO–Al$_2$O$_3$ (BF)	24	17.09	19.37	—	—	2×10^{14}	6.7	0.12	0.88	33

Table 8.A8. Kinetic parameters for n-pentane hydrogenolysis over metal catalysts

Catalyst	T (°C)	log (HC)	log (H₂)	E_a (kcal/mol)	log A	Rate (molecules/cm²/sec)	−log RP	S_{C_4}	S_{C_3}	References
						at 100°C				
0.5% Rh/Al₂O₃	100	17.51	18.51	—	—	3×10^{11}	9.9	~0.12	~0.88	42
1.9% Rh/Al₂O₃	100	17.51	18.51	—	—	2×10^{11}	10.1	~0.15	~0.85	42
8.7% Rh/Al₂O₃	100	17.51	18.51	—	—	9×10^{10}	10.4	~0.60	~0.40	42
12.4% Rh/Al₂O₃	100	17.51	18.51	—	—	7×10^{10}	10.6	~0.66	~0.34	42
						at 185°C				
Rh, powder	130–170	17.51	18.51	28	28.2	9×10^{14}	6.4	—	—	9
5.5% Rh/Al₂O₃	110–140	$17.51^{1.0}$	$18.51^{-1.6}$	27 / 17	{ 25.1 for C₃ + C₂ / 21.0 for C₄ + C₁ }	7×10^{12}	8.5	~0.8	~0.2	43
Os/Al₂O₃	185	18.35	19.35	—	—	2×10^{13}	9.0	—	—	44
0.6% Re/Al₂O₃	185	18.35	19.35	—	—	4×10^{12}	9.6	~0.14	~0.20	44
0.6% Ir/Al₂O₃	185			—	—	1×10^{12}	10.2	0.1	0.84	44
0.6% Re–Ir/Al₂O₃, bimetallic										
(X_Re = 0.88)	185	18.35	19.35	—	—	6×10^{12}	9.4	—	—	44
(X_Re = 0.45)	185	18.35	19.35	—	—	8×10^{12}	9.3	~0.22	~0.56	44
(X_Re = 0.10)	185	18.35	19.35	—	—	4×10^{12}	9.6	—	—	44

(continued)

425

Table 8.A8.—Continued

Catalyst	T (°C)	log (HC)	log (H_2)	E_a (kcal/mol)	log A	Rate (molecules/cm²/sec) at 250°C	$-\log RP$	S_{C_4}	S_{C_3}	References
5% Fe/SiO₂	375–450	$18.79^{0.5}$	$20.09^{-1.6}$	23	22.8	2×10^{13}	9.5	0.0	~0.08	45
5% Co/SiO₂	275–375	$18.79^{0.9}$	$20.09^{-1.5}$	31	27.2	2×10^{14}	8.4	~0.12	~0.1	45
5% Ni/SiO₂	175–375	$18.79^{0.9}$	$20.09^{-1.6}$	31	27.4	3×10^{14}	8.2	~.66	~.16	45
5% Ru/SiO₂	230–290	$18.79^{0.9}$	$20.09^{-1.6}$	29	29.4	2×10^{17}	5.4	—	—	45
Ru/C	250–325	$18.79^{1.0}$	$20.09^{-1.5}$	37	32.1	5×10^{16}	6.0	~0.06	~0.14	45
5% Rh/SiO₂	250–325	$18.79^{1.0}$	$20.09^{-1.3}$	30	28.0	3×10^{15}	7.2	~0.04	~0.28	45
Rh/C	290–375	$18.79^{1.0}$	$20.09^{-1.3}$	31	27.8	7×10^{14}	7.8	~0.08	~0.30	45
5% Pd/SiO₂	370–480	$18.79^{0.9}$	$20.09^{-1.4}$	49	31.5	1×10^{11}	11.6	~0.62	~0.36	45
Pd/Al₂O₃	250	18.35	19.35	—	—	8×10^{9}	11.6	—	—	44
Re, film	190–330	16.51	17.51	~0	~12	$\sim7 \times 10^{11}$	8.5	~0.56	~0.4	37
5% Ir/SiO₂	275–360	$18.79^{1.0}$	$20.09^{-1.5}$	~32	27.3	9×10^{13}	8.7	~0.08	~0.38	45
0.25% Ir/Al₂O₃	250	18.35	19.35	—	—	4×10^{14}	7.6	—	—	44
0.25% Ir + Cu/Al₂O₃, bimetallic										
(X_{Cu} = 0.20)	250	18.35	19.35	—	—	4×10^{13}	8.6	—	—	44
(X_{Cu} = 0.50)	250	18.35	19.35	—	—	3×10^{12}	9.8	—	—	44
(X_{Cu} = 0.80)	250	18.35	19.35	—	—	4×10^{11}	10.6	—	—	44
Ir, powder	210–270	17.51	18.51	25	22.7	2×10^{12}	10.2	—	—	9
5% Pt/SiO₂	350–500	$18.79^{0.7}$	$20.09^{-1.4}$	28	27.0	2×10^{15}	7.3	0.44	0.56	45
Pt/Al₂O₃	250	18.35	19.35	—	—	4×10^{10}	11.6	—	—	44

Catalyst										Ref.
0.33% Ru/Al_2O_3	450	18.61	19.31	~27	~22	7×10^{13}	8.6	—	—	46
0.86% Ru/Al_2O_3	490	18.61	19.31	—	—	1×10^{14}	8.4	—	—	46
Rh, film	320	16.51	17.51	—	—	$\sim\!1 \times 10^{11}$	9.3	100% C_1		47
Rh-Cu, films										
($X_{Cu} = 0.06$)	330	16.51	17.51	—	—	$\sim\!9 \times 10^{10}$	9.4	—	—	47
($X_{Cu} = 0.92$)	310	16.51	17.51	—	—	$\sim\!2 \times 10^{10}$	10.1	—	—	47
($X_{Cu} = 1.0$)	310	16.51	17.51	—	—	$\sim\!4 \times 10^{9}$	10.8	—	—	47
Rh-Ag, film										
($X_{Ag} = 0.19$)	320	16.51	17.51	~20	~18	$\sim\!2 \times 10^{10}$	10.1	—	—	47
Rh-Sn, film										
($X_{Sn} = 0.17$)	280	16.51	17.51	—	—	$\sim\!1 \times 10^{11}$	9.2	—	—	47
Rh-Au, film										
($X_{Au} = 0.23$)	320	16.51	17.51	29	20	$\sim\!4 \times 10^{9}$	10.8	—	—	47
Re-Au, films										
($X_{Au} = 0.14$)	280	16.51	17.51	~5	—	$\sim\!3 \times 10^{11}$	8.8	~0.5	~0.5	37
($X_{Au} = 0.68$)	290	16.51	17.51	~15	—	$\sim\!2 \times 10^{10}$	10.1	~0.0	~0.64	37
($X_{Au} = 0.93$)	290	16.51	17.51	~14	—	$\sim\!2 \times 10^{10}$	10.1	~0.0	~0.98	37
16% Pt/SiO_2 (D ~ 0.3)	312	18.39	19.34	~28	~21	$\sim\!3 \times 10^{10}$	11.8	~0.14	~0.72	48
16% Pt-Au/SiO_2, bimetallic (D ~ 0.3)										
($X_{Au} = 0.975$)	370	18.39	19.34	~39	~22	$\sim\!3 \times 10^{8}$	13.8	~0.5	~0.5	48
($X_{Au} = 0.875$)	350	18.39	19.34	~45	~25	$\sim\!2 \times 10^{9}$	13.0	~0.4	~0.45	48
1.3% Pt/SiO_2 (14 Å)	283	$18.03^{0.4}$	$19.05^{-0.8}$	—	—	2×10^{12}	9.6	—	—	49
8.2% Pt/SiO_2 (195 $\varkappa$)	283	$18.03^{0.4}$	$19.05^{-0.8}$	—	—	4×10^{11}	10.3	—	—	49

427

Table 8.A9. Kinetic parameters for isopentane hydrogenolysis over metal catalysts

Catalyst	T (°C)	log (HC)	log (H_2)	E_a (kcal/mol)	log A	Rate (molecules/cm²/sec)	$-\log RP$	S_{iC_4}	S_{nC_3}	References
0.5% Ru/Al$_2$O$_3$	110	18.35[0]	19.35[1]	43	37.5	1×10^{13}	9.1	~0.78	~0.1	34
Pd, film	310	17.10	18.16	—	—	2×10^{12}	8.7	~0.32	~0.5	10
2% Pt/Al$_2$O$_3$	300	18.39	19.35	—	—	2×10^{13}	9.0	—	—	15
Pt/Al$_2$O$_3$	300	18.39	19.35	—	—	2×10^{13}	8.9	~0.76	—	36
Pt, film	278	17.10	18.16	—	—	3×10^{12}	8.6	—	~0.38	10

Table 8.A10. Kinetic parameters for neopentane hydrogenolysis over metal catalysts

Catalyst	T (°C)	log (HC)	log (H$_2$)	E_a (kcal/mol)	log A	Rate (molecules/cm^2/sec)	$-$log RP	S_{iC_4}	S_{C_3}	References
0.5% Ru/Al$_2$O$_3$	125–155	18.35$^{0.9}$	19.35$^{-0.9}$	43	35.7	4×10^{13}	at 150°C 8.6	~0.16	~0.08	34
Rh, powder	110–150	17.51	18.51	40	33.4	8×10^{12}	8.5	—	—	9
Ni, film	220–265	17.10	18.16	32	26.3	1×10^{13}	at 250°C 8.0	—	—	7
5% Ru/SiO$_2$	160–180	18.35	19.35	36	30	1×10^{15}	7.2	~1.0	—	50
5% Rh/SiO$_2$	170–200	18.35	19.35	53	37	8×10^{14}	7.3	~1.0	—	50
5.5% Rh/Al$_2$O$_3$	160–200	17.51$^{1.0}$	18.51$^{-1.5}$	20	22.6	2×10^{14}	8.2	—	—	43
W, film	200–220	17.10	18.16	11	17.5	9×10^{12}	8.0	—	—	7
10% Os/SiO$_2$	130–180	18.35	19.35	32	28.4	1×10^{15}	7.1	~1.0	—	50
10% Ir/SiO$_2$	180–200	18.35	19.35	46	33	6×10^{13}	8.4	~1.0	—	50
1% Pt/spheron	310–370	18.35	19.35	59	33	2×10^{8}	13.9	~1.0	—	50
3% Pt/Na–Y (10 Å)	200–240	18.07	19.37	37	27.6	2×10^{12}	9.6	~0.8	~0.08	51
3% Pt/Ca–Y (10 Å)	200–240	18.07	19.37	35	27.3	6×10^{12}	9.2	~0.6	~0.2	51
3% Pt/LaY (10 Å)	200–240	18.07	19.37	33	26.9	1.6×10^{13}	8.7	~0.5	~0.34	51
2% Pt/SiO$_2$ (12 Å)	260–300	18.07	19.37	35	26.4	7×10^{11}	10.1	~0.5	~0.1	51
0.9% Pt/SiO$_2$ (40 Å)	280–320	18.07	19.37	28	23.8	1.5×10^{12}	9.8	~.6	~0.1	51
0.9% Pt/SiO$_2$ (70 Å)	280–320	18.07	19.37	25	22.5	1.3×10^{12}	9.8	~0.5	~0.2	51
2.5% Pt/SiO$_2$	190–360	17.91$^{1.0}$	19.39$^{1.0-0.3}$	35	28.1	4×10^{13}	8.2	—	—	52
6.3% Pt–Mo/SiO$_2$ (X_{Mo} = 0.47)	190–360	17.91$^{1.0}$	19.39$^{1.0-0}$	16	22.4	5×10^{15}	6.1	—	—	52
6.8% Pt–W/SiO$_2$ (X_W = 0.6)	190–360	17.91$^{1.0}$	19.39$^{1.0-0}$	21	23.6	7×10^{14}	6.9	—	—	52
Pt, film	240–295	17.10	18.16	21	20.8	1×10^{12}	8.9	~0.7	—	10
Pt–Pd, films (X_{Pd} = 0.0)	250	16.34	17.34	—	—	9×10^{10}	9.3	~0.96	—	53
(X_{Pd} = 0.2)	250	16.34	17.34	—	—	2×10^{10}	9.9	~0.6	~0.4	53

(continued)

Table 8.A10.—Continued

Catalyst	T (°C)	log (HC)	log (H$_2$)	E_a (kcal/mol)	log A	Rate (molecules/cm^2/sec)	$-\log RP$	S_{iC_4}	S_{C_3}	References
							at 250°C			
(X_{Pd} = 0.3)	250	16.34	17.34	—	—	2×10^{11}	8.9	~0.9	~0.1	53
(X_{Pd} = 0.54)	250	16.34	17.34	—	—	1×10^{12}	8.1	~0.5	~0.3	53
(X_{Pd} = 0.8)	250	16.34	17.34	—	—	2×10^{11}	8.9	~0.7	~0.3	53
(X_{Pd} = 1.0)	250	16.34	17.34	—	—	2×10^{12}	8.0	~0.7	~0.2	53
							at indicated T (°C)			
Pd, film	310	17.10	18.16	~0	—	9×10^{12}	8.0	~0.7	~0.2	10
Pd, film	265–310	17.10	18.16	—	—	3×10^{12}	8.5	—	—	38
2% Pt/Al$_2$O$_3$	300	18.39	19.35	—	—	1×10^{13}	9.2	—	—	15
Pt/Al$_2$O$_3$	300	18.39	19.35	—	—	8×10^{12}	9.4	—	—	36
1% Pt/C	370	18.35	19.35	—	—	5×10^{13}	8.5	~1.0	—	54
6% Pt–6% Fe/c	370	18.35	19.35	—	—	1.5×10^{12}	10.0	~1.0	—	54
0.6% Pt/γ-Al$_2$O$_3$	307	18.35	19.35	—	—	3×10^{12}	9.8	~1.0	—	55
2% Pt/γ-Al$_2$O$_3$	307	18.35	19.35	—	—	4×10^{12}	9.7	~1.0	—	55
4.3% Pt/SiO$_2$	307	18.35	19.35	—	—	2×10^{13}	8.9	~1.0	—	55
1% Pt/spheron	307	18.35	19.35	—	—	6×10^{10}	11.4	~1.0	—	55
2% Pt/η-Al$_2$O$_3$	307	18.35	19.35	—	—	7×10^{12}	9.4	~1.0	—	55
Pt, powder	307	18.35	19.35	—	—	1.2×10^{11}	11.1	~1.0	—	55
5% Pt/Ca–Y (D = 1.0)	272	18.35	19.35	—	—	1.8×10^{13}	8.9	~1.0	—	56
2% Pt/η-Al$_2$O$_3$ (D = 0.64)	272	18.35	19.35	—	—	4×10^{11}	10.7	~1.0	—	56
2% Pt/η-Al$_2$O$_3$ (D = 0.08)	272	18.35	19.35	—	—	6×10^{11}	10.4	~1.0	—	56
Pt, film clean	360	17.02	18.02	—	—	7×10^{13}	7.0	—	—	20
steady state	360	17.02	18.02	—	—	4×10^{12}	8.3	—	—	20
Au, powder	450	18.35	19.35	51	20	~10^5	17	—	—	50

Table 8.A11. Kinetic parameters for cyclopentane ring opening and hydrogenolysis over metal catalysts

Catalyst	T (°C)	log (HC)	log (H$_2$)	E_a (kcal/mol)	log A	Rate (molecules/cm^2/sec)	$-\log RP$	S_{C_5}	References
						at 250°C			
Re/Al$_2$O$_3$	240	18.39	19.35	12	17.9	8×10^{12}	9.4	~0.5	35
Os/Al$_2$O$_3$	240	18.39	19.35	18	21.8	3×10^{14}	7.9	—	35
Ir/Al$_2$O$_3$	240	18.39	19.35	28	27.6	8×10^{15}	6.4	~1.0	35
Pt/Al$_2$O$_3$	240	18.39	19.35	35	27.7	1.4×10^{13}	9.1	~1.0	35
Pt–Re/Al$_2$O$_3$, bimetallic									
(X_{Re} = 0.25)	240	18.39	19.35	40	29.6	8×10^{12}	9.4	—	35
(X_{Re} = 0.50)	240	18.39	19.35	26	24.2	2×10^{13}	9.0	—	35
(X_{Re} = 0.75)	240	18.39	19.35	23	24.8	1×10^{15}	7.1	—	35
						at indicated T (°C)			
Rh, powder	150	17.51	18.51	18	21.2	9×10^{11}	9.4	—	9
1% Pd/Al$_2$O$_3$ ($D = 0.3$–0.8)	290	18.51	19.33	40	28.4	9×10^{12}	9.4	1.0	57
0.25% Pd/Al$_2$O$_3$ ($D = 0.8$–1.0)	290	18.51	19.33	40	28.6	1.4×10^{13}	9.2	1.0	57
0.25–8% Pd/Al$_2$O$_3$	290	18.51	19.33	40	28.6	$9\text{-}13 \times 10^{12}$	9.3	1.0	58
0.7% Pd/SiO$_2$ ($D = 0.04$)	290	18.51	19.33	—	—	3×10^{12}	9.9	1.0	58
1.3% Pd/SiO$_2$ ($D = 0.32$)	290	18.51	19.33	—	—	2×10^{13}	9.1	1.0	58
Pd, film	300	16.41	17.41 (D$_2$)	~28	~23	$\sim 1 \times 10^{12}$	8.3	~0.95	59
Pd–Au, films									
(X_{Au} = 0.21)	330	16.41	17.41 (D$_2$)	—	—	$\sim 2 \times 10^{11}$	9.0	~1.0	59
(X_{Au} = 0.34)	310	16.41	17.41 (D$_2$)	—	—	$\sim 9 \times 10^{10}$	9.3	~1.0	59
(X_{Au} = 0.86)	400	16.41	17.41 (D$_2$)	—	—	$\sim 3 \times 10^{12}$	7.8	~0.98	59
(X_{Au} = 1.0)	500	16.41	17.41 (D$_2$)	—	—	$\sim 4 \times 10^{11}$	8.7	~0.85	59
2% Pt/SiO$_2$ ($D = 0.5$–1.0)	~300	~16.2	~17.8	—	—	$6\text{-}9 \times 10^{12}$	7.1	—	60

Table 8.A12. Kinetic parameters for *n*-hexane hydrogenolysis over metal catalysts

Catalyst	T (°C)	log (HC)	log (H$_2$)	E_a (kcal/mol)	Rate (molecules/cm^2/sec)	$-\log RP$	S_{C_5}	S_{C_4}	S_{C_3}	References
						at indicated T (°C)				
2.9% Ni/SiO$_2$ ($D = 0.15$)	330	18.69	19.29	—	~5 × 10^9	12.8	97% C$_1$ (spanning S_{C_5}–S_{C_4})		—	61
5% Ni/SiO$_2$-Al$_2$O$_3$ ($D = 0.15$)	330	18.69	19.29	—	~2 × 10^7	15.3	—	—	—	61
7% Ni/SiO$_2$-Al$_2$O$_3$ ($D = 0.15$)	350	18.69	19.29	—	~1 × 10^8	14.5	—	—	—	61
16% Ni/SiO$_2$-Al$_2$O$_3$ ($D = 0.15$)	275	18.69	19.29	—	~2 × 10^8	14.3	—	—	—	61
3% Ni/Li–Y ($D = 0.3$)	300	18.42	19.34	—	~6 × 10^{12}	9.5	~0.3	~0.2	~0.1	14
Ni–Cu, powders										
($X_{Cu} = 0.0$)	330	18.12	19.36	44	1 × 10^{14}	7.8	~0.4	~0.2	~0.1	62
($X_{Cu} = 0.05$)	330	18.12	19.36	44	3 × 10^{11}	10.5	~0.5	~0.2	~0.1	62
($X_{Cu} = 0.23$)	330	18.12	19.36	47	3 × 10^{11}	10.5	~0.7	~0.2	—	62
($X_{Cu} = 0.47$)	330	18.12	19.36	55	~5 × 10^{10}	11.2	~0.4	~0.2	~0.4	62
Cu/SiO$_2$	325	17.74	18.69	~18	~2 × 10^{14}	7.2	~0.1	~0.1	—	63
Rh/SiO$_2$	175	17.74	18.69	~22	~4 × 10^{14}	6.9	~0.5	~0.4	~0.1	63
Rh–Cu/SiO$_2$ ($X_{Cu} = 0.85$)	300	17.74	18.69	~18	~2 × 10^{13}	8.2	~0.3	~0.3	~0.2	63
Rh–Cu, films										
($X_{Cu} = 0.0$)	240	16.51	17.51	~12	~3 × 10^{11}	8.9	all C$_1$ (spanning S_{C_5}–S_{C_3})			47
($X_{Cu} = 0.03$)	280	16.51	17.51	~23	~1 × 10^{10}	10.3	67% C$_1$ (spanning S_{C_5}–S_{C_3})			47
($X_{Cu} = 0.58$)	270	16.51	17.51	~15	~9 × 10^9	10.4	all C$_1$ (spanning S_{C_5}–S_{C_3})			47
($X_{Cu} = 0.92$)	280	16.51	17.51	~20	~1 × 10^{10}	10.2	all C$_1$ (spanning S_{C_5}–S_{C_3})			47
Rh–Cu, films										
($X_{Cu} = 1.0$)	310	16.51	17.51	~10	~2 × 10^{10}	10.0	all C$_1$ (spanning S_{C_5}–S_{C_3})			47

Rh–Sn, films									
($X_{Sn} = 0.03$)	290	16.51	17.51	~20	~3 × 10^{10}	9.9	—	all C$_1$	—	47
($X_{Sn} = 0.36$)	320	16.51	17.51	~14	~2 × 10^{10}	10.0	—	all C$_1$	—	47
Rh–Au, film									
($X_{Au} = 0.87$)	300	16.51	17.51	~20	~1 × 10^{10}	10.3	—	all C$_1$	—	47
Rh–Ag, film									
($X_{Ag} = 0.26$)	280	16.51	17.51	~16	~1 × 10^{10}	10.3	—	all C$_1$	—	47
10% Pd/Al$_2$O$_3$ ($D = 0.3$)	350	16.99	19.39	—	~4 × 10^{11}	9.2	~0.6	~0.1	~0.1	64
10% Pd–Au/Al$_2$O$_3$, bimetallic ($D = 0.3$)										37
($X_{Au} = 0.5$)	350	16.99	19.39	—	~9 × 10^{10}	9.9	~0.7	~0.1	~0.1	64
($X_{Au} = 0.65$)	350	16.99	19.39	—	~2 × 10^{9}	11.5	~0.5	~0.2	~0.3	64
Ta, film	370	16.51	17.51	—	~5 × 10^{10}	9.6	—	—	—	37
Ta–Au, film									
($X_{Au} = 0.47$)	360	16.51	17.51	—	~4 × 10^{10}	9.7	—	—	—	37
Re, film	180	16.51	17.51	~0	~3 × 10^{12}	7.9	~0.4	~0.2	~0.2	37
Re–Au, film									
($X_{Au} = 0.21$)	340	16.51	17.51	—	~3 × 10^{11}	8.9	~0.1	~0.2	~0.3	37
Ir–Au, films									
($X_{Au} = 0.0$)	240	16.44	17.44	—	~6 × 10^{11}	8.5	~0.3	~0.3	~0.2	65
($X_{Au} = 0.86$)	360	16.44	17.44	—	~2 × 10^{10}	10.0	—	—	~0.1	65
($X_{Au} = 0.94$)	400	16.44	17.44	—	~1 × 10^{10}	10.3	—	—	~0.2	65
0.2% Pt/Al$_2$O$_3$ (10 Å)	300	18.3	19.35	—	9 × 10^{12}	9.2	~0.3	~0.3	~0.3	66
0.6% Pt/Al$_2$O$_3$ (10 Å)	300	18.3	19.35	—	7 × 10^{12}	9.3	~0.3	~0.3	~0.3	66
1.7% Pt/Al$_2$O$_3$ (16 Å)	300	18.3	19.35	—	4 × 10^{12}	9.5	~0.3	~0.3	~0.3	66
4.7% Pt/Al$_2$O$_3$ (16 Å)	300	18.3	19.35	—	4 × 10^{12}	9.5	~0.3	~0.3	~0.3	66
9.5% Pt/Al$_2$O$_3$ (55 Å)	300	18.3	19.35	—	4 × 10^{12}	9.5	~0.3	~0.3	~0.3	66
16% Pt/SiO$_2$ ($D = 0.3$)	300	18.14	19.37	~45	~1 × 10^{10}	12.0	~0.3	~0.3	~0.4	48

(continued)

Table 8.A12.—Continued

Catalyst	T (°C)	log (HC)	log (H₂)	E_a (kcal/mol)	Rate (molecules/cm²/sec)	$-\log RP$	S_{C_3}	S_{C_4}	S_{C_5}	References
						at indicated T (°C)				
16% Pt–Au/SiO₂, bimetallic ($D = 0.3$)										
($X_{Au} = 0.875$)	370	18.14	19.37	—	$\sim 6 \times 10^{8}$	13.2	~0.3	~0.3	~0.3	48
($X_{Au} = 0.92$)	320	18.14	19.37	—	$\sim 2 \times 10^{8}$	13.7	~0.4	—	~0.1	48
($X_{Au} = 0.99$)	390	18.14	19.37	—	$\sim 7 \times 10^{8}$	13.1	~0.2	~0.4	~0.3	48
Pt–Cu/SiO₂, bimetallic										
($X_{Cu} = 0.0$)	300	18.12	19.36	30–35	$\sim 6 \times 10^{11}$	10.2	~0.5	~0.3	~0.3	67
($X_{Cu} = 0.15$)	300	18.12	19.36	30–35	$\sim 4 \times 10^{11}$	10.3	~0.5	~0.3	~0.3	67
($X_{Cu} = 0.25$)	300	18.12	19.36	30–35	$\sim 2 \times 10^{11}$	10.6	—	—	—	67
($X_{Cu} = 0.85$)	300	18.12	19.36	30–35	$\sim 4 \times 10^{10}$	11.3	~0.5	~0.2	~0.4	67
($X_{Cu} = 1.0$)	300	18.12	19.36	30–35	$\sim 5 \times 10^{10}$	11.2	~0.4	~0.2	~0.4	67
Au/SiO₂	300	17.74	18.69	—	6×10^{14}	6.8	~0.4	~0.4	~0.3	63
Sn/SiO₂	300	17.74	18.69	~18	1×10^{12}	9.4	100% C₁			63
Pt–Sn/SiO₂										
($X_{Sn} = 0.05$)	340	17.74	18.69	—	9×10^{12}	8.6	~0.9	—	~0.1	63
Pt–Au/SiO₂										
($X_{Au} = 0.95$)	316	17.74	18.69	~28	7×10^{12}	8.7	~0.2	~0.3	~0.4	63
Pt–Au/SiO₂										
($X_{Au} = 0.5$)	320	17.74	18.69	~28	4×10^{13}	8.0	~0.2	~0.2	~0.4	63
Pt, film (15 Å)	273	17.47	18.47	≤20	4×10^{12}	8.7	~0.3	~0.3	~0.4	68
Pt, film (20 Å)	273	17.47	18.47	≤20	1.5×10^{12}	9.1	~0.3	~0.3	~0.4	68
Pt, film (36 Å)	273	17.47	18.47	≤20	6×10^{11}	9.6	~0.3	~0.4	~0.3	68
Pt, film (>100 Å)	273	17.47	18.47	≤20	4×10^{11}	9.7	~0.5	~0.2	~0.3	68

Table 8.A13. Kinetic parameters for 2-methylpentane hydrogenolysis over metal catalysts

Catalyst	T (°C)	log (HC)	log (H_2)	Rate (molecules/cm^2/sec)	$-\log RP$	S_{iC_5}	S_{nC_5}	S_{iC_4}	References
10% Pd/Al$_2$O$_3$ ($D = 0.3$)	350	17.11	19.39	$\sim 3 \times 10^{11}$	9.5	~ 0.3	~ 0.45	~ 0.05	64
10% Pd–Au/Al$_2$O$_3$, bimetallic									
($D = 0.3$)									
($X_{Au} = 0.5$)	350	17.11	19.39	$\sim 1 \times 10^{10}$	10.9	~ 0.4	~ 0.4	~ 0.1	64
($X_{Au} = 0.65$)	350	17.11	19.39	$\sim 6 \times 10^{9}$	11.2	~ 0.3	~ 0.4	~ 0.1	64
Pt/Al$_2$O$_3$	300	18.39	19.35	3×10^{13}	8.7	—	—	—	36
Pt, film (20 Å)	273	17.47	18.47	5×10^{12}	7.8	~ 0.3	~ 0.2	~ 0.3	68
Pt, film (38 Å)	273	17.47	18.47	2×10^{12}	8.2	~ 0.25	~ 0.15	~ 0.3	68
Pt, film (58 Å)	273	17.47	18.47	1.3×10^{12}	8.4	~ 0.2	~ 0.15	~ 0.3	68
Pt, film (>100 Å)	273	17.47	18.47	5×10^{11}	8.6	~ 0.1	~ 0.15	~ 0.4	68

Table 8.A14. Kinetic parameters for 3-methylpentane hydrogenolysis over metal catalysts

Catalyst	T (°C)	log (HC)	log (H₂)	Rate (molecules/cm²/sec)	$-\log RP$	S_{iC_5}	S_{nC_5}	S_{nC_4}	References
Re, film	295	16.51	17.51	$\sim2 \times 10^{11}$	9.0	—	—	—	37
Re–Au, film ($X_{Au} = 0.27$)	330	16.51	17.51	$\sim2 \times 10^{11}$	9.1	—	—	—	37
Ir–Au, film									
($X_{Au} = 0.0$)	400	16.44	17.44	$\sim4 \times 10^{11}$	8.7	—	—	—	69
($X_{Au} = 0.4$)	400	16.44	17.44	$\sim1 \times 10^{11}$	9.3	—	—	—	69
($X_{Au} = 0.8$)	400	16.44	17.44	$\sim1 \times 10^{11}$	9.3	—	—	—	69
($X_{Au} = 1.0$)	420	16.44	17.44	$\sim1 \times 10^{11}$	9.3	—	—	—	69
Pt/Al₂O₃	300	18.39	19.35	4×10^{13}	8.6	—	—	—	36
Pt, film (20 Å)	273	17.47	18.47	1.5×10^{13}	8.1	~0.65	~0.2	~0.2	68
Pt, film (58 Å)	273	17.47	18.47	9×10^{12}	8.4	~0.6	~0.15	~0.3	68
Pt, film (>100 Å)	273	17.47	18.47	6×10^{12}	8.5	~0.4	~0.1	~0.4	68

Table 8.A15. Kinetic parameters for cyclohexane hydrogenolysis over metal catalysts

Catalyst	T (°C)	log (HC)	log (H_2)	E_a (kcal/mol)	Rate (molecules/cm²/sec)	$-\log RP$	%C	References
Ni, powder	200–300	17.47	18.47 (D_2)	21	$\sim 4 \times 10^{12}$	8.7	—	70
1% Ru/SiO$_2$	316	18.62	19.31	—	6×10^{13}	8.7	≥80	3
Ru, powder ($D = 6 \times 10^{-4}$)	316	18.62	19.31	—	$\sim 1.5 \times 10^{14}$	8.3	≥90	71
5% Ru/SiO$_2$ ($D = 0.24$)	316	18.62	19.31	—	$\sim 8 \times 10^{13}$	8.6	≥90	71
1% Ru/SiO$_2$ ($D = 0.41$)	316	18.62	19.31	—	$\sim 6 \times 10^{13}$	8.7	≥90	71
0.1% Ru/SiO$_2$ ($D = 1.0$)	316	18.62	19.31	—	$\sim 2 \times 10^{13}$	9.2	≥90	71
Ru–Cu, powder ($X_{Cu} = 0.0$)	316	18.62	19.31	—	$\sim 5 \times 10^{13}$	8.8	≥90	8
($X_{Cu} = 0.005$)	316	18.62	19.31	—	$\sim 2 \times 10^{12}$	10.2	≥90	8
($X_{Cu} = 0.011$)	316	18.62	19.31	—	$\sim 4 \times 10^{11}$	10.9	≥90	8
1% Os/SiO$_2$	316	18.62	19.31	—	1.2×10^{14}	8.4	≥80	3
1.3% Os–Cu/SiO$_2$ ($X_{Cu} = 0.5$)	316	18.62	19.31	—	$\sim 1 \times 10^{13}$	9.5	≥80	3
1.6% Os–Cu/SiO$_2$ ($X_{Cu} = 0.67$)	316	18.62	19.31	—	$\sim 1 \times 10^{13}$	9.5	≥80	3
Pt(111)	300	17.69	18.51	~25	$2.2 \times 10^{12}\,n\text{-}C_6$ $1.7 \times 10^{12}\,C_1\text{-}C_3$	9.2 9.3	—	72
Pt(557)	300	17.69	18.51	~25	$1.0 \times 10^{12}\,n\text{-}C_6$ $1.0 \times 10^{12}\,C_1\text{-}C_3$	9.5 9.5	—	72
Pt(10, 8, 7)	300	17.69	18.51	~25	$1.2 \times 10^{12}\,n\text{-}C_6$ $1.0 \times 10^{12}\,C_1\text{-}C_3$	9.5 9.5	—	72
Pt(25, 10, 7)	300	17.69	18.51	~25	$1.0 \times 10^{12}\,n\text{-}C_6$ $7 \times 10^{11}\,C_1\text{-}C_3$	9.5 9.7	—	72

Table 8.A16. Kinetic parameters for methylcyclopentane ring opening over metal catalysts

Catalyst	T (°C)	log (HC)	log (H$_2$)	Rate (molecules/cm^2/sec)	$-\log RP$	S_{nC_6}	S_{2MP}	S_{3MP}	References
Ni–Cu, powders									
($X_{Cu} = 0.0$)	280	18.12	19.36	6×10^{14}	7.2	0.08	0.38	0.54	73
($X_{Cu} = 0.05$)	280	18.12	19.36	$\sim 2 \times 10^{12}$	9.7	0.06	0.56	0.38	73
1% Rh–Co/SiO$_2$									
($X_{Co} = 0.5$)	260	18.95	19.79	$\sim 6 \times 10^{13}$	9.0	0.20	0.52	0.28	74
Ir–Au, films									
($X_{Au} = 0.0$)	335	16.44	17.44	$\sim 5 \times 10^{10}$	9.6	—	—	—	69
($X_{Au} = 0.73$)	370	16.44	17.44	$\sim 6 \times 10^{10}$	9.5	—	—	—	69
0.3% Pt/Al$_2$O$_3$ ($D = 0.9$)	470	19.94	20.63	$\sim 4 \times 10^{14}$	9.2	0.67	0.20	0.13	75
0.7% Pt/SiO$_2$ ($D = 0.9$)	485	19.92	20.61$^{0.7}$	$\sim 9 \times 10^{14}$	8.8	0.38	0.40	0.22	76
0.65% Pt/Al$_2$O$_3$ ($D = 0.9$)	485	19.92	20.61$^{0.7}$	$\sim 2 \times 10^{15}$	8.5	—	—	—	76
10% Pt/Al$_2$O$_3$ ($D = 0.3$)	230	19.12	18.17	$\sim 2 \times 10^{11}$	11.7	—	0.61	0.39	77
10% Pt/Al$_2$O$_3$ ($D = 0.3$)	230	18.69	19.29	$\sim 1 \times 10^{12}$	10.5	—	0.77	0.23	77
Pt, film (20 Å)	273	17.47	18.47	2.4×10^{13}	7.9	0.31	0.51	0.18	68
Pt, film (40 Å)	273	17.47	18.47	1.4×10^{13}	8.2	0.18	0.60	0.22	68
Pt, film (>100 Å)	273	17.47	18.47	8×10^{12}	8.4	0.02	0.64	0.34	68

Table 8.A17. Kinetic parameters for methylcyclopentane hydrogenolysis over metal catalysts

Catalyst	T (°C)	log (HC)	log (H_2)	Rate (molecules/cm²/sec)	$-\log RP$	%C_1	References
Ni–Cu, powders							
($X_{Cu} = 0.0$)	280	18.12	19.36	$\sim 4 \times 10^{14}$	7.4	59	73
($X_{Cu} = 0.05$)	280	18.12	19.36	$\sim 1 \times 10^{12}$	9.9	59	73
1% Rh–Co/SiO$_2$							
($X_{Co} = 0.5$)	260	18.95	19.79	$\sim 3 \times 10^{12}$	10.3	—	74
Rh–Cu, films							
($X_{Cu} = 0.0$)	310	16.51	17.51	$\sim 2 \times 10^{11}$	9.0	75	47
($X_{Cu} = 0.03$)	350	16.51	17.51	$\sim 4 \times 10^{10}$	9.7	66	47
($X_{Cu} = 0.06$)	375	16.51	17.51	$\sim 4 \times 10^{10}$	9.7	70	47
($X_{Cu} = 0.95$)	300	16.51	17.51	$\sim 2 \times 10^{10}$	10.0	100	47
($X_{Cu} = 1.0$)	200	16.51	17.51	$\sim 4 \times 10^{9}$	10.7	—	47
Re–Au, films							
($X_{Au} = 0.0$)	295	16.51	17.51	$\sim 1 \times 10^{11}$	9.2	—	37
($X_{Au} = 0.22$)	340	16.51	17.51	$\sim 6 \times 10^{11}$	8.6	—	37
($X_{Au} = 0.47$)	375	16.51	17.51	$\sim 1 \times 10^{11}$	9.3	—	37
Ir–Au, films							
($X_{Au} = 0.0$)	335	16.44	17.44	$\sim 9 \times 10^{10}$	9.2	—	69
($X_{Au} = 0.73$)	370	16.44	17.44	$\sim 4 \times 10^{10}$	9.6	—	69
0.3% Pt/Al$_2$O$_3$							
($D = 0.9$)	470	19.94	20.63	$\sim 4 \times 10^{13}$	10.2	—	75
Pt, film (20 Å)	273	17.47	18.47	2×10^{12}	9.0	—	68
Pt, film (40 Å)	273	17.47	18.47	4×10^{11}	9.7	—	68
Pt, film (>100 Å)	273	17.47	18.47	4×10	9.7	—	68

Table 8.A18. Kinetic parameters for benzene hydrogenolysis over metal catalysts

Catalyst	T (°C)	log (HC)	log (H$_2$)	E_a (kcal/mol)	log A	Rate (molecules/cm^2/sec) at 227°C	$-$log RP	References
1% Ru/SiO$_2$	130–170	16–8 (pulse)	19.6	30	27.3	2×10^{14}	6.4	78
1% Ru/Al$_2$O$_3$	145–190	16–8 (pulse)	19.6	30	27.4	2×10^{14}	6.3	78
1% Tc/SiO$_2$	170–235	16–8 (pulse)	19.6	29	25.3	5×10^{12}	7.9	78
1% Tc/Al$_2$O$_3$	170–235	16–8 (pulse)	19.6	29	25.6	1×10^{13}	7.7	78
1% Re/SiO$_2$	205–235	16–8 (pulse)	19.6	32	26.2	2×10^{12}	8.4	78
k% Re/Al$_2$O$_3$	200–250	16–8 (pulse)	19.6	32	25.7	7×10^{11}	8.8	78

Table 8.A19. Kinetic parameters for n-heptane hydrogenolysis over metal catalysts

Catalyst	T (°C)	log (HC)	log (H$_2$)	E_a (kcal/mol)	log A	Rate (molecules/cm^2/sec)	$-\log RP$	S_{C_6}	S_{C_5}	S_{C_4}	References
							at 205°C				
Pd, powder	250–350	18.61	19.31	~28	~22.4	5×10^{9}	12.7	0.92	0.08	—	79
Rh, powder	100–125	18.61	19.31	~35	~31.1	1.5×10^{15}	7.2	0.82	0.10	0.06	79
Ru, powder	80–115	18.61	19.31	~35	~32.1	1.7×10^{16}	6.2	~0.5	~0.2	~0.25	79
Ir, powder	130–200	18.61	19.31	~30	~28.8	1.5×10^{15}	7.2	~0.3	~0.3	~0.3	79
Pt, powder	250–350	18.61	19.31	~27	23.0	5×10^{10}	11.7	~0.3	~0.2	~0.3	79
							at indicated T (°C)				
0.3% Pt/Al$_2$O$_3$ ($D = 0.9$)	470	19.94	20.63	—	—	$\sim 2 \times 10^{14}$	9.4	0.15	—	—	74
Pt/Al$_2$O$_3$	300	18.39	19.35	—	—	6×10^{12}	9.4	—	—	—	36
Pt(111)	300	17.69	19.19	—	—	7×10^{12}	8.6	~0.3	~0.3	~0.3	72
Pt(557)	300	17.69	19.19	—	—	6×10^{12}	8.7	~0.3	~0.3	~0.3	72
Pt(10, 8, 7)	300	17.69	19.19	—	—	1.2×10^{13}	8.4	~0.3	~0.3	~0.3	72
Pt(25, 10, 7)	300	17.69	19.19	—	—	2×10^{13}	8.2	~0.3	~0.3	~0.3	72
1% Pt/Al$_2$O$_3$ ($D = 0.8$)	420	14.99	—	21	~19	$\sim 2 \times 10^{12}$	6.5	all C$_1$			80

Table 8.A20. Kinetic parameters for toluene hydrodealkylation and hydrogenolysis over metal catalysts

Catalyst	T (°C)	log (HC)	log (H$_2$)	E_a (kcal/mol)	log A	Rate (molecules/cm^2/sec)	$-\log RP$	S_B	References
						at 380°C			
5% Ni/Al$_2$O$_3$	290	$18.39^{0.3}$	$18.95^{-0.2}$	31	25.2	5×10^{14}	7.5	0.94	81
1% Ru/Al$_2$O$_3$	380	$18.39^{0.2}$	$18.95^{1.0}$	33	24.4	3×10^{13}	8.8	0.82	81
Ru/Al$_2$O$_3$	~380	~18.3	~19.3	29	21.9	2×10^{12}	9.9	0.80	82
1% Rh/Al$_2$O$_3$	320	$18.39^{0.2}$	$18.95^{0.2}$	32	25.1	2×10^{14}	8.0	0.98	81
Rh/Al$_2$O$_3$	~380	~18.3	~19.3	30	24.1	1×10^{14}	8.1	0.96	82
1% Pd/Al$_2$O$_3$	450	$18.39^{0.5}$	$18.95^{-0.4}$	39	25.0	1.1×10^{12}	10.2	1.0	81
Pd/Al$_2$O$_3$	~380	~18.3	~19.3	37	24.6	2×10^{12}	9.9	0.99	82
10% Re/Al$_2$O$_3$	450	$18.39^{-0.2}$	$18.95^{1.7}$	33	22.7	5×10^{11}	10.6	0.85	81
2% Os/Al$_2$O$_3$	380	$18.39^{0.0}$	$18.95^{1.2}$	25	22.1	4×10^{13}	8.6	0.93	81
2% Ir/Al$_2$O$_3$	340	$18.39^{0.2}$	$18.95^{0.5}$	28	23.3	1×10^{14}	8.2	0.95	81
2% Pt/Al$_2$O$_3$	380	$18.39^{0.5}$	$18.95^{-0.1}$	34	24.0	5×10^{12}	9.5	0.96	81
Pt/Al$_2$O$_3$	~380	~18.3	~19.3	33	22.5	3×10^{11}	10.7	0.98	82
						at indicated T (°C)			
0.6% Rh/γ-Al$_2$O$_3$	475	18.35	19.35	13	17.7	9×10^{13}	8.3	0.9	83
0.6% Rh/α-Al$_2$O$_3$	475	18.35	19.35	11	16.8	4×10^{13}	8.6	0.9	83
0.6% Rh/glass	475	18.35	19.35	13	17.9	1.3×10^{14}	8.1	0.9	83
0.6% Rh/SiO$_2$–Al$_2$O$_3$	475	18.35	19.35	10	16.0	1.3×10^{13}	9.1	0.9	83
0.6% Rh/SiO$_2$	475	18.35	19.35	12	17.2	5×10^{13}	8.5	0.9	83
0.6% Rh/MgO	475	18.35	19.35	31	20.6	4×10^{11}	10.6	0.9	83
0.6% Rh/ZnO	475	18.35	19.35	38	24.1	1.3×10^{13}	9.1	0.9	83
0.6% Rh/α-Cr$_2$O$_3$	475	18.35	19.35	19	19.2	5×10^{13}	8.5	0.9	83
10% Pt/glass	76	18.29	19.36	8	18.7	5×10^{13}	8.4	1.0	84

Table 8.A21. Kinetic parameters for other hydrogenolysis reactions over metal catalysts

Hydrocarbon	Catalyst	T (°C)	log (HC)	log (H$_2$)	E_a (kcal/mol)	Rate (molecules/cm^2/sec)	$-\log RP$	Selectivity	References
⬚	Ni, film	85–200	17.21	18.29	8	—	—	10% nC_5 90% iC_5	85
	Pd, film	200	17.21	18.29	19	4×10^{10}	10.5	28% nC_5 72% iC_5	85
	Pt, film	50–150	17.21	18.29	17	—	—	33% nC_5 67% iC_5	85
	Pt/Al$_2$O$_3$	300	18.39	19.35	—	2×10^{13}	8.9		36
✕	3% Pt/La–Y (10 Å)	200	18.07	19.37	12	$\sim 3 \times 10^{12}$	9.4		51
	3% Pt/Ca–Y (10 Å)	200	18.07	19.37	—	$\sim 2 \times 10^{12}$	9.6		51
	3% Pt/Na–Y (10 Å)	200	18.07	19.37	18	$\sim 1.5 \times 10^{12}$	9.7		51
	Pt/Al$_2$O$_3$	300	18.39	19.35	—	3×10^{13}	8.7		36
	Pt/Al$_2$O$_3$	300	18.39	19.35	—	1×10^{13}	9.1		36
	Pt/Al$_2$O$_3$	300	18.39	19.35	—	2×10^{13}	8.9		36
	Pt/Al$_2$O$_3$	300	18.39	19.35	—	2×10^{13}	8.8		36
	Pt/Al$_2$O$_3$	300	18.39	19.35	—	5×10^{13}	8.5		36
	Pt/Al$_2$O$_3$	300	18.39	19.35	—	6×10^{12}	9.4		36
	Pt/Al$_2$O$_3$	300	18.39	19.35	—	2×10^{13}	8.8		36
	Pt/Al$_2$O$_3$	300	18.39	19.35	—	2×10^{13}	8.9		36
	Rh–Cu, films ($X_{Cu} = 0.0$)	260	16.51	17.51	—	$\sim 4 \times 10^{11}$	8.9	13% ring opening	47
	($X_{Cu} = 0.88$)	250	16.51	17.51	—	$\sim 5 \times 10^{10}$	9.8	8% ring opening	47
	10% Pt/glass	76	18.29	19.36	8	5×10^{13}	8.4	$S_B = 1.0$	84
	10% Pt/glass	76	18.29	19.36	17	3×10^{13}	8.7		84
	10% Pt/glass	76	18.29	19.36	28	4×10^{12}	9.4		84

Table 8.A22. Kinetic parameters for cracking reactions over nickel powder [13]

Hydrocarbon	T (°C)	log (HC)	E_a (kcal/mol)	log A	Rate (molecules/cm²/sec) at 300°C	$-\log RP$
△	315	18.21	38	25.5	1.3×10^{11}	11.2
C_2H_4	315	18.21	37	24.6	4×10^{10}	11.7
C_2H_2	315	18.21	40	25.6	3×10^{10}	11.9
	315	18.21	38	25.0	4×10^{10}	11.6
	315	18.21	44	27.2	3×10^{10}	11.6
⬡	315	18.21	57	30.5	8×10^{8}	13.2

Table 8.B1. Kinetic parameters for ethylene hydrogenation over metal catalysts

Catalyst	T (°C)	log (HC)	log (H$_2$)	E_a (kcal/mol)	log A	Rate (molecules/cm^2/sec)	$-$log RP	References
						at 0°C		
Cr, film	0	18.12	18.12	—	—	~3 × 10^{13}	8.7	86
Fe, film	0	18.12	18.12	—	—	~6 × 10^{14}	7.4	86
Ni, film	0	18.12	18.12	—	—	~2 × 10^{15}	6.9	86
Ni, film	0–72	17.91	17.91	10.5	22.5	1.4 × 10^{14}	7.8	87
Ni–Cu, films								
(X_{Cu} = 0.18)	−12 to −25	17.91	17.91	8.6	22.0	1.4 × 10^{15}	6.8	87
(X_{Cu} = 0.6)	−11 to −23	17.91	17.91	10.5	23.2	7 × 10^{14}	7.1	87
(X_{Cu} = 0.98)	20–69	17.91	17.91	9.8	22.3	3 × 10^{14}	7.4	87
Tc, film	0	18.12	18.12	—	—	~3 × 10^{13}	8.7	86
Rh, film	0	18.12	18.12	—	—	~3 × 10^{17}	4.7	86
Pd, film	0	18.12	18.12	—	—	~3 × 10^{16}	5.7	86
Pd, foil	0	17.51	17.51	—	—	4 × 10^{15}	6.0	88
Pd, foil	0	17.51	17.51	—	—	8 × 10^{14}	6.7	88
Ta, film	0	18.12	18.12	—	—	~3 × 10^{13}	8.7	86
W, film	0	18.12	18.12	—	—	~3 × 10^{13}	8.7	86
Pt, film	0	18.12	18.12	—	—	~1 × 10^{16}	6.2	86
						at 25°C		
Ni(100)	25	17.17^0	18.17	—	—	≤2 × 10^{13}	≤7.9	89
Ni(110)	25	17.17^0	18.17	—	—	3 × 10^{14}	6.8	89
Ni(111)	25	17.17^0	18.17	—	—	1 × 10^{15}	6.3	89
Ni(111) crystallites	25	17.17^0	18.17	—	—	1.5 × 10^{15}	6.1	89
0.05% Pt/SiO$_2$	−70 to −100	17.87	18.69	9	22.0	3 × 10^{15}	6.5	90
0.05% Pt/SiO$_2$ + Al$_2$O$_3$	−70 to −100	17.87	18.69	9	22.2	4 × 10^{15}	6.3	90
0.05% Pt/SiO$_2$, preoxidized	−70 to −100	18.87	18.69	9	22.4	7 × 10^{15}	6.1	90

(continued)

Table 8.B1.—Continued

Catalyst	T (°C)	log (HC)	log (H_2)	E_a (kcal/mol)	log A	Rate (molecules/cm²/sec) at indicated T (°C)	$-$log RP	References
Ni–Cu, film ($X_{Cu} = 0.03$)	-40	$19.21^{0.0}$	$19.21^{1.0}$	—	—	9×10^{11}	11.3	91
Ni(100)	90	18.35	19.35	~4	~16	4×10^{13}	8.8	92
Ni(110)	90	18.35	19.35	~6	~17	2×10^{13}	9.1	92
Ni(111)	90	18.35	19.35	~8	~18	3×10^{13}	8.9	92
Ni(321)	90	18.35	19.35	~5	~17	9×10^{13}	8.5	92
Ni, ribbon	200	18.09	19.37	~12	~18	$\sim6 \times 10^{12}$	9.4	93
Ni, ribbon	200	18.09 (+1.3 ppm O_2)	19.37			$\sim9 \times 10^{14}$	7.2	93
Ni, ribbon	310	18.09 (+22.6 ppm O_2)	19.37	—	—	$\sim2 \times 10^{15}$	6.9	94
Ni–Au, films								
($X_{Au} = 0.16$)	180	$17.91^{1.0}$	$17.91^{1.0}$	4.2	16.9	8×10^{14}	7.1	87
($X_{Au} = 0.55$)	180	$17.91^{1.0}$	$17.91^{1.0}$	3.6	16.4	5×10^{14}	7.3	87
($X_{Au} = 1.0$)	486	$17.91^{1.0}$	$17.91^{1.0}$	—	—	6×10^{12}	9.2	87
Ni–Pd, films								
($X_{Pd} = 0.0$)	-100	17.68	18.13			$\sim4 \times 10^{12}$	9.2	95
($X_{Pd} = 0.2$)	-100	17.68	18.13			$\sim3 \times 10^{13}$	8.3	95
($X_{Pd} = 0.5$)	-100	17.68	18.13			$\sim5 \times 10^{13}$	8.1	95
($X_{Pd} = 0.75$)	-100	17.68	18.13			$\sim4 \times 10^{13}$	8.2	95
($X_{Pd} = 1.0$)	-100	17.68	18.13			$\sim2 \times 10^{13}$	8.5	95
Cu, film	150	$17.91^{1.0}$	$17.91^{1.0}$	12	20.4	2×10^{14}	7.6	87
Pt, wire	120	17.93	18.57	10	23.1	4×10^{17}	4.5	96
0.05% Pt/SiO₂ + Al₂O₃	100	17.87	18.69	17–20	—	2×10^{16}	5.7	97
0.05% Pt/SiO₂	100	17.87	18.69	17–20	—	1×10^{15}	7.0	97
0.05% Pt/SiO₂	40	17.87	18.69	16	24.3	1.5×10^{13}	8.8	98
0.54% Pt/Na-Y	-84	17.87	18.69	—	—	8×10^{12}	9.1	56
0.59% Pt/Ca-Y	-84	17.87	18.69	—	—	4×10^{13}	8.4	56
0.6% Pt/Mg-Y	-84	17.87	18.69	—	—	4×10^{13}	8.4	56
0.5% Pt/La-Y	-84	17.87	18.69	—	—	3×10^{13}	8.5	56
0.53% Pt/SiO₂	-84	17.87	18.69	—	—	1×10^{13}	9.0	56
1.2–12% Pt/SiO₂	-80	17.79	19.38	10–12	24–26	2–8×10^{12}	9.0	99

Table 8.B2. Kinetic parameters for hydrogenation reactions of terminal olefins

Hydrocarbon	Catalyst	T (°C)	log (HC)	log (H$_2$)	E_a (kcal/mol)	log A	Rate (molecules/cm^2/sec)	$-\log RP$	References
C$_3$H$_6$	7–81% Pt/SiO$_2$	−57	18.09	19.37	~10	~24.2	6–15 × 10^{13}	7.9	32
C$_4$H$_8$	Pt(223)	25	17.36	18.36	—	—	1.3 × 10^{16}	5.2	100
C$_4$H$_8$	3.2% Ni/mordenite ($D = 0.2$)	80	18.91	19.22	~12	~22.2	~7 × 10^{14}	8.2	101
C$_5$H$_{10}$	10^{-2}% Au/SiO$_2$	120	16.62	19.39	—	—	4 × 10^{11}	8.8	102
C$_5$H$_{10}$	7 × 10^{-2}% Au/SiO$_2$	120	16.62	19.39	—	—	8 × 10^{10}	9.5	102
C$_5$H$_{10}$	0.25% Au/SiO$_2$	120	16.62	19.39	—	—	2 × 10^{10}	10.2	102
C$_5$H$_{10}$	1% Au/SiO$_2$	120	16.62	19.39	—	—	8 × 10^{8}	11.6	102
C$_5$H$_{10}$	0.1–1% Au/Al$_2$O$_3$	120	16.62	19.39	—	—	4 × 10^{8}	11.9	102

447

Table 8.B3. Kinetic parameters for benzene hydrogenation over metal catalysts

Catalyst	T (°C)	log (HC)	log (H$_2$)	E_a (kcal/mol)	log A	Rate (molecules/cm^2/sec) at 25°C	$-$log RP	References
4.3–79% Ni/SiO$_2$	60	18.58	19.31	14	22.1	4–9 $\times$ 10^{11}	10.4	103
Ni–Al, powders (~50% by wt Al)								
(X$_{Al}$ = 0.44)	27–38	18.49	19.33	13	21.6	1 $\times$ 10^{12}	10.3	104
(X$_{Al}$ = 0.58)	27–38	18.49	19.33	13	21.3	7 $\times$ 10^{11}	10.5	104
(X$_{Al}$ = 0.79)	27–38	18.49	19.33	12	20.4	5 $\times$ 10^{11}	10.6	104
4.3–52% Ni/SiO$_2$ (15–51 Å)	25	18.37	19.29	—	—	1–1.7 $\times$ 10^{13}	9.0	105
10% Ni/SiO$_2$	30	18.36$^{0.2}$	19.29	—	—	~2 $\times$ 10^{13}	8.9	106
Ni–Ti–Al, powders (~50% by wt Al)								
(X$_{Ti}$ = 0.02)	27–38	17.67	19.38	12	20.8	1 $\times$ 10^{12}	9.5	107
(X$_{Ti}$ = 0.07)	27–38	17.67	19.38	13	21.6	1 $\times$ 10^{12}	9.5	107
(X$_{Ti}$ = 0.17)	27–38	17.67	19.38	13	21.5	1 $\times$ 10^{12}	9.5	107
Ni–V–Al, powders (~50% by wt Al)								
(X$_V$ = 0.03)	27–38	17.67	19.38	12	20.9	1 $\times$ 10^{12}	9.5	107
(X$_V$ = 0.14)	27–38	17.67	19.38	13	21.5	1 $\times$ 10^{12}	9.5	107
Ni–Cr–Al, powders (~50% by wt Al)								
(X$_{Cr}$ = 0.03)	27–38	17.67	19.38	13	21.6	1 $\times$ 10^{12}	9.5	107
(X$_{Cr}$ = 0.11)	27–38	17.67	19.38	12	20.7	9 $\times$ 10^{11}	9.6	107
Ni–Fe–Al, powders (~50% by wt Al)								
(X$_{Fe}$ = 0.1)	27–38	17.67	19.38	13	21.5	1 $\times$ 10^{12}	9.5	107
(X$_{Fe}$ = 0.41)	27–38	17.67	19.38	12	20.3	4 $\times$ 10^{11}	9.9	107
(X$_{Fe}$ = 0.61)	27–38	17.67	19.38	12	20.1	2 $\times$ 10^{11}	10.2	107
Ni–Co–Al, powders (~50% by wt Al)								
(X$_{Co}$ = 0.10)	27–38	17.67	19.38	12	20.8	1 $\times$ 10^{12}	9.5	107
(X$_{Co}$ = 0.40)	27–38	17.67	19.38	12	20.6	7 $\times$ 10^{11}	9.7	107

(continued)

Ni–Mo–Al, powders (~50% by wt Al)								
(X_{Mo} = 0.02)	27–38	17.67	19.38	11	20.1	1×10^{12}	9.5	107
(X_{Mo} = 0.37)	27–38	17.67	19.38	12	20.9	1×10^{12}	9.5	107
5.4–75% Ni/Al₂O₃	0–70	17.47	18.61	10	18.6	3×10^{11}	9.8	108
0.6% Re/Al₂O₃	30	17.78	19.38	—	—	$\sim 1 \times 10^{13}$	8.6	44
Os/Al₂O₃	30	17.78	19.38	—	—	$\sim 1 \times 10^{15}$	6.6	44
0.6% Ir/Al₂O₃	30	17.78	19.38	—	—	$\sim 4 \times 10^{13}$	8.0	44
0.6% Ir–Re/Al₂O₃, bimetallic								44
(X_{Re} = 0.1)	30	17.78	19.38	—	—	$\sim 6 \times 10^{13}$	7.9	44
(X_{Re} = 0.45)	30	17.78	19.38	—	—	$\sim 6 \times 10^{13}$	7.9	44
(X_{Re} = 0.86)	30	17.78	19.38	—	—	$\sim 4 \times 10^{13}$	8.0	44
Pt/Na–Y (20 Å)	25	$18.36^{0.0}$	19.35	11	21.9	8×10^{13}	8.3	109
Pt/Na–Y (10 Å)	25	$18.36^{0.0}$	19.35	11	21.6	4×10^{13}	8.6	109
3% Pt/Al₂O₃	25	$18.36^{0.0}$	19.35	—	—	5×10^{13}	8.5	109
10% Pt/glass	63	18.29	19.36	18	24.3	1.5×10^{11}	10.6	84
0.2–10% Pt/Al₂O₃	50	18.26	19.36	10	21.0	8×10^{13}	8.2	110
2.0–2.8% Pt/SiO₂, fired at 100–500°C	25	17.87	19.37	—	—	$3\text{–}5 \times 10^{13}$	8.1	111
2.0% Pt/SiO₂, fired at 600°C	25	17.87	19.37	—	—	$\leq 5 \times 10^{11}$	≤10	111
							at 100°C	
10% Co/SiO₂	70–175	17.99	19.09	14	16.5	1.1×10^{13}	8.8	112
Ni/SiO₂	70–140	18.99	20.14	14	22.6	3×10^{14}	8.4	113
Ni, powder	100	$18.44^{0.0}$	19.34	—	—	4×10^{13}	8.6	114
10–85% Ni/Cr₂O₃–Al₂O₃	100	$18.44^{0.0}$	19.34	—	—	4×10^{13}	8.6	114
7–75% Ni/Al₂O₃	100	$18.44^{0.0}$	19.34	—	—	4×10^{13}	8.6	114
9–20% Ni/SiO₂	100	$18.44^{0.0}$	19.34	—	—	4×10^{13}	8.6	114
3–70% Ni/Cr₂O₃	100	$18.44^{0.0}$	19.34	—	—	5×10^{13}	8.5	114
11% Ni/SiO₂	60	18.36	$19.29^{0.6}$	13	21.7	3×10^{14}	7.7	115

Table 8.B3.—Continued

Catalyst	T (°C)	log (HC)	log (H_2)	E_a (kcal/mol)	log A	Rate (molecules/cm²/sec) at 100°C	$-\log RP$	References
10% Ni/SiO$_2$	70–175	17.99	19.09	6	21.4	1.5×10^{13}	8.7	112
1–5% Ni/SiO$_2$	80–160	17.99$^{0.2}$	19.09$^{1.5}$	6	16.7	1.6×10^{13}	8.7	116
10% Ni/SiO$_2$	80–160	17.99$^{0.2}$	19.09$^{1.5}$	14	21.2	1.1×10^{13}	8.9	116
1% Ni/SiO$_2$–Al$_2$O$_3$	80–160	17.99$^{0.2}$	19.09$^{1.5}$	—	—	$\leq 10^{11}$	—	116
5% Ni/SiO$_2$–Al$_2$O$_3$	80–160	17.99$^{0.2}$	19.09$^{1.5}$	8	17.1	3×10^{12}	9.4	116
10% Ni/SiO$_2$–Al$_2$O$_3$	80–160	17.99$^{0.2}$	19.09$^{1.5}$	14	21.0	7×10^{12}	9.0	116
1% Tc/SiO$_2$	110–160	16.8	19.6	11	18.4	9×10^{11}	8.7	78
1% Tc/Al$_2$O$_3$	110–170	16.8 (pulse)	19.6	9	17.7	5×10^{12}	8.0	78
1% Ru/SiO$_2$	95–130	16.8 (pulse)	19.6	7	18.2	1×10^{14}	6.7	78
1% Ru/Al$_2$O$_3$	95–145	16.8 (pulse)	19.6	8	18.7	1×10^{14}	6.7	78
Pd/Al$_2$O$_3$	70–140	18.99	20.14	14	22.8	4×10^{13}	9.2	113
Pd/Al$_2$O$_3$	100	17.78	19.38	—	—	4×10^{13}	8.0	44
0.5–2.5% Pd/SiO$_2$, (reduced at 300°C)	100	17.67	19.38	—	—	2×10^{13}	8.2	117
0.5–2.5% Pd/SiO$_2$, (reduced at 450°C)	100	17.67	19.38	—	—	7×10^{12}	8.6	117
1% Pd/SiO$_2$	105–150	16.8 (pulse)	19.6	9.5	18.2	4×10^{12}	8.1	78
1% Pd/Al$_2$O$_3$	105–160	16.8 (pulse)	19.6	9.5	17.5	2×10^{12}	8.4	78
Pd, film	80–250	16.33	17.54 (D$_2$)	5.1	17.4	3×10^{14}	5.7	118
Pd–Au, films ($X_{Au} = 0.17$)	80–250	16.33	17.54 (D$_2$)	4.3	16.2	5×10^{13}	6.5	118
($X_{Au} = 0.38$)	80–250	16.33	17.54 (D$_2$)	6.8	17.2	2×10^{13}	6.9	118

Catalyst	Temp. range							Ref.
1% Re/SiO₂	130–190	16.8 (pulse)	19.6	11	17.9	3×10^{11}	9.2	78
1% Re/Al₂O₃	140–185	16.8 (pulse)	19.6	11	17.5	1×10^{11}	9.7	78
Os/Al₂O₃	100	18.09	19.37	—	—	1.7×10^{14}	7.7	35
Ir/Al₂O₃	100	18.09	19.37	—	—	8×10^{14}	7.0	35
0.25% Ir/Al₂O₃	100	17.78	19.38	—	—	5×10^{14}	6.9	44
0.25% Ir + Cu/Al₂O₃, bimetallic								
(X_{Cu} = 0.2)	100	17.78	19.38	—	—	3×10^{14}	7.2	44
(X_{Cu} = 0.5)	100	17.78	19.38	—	—	1×10^{14}	7.6	44
(X_{Cu} = 0.8)	100	17.78	19.38	—	—	5×10^{13}	7.9	44
Pt/Al₂O₃	70–140	18.99	20.14	13	22.4	7×10^{14}	8.0	113
Pt–Ir/Al₂O₃ (X_{Ir} = 0.5)	105	18.55[0]	19.30	—	—	7×10^{13}	8.5	119
0.5% Pt/Al₂O₃	40–90	18.26	19.23	10	20.2	2×10^{14}	7.8	120
0.4% Pt–0.5% Sn/Al₂O₃ (coimpreg.)	40–90	18.26	19.23	10	19.2	2×10^{13}	8.8	120
0.4% Pt–0.7% Sn/Al₂O₃ (Pt(II)Cl₂(SnCl₃)₂)⁻²	40–90	18.26	19.23	10	20.0	1.5×10^{14}	7.9	120
Pt/Al₂O₃	100	18.09	19.37	—	—	5×10^{14}	7.2	35
Pt–Re/Al₂O₃, bimetallic								
(X_{Re} = 0.0)	100	18.09	19.37	—	—	5×10^{13}	8.3	35
(X_{Re} = 0.2)	100	18.09	19.37	—	—	6×10^{13}	8.2	35
(X_{Re} = 0.5)	100	18.09	19.37	—	—	3×10^{13}	8.5	35
(X_{Re} = 0.6)	100	18.09	19.37	—	—	7×10^{13}	8.1	35
10% Pt/SiO₂	70–175	17.99	19.09	6	21.4	1.5×10^{13}	8.7	112
Pt/Al₂O₃	100	17.78	19.38	—	—	5×10^{14}	6.9	44
1% Pt/SiO₂	100–150	16.8 (pulse)	19.6	7	18.2	1×10^{14}	6.7	78
1% Pt/Al₂O₃	110–165	16.8 (pulse)	19.6	8	18.3	4×10^{13}	7.1	78

(continued)

Table 8.B3.—Continued

Catalyst	T (°C)	log (HC)	log (H$_2$)	E_a (kcal/mol)	log A	Rate (molecules/cm^2/sec) at indicated T (°C)	$-$log RP	References
3% Ni/Na–Y	150	18.16	19.37	—	—	5×10^{12}	9.3	14
3% Ni/Li–Y	150	18.16	19.37	—	—	5×10^{12}	9.3	14
3% Ni/Ca–Y	150	18.16	19.37	—	—	4×10^{12}	9.4	14
3% Ni/Mg–Y	150	18.16	19.37	—	—	4×10^{12}	9.4	14
5% Ni/SiO$_2$	150	18.16	19.37	—	—	1.6×10^{14}	7.8	14
5% Ni/SiO$_2$–Al$_2$O$_3$	150	18.16	19.37	—	—	1.3×10^{13}	8.9	14
Ni/Cr$_2$O$_3$	170	18.09	19.35	11	19.2	6×10^{13}	8.1	121
8% Ni/SiO$_2$–Al$_2$O$_3$	200	~18 (pulse)	19.3	—	—	8×10^{14}	—	16
+ 0.44% Na$^+$	200	~18 (pulse)	19.3	—	—	1.1×10^{15}	—	16
+ 2.4% Na$^+$	200	~18 (pulse)	19.3	—	—	3×10^{14}	—	16
Ni, film	150	17.27	19.02$^{1-2}$	12	20.4	2×10^{14}	6.9	122
Ni–Cu, film (X_{cu} = 0.15–1.0)	150	17.27	19.02$^{1-2}$	25	26.3	4×10^{13}	7.6	122
Ni(100)	170	17.17	18.17	—	—	1.5×10^{13}	7.8	89
Ni(110)	170	17.17	18.17	—	—	2.5×10^{13}	7.6	89
Ni(111)	170	17.17	18.17	—	—	1.6×10^{13}	7.8	89
Ni(111) crystallites	170	17.17	19.36	11	18.6	1.5×10^{13}	7.8	89
9.1% Ru/Na–A	80	18.62	19.31	—	—	3×10^{12}	10.0	123
14–24% Ru/Na–X	80	18.62	19.31	—	—	2×10^{13}	9.2	123
4–5% Ru/Na–Y	80	18.62	19.31	—	—	4×10^{13}	8.9	123

Catalyst	Temp						Ref
0.9% Ru/Na-L	80	18.62	—	—	1×10^{14}	8.5	123
4.3% Ru/Na-L	80	18.62	—	—	1.4×10^{13}	9.3	123
2.4% Ru/mordenite	80	18.62	—	—	5×10^{13}	8.8	123
0.25–8% Pd/Al$_2$O$_3$	140	18.26	—	—	$1–1.4 \times 10^{14}$	8.0	58
0.7–1.3% Pd/SiO$_2$	140	18.26	—	—	$4–6 \times 10^{13}$	8.3	58
0.2–1% Pd/Al$_2$O$_3$	140	18.26	—	—	8×10^{12}	9.2	124
2.1% Pd/Al$_2$O$_3$	150	18.07$^{0.0}$	—	—	3×10^{13}	8.0	125
2.2% Pd/Al$_2$O$_3$	150	17.85	—	—	$1.5–5 \times 10^{15}$	5.6	126
0.1% Pt/Al$_2$O$_3$ ($D = 0.9$)	85	18.87	—	—	$\sim 1 \times 10^{14}$	8.7	127
+ 0.3% SO$_4^{2-}$	85	18.87	—	—	$\sim 6 \times 10^{13}$	8.9	127
0.1% Pt/Al$_2$O$_3$ ($D = 0.9$) + 0.6% SO$_4^{2-}$	85	18.87	—	—	$\sim 2 \times 10^{12}$	10.4	127
0.7% Pt/SiO$_2$–Al$_2$O$_3$ ($D = 0.1–0.5$)	120	18.44	—	—	4×10^{15}	6.6	128
0.7% Pt/Al$_2$O$_3$ ($D = 0.2–0.8$)	120	18.44	—	—	5×10^{15}	6.6	128
0.4% Pt/Al$_2$O$_3$ ($D = 0.8$)	150	18.55$^{0.0}$	—	—	$\sim 9 \times 10^{14}$	7.4	129
0.6% Pt/Al$_2$O$_3$ ($D = 0.9$)	150	18.55	—	—	$\sim 1 \times 10^{14}$	8.4	130
+ 0.2% Na$^+$	150	18.55	—	—	$\sim 7 \times 10^{13}$	8.6	130
+ 0.6% Na$^+$	150	18.55	—	—	$\sim 5 \times 10^{13}$	8.7	130
0.2–1.0% Pt/Al$_2$O$_3$	50	18.26	—	—	5×10^{13}	8.4	124
1% Pt-Pd/Al$_2$O$_3$, bimetallic ($X_{Pd} = 0.2–0.8$)	50	18.26	—	—	1.5×10^{13}	8.9	124
3.9% Pt/SiO$_2$	20	18.21	—	—	3×10^{13}	8.6	131
5% Pt–5% Mo/SiO$_2$	20	18.21	—	—	2×10^{14} cm^{-2} Pt	7.8	131

Table 8.B4. Kinetic parameters for other hydrogenation reactions catalyzed by metals

Hydrocarbon	Catalyst	T (°C)	log (HC)	log (H$_2$)	E_a (kcal/mol)	log A	Rate (molecules/cm^2/sec)	$-\log RP$	Selectivity	References
H$_2$C=C=CH$_2$	Co/SiO$_2$	100	18.21	18.51$^{1.0}$	9	~20	~1×10^{15}	7.2	~75% C$_3$H$_6$	132
	Ni/SiO$_2$	100	18.21	18.51$^{1.0}$	11	~22	~3×10^{15}	6.7	~100% C$_3$H$_6$	132
	Ru/SiO$_2$	100	18.21	18.51$^{1.0}$	5	~18	~1×10^{15}	7.2	~75% C$_3$H$_6$	132
	Rh/SiO$_2$	100	18.21	18.51$^{1.0}$	8	~21	~2×10^{16}	5.9	~75% C$_3$H$_6$	132
	Pd/SiO$_2$	100	18.21	18.51$^{1.0}$	8	~22	~1×10^{17}	5.2	~75% C$_3$H$_6$	132
	Ir/SiO$_2$	100	18.21	18.51$^{1.0}$	9	~20	~3×10^{14}	7.7	~75% C$_3$H$_6$	132
	Pt/SiO$_2$	100	18.21	18.51$^{1.0}$	16	~25	~1×10^{16}	6.2	~75% C$_3$H$_6$	132
(structure)	3.2% Ni/mordenite ($D = 0.2$)	80	18.91	19.22	—	—	~2×10^{14}	8.5		101
(structure)	3.2% Ni/mordenite ($D = 0.2$)	80	18.91	19.22	—	—	~3×10^{14}	8.3		101
(structure)	4.3–79% Ni/SiO$_2$	100	18.79	19.26	1	—	7–15×10^{13}	8.5	33–45% 1-C$_4$H$_8$ 67–55% trans-2-C$_4$H$_8$	103
H$_3$CC≡CCH$_3$	Pd–Au, powders— ($X_{Au} = 0.0$)	−100–10	17.46	17.86	13	26.7	1.7×10^{17}	4.2	at 25°C	133
	($X_{Au} = 0.25$)	60–120	17.46	17.86	14	24.4	1.6×10^{14}	7.2		133

	Temp (°C)								Ref.
$(X_{Au} = 0.40)$	110–170	17.46	17.86	13	23.0	3×10^{13}	7.9		133
$(X_{Au} = 0.85)$	200–260	17.46	17.86	14	22.0	6×10^{11}	9.6		133
$(X_{Au} = 0.94)$	300–350	17.46	17.86	14	20.8	4×10^{10}	10.8		133
0.1–2% Pt/SiO_2	20	$\sim17.39^{0.0}$	19.39	—	—	1.5×10^{16}	5.6		134
0.3–1.2% Pt/Al_2O_3	20	$\sim17.39^{0.0}$	19.39	—	—	1.0×10^{16}	5.8		134
0.6–3.8% Pd/SiO_2	0	$17.39^{0.0}$	$18.41^{0.5}$	9.3	22.4	1.1×10^{15}	6.2		135
0.5% Pt/Ca–Y	303	20.07	20.84	—	—	$\sim1 \times 10^{16}$	7.9		136
0.4–3.7% Pt/SiO_2	22	$17.61^{0.0}$	$18.39^{0.5-0.8}$	8	21.5	4×10^{15}	5.9		137
Pt(223)	25	17.36	18.36	5	19.3	4×10^{15}	5.6		100
4–39% Ni/SiO_2	90	18.55	19.32	15	21.8	$3-6 \times 10^{12}$	9.6		103
51% Ni/SiO_2	90	18.55	19.32	—	—	9×10^{12}	9.4		103
79% Ni/SiO_2	90	18.55	19.32	—	—	1.1×10^{13}	9.3		103
10% Pt/glass	63	18.29	19.36	17	23.2	1.5×10^{12}	9.9		84
Pd–Au, films									
$(X_{Au} = 0.0)$	100	16.40	17.54	—	—	1.8×10^{14}	5.9	68% trans	118
$(X_{Au} = 0.07)$	100	16.40	17.54	—	—	1.5×10^{14}	6.0	59% trans	118
$(X_{Au} = 0.18)$	100	16.40	17.54	—	—	8×10^{13}	6.3	60% trans	118
$(X_{Au} = 0.37)$	100	16.40	17.54	—	—	3×10^{13}	6.7	63% trans	118
10% Pt/glass	76	18.29	19.36	28	29.0	4×10^{11}	10.4		84
10% Pt/glass	76	18.29	19.36	31	29.4	6×10^{10}	11.3		84

Table 8.C1. Kinetic parameters for cyclohexane dehydrogenation to benzene over metal catalysts

Catalyst	$T(C°)$	$\log$ (HC)	$\log$ (H$_2$)	E_a (kcal/mol)	$\log A$	Rate (molecules/cm^2/sec)	$-\log RP$	References
Ni–Cu, powders								
($X_{Cu} = 0.0$)	316	18.62	19.31	—	—	$\sim 2 \times 10^{14}$	8.2	5
($X_{Cu} = 0.2$–0.9)	316	18.62	19.31	—	—	$\sim 4 \times 10^{14}$	7.9	5
($X_{Cu} = 1.0$)	316	18.62	19.31	—	—	$\sim 4 \times 10^{12}$	8.9	5
Ni, powder	300	17.47	18.47 (D$_2$)	21	20.7	6×10^{12}	8.5	70
1% Cu/SiO$_2$	316	18.62	19.31	—	—	$\sim 1 \times 10^{12}$	10.5	3
0.81% Pd/Al$_2$O$_3$	225	19.05	19.12	16	21.3	$\sim 2 \times 10^{14}$	8.6	138
0.2–1% Pd/Al$_2$O$_3$	190	18.21	18.46	16	20.8	2×10^{13}	8.7	139
1% Ru/SiO$_2$	316	18.62	19.31	—	—	$\sim 2 \times 10^{14}$	8.2	3
Ru, powder	316	18.62	19.31	—	—	$\sim 5 \times 10^{13}$	8.8	71
5% Ru/SiO$_2$	316	18.62	19.31	—	—	$\sim 7 \times 10^{13}$	8.6	71
0.1% Ru/SiO$_2$	316	18.62	19.31	—	—	$\sim 1.1 \times 10^{14}$	8.4	71
1% Ru–0.66% Cu/SiO$_2$	316	18.62	19.31	—	—	$\sim 2 \times 10^{14}$	8.2	71
Ru–Cu, powders								
($X_{Cu} = 0.0$)	316	18.62	19.31	—	—	$\sim 2 \times 10^{13}$	9.2	8
($X_{Cu} = 0.005$)	316	18.62	19.31	—	—	$\sim 9 \times 10^{12}$	9.5	8
($X_{Cu} = 0.011$)	316	18.62	19.31	—	—	$\sim 1 \times 10^{13}$	9.5	8
1% Os/SiO$_2$	316	18.62	19.31	—	—	$\sim 2 \times 10^{14}$	8.2	3
1% Os–0.33% Cu/SiO$_2$	316	18.62	19.31	—	—	$\sim 2 \times 10^{14}$	8.2	3
0.06% Pt/Al$_2$O$_3$ ($D = 0.8$)								
initial	150	19.39	0.0	19	22.2	$\sim 3 \times 10^{12}$	9.6	140
steady state	150	19.39	0.0	—	—	$\sim 2 \times 10^{12}$	9.8	140
0.55% Pt/Al$_2$O$_3$	350	18.79	19.27	—	—	4×10^{15}	7.0	141
0.1–2% Pt/Al$_2$O$_3$	300	18.65	19.30	—	—	4×10^{15}	6.8	142
2% Pt/Al$_2$O$_3$ ($D = 0.8$)	315	18.61	19.31	—	—	$\sim 3 \times 10^{15}$	7.0	143
Pt/Al$_2$O$_3$	250	18.55	19.30	17	23.3	1.9×10^{16}	6.1	119

Catalyst								Ref.
Pt-Ir/Al₂O₃	250	18.55	19.30	17	23.2	1.4×10^{16}	6.2	119
16% Pt/C (D = 0.2)	271	18.35	—	20	~21.7	$\sim 5 \times 10^{13}$	8.5	144
0.2–0.66 Pt/SiO₂Al₂O₃ (D = 0.9)								
(13% Al₂O₃)	230	18.28	19.36	—	—	$\sim 2 \times 10^{13}$	8.8	145
(1.9% Al₂O₃)	230	18.28	19.36	—	—	$\sim 8 \times 10^{13}$	8.2	145
(0.5% Al₂O₃)	230	18.28	19.36	—	—	$\sim 1 \times 10^{14}$	8.1	145
0.2–1% Pt/Al₂O₃	190	18.21	18.46	16	22.5	1.0×10^{15}	7.0	139
0.2% Pt–0.8% Pd/Al₂O₃	190	18.21	18.46	15	21.5	3×10^{14}	7.6	139
0.4% Pt–0.6% Pd/Al₂O₃	190	18.21	18.46	15	21.3	1.8×10^{14}	7.8	139
0.8% Pt–0.2% Pd/Al₂O₃	190	18.21	18.46	15	21.4	2×10^{14}	7.7	139
Pt(111)	260	17.69	18.51	~18	~23	1.2×10^{16}	5.5	72
Pt(557)	260	17.69	18.51	~19	~23	5×10^{15}	5.8	72
Pt(25, 10, 7)	260	17.69	18.51	~18	~23	2×10^{16}	5.2	72
Pt, powder	300	17.47	18.47 (D₂)	19	20.7	3×10^{13}	7.8	70
2–2.4% Pt/SiO₂ (D = 0.05–0.98)	300	16.21	17.84	19	~21	1.3×10^{14}	5.9	60
2% Pt/SiO₂ (D = 0.98, preoxidized)	300	16.21	17.84	—	—	6×10^{14}	5.3	60
2% Pt/SiO₂ (D = 0.5, preoxidized)	300	16.21	17.84	—	—	1.7×10^{14}	5.8	60
Dehydrogenation to Cyclohexene								
Pt(111)	300	17.69	18.51	—	—	1.0×10^{13}	8.5	72
Pt(557)	300	17.69	18.51	—	—	5×10^{13}	7.8	72
Pt(25, 10, 7)	300	17.69	18.51	—	—	3×10^{12}	9.0	72

457

Table 8.C2. Kinetic parameters for other dehydrogenation reactions catalyzed by metals

Hydrocarbon	Catalyst	T (°C)	log (HC)	log (H$_2$)	E_a (kcal/mol)	log A	Rate (molecules/cm^2/sec)	$-\log RP$	S_D	References
	Pt–Au, powders									
	($X_{Au} = 0.0$)	360	$17.99^{1.0}$	$19.67^{-1.1}$	30	23.8	4×10^{13}	9.4	—	146
	($X_{Au} = 0.86$)	360	$17.99^{1.0}$	$19.67^{-0.5}$	27	21.1	8×10^{11}	10.0	—	146
	($X_{Au} = 0.93$)	360	17.99	19.67	—	—	3×10^{11}	10.5	—	146
	($X_{Au} = 0.975$)	360	$17.99^{1.0}$	$19.67^{-0.5}$	27	20.4	1.5×10^{11}	10.8	—	146
	($X_{Au} = 0.995$)	360	$17.99^{1.0}$	$19.67^{-0.5}$	27	19.7	3×10^{10}	11.5	—	146
	($X_{Au} = 1.0$)	360	17.99	19.67	—	—	4×10^{9}	12.4	—	146
	W, film	425	16.51	17.51	—	—	$\sim 2 \times 10^{12}$	8.1	~−0.9	37
	Re–Au, films									
	($X_{Au} = 0.29$)	330	16.51	17.51	~13	~16	$\sim 2 \times 10^{11}$	9.1	~−0.7	37
	($X_{Au} = 0.49$)	350	16.51	17.51	—	—	$\sim 5 \times 10^{10}$	9.7	~−0.6	37
	0.33–0.86% Ru/Al$_2$O$_3$	490	18.61	19.31	~27	~22.2	$2\text{–}3 \times 10^{14}$	8.0	—	46
	1.3% Pt/C (20 Å)	460	18.89	19.23	—	—	2×10^{16}	6.4	—	147
	2.8% Pt/C (50 Å)	460	18.89	19.23	—	—	7×10^{15}	6.9	—	147
	Pt(223)	150	17.36	18.36	9	~19.4	6×10^{14}	6.4	~−0.01	100
	16% Pt/C ($D = 0.2$)	263	18.35	—	15	~19.0	$\sim 9 \times 10^{12}$	9.2	—	144
	Re/Al$_2$O$_3$	310	19.39	0.0	—	—	3×10^{12}	10.7	—	35
	Os/Al$_2$O$_3$	310	19.39	0.0	—	—	3×10^{12}	10.7	—	35
	Ir/Al$_2$O$_3$	310	19.39	0.0	—	—	4×10^{12}	10.5	—	35
	Pt/Al$_2$O$_3$	310	19.39	0.0	—	—	8×10^{13}	9.2	—	35
	Pt–Re/Al$_2$O$_3$, bimetallic									
	($X_{Re} = 0.2$)	310	19.39	0.0	—	—	5×10^{13}	9.4	—	35
	($X_{Re} = 0.75$)	310	19.39	0.0	—	—	2×10^{13}	9.8	—	35

Table 8.D1. Kinetic parameters for n-butane isomerization over metal catalysts

Catalyst	T (°C)	log (HC)	log (H$_2$)	E_a (kcal/mol)	log A	Rate (molecules/cm^2/sec)	$-\log RP$	S_1	References
Pd, film	300	17.10	18.16	~38	25.4	1×10^{11}	10.0	~0.01	10
Rh, powder	130	17.51	18.51	—	—	$\sim 2 \times 10^{11}$	10.1	~0.02	9
Ir, powder	200	17.51	18.51	—	—	$\sim 2 \times 10^{10}$	11.1	~0.02	9
2% Pt/Al$_2$O$_3$	300	18.39	19.35	—	—	1.1×10^{13}	9.3	~0.49	15
Pt/Al$_2$O$_3$	300	18.39	19.35	—	—	3×10^{11}	10.9	~0.03	36
Pt, powder	370	17.51$^{1.0}$	18.81$^{-1.7}$	24	19.2	1.2×10^{11}	10.3	~0.15	19
Pt, film	300	17.10	18.16	~21	20.1	1.4×10^{12}	8.9	~0.11	10
Pt(111), film	320	17.10	18.16	—	—	1.7×10^{12}	8.8	~0.12	10
Pt(100), film	300	17.10	18.16	—	—	4×10^{12}	8.4	~0.16	10
Pt, film	360	17.02	18.02	22	20.2	4×10^{12}	8.3	~0.38	20

Table 8.D2. Kinetic parameters for isobutane isomerization over metal catalysts

Catalyst	T (°C)	log (HC)	log (H$_2$)	E_a (kcal/mol)	log A	Rate (molecules/cm^2/sec)	$-\log RP$	S_1	References
Pd, film	300	17.10	18.16	~21	19.5	4×10^{11}	9.4	~0.03	10
Pd, film	310	17.10	18.16	21	19.3	3×10^{11}	9.5	~0.05	38
Ta, film	200	16.51	17.51	~5	~14	$\sim 1 \times 10^{11}$	9.4	~0.05	37
W, film	155	16.51	17.51	~12	~18	$\sim 4 \times 10^{11}$	8.8	~0.12	37
Re, film	200	16.51	17.51	—	—	$\sim 2 \times 10^{11}$	9.1	~0.17	37
Pt, Al$_2$O$_3$	300	18.39	19.35	—	—	5×10^{11}	10.6	~0.06	36
Pt, powder	375	17.51$^{0.9}$	18.81$^{-2.0}$	29	21.3	4×10^{11}	9.8	~0.65	19
Pt, film	300	17.10	18.16	~21	20.7	6×10^{12}	8.2	~0.50	10
Pt, film	300	17.10	18.16	~21	20.5	2×10^{12}	8.3	~0.67	10
Pt(111), film	300	17.10	18.16	19	19.4	2×10^{12}	8.7	0.83	10
Pt(100), film	300	17.10	18.16	—	—	9×10^{12}	8.0	~0.59	10
Pt, film									
clean	360	17.02	18.02	—	—	3×10^{13}	7.5	~0.64	20
steady state	360	17.02	18.02	—	—	$\sim 3 \times 10^{12}$	8.5	~0.32	20

Table 8.D3. Kinetic parameters for *n*-pentane isomerization over metal catalysts

Catalyst	T (°C)	log (HC)	log (H$_2$)	E_a (kcal/mol)	log A	Rate (molecules/cm^2/sec)	$-\log RP$	S_1	References
5% Pd/SiO$_2$	400	19.09$^{0.9}$	20.39$^{-1.4}$	49	29.0	1.5×10^{13}	9.8	~0.26	45
Pd/Al$_2$O$_3$	250	18.35	19.35	—	—	$\sim 2 \times 10^{9}$	13.0	~0.10	44
Rh, film	270	16.51	17.51	—	—	$\sim 8 \times 10^{9}$	12.4	~0.01	47
0.25% Ir/Al$_2$O$_3$	250	18.35	19.35	—	—	$\leq 4 \times 10^{11}$	10.7	≤0.01	44
0.25% Ir + Cu/Al$_2$O$_3$									
(X_{Cu} = 0.4)	250	18.35	19.35	—	—	$\sim 3 \times 10^{11}$	10.8	~0.06	44
(X_{Cu} = 0.6)	250	18.35	19.35	—	—	$\sim 2 \times 10^{11}$	11.0	~0.15	44
(X_{Cu} = 0.9)	250	18.35	19.35	—	—	$\sim 1 \times 10^{11}$	11.3	~0.30	44
5% Pt/SiO$_2$	400	19.09$^{1.0}$	20.39$^{-2.7}$	53	31.1	1.0×10^{14}	9.0	~0.15	45
Pt/C	400	19.09$^{1.0}$	20.39$^{-2.2}$	57	32.2	7×10^{13}	9.1	~0.15	45
0.6% Pt/Al$_2$O$_3$(BF) (D = 0.9)	450	18.79	19.27	—	—	$\sim 2 \times 10^{15}$	7.4	—	130
0.6% Pt-0.5% Na$^+$/Al$_2$O$_3$ (BF) (D = 0.9)	450	18.79	19.27	—	—	$\sim 2 \times 10^{14}$	8.4	—	130
0.6% Pt-2.5% Na$^+$/Al$_2$O$_3$ (D = 0.9)	450	18.79	19.27	—	—	$\sim 5 \times 10^{13}$	9.0	—	130
16% Pt/SiO$_2$ (D = 0.3)	312	18.39	19.34	—	—	$\sim 8 \times 10^{10}$	11.4	~0.67	48
16% Pt-Au/SiO$_2$, bimetallic (D = 0.3)									
(X_{Au} = 0.875)	350	18.39	19.34	—	—	$\sim 5 \times 10^{9}$	12.6	~0.26	48
(X_{Au} = 0.975)	370	18.39	19.34	~34	~22	$\sim 1 \times 10^{10}$	12.3	~0.96	48
(X_{Au} = 0.99)	343	18.39	19.34	~23	~17	$\sim 2 \times 10^{9}$	13.0	1.0	48
1.3% Pt/SiO$_2$ (D = 0.56)	283	18.03$^{0.6}$	19.05$^{-1.8}$	—	—	8×10^{11}	10.0	~0.29	49
8.2% Pt/SiO$_2$ (D = 0.07)	283	18.03$^{0.6}$	19.05$^{-1.8}$	—	—	1×10^{12}	9.9	~0.74	49

Table 8.D4. Kinetic parameters for neopentane isomerization over metal catalysts

Catalyst	T (°C)	log (HC)	log (H_2)	E_a (kcal/mol)	log A	Rate (molecules/cm²/sec)	$-\log RP$	S_I	References
10% Ir/SiO₂	200	18.35	19.35	50	34	1.1×10^{11}	7.2	~0.13	50
Pt/Al₂O₃	300	18.39	19.35	—	—	5×10^{11}	10.6	~0.05	36
1% Pt/spheron	300	18.35	19.35	49	~31	$\sim 2 \times 10^{12}$	10.0	~0.97	50
0.6% Pt/Al₂O₃	307	18.35	19.35	—	—	4×10^{12}	9.6	~0.36	55
2% Pt/Al₂O₃	307	18.35	19.35	—	—	8×10^{12}	9.3	~0.29	55
4.3% Pt/SiO₂	307	18.35	19.35	—	—	6×10^{12}	9.5	~0.23	55
1% Pt/spheron	307	18.35	19.35	—	—	1.7×10^{12}	10.0	~0.96	55
Pt, powder	307	18.35	19.35	—	—	1.1×10^{12}	10.2	~0.90	55
6% Pt–6% Fe/C	370	18.35	19.35	—	—	4×10^{12}	9.6	~0.71	54
1% Pt/C	370	18.35	19.35	—	—	1.3×10^{14}	8.1	~0.72	54
3% Pt/Na–Y (10 Å)	225	18.07	19.37	~37	~26	3×10^{10}	11.5	~0.11	51
3% Pt/Ca–Y (10 Å)	225	18.07	19.37	~35	~25	3×10^{10}	11.5	~0.03	51
0.9% Pt/SiO₂ (40 Å)	315	18.07	19.37	~28	~24	2.6×10^{13}	8.5	~0.47	51
0.9% Pt/SiO₂ (70 Å)	315	18.07	19.37	~25	~23	1.8×10^{13}	8.7	~0.58	51
2% Pt/SiO₂ (12 Å)	315	18.07	19.37	~35	~26	2.2×10^{13}	8.6	~0.40	51
2.5% Pt/SiO₂	300	$17.91^{1.0}$	$19.39^{1.0-0.1}$	34	27.4	2.6×10^{14}	7.4	~0.22	52
2.8% Pt–4% W/SiO₂	300	$17.91^{1.0}$	$19.39^{1.0-0.1}$	—	—	9×10^{13}	7.8	~0.02	52
Pt, film	275	17.10	18.16	—	—	$\sim 5 \times 10^{12}$	8.3	~0.76	38
Pt, film	275	17.10	18.16	~21	21.0	5×10^{12}	8.3	~0.59	10
Au, powder	450	18.35	19.35	48	~25	$\sim 4 \times 10^{10}$	10.6	~0.99	50

Table 8.D5. Kinetic parameters for *n*-hexane isomerization over metal catalysts

Catalyst	T (°C)	log (HC)	log (H$_2$)	E_a (kcal/mol)	Rate (molecules/cm^2/sec)	$-\log RP$	S_I	S_{2MP}	S_{3MP}	References
5% Ni/SiO$_2$–Al$_2$O$_3$ (BF) (D = 0.3)	330	18.69	19.29	—	$\sim2 \times 10^8$	14.2	0.86	—	—	61
7% Ni/SiO$_2$–Al$_2$O$_3$ (BF) (D = 0.3)	350	18.69	19.29	—	$\sim4 \times 10^9$	12.9	0.95	—	—	61
16% Ni/SiO$_2$–Al$_2$O$_3$ (BF) (D = 0.3)	275	18.69	19.29	—	$\sim1 \times 10^8$	14.5	0.19	—	—	61
3% Ni/Li-Y (BF) (D = 0.3)	300	18.42	19.34	—	$\sim4 \times 10^{11}$	10.6	~-0.07	0.66	0.34	14
Ni–Cu, powders										
(X$_{Cu}$ = 0.0)	330	18.12	19.36	~44	$\sim6 \times 10^{12}$	9.2	~-0.04	0.7	0.3	62
(X$_{Cu}$ = 0.05)	330	18.12	19.36	~44	$\sim2 \times 10^{10}$	11.2	~-0.08	0.66	0.34	62
(X$_{Cu}$ = 0.47)	330	18.12	19.36	~55	$\sim7 \times 10^{10}$	11.1	~-0.61	0.78	0.22	62
Mo/MoO$_2$	330	18.69	19.29	—	—	—	~-0.98	—	—	61
Rh–Cu/SiO$_2$, bimetallic										
(X$_{Cu}$ = 0.0)	175	17.74	18.69	—	$\sim4 \times 10^{13}$	8.0	~-0.08	0.9	0.1	63
(X$_{Cu}$ = 0.85)	300	17.74	18.69	—	$\sim4 \times 10^{12}$	9.0	~-0.08	0.66	0.33	63
(X$_{Cu}$ = 1.0)	325	16.99	19.39	—	$\sim2 \times 10^{12}$	9.3	≤0.01	0.66	0.33	63
10% Pd/Al$_2$O$_3$ (D = 0.3)	350	16.99	19.39	—	$\sim6 \times 10^{10}$	10.0	~-0.06	0.6	0.4	64
10% Pd–Au/Al$_2$O$_3$, bimetallic (D = 0.3)										
(X$_{Au}$ = 0.5)	350	16.99	19.39	—	$\sim1.3 \times 10^{10}$	10.7	~-0.09	0.64	0.36	64
(X$_{Au}$ = 0.35)	350	16.99	19.39	—	$\sim2 \times 10^8$	12.5	~-0.04	0.6	0.4	64
Pd/Na-Y (BF)	350	18.69	19.29	—	—	—	~-0.91	—	—	61
Re, film	180	16.51	17.51	0	$\sim7 \times 10^{10}$	9.5	~-0.03	1.0	—	37
0.5% Pt/Al$_2$O$_3$ (D = 0.2–0.6)	490	19.67	20.27	—	$1-1.5 \times 10^{15}$	8.3	~-0.65	0.67	0.33	148

(continued)

463

Table 8.D5. —Continued

Catalyst	T (°C)	log (HC)	log (H₂)	E_a (kcal/mol)	Rate (molecules/cm²/sec)	−log RP	S_1	S_{2MP}	S_{3MP}	References
10% Pt/Al₂O₃ (D = 0.2–0.6)	490	19.67	20.27	—	$1-1.5 \times 10^{15}$	8.3	~0.65	0.67	0.33	148
Pt/Al₂O₃	330	18.69	19.29	—	—	—	0.93	—	—	61
0.2–9.5% Pt/Al₂O₃ (D = 0.2–1.0)	300	18.3	19.35	—	$6-8 \times 10^{12}$	9.2	0.3–0.6	0.74	0.26	66
Pt–Au, powders										
(X_Au = 0.0)	360	17.77	19.67	—	$\sim 4 \times 10^{11}$	10.0	—	1.0	—	146
(X_Au = 0.86)	360	17.77	19.67	—	$\sim 2 \times 10^{7}$	14.3	—	1.0	—	146
(X_Au = 0.93)	360	17.77	19.67	—	$\sim 1 \times 10^{6}$	15.6	—	1.0	—	146
16% Pt/SiO₂ (D = 0.3)	295	18.14	19.37	45	$\sim 2 \times 10^{10}$	11.7	0.58	0.63	0.37	48
16% Pt–Au/SiO₂, bimetallic (D = 0.03)										
(X_Au = 0.875)	370	18.14	19.37	—	$\sim 3 \times 10^{9}$	12.5	0.50	0.23	0.77	48
(X_Au = 0.92)	320	18.14	19.37	—	$\sim 2 \times 10^{9}$	12.7	0.90	0.41	0.59	48
(X_Au = 0.99)	390	18.14	19.37	—	$\sim 5 \times 10^{9}$	12.3	0.76	0.41	0.59	48
Pt–Cu/SiO₂, bimetallic										
(X_Cu = 0.0)	300	18.12	19.36	~60	$\sim 2 \times 10^{12}$	9.6	~0.63	0.66	0.33	67
(X_Cu = 0.39)	300	18.12	19.36	~60	$\sim 8 \times 10^{10}$	11.0	~0.30	—	—	67
(X_Cu = 0.98)	300	18.12	19.36	~60	$\sim 1 \times 10^{10}$	11.9	~0.08	0.6	0.4	67
Pt–Sn/SiO₂										
(X_Sn = 0.0)	300	17.74	18.69	—	1.1×10^{15}	6.5	0.63	0.65	0.35	63
(X_Sn = 0.05)	340	17.74	18.69	—	1.7×10^{12}	9.3	0.12	0.84	0.16	63
Pt–Au/SiO₂										
(X_Au = 0.5)	320	17.74	18.69	—	5×10^{13}	7.9	0.50	0.55	0.45	63
(X_Au = 0.95)	320	17.74	18.69	—	8×10^{11}	9.7	0.08	0.93	0.07	63
Pt, film (15 Å)	273	17.47	18.47	—	2.4×10^{13}	7.9	0.84	—	—	68
Pt, film (20 Å)	273	17.47	18.47	—	1.8×10^{13}	8.0	0.89	—	—	68
Pt, film (38 Å)	273	17.47	18.47	—	8×10^{12}	8.4	0.94	—	—	68
Pt, film (58 Å)	273	17.47	18.47	—	4×10^{12}	8.7	0.92	—	—	68
Pt, film (>100 Å)	273	17.47	18.47	—	1.3×10^{12}	8.2	0.78	—	—	68

Table 8.D6. Kinetic parameters for 2-methylpentane isomerization over metal catalysts

Catalyst	T (°C)	log (HC)	log (H$_2$)	Rate (molecules/cm^2/sec)	$-\log RP$	S_I	S_{nC_6}	S_{3MP}	References
10% Pd/Al$_2$O$_3$ ($D = 0.3$)	350	17.11	19.39	$\sim 1 \times 10^{11}$	9.9	0.16	~ 0.45	~ 0.55	64
10% Pd–Au/Al$_2$O$_3$, bimetallic ($D = 0.3$)									
($X_{Au} = 0.5$)	350	17.11	19.39	$\sim 3 \times 10^{9}$	11.4	0.20	~ 0.5	~ 0.5	64
($X_{Au} = 0.65$)	350	17.11	19.39	$\sim 4 \times 10^{19}$	11.3	0.25	~ 0.5	~ 0.5	64
Pt/Al$_2$O$_3$	300	18.39	19.35	1.0×10^{13}	9.2	0.23	—	—	36
Pt, film (20 Å)	273	17.47	18.47	5×10^{14}	6.6	~ 0.94	—	—	68
Pt, film (38 Å)	273	17.47	18.47	4×10^{14}	6.7	~ 0.96	—	—	68
Pt, film (58 Å)	273	17.47	18.47	1.5×10^{14}	7.1	~ 0.95	—	—	68
Pt, film (>100 Å)	273	17.47	18.47	6×10^{13}	7.5	~ 0.92	—	—	68

Table 8.D7. Kinetic parameters for 3-methylpentane isomerization over metal catalysts

Catalyst	T (°C)	log (HC)	log (H$_2$)	Rate (molecules/cm^2/sec)	$-\log RP$	S_I	S_{nC_6}	S_{2MP}	References
Ir–Au, films									
(X_{Au} = 0.0)	400	16.44	17.44	~8 × 10^{10}	9.4	~0.12	0.40	0.6	69
(X_{Au} = 0.6)	400	16.44	17.44	~4 × 10^{10}	9.7	~0.06	—	1.0	69
(X_{Au} = 0.19)	400	16.44	17.44	~1 × 10^{10}	10.3	~0.03	—	1.0	69
Pt/Al$_2$O$_3$	300	18.39	19.35	4 × 10^{13}	8.6	0.42	—	—	36
Pt, film (20 Å)	273	17.47	18.47	7 × 10^{13}	7.5	0.83	—	—	68
Pt, film (40 Å)	273	17.47	18.47	3 × 10^{14}	6.8	0.96	—	—	68
Pt, film (58 Å)	273	17.47	18.47	2 × 10^{14}	7.0	0.96	—	—	68
Pt, film (>100 Å)	273	17.47	18.47	7 × 10^{13}	7.5	0.92	—	—	68

Table 8.D8. Kinetic parameters for n-heptane isomerization over metal catalysts

Catalyst	T (°C)	log (HC)	log (H$_2$)	Rate (molecules/cm^2/sec)	$-\log RP$	S_I	References
Ru, powder	88	18.61	19.31	~1 × 10^{10}	12.4	~0.07	79
Rh, powder	113	18.61	19.31	~2 × 10^{10}	12.1	~0.07	79
Pd, powder	300	18.61	19.31	~4 × 10^{10}	11.8	~0.06	79
Ir, powder	125	18.61	19.31	~4 × 10^{11}	10.8	~0.13	79
Pt, powder	275	18.61	19.31	~2 × 10^{12}	10.1	~0.47	79
0.3% Pt/Al$_2$O$_3$–(BF) (D = 0.9)	471	19.94	20.63	~1 × 10^{15}	8.7	0.82	75
Pt/Al$_2$O$_3$	300	18.39	19.35	~6 × 10^{11}	10.4	~0.05	36

Table 8.D9. Kinetic parameters for other isomerization reactions catalyzed by metals

Hydrocarbon	Catalyst	T (°C)	log (HC)	log (H$_2$)	Rate (molecules/cm^2/sec)	−log RP	S_I	References
dbm [structure]	3.2% Ni/mordenite ($D = 0.2$)	80	18.91	19.22	$\sim 2 \times 10^{14}$ $E_a \sim 12$ kcal/mol	8.5	~0.2 36% cis	101
	Pt(223)	25	17.36	18.36	2×10^{15}	6.0	~0.15 60% cis	100
[structure]	2% Pt/Al$_2$O$_3$	300	18.39	19.35	1.6×10^{13}	9.0	0.48	15
	Pt, film	278	17.10	18.16	7×10^{11}	9.1	0.27	10
[structure]	Pt/Al$_2$O$_3$	300	18.39	19.35	3×10^{11}	10.7	0.02	36
	3% Pt/La–Y (BF)	200	18.07	19.37	$\sim 1.5 \times 10^{13}$	8.7	~0.6	51
	3% Pt/Ca–Y (BF)	200	18.07	19.37	$\sim 3 \times 10^{12}$	9.4	~0.61	51
	3% Pt/Na–Y (BF)	200	18.07	19.37	$\sim 8 \times 10^{11}$	10.0	~0.35	51
[structure]	Pt/Al$_2$O$_3$	300	18.39	19.35	5×10^{11}	10.5	~0.02	36
[structure]	Pt/Al$_2$O$_3$	300	18.39	19.35	7×10^{11}	10.3	~0.06	36
[structure]	Pt/Al$_2$O$_3$	300	18.39	19.35	3×10^{13}	8.7	~0.47	36
[structure]	Pt/Al$_2$O$_3$	300	18.39	19.35	3×10^{13}	8.7	~0.52	36
[structure]	Pt/Al$_2$O$_3$	300	18.39	19.35	3×10^{13}	8.7	~0.37	36
[structure]	Pt/Al$_2$O$_3$	300	18.39	19.35	2×10^{11}	10.9	~0.01	36
cis- epimerization [structure]	4.8% Pd/C	73	17.12	18.03	8×10^{10}	10.0	$E_a = 18$ kcal/mol	149
	19% Pd/C	73	17.12	18.03	2×10^{10}	10.6	$E_a = 20$ kcal/mol	149

467

Table 8.E1. Kinetic parameters for n-pentane dehydrocyclization over metal catalysts

Catalyst	T (°C)	log (HC)	log (H_2)	Rate (molecules/cm²/sec)	$-\log RP$	S_C	References
Rh, film	320	16.51	17.51	$\sim 1 \times 10^{10}$	10.4	~0.07	47
Rh–Cu, films							
($X_{Cu} = 0.06$)	330	16.51	17.51	$\sim 2 \times 10^{10}$	10.1	~0.15	47
($X_{Cu} = 0.92$)	310	16.51	17.51	$\sim 4 \times 10^{10}$	9.8	~0.78	47
Rh–Au, films							
($X_{Au} = 0.23$)	320	16.51	17.51	$\sim 4 \times 10^{9}$	10.8	~0.5	47
Rh–Sn, films							
($X_{Sn} = 0.03$)	260	16.51	17.51	$\sim 6 \times 10^{10}$	9.6	~0.55	47
($X_{Sn} = 0.17$)	280	16.51	17.51	$\sim 8 \times 10^{10}$	9.5	~0.34	47
Pd/Al_2O_3	250	18.35	19.35	$\sim 3 \times 10^{9}$	12.7	~0.33	44
0.25% Ir + Cu/Al_2O_3							
($X_{Cu} = 0.5$)	250	18.35	19.35	$\sim 6 \times 10^{10}$	11.4	~0.02	44
($X_{Cu} = 0.9$)	250	18.35	19.35	$\sim 3 \times 10^{10}$	11.7	~0.08	44
Pt/Al_2O_3	250	18.35	19.35	$\sim 3 \times 10^{11}$	10.7	~0.50	44
16% Pt/SiO_2 ($D = 0.3$)	312	18.39	19.34	$\sim 1 \times 10^{10}$	12.2	~0.08	48
16% Pt–Au/SiO_2, bimetallic ($D = 0.3$) ($X_{Au} = 0.875$)	350	18.39	19.34	$\sim 8 \times 10^{9}$	12.3	~0.61	48

Table 8.E2. Kinetic parameters for n-hexane dehydrocyclization over metal catalysts

Catalyst	T (°C)	log (HC)	log (H$_2$)	Rate (molecules/cm^2/sec)	$-\log RP$	S_C	S_{MCP}	S_{B_2}	References
Ni–Cu, powders									
($X_{Cu} = 0.0$)	330	18.12	19.36	~1.5 × 10^{12}	9.8	~0.014	~0.35	~0.65	62
($X_{Cu} = 0.05$)	330	18.12	19.36	~4 × 10^9	12.3	~0.01	~0.5	~0.5	62
($X_{Cu} = 0.23$)	330	18.12	19.36	~8 × 10^9	12.0	~0.03	~0.75	~0.25	62
($X_{Cu} = 0.47$)	330	18.12	19.36	~9 × 10^9	12.0	~0.07	~0.8	~0.2	62
Rh–Cu/SiO$_2$, bimetallic									
($X_{Cu} = 0.85$)	300	17.74	18.69	~1 × 10^{13}	8.6	~0.24	~0.1	~0.9	63
($X_{Cu} = 1.0$)	325	17.74	18.69	~2 × 10^{13}	8.3	~0.08	~0	~1.0	63
Rh, film	240	17.74	18.69	~1 × 10^{11}	10.6	~0.25	—	—	47
Rh–Cu, films									
($X_{Cu} = 0.03$)	280	17.74	18.69	~5 × 10^{10}	10.9	~0.77	—	—	47
($X_{Cu} = 0.58$)	270	17.74	18.69	~1 × 10^{10}	11.6	~0.60	—	—	47
($X_{Cu} = 0.92$)	280	17.74	18.69	~9 × 10^9	11.6	~0.41	—	—	47
Rh–Sn, films									
($X_{Sn} = 0.03$)	290	17.74	18.69	~4 × 10^{10}	11.0	~0.60	—	—	47
($X_{Sn} = 0.36$)	320	17.74	18.69	~8 × 10^{10}	10.7	~0.84	—	—	47
Rh–Au, film									
($X_{Au} = 0.87$)	300	17.74	18.69	~2 × 10^{10}	11.3	~0.75	—	—	47
Rh–Ag, film									
($X_{Ag} = 0.26$)	280	17.74	18.69	~1 × 10^{10}	11.6	~0.50	—	—	47
10% Pd/Al$_2$O$_3$ ($D = 0.3$)	350	16.99	19.39	~2 × 10^{11}	9.5	~0.22	~0.65	~0.35	64
10% Pd–Au/Al$_2$O$_3$, bimetallic ($D = 0.3$)									
($X_{Au} = 0.5$)	350	16.99	19.39	~5 × 10^{10}	10.1	~0.21	~0.6	~0.4	64
($X_{Au} = 0.65$)	350	16.99	19.39	~1.3 × 10^9	11.7	~0.10	~0.6	~0.4	64

(continued)

Table 8.E2.—Continued

Catalyst	T (°C)	log (HC)	log (H₂)	Rate (molecules/cm²/sec)	−log RP	S_C	S_{MCP}	S_{Bz}	References
Re, film	300	16.51	17.51	$\sim 5 \times 10^{10}$	9.6	~0.14	—	—	37
Re–Au, film ($X_{Au} = 0.79$)	355	16.51	17.51	$\sim 1 \times 10^{11}$	9.3	~0.64	—	—	37
W, film	370	16.51	17.51	$\sim 1.4 \times 10^{11}$	9.2	~1.0	—	—	37
W–Au, film ($X_{Au} = 0.33$)	410	16.51	17.51	$\sim 2 \times 10^{11}$	9.0	~1.0	—	—	37
Ta, film	375	16.51	17.51	$\sim 4 \times 10^{11}$	8.7	~0.9	—	—	37
Ta–Au, film ($X_{Au} = 0.47$)	365	16.51	17.51	$\sim 5 \times 10^{11}$	8.6	~0.9	—	—	37
Ir–Au, films ($E_a = 10\text{--}15$ kcal/mol)									
($X_{Au} = 0.0$)	390	16.44	17.44	$\sim 8 \times 10^{10}$	9.4	~0.33	—	—	65
($X_{Au} = 0.86$)	360	16.44	17.44	$\sim 3 \times 10^{10}$	9.8	~0.65	—	—	65
($X_{Au} = 0.94$)	400	16.44	17.44	$\sim 5 \times 10^{10}$	9.6	~0.81	—	—	65
0.5% Pt/Al₂O₃ ($D = 0.2\text{--}0.65$)	490	19.67	20.27	$2\text{--}3 \times 10^{14}$	9.0	≤0.3	~0.5	~0.5	148
10% Pt/Al₂O₃ ($D = 0.2\text{--}0.65$)	490	19.67	20.27	$2\text{--}3 \times 10^{14}$	9.0	≤0.3	~0.5	~0.5	148

0.2% Pt/Al$_2$O$_3$ (D = 1.0)	300	18.3	19.35	9×10^{12}	9.2	~0.34	1.0	—	66
1.7% Pt/Al$_2$O$_3$	300	18.3	19.35	1.6×10^{12}	9.9	~0.13	1.0	—	66
9.5% Pt/Al$_2$O$_3$ (D = 0.2)	300	18.3	19.35	8×10^{11}	10.2	~0.07	1.0	—	66
Pt–Cu/SiO$_2$, bimetallic (E_a = 30–40 kcal/mol)									
(X_{Cu} = 0.0)	300	18.12	19.36	$\sim 3 \times 10^{11}$	10.5	~0.1	~0.9	~0.1	67
(X_{Cu} = 0.39)	300	18.12	19.36	$\sim 1 \times 10^{11}$	11.0	~0.4	—	—	67
(X_{Cu} = 0.71)	300	18.12	19.36	$\sim 5 \times 10^{10}$	11.3	~0.45	—	—	67
(X_{Cu} = 0.98)	300	18.12	19.36	$\sim 4 \times 10^{10}$	11.4	~0.33	~0.6	~0.4	67
16% Pt/SiO$_2$ (D = 0.3)	295	18.14	19.37	$\sim 4 \times 10^{9}$	12.4	~0.12	0.78	0.22	48
16% Pt–Au/SiO$_2$, bimetallic (D = 0.3)									
(X_{Au} = 0.875)	370	18.14	19.37	$\sim 2 \times 10^{9}$	12.7	~0.40	0.88	0.12	48
(X_{Au} = 0.99)	390	18.14	19.37	$\sim 9 \times 10^{8}$	13.0	~0.14	0.77	0.23	48
Pt/SiO$_2$	300	17.74	18.69	4×10^{13}	8.0	~0.02	0.95	0.05	63
Pt–Sn/SiO$_2$ (X_{Sn} = 0.05)	340	17.74	18.69	4×10^{12}	9.0	~0.25	0.86	0.14	63
Pt–Au/SiO$_2$, bimetallic									
(X_{Au} = 0.5)	316	17.74	18.69	1×10^{13}	8.6	~0.10	0.5	0.5	63
(X_{Au} = 0.95)	320	17.74	18.69	2×10^{12}	9.3	~0.24	1.0	—	63

Table 8.E3 Kinetic parameters for other dehydrocyclization reactions catalyzed by metals

Hydrocarbon	Catalyst	T (°C)	log (HC)	log (H₂)	Rate (molecules/cm²/sec)	-log RP	S_C	References
	10% Pd/Al₂O₃ (D = 0.3)	350	17.11	19.39	~3 × 10¹¹	9.5	~0.41	64
	10% Pd–Au/Al₂O₃, bimetallic (D = 0.3)							
	(X_{Au} = 0.5)	350	17.11	19.39	~7 × 10⁹	11.0	~0.38	64
	(X_{Au} = 0.65)	350	17.11	19.39	~6 × 10⁹	11.1	~0.42	64
	Pt/Al₂O₃	300	18.39	19.35	4 × 10¹²	9.6	0.10	36
	Re, film	380	16.51	17.51	~5 × 10¹⁰	9.6	~0.28	37
	Re–Au, film (X_{Au} = 0.27)	330	16.51	17.51	~8 × 10¹⁰	9.4	~0.28	37
	Ir–Au, films (X_{Au} = 0.0)	400	16.44	17.44	~3 × 10¹¹	8.8	~0.38	69
	(X_{Au} = 0.4)	400	16.44	17.44	~5 × 10¹¹	8.6	~0.80	69
	(X_{Au} = 0.81)	400	16.44	17.44	~3 × 10¹¹	8.8	~0.66	69
	Pt/Al₂O₃	300	18.39	19.35	~1 × 10¹³	9.2	0.11	36

Pd, powder	300	18.61	19.31	$\sim 2 \times 10^{10}$	12.1	~ 0.03	79
0.3% Pt/Al$_2$O$_3$ ($D = 0.9$)	471	19.94	20.63	$\sim 7 \times 10^{13}$	9.9	~ 0.04	75
Pt, powder	275	18.61	19.31	$\sim 8 \times 10^{11}$	10.5	~ 0.16	79
Pt/Al$_2$O$_3$	300	18.39	19.35	6×10^{12}	9.4	~ 0.45	36
Pt(111)	300	17.69	19.19	1.8×10^{13} $E_a \sim 44$ kcal/mol	8.2	$\sim 0.1-0.2$	72
Pt(557)	300	17.69	19.19	2×10^{13} $E_a \sim 34$ kcal/mol	8.2	$\sim 0.1-0.2$	72
Pt(10, 8, 7)	300	17.69	19.19	2×10^{13}	8.2	$\sim 0.1-0.2$	72
Pt(25, 10, 7)	300	17.69	19.19	1.0×10^{13} $E_a \sim 34$ kcal/mol	8.5	$\sim 0.1-0.2$	72
1% Pt/Al$_2$O$_3$ ($D = 0.8$)	420	14.99	—	$\sim 1 \times 10^{15}$ to To$_1$, $E_a \sim 10$ kcal/mol	3.8	~ 0.83	80
1% Pt/Al$_2$O$_3$ ($D = 0.8$)	420	14.99	—	$\sim 2 \times 10^{14}$ to B$_2$, $E_a \sim 21$ kcal/mol	4.5	~ 0.16	80
Pt/Al$_2$O$_3$	300	18.39	19.35	6×10^{12}	9.4	0.11	36
Pt/Al$_2$O$_3$	300	18.39	19.35	1×10^{13}	9.2	0.17	36

Table 8.E4. Kinetic parameters for hydro- and dehydroisomerization reactions catalyzed by metals

Hydrocarbon	Catalyst	T (°C)	log (HC)	log (H$_2$)	Rate (molecules/cm^2/sec)	$-\log RP$	S_{DI}	References
	0.5% Pt/Ca–Y (BF) ($D = 0.8$)	303	19.96	$20.81^{-0.5}$	$\sim 4 \times 10^{14}$ ($E_a \sim 30$ kcal/mol)	9.2	~0.04	136
	0.5% Pt/Ca–Y (BF) ($D = 0.8$)	288	20.07	$20.84^{-0.3}$	$\sim 2 \times 10^{14}$ ($E_a \sim 33$ kcal/mol)	9.6	1.0 all MCP	150
	Ni–Cu, powder ($X_{Cu} = 0.05$)	280	18.12	19.36	$\sim 8 \times 10^{10}$	11.1	~0.03	73
	Rh, film	310	16.51	17.51	$\sim 1 \times 10^{10}$	10.3	~0.06	47
	Rh–Cu, films ($X_{Cu} = 0.03$)	350	16.51	17.51	$\sim 1 \times 10^{10}$	10.3	~0.27	47
	($X_{Cu} = 0.06$)	375	16.51	17.51	$\sim 5 \times 10^{10}$	9.6	~0.57	47
	1% Rh–Co/SiO$_2$ ($X_{Co} = 0.5$)	260	18.95	19.79	$\sim 4 \times 10^{12}$	10.2	~0.07	74
	Re, film	333	16.51	17.51	$\sim 9 \times 10^{10}$	9.4	~0.45	37
	Re–Au, films ($X_{Au} = 0.22$)	380	16.51	17.51	$\sim 3 \times 10^{11}$	8.9	~0.22	37
	($X_{Au} = 0.47$)	315	16.51	17.51	$\sim 1.4 \times 10^{11}$	9.2	~0.95	37
	Ir, film	335	16.44	17.44	$\sim 2 \times 10^{11}$	9.0	~0.57	69
	Ir–Au, film ($X_{Au} = 0.73$)	370	16.44	17.44	$\sim 1 \times 10^{11}$	9.3	~0.47	69
	0.3% Pt/Al$_2$O$_3$ (BF) ($D = 0.9$)	471	19.94	20.63	$\sim 2 \times 10^{14}$	9.5	~0.3	75
	Pt, film (40 Å)	273	17.47	18.47	$\sim 1 \times 10^{14}$	9.8	~0.03	68

1. J.H. Sinfelt, Adv. Catal. **23**, 91 (1973); J.H. Sinfelt, Catal. Rev. **9**, 147 (1974), and references cited therein.
2. J.H. Sinfelt, W.F. Taylor, and D.J.C. Yates, J. Phys. Chem. **69**, 95 (1965).
3. J.H. Sinfelt, J. Catal. **29**, 308 (1973).
4. D.J.C. Yates and J.H. Sinfelt, J. Catal. **8**, 348 (1967).
5. J.H. Sinfelt, J.L. Carter, and D.J.C. Yates, J. Catal. **24**, 283 (1972).
6. L. Babernics, L. Guczi, K. Matusek, A. Sárkány, and P. Tétényi, Proc. 6th Int. Congr. Catal., Lond., 456 (1976).
7. J.R. Anderson and B.G. Baker, Proc. Roy. Soc. (Lond.) **A271**, 402 (1963).
8. J.H. Sinfelt, Y.L. Lam, J.A. Cusumano, and A.E. Barnett, J. Catal. **42**, 227 (1976).
9. A. Sárkány, K. Matusek, and P. Tétényi, Faraday Trans. **73**, 1699 (1977).
10. J.R. Anderson and N.R. Avery, J. Catal. **5**, 446 (1966).
11. D.C. Tajbl, Ind. Eng. Chem. Proc. Res. Develop. **8**, 365 (1969).
12. R.S. Hansen, private communication.
13. J. Freel and A.K. Galwey, J. Catal. **10**, 277 (1968).
14. J.T. Richardson, J. Catal. **21**, 122 (1971).
15. G. Leclercq, L. Leclercq, and R. Maurel, J. Catal. **44**, 68 (1976).
16. C.P. Huang and J.T. Richardson, J. Catal. **52**, 332 (1978).
17. T.J. Plunkett and J.K.A. Clarke, Faraday Trans. **68**, 500 (1972).
18. Y.I. Yermakov, B.N. Kuznetsov, and Y.A. Ryndin, J. Catal. **42**, 73 (1976).
19. L. Guczi, A. Sárkány, and P. Tétényi, Faraday Trans. **70**, 1971 (1974).
20. R.S. Dowie, D.A. Whan, and C. Kemball, Faraday Trans. **68**, 2150 (1972).
21. C. Naccache, N. Kaufherr, M. Dufaux, J. Pandiera, and B. Imelik, in *Molecular Sieves*, vol. 2, J.R. Katzer, ed., American Chemical Society, Washington, D.C., 1977, p. 538.
22. D.W. McKee and F.J. Norton, J. Catal. **3**, 252 (1964).
23. C.M. Machiels and R.B. Anderson, J. Catal. **58**, 253 (1979).
24. F.E. Shephard, J. Catal. **14**, 148 (1969).
25. L. Guczi, A. Sárkány, and P. Tétényi, Proc. 5th Int. Congr. Catal., Miami, paper 78, p. 1111 (1972).
26. J.R. Anderson and N.R. Avery, J. Catal. **8**, 48 (1967).
27. T.S. Sridhar and D.M. Ruthven, J. Catal. **24**, 153 (1972); A. Verma and D.M. Ruthven, J. Catal. **46**, 160 (1977); A. Verma and D.M. Ruthven, J. Catal. **19**, 401 (1970).
28. R.A. Dalla Betta, J.A. Cusumano, and J.H. Sinfelt, J. Catal. **19**, 343 (1970).
29. D.R. Kahn, E.E. Petterson, and G.A. Somorjai, J. Catal. **34**, 294 (1974).
30. M. Boudart, A. Aldag, J.E. Benson, N.A. Dougharty, and G.C. Harkins, J. Catal. **6**, 92 (1966).
31. N.A. Dougharty, Ph.D. thesis, University of California, Berkeley, 1964.
32. P.H. Otero-Schipper, W.A. Wachter, J.B. Butt, R.L. Burwell, Jr., and J.B. Cohen, J. Catal. **50**, 494 (1977).
33. P.A. Camagnon, C. Hoang-Van, and J.C. Teichner, Proc. 6th Int. Congr. Catal., Lond., 117 (1976).
34. J.C. Kempling and R.B. Anderson, Proc. 5th Int. Congr. Catal., Miami, 1099 (1972); Ind. Eng. Chem. Proc. Res. Develop. **11**, 146 (1972).
35. C. Bolivar, H. Charcosset, R. Frety, L. Tournayan, C. Betizeau, G. Leclercq, and M. Maurel, J. Catal. **45**, 179 (1976).
36. C. Leclercq, L. Leclercq, and R. Maurel, J. Catal. **50**, 87 (1977).
37. J.K.A. Clarke and J.F. Taylor, Faraday Trans. **71**, 2063 (1975).

38. J.R. Anderson and N.R. Avery, J. Catal. **2**, 542 (1963).
39. J. Datka, P. Gallezot, J. Massardier, and B. Imelik, in Proc. 5th Ibero-Am. Symp. Catal., Lisbon (1976).
40. P.H. Otero-Schipper, W.A. Wachter, J.B. Butt, R.L. Burwell, Jr., and J.B. Cohen, J. Catal. **53**, 414 (1978).
41. J.C. Schlatter and M. Boudart, J. Catal. **25**, 93 (1972).
42. H.C. Yao, Y.F. Yu Yao, and K. Otto, J. Catal. **56**, 21 (1979).
43. H.C. Yao and M. Shelef, J. Catal. **56**, 12 (1979).
44. J.P. Brunelle, R.E. Montarnal, and A.A. Sugier, Proc. 6th Int. Congr. Catal., Lond., 844 (1976).
45. E. Kikuchi, M. Tsurumi, and Y. Morita, J. Catal. **22**, 226 (1971).
46. V. Ragaini, L. Forni, and Le Van Mao, J. Catal. **37**, 339 (1975).
47. A. Péter and J.K.A. Clarke, Faraday Trans. **72**, 1201 (1976).
48. J.R.H. van Schaik, R.P. Dessing, and V. Ponec, J. Catal. **38**, 273 (1975).
49. J.P. Brunelle, A. Sugier, and J.F. LePage, J. Catal. **43**, 273 (1976).
50. M. Boudart and L.D. Ptak, J. Catal. **16**, 90 (1970).
51. K. Foger and J.R. Anderson, J. Catal. **54**, 318 (1978).
52. B.N. Kuznetsov, Yu.I. Yermakov, M. Boudart, and J.P. Collman, J. Mol. Catal. **4**, 49 (1978).
53. Z. Karpiński and T. Kóscielski, J. Catal. **56**, 430 (1979).
54. C.H. Bartholomew and M. Boudart, J. Catal. **25**, 173 (1972).
55. M. Boudart, A.W. Aldag, L.D. Ptak, and J.E. Benson, J. Catal. **11**, 35 (1968).
56. R.A. Dalla Betta and M. Boudart, Proc. 5th Int. Congr. Catal., Miami, paper 96, p. 1329 (1972).
57. S. Fuentes and F. Figueras, J. Catal. **54**, 397 (1978).
58. S. Fuentes and F. Figueras, Faraday Trans. **74**, 174 (1978).
59. J.K.A. Clarke and J.F. Taylor, Faraday Trans. **72**, 917 (1976).
60. A.N. Mitrofanova, V.S. Boronin, and O.M. Poltorak, Russ. J. Phys. Chem. **46**, 32 (1972), and references cited therein.
61. R. Burch, J. Catal. **58**, 220 (1979).
62. V. Ponec and W.M.H. Sachtler, Proc. 5th Int. Congr. Catal., Miami, paper 43, p. 645 (1972).
63. J.K.A. Clarke, T. Manninger, and T. Baird, J. Catal. **54**, 230 (1978).
64. A. O'Cinneide and F.G. Gault, J. Catal. **37**, 311 (1975).
65. T.J. Plunkett and J.K.A. Clarke, J. Catal. **35**, 330 (1974).
66. E. Santacessaria, D. Gelosa, S. Carra, and T. Adami, Ind. Eng. Chem. Prod. Res. Develop. **17**, 68 (1978).
67. H.C. DeJongste, F.J. Kuijers, and V. Ponec, Proc. 6th Int. Congr. Catal., London, 1976, p. 915.
68. J.R. Anderson and Y. Shimoyama, Proc. 5th Int. Congr. Catal., Miami, paper 47, p. 695 (1972).
69. Z. Karpiński and J.K.A. Clarke, Faraday Trans. **71**, 2310 (1975).
70. A. Sárkány, I. Guczi, and T. Tétényi, J. Catal. **39**, 181 (1975).
71. Y.L. Lam and J.H. Sinfelt, J. Catal. **42**, 319 (1976).
72. W.D. Gillespie and G.A. Somorjai, J. Cat., to be published (1981).
73. A. Roberti, V. Ponec, and W.M.H. Sachtler, J. Catal. **28**, 381 (1973).
74. J.R. Anderson and D.E. Mainwaring, J. Catal. **35**, 162 (1974).
75. J.H. Sinfelt, H. Hurwitz, and J.C. Rohrer, J. Catal. **1**, 481 (1962).
76. J.G. Brandenberger, W.L. Callender, and W.K. Meerbott, J. Catal. **42**, 282 (1976).
77. C. Corolleur, F.G. Gault, and L. Beránek, React. Kinet. Catal. Lett. **5**, 459 (1976).
78. H. Kubicka, J. Catal. **12**, 223 (1968).

79. J.L. Carter, J.A. Cusumano, and J.H. Sinfelt, J. Catal. **20**, 223 (1971).
80. A.V. Sklyarov, O.V. Kyrlov, and G. Keulks, Kinet. Katal. **18**, 1213 (1977).
81. D.G. Grenoble, J. Catal. **56**, 32 (1979).
82. V.N. Mozhaiko, G.L. Rabinovich, G.N. Maslyanskii, and L.P. Erdyakova, Neftekhimiya **15**, 95 (1975).
83. K. Kochloefl, Proc. 6th Int. Congr. Catal., Lond., 1122 (1976).
84. G. Lietz and J. Völter, J. Catal. **45**, 121 (1976).
85. G. Marie, G. Plouidy, J.C. Prudhomme, and F.G. Gault, J. Catal. **4**, 556 (1967).
86. O. Beek, Rev. Mod. Phys. **17**, 61 (1945), and references cited therein.
87. J.S. Campbell and P.H. Emmett, J. Catal. **7**, 252 (1967).
88. I. Yasumori, H. Shinohara, and Y. Inoue, Proc. 5th Int. Congr. Catal., Miami, paper 52, p. 771 (1972).
89. G. Dalmai-Imelik and J. Massardier, Proc. 6th Int. Congr. Catal. Lond., 90 (1976).
90. J.C. Schlatter and M. Boudart, J. Catal. **24**, 482 (1972).
91. A. Frackiewicz and Z. Karpiński, Proc. 5th Int. Congr. Catal., Miami, paper 42, p. 635 (1972).
92. R.E. Cunningham and A.T. Gwathmey, Adv. Catal. **9**, 25 (1957).
93. P. Pareja, A. Amariglio, and H. Amariglio, J. Catal. **36**, 379 (1975).
94. P. Pareja, A. Amariglio, and H. Amariglio, React. Kinet. Catal. Lett. **4**, 459 (1976).
95. R.L. Moss, D. Pope, and H.R. Gibbens, J. Catal. **46**, 204 (1977).
96. V.B. Kazanskii and V.P. Strunin, Kinet. Katal. **1**, 553 (1960).
97. J.H. Sinfelt and P.J. Lucchesi, J. Am. Chem. Soc. **85**, 3365 (1963).
98. J.H. Sinfelt, J. Phys. Chem. **68**, 856 (1964).
99. T.A. Dorling, M.J. Eastlake, and R.L. Moss, J. Catal. **14**, 23 (1969).
100. S.M. Davis and G.A. Somorjai, J. Catal. **65**, 78 (1980).
101. P. Chatoransky, Jr., and W.L. Kranich, J. Catal. **21**, 1 (1971).
102. P.A. Sermon, G.C. Bond, and P.B. Wells, Faraday Trans. **75**, 2385 (1979).
103. R.A. Ross, G.D. Martin, and W.G. Cook, Ind. Eng. Chem. Prod. Res. Develop. **14**, 151 (1975).
104. G.D. Lyubarskii, L.I. Ivanovskaya, and G.G. Isaeva, Kinet. Katal. **1**, 235 (1960).
105. J.W.E. Coenen, R.Z.C. Van Meerten, and H.Th. Rijnten, Proc. 5th Int. Congr. Catal., Miami, 1972, paper 45, p. 671.
106. R.Z.C. Van Meerten, A.C.M. Verhaak, and J.W.E. Coenen, J. Catal. **44**, 217 (1976).
107. G.D. Lyubarskii, L.I. Ivanovskaya, G.L. Isaeva, D.I. Lanier, and N.M. Kogan, Kinet. Katal. **1**, 358 (1960).
108. G.M. Dixon and K. Singh, Faraday Trans. **65**, 1129 (1969).
109. P. Gallezot, V. Datka, J. Massardier, M. Primet, and B. Imelik, Proc. 6th Int. Congr. Catal., Lond., 696 (1976).
110. J.M. Bassett, G. Dalmai-Imelik, M. Primet, and R. Mutin, J. Catal. **37**, 22 (1975).
111. T.A. Dorling and R.L. Moss, J. Catal. **5**, 111 (1966).
112. W.F. Taylor, J. Catal. **9**, 99 (1967).
113. P.C. Aben, J.C. Platteeuw, and B. Southamer, Proc. 4th Int. Congr. Catal., Moscow, 1968, paper 31.
114. M.S. Borisova, V.A. Dzisko, and Yu.O. Bulgakova, Kinet. Katal. **12**, 344 (1971).
115. R.Z.C. Van Meerten and J.W. E. Coenen, J. Catal. **37**, 37 (1975).
116. W.F. Taylor and H.K. Staffin, Faraday Trans. **63**, 2309 (1967).
117. R.L. Moss, D. Pope, B.J. Davis, and D.H. Edwards, J. Catal. **58**, 206 (1979).
118. A.O. Cinneide and J.K.A. Clarke, J. Catal. **26**, 233 (1972).
119. A.V. Ramaswamy, P. Ratnasamy, S. Sivasanker, and A.J. Leonard, Proc. 6th Int. Congr. Catal., Lond., 855 (1976).

120. A. Compero, M. Ruiz, and R. Gómez, React. Kinet. Catal. Lett. **5**, 177 (1976).
121. Yu. S. Snagovskii, G.D. Lyubarskii, and G.M. Ostrovskii, Kinet. Katal. **7**, 232 (1964).
122. P. Van Der Plank and W.M.H. Sachtler, J. Catal. **12**, 35 (1968).
123. B. Coughlan, S. Narayanan, W.A. McCann, and W.M. Carroll, J. Catal. **49**, 97 (1977).
124. R. Gómez, S. Fuentes, F.J. Fernández del Valle, A. Campero, and J.M. Ferreira, J. Catal. **38**, 47 (1975).
125. M.A. Vannice and W.C. Neikam, J. Catal. **23**, 401 (1971).
126. K.M. Sancier, J. Catal. **23**, 404, (1971).
127. R. Maurel, G. Leclercq, and J. Barbier, J. Catal. **37**, 324 (1975).
128. R. Ratnasamy, J. Catal. **31**, 466 (1973).
129. G.N. Maslyanskii, B.B. Zharkov, and A.Z. Rubinov, Kinet. Katal. **12**, 699 (1971).
130. N.R. Bursian, S.B. Kogan, N.M. Dvorova, L.P. Erdyakova, I.A. Korchagina, and N.K. Volnykhina, Kinet. Katal. **18**, 197 (1977).
131. Yu.I. Ermakov, B.N. Kuznetsov, Yu.A. Ryndin, and V.K. Duplyakin, Kinet. Katal. **15**, 978 (1974).
132. C.P. Khulbe and R.S. Mann, Proc. 6th Int. Congr. Catal., Lond., 447 (1976).
133. H.G. Rushford and D.A. Whan, Faraday Trans. **67**, 3377 (1971).
134. R.L. Burwell Jr., H.H. Kung, and R.J. Pellet, Proc. 6th Int. Congr. Catal., Lond., 108 (1976).
135. E.E. Gonzo and M. Boudart, J. Catal. **52**, 462 (1978).
136. V.I. Garanin, U.M. Kurkchi, and Kh.M. Minachev, Kinet. Katal. **9**, 889 (1968).
137. E. Segal, R.J. Madon, and M. Boudart, J. Catal. **52**, 45 (1978).
138. R. Maatman, W. Ribbens, and B. Vonk, J. Catal. **31**, 384 (1973).
139. J. Haro, R. Gómez, and J.M. Ferreira, J. Catal. **45**, 326 (1976).
140. R.W. Maatman, P. Mahaffy, P. Hoekstra, and C. Addink, J. Catal. **23**, 105 (1971).
141. M. Kraft and H. Spindler, Proc. 4th Int. Congr. Catal., Moscow, paper 69 (1968).
142. N.M. Zaidman, V.A. Dzis'ko, A.P. Karnaukhov, N.P. Krasilenko, N.G. Koroleva, and G.P. Vishnyakova, Kinet. Katal. **9**, 709 (1968).
143. J.A. Cusumano, G.W. Debinski, and J.H. Sinfelt, J. Catal. **5**, 471 (1966).
144. I. Horescu and A.P. Rudenko, Russ. J. Phys. Chem. **44**, 1601 (1970).
145. F. Figureras, B. Mencier, L. DeMorgues, C. Naccache, and Y. Trambouze, J. Catal. **19**, 315 (1970).
146. P. Biloen, F.M. Dautzenberg, and W.M.H. Sachtler, J. Catal. **50**, 77 (1977).
147. N. Nakamura, M. Yamoda, and A. Amano, J. Catal. **39**, 125 (1975).
148. F.M. Dautzenberg and J.C. Platteeuw, J. Catal. **19**, 41 (1970).
149. J.K.A. Clarke, E. McMahon, and A.D. O'Cinneide, Proc. 5th Int. Congr. Catal., Miami, paper 46, p. 685 (1972).
150. U.M. Kurkchi, V.I. Garanin, and Kh.M. Minachev, Kinet. Katal. **9**, 472 (1968).

9. Hydrocarbon Conversion on Platinum

Introduction

Platinum is one of the most versatile, all-purpose, heterogeneous metal catalysts. It is employed under reducing conditions (in the presence of excess hydrogen) for the conversion of aliphatic straight-chain hydrocarbons to aromatic molecules (dehydrocyclization) and to branched molecules (isomerization), and for hydrogenation on a large scale in the chemical and petroleum-refining industries.[1] It is used as an oxidation catalyst for ammonia oxidation, an important step in the process of producing fertilizers.[2] Platinum is the catalyst for the oxidation of carbon monoxide and unburned hydrocarbons in the control of car emissions.[3] Platinum is perhaps the most widely used and most active electrode for catalyzed reactions in electrochemical cells.[4] Its chemical stability in both oxidizing and reducing conditions makes this metal an ideal catalyst in many applications. Mined mostly in South Africa and in the USSR, platinum, along with rhodium (which occurs as an impurity in platinum ores), is very rare and therefore expensive. Its regeneration and recovery must be an important part of any technology that uses this metal.

For this reason it is of considerable importance to scrutinize the catalytic activity of platinum on the atomic scale, to learn what makes this metal so versatile as a catalyst and so selective for important catalytic transformations after suitable preparation. Once the elements of catalytic activity are revealed, it should be possible to use this metal more economically or perhaps to find ways to synthesize new catalyst systems to substitute for this excellent but rare catalyst.

Let us concentrate on the atomic-scale study of the platinum surface under reducing conditions used during hydrocarbon conversion reactions. In this circumstance H—H, C—H, and C—C bond-breaking processes are essential. In Figure 9.1 the various hydrocarbon conversions of interest are listed. Dehydrogenation involves C—H bond breaking only, while hydrogenolysis necessitates the breaking of C—C bonds. Dehydrocyclization must involve the complex process of dehydrogenation and ring closure.

Figure 9.1 shows several reactions that are all catalyzed by platinum. The simpler hydrogenation and dehydrogenation reactions have turnover frequencies in the range 0.1 to 10 sec^{-1} under the usual conditions (400 to 600 K, atmospheric pressures of reactant and excess hydrogen)

HYDROGENATION

-DEHYDROGENATION

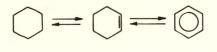

HYDROGENOLYSIS

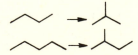

ISOMERIZATION

DEHYDROCYCLIZATION

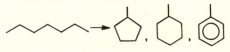

Figure 9.1. Several competing hydrocarbon reactions that occur on platinum catalyst surfaces.

that are employed in the chemical technology.[5] However, platinum is really noted for being an excellent catalyst for the more complex reactions of dehydrocyclization (for example, *n*-heptane to toluene) and isomerization (for example, *n*-pentane to 2-methyl butane) that have turnover frequencies of about 10^{-4} to 10^{-2} sec^{-1} under similar experimental conditions to those used to carry out the more facile reactions.[6] One of the key questions in the molecular-scale study of the hydrocarbon catalysis of platinum is the way this metal catalyzes the complex, low-turnover frequency reactions selectively while blocking the simpler, high-rate dehydrogenation and hydrogenation reactions and the slower, but unwanted, hydrogenolysis reaction. This happens after suitable preparation of the platinum catalyst prior to exposure to the reaction mixture. Boudart et al.[7] reported a 27-fold increase in the selectivity for the isomerization of neopentane as compared to its hydrogenolysis when the supported Pt catalyst was pretreated at high temperature (900°C). When prepared differently, platinum can be an excellent and selective hydrogenolysis or dehydrogenation catalyst as well.[1] By depositing a partial monolayer of strongly bound oxygen by heat treatment in oxygen, one can increase the hydrogenolysis rate of *n*-heptane or other alkanes by tenfold or more.[8] Several systematic studies have led to proposals of reaction pathways for the complex isomerization and dehydrocyclization reactions.[9] Anderson and Avery[10] have studied *n*-butane, isobutane, neopentane, and isopentane over evaporated platinum films. They report evidence indicating the importance of multiple bonds between adsorbed molecules and the metal surface. They suggest the presence of diadsorbed and triadsorbed intermediates. Intramolecular rearrangements by bond shifts are proposed as leading to the formation of the isomers.[11] Csicsery and Burnett[12] have suggested that isomerization and dehydrocyclization both involve the formation of cyclic intermediates. Several other groups[9,13–15] have come to the same conclusion, with the necessary presence of a C_5 cyclic intermediate gaining the greatest acceptance. In Figure 9.2 different paths for the isomerization of *n*-heptane are shown, all involving cyclic intermediates. Dehydrogenation of the cyclohexane intermediate leads to the formation of aromatic molecules, while ring-opening hydrogenolysis reactions lead to the formation of isomers of *n*-heptane.

While some of the investigators have found evidence for the predominance of the bond-shift mechanism, and others have proposed the cyclic ring-opening mechanism operating exclusively, all of them seem to agree that the isomerization and dehydrocyclization reactions are

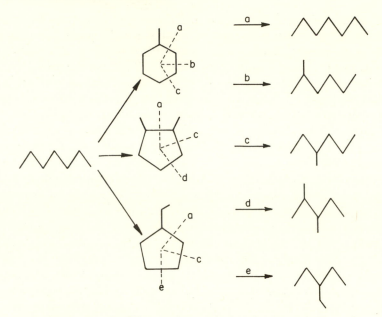

Figure 9.2. Possible reaction path for the isomerization of *n*-heptane via five- or six-member-ring intermediates.

structure-sensitive. That is, the reaction rates and the selectivity change markedly with changing catalyst particle size. This type of behavior was classified by Boudart,[16] and recently and more elaborately by Monogue and Katzer[17] and Blakely and Somorjai.[18] Boudart's classification separates reactions into two groups. In the first the specific reaction rate depends on the mean particle size of the catalyst; in the second the specific reaction rate is independent of the mean particle size of the catalyst. Changes in the mean metal particle size are generally accomplished by sintering the catalyst and imply changes in the surface structure of the metal particle, although how the surface structure changes is unknown.

A number of other reactions besides the isomerization reactions just described have been categorized. The dehydrogenation of cyclohexane[19] and the hydrogenation of benzene[20] are independent of the metal particle size on supported catalysts in polycrystalline foils. The hydrogenation of cyclopropane to propane is structure-insensitive on both supported catalysts[21] and single-crystal catalysts.[22] The hydrogen–deuterium exchange has been found to be structure-insensitive on supported catalysts at high pressures, but Bernasek and Somorjai found the reaction to be structure-sensitive on single-crystal catalysts at low pressure.[23] Possibly all the structure-insensitive reactions proceed through a structure-sensitive step, but the sensitive step is not rate-limiting in the conditions of the experiment.

These macroscopic reaction rate and selectivity studies have been very valuable in pointing out important features of hydrocarbon catalysis by platinum. The sensitivity of the rates and product distribution to the surface structure of the catalyst is revealed. The importance of additives such as oxygen in influencing the reaction mechanism has also been uncovered.[24]

One important aspect of catalysis research involves studies of catalyst deactivation and regeneration. A typical catalytic re-forming unit that uses dispersed platinum catalyst to carry out the dehydrocyclization and isomerization functions may contain 200,000 lb of catalyst pellets that are impregnated with about 1000 lb of suitably prepared platinum or bimetallic catalysts (Pt–Re, Pt–Sn, etc.). As the catalyst begins to deactivate, as indicated by the decreasing conversion rates, the temperature of the reactor is increased to maintain the reaction rates. Increasing temperatures may also lead to increasing deactivation-rate processes and finally to the complete deactivation of the catalyst, due to carbon deposits or to the deposition of impurities from the reactant feed. At this point the catalyst must be regenerated using oxidation heat treatments followed by reduction. It is essential that the catalyst retain its active surface structure and large surface area during all these chemical processes to avoid the need for frequent and very costly replacement of the catalyst bed. Thus the kinetics of metal particle growth and redispersion are subjects of intense investigation in the field of catalysis.[25]

During the past decade, we have studied several hydrocarbon reactions on platinum single-crystal surfaces of various atomic structures to identify the active sites where H—H, C—H, and C—C bond-breaking reactions occur.[26] These included the exchange of H_2 and D_2; the dehydrogenation and hydrogenolysis of cyclohexane; the dehydrogenation, hydrogenation, and hydrogenolysis of cyclohexene; and the dehydrocyclization and hydrogenolysis of n-heptane. These reactions were carried out at both low pressures ($<10^{-4}$ torr total pressure) and high pressures (>10 torr). At low pressures the initial rates and product distributions were those characteristic of the clean metal surface. At high pressures the surface became covered with a near-monolayer of carbonaceous deposit within seconds, and then the reaction rates and product distributions were characteristic of the covered platinum surface.

From these studies came several significant discoveries. One was the identification of active sites, atomic steps, and kinks for the H—H, C—H, and C—C bond-breaking reactions. In addition, we have uncovered the importance of active carbonaceous monolayer deposits that form on the platinum surfaces in controlling the selectivity during hydrocarbon conversion. Additives such as oxygen and chlorine, together with surface irregularities, steps, and kinks, were also found to be essential to produce the reaction selectivities for which platinum is noted.[27] As a result of these studies, an atomic-scale picture of the working platinum catalyst emerges. Two important elements that play key roles in catalysis by platinum can be identified:[28]

1. Surface irregularities, steps, and kinks have unique activities in the breaking of H—H, C—H, and C—C bonds. Their availability on a well-prepared surface or their deactivation by impurity blocking can markedly change the selectivity in hydrocarbon conversion reactions.

2. Additives present on the surface because of catalyst pretreatment (by oxygen or chlorine, for example) or because they are deposited from the reaction mixture (carbonaceous overlayer) are essential chemical components of the working platinum catalyst, influencing both its selectivity and its activity. Additives may also cause structural reconstruction, thereby drastically altering catalyst behavior.

Structures of Clean, Single-Crystal Platinum Surfaces with Distinguishable Catalytic Activities

There are three types of platinum surfaces with readily distinguishable chemical activities. These are shown in Figure 9.3. The low-Miller-index platinum (111) surface has a very low density of surface imperfections as compared to the number of surface atoms on its terraces (1.5×10^{15} atm/cm^2). The high-Miller-index surface, of which the Pt(557) and Pt(679) crystal faces are examples, is characterized by stable step structure; 10 to 40 percent of the surface atoms are located at steps of monatomic height. The Pt(679) surface has, in addition to high step density, a high density of kinks in the steps. These are sites of even lower coordination number than step atom sites.

By cutting the single crystal in well-defined directions with respect to the (111) or (100) low-Miller-index crystal faces, surfaces with different terrace widths and kink densities in the steps can be prepared. These high-Miller-index surfaces, their structure and preparation, were discussed in detail in Chapter 3. Low-energy electron diffraction identifies the terrace width and the step height as long as the step structure is periodic.

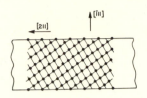

a) Pt - (Ī11)

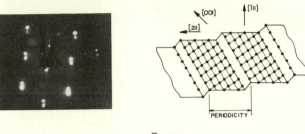

b) Pt - (557)

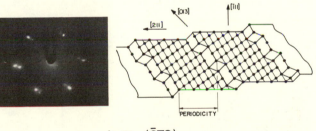

c) Pt - (679)

Figure 9.3. LEED pattern and schematic representations of the surface configurations of platinum single-crystal surfaces. (a) Pt(111) containing fewer than 10^{12} defects/cm². (b) Pt(557) crystal face containing 2.5×10^{14} step atoms/cm² with an average spacing between steps of six atoms. (c) Pt(679) crystal surface containing 2.3×10^{14} step atoms/cm² and 7×10^{14} kink atoms/cm², with an average spacing between steps of seven atoms, and between kinks of three atoms.

Structure-Sensitive Deactivation of Clean Platinum Surfaces during Hydrocarbon Reactions

Several catalyzed hydrocarbon reactions can be studied at low pressures (10^{-8} to 10^{-4} torr), starting with clean platinum crystal faces. In Figure 9.4 the turnover numbers for the dehydrogenation of cyclohexene are plotted as a function of reaction time for the flat (111), a stepped surface, and a kinked surface.[29] The rates increase as a function of time, reach a maximum value, then drop rapidly as deactivation commences. The rates of deactivation are most rapid for the flat (111) crystal face, followed by the stepped surface; the kinked surface deactivates at the slowest rate [(111) > stepped > kinked]. Interestingly, the kinked surface is not only the slowest to poison but also the most active. In fact, the activity follows the opposite trend, kinked > stepped > (111), for this reaction on a clean surface.

The reason for the catalyst deactivation at low pressures is the buildup

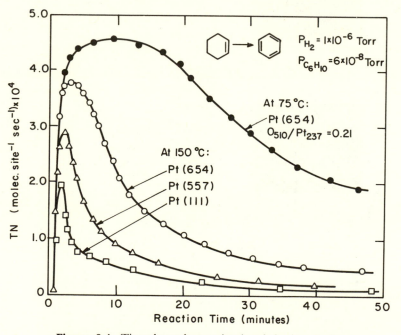

Figure 9.4. Time-dependent poisoning behavior during cyclohexene dehydrogenation to benzene at constant pressure (6 × 10^{-8} torr hydrocarbon, 1 × 10^{-6} torr hydrogen). The effects of surface structure and temperature and the presence of surface oxygen are shown.

of a carbonaceous deposit on all three crystal faces that can be monitored by Auger electron spectroscopy (AES). This is shown in Figure 9.5, where both the rate of reaction and the carbon-to-platinum Auger peak ratios are plotted as a function of time. Clearly, the buildup of the carbonaceous deposit over 50 percent of a monolayer can be correlated with the rapid deactivation of the platinum surface.

Why is the most reactive kinked platinum surface poisoned at the slowest rate, while the least active Pt(111) crystal face is poisoned at the fastest rate? It appears that hydrogen binds more strongly at the kink and step sites than on the terrace sites. This is demonstrated by the thermal desorption spectra of hydrogen from the clean platinum crystal faces of different surface structure (Figure 6.2). The (111) face exhibits one thermal desorption peak that appears near 700 K; the stepped Pt surface, two thermal desorption peaks, one of which is at a higher temperature than the one on the (111) crystal face. The kinked surface exhibits three thermal desorption peaks, one of which is of the highest desorption temperature found for any platinum surface.

The strong hydrogen-bonding and retention capability of the kinked Pt surfaces appears to slow down the deactivation due to carbon deposition. The other surfaces, which do not retain hydrogen so strongly,

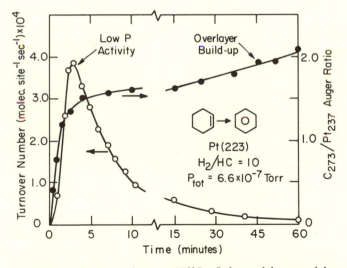

Figure 9.5. Comparison at 150°C of the cyclohexene dehydrogenation rate over Pt(223) at low pressures, with the simultaneous buildup of the irreversibly chemisorbed carbonaceous overlayer, C_{273}/Pt_{237} ratio of 2.8 corresponds to monolayer coverage.

poison much faster. There could be other reasons for the structure sensitivity of deactivation that are related to the structure of the carbonaceous deposit as it nucleates and grows differently on the different platinum crystal faces. The structure and bonding of this carbon deposit are under intensive investigation at present.

Structure Sensitivity of the H—H, C—H, and C—C Bond Breaking of Clean Platinum Crystal Surfaces

One of the striking confirmations of the importance of surface structure in chemical activity comes from molecular beam–surface scattering studies.[30] A molecular beam containing a mixture of H_2 and D_2 is directed at a step surface of a platinum crystal. The probability of forming HD exhibits a marked dependence on the orientation of the step with respect to the incident beam. This is shown in Figure 7.13. As the angle of incidence is varied by the rotation of the crystal, the rate of HD production drops by 50 percent over a wide variety of reaction conditions (curve A). The reaction probability is highest when the beam strikes the open edge of the step, and lowest when the bottom of the step is shadowed. When the H_2–D_2 mixed beam is incident parallel to the steps at all angles of crystal rotation, the rate of HD production is independent of the angle of incidence (curve B). These results and a simple calculation indicate that the area between the platinum step and the terrace is about seven times more active for the dissociative adsorption of the hydrogen molecules than are the platinum terrace atoms.

The reason for this different activity at the step and terrace atom sites on terraced platinum surfaces is revealed by detailed kinetic analysis of the molecular-beam-scattering data.[31] The dissociative adsorption of hydrogen appears to be an activated process on the Pt(111) surface, with a barrier height of 0.5 to 1.5 kcal/mol. On the other hand, the atomization of hydrogen requires no activation energy on the stepped surfaces of platinum. This alone accounts for the different reaction probabilities of the two surface sites.

Although the dissociation probability of the hydrogen molecule on the stepped surfaces is appreciably higher than on the Pt(111) surface, after dissociation has taken place the kinetics and mechanism of HD formation are identical on both stepped and low-Miller-index platinum surfaces.

The recombination of H and D atoms to form HD follows on both surfaces a parallel mechanism, with one of the branches operative in the entire temperature range studied, 25 to 800°C. This branch has an activation energy and pseudo-first-order preexponential of $E = 13.0$ kcal/mol and $A_1 = 8 \times 10^4$/sec^{-1} for the stepped surface, and $E_1 = 15.6$ kcal/mol and $A_1 = 2.7 \times 10^5$/sec^{-1} for the Pt(111) surface.

For temperatures above 300°C the second branch is observed, but the values of the preexponential factors and activation energies could not be uniquely determined for either of the two crystals. Below 300°C a third process appears in series with the first branch. Thus the mechanism of H_2–D_2 exchange is quite complex at the low pressures studied by the molecular beam techniques and certainly is structure-sensitive. The structure sensitivity disappears at high surface coverages of hydrogen that is obtained at higher pressures as the reaction mechanism changes with coverage.

Let us concentrate on the reactivity of these irregularities, steps, and kinks on the platinum surfaces for breaking C—H and C—C bonds. For this purpose the dehydrogenation of cyclohexane and cyclohexene, which involves only C—H bond breaking, was monitored as a function of changing step and kink densities. Simultaneously, the hydrogenolysis of cyclohexane to form n-hexane was also monitored.[18] The latter reaction involved C—C bond scission. By determining the ratio of benzene to n-hexene, one could find out the efficiency for C—H and C—C bond breaking respectively on a given surface.

During all these low-pressure studies,[18,32] the optimum rates were determined and reported as obtained before the onset of deactivation. Thus the reactivities reported are the properties of clean platinum.

The results of studies of these dehydrogenation and hydrogenolysis reactions on a series of platinum surfaces with varying step concentrations are shown in Figures 9.6 and 9.7. It should be noted that the rate of dehydrogenation of cyclohexene is about two orders of magnitude faster at these low pressures than the rate of hydrogenation of cyclohexane.[18,32] As shown in Figure 9.6a, the rate of cyclohexene dehydrogenation increases with increasing step density. This indicates that the steps are indeed active in breaking the C—H bonds in these molecules. It should be recalled from our previous studies of H_2–D_2 exchange that step surfaces were also active in breaking H—H bonds. In Figure 9.6b the kink density is varied while step density is constant.[18,32] Although perhaps there is a small increase in the rate of dehydrogenation, the results are inconclusive because of the large scattering in the data. Figure 9.7, however, reveals significant differences between the chemistry of stepped and kinked surfaces. The dehydrogenation of cyclohexane is step-density-independent. This reaction, which is slow compared with the dehydrogenation of cyclohexene, appears to be structure-insensitive.

489

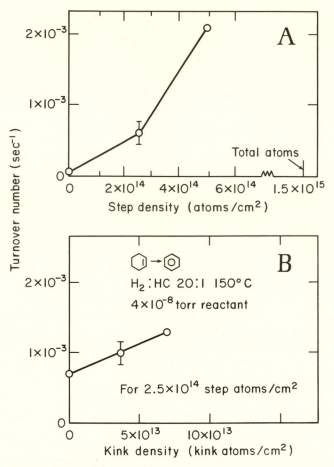

Figure 9.6. Cyclohexene dehydrogenation to benzene as a function of (A) step density, and (B) kink density.

Some degree of hydrogenolysis is detectable, increasing slowly with step density. However, on kinked surfaces the hydrogenolysis rate appears to be an order of magnitude larger than on stepped surfaces. This is clearly detectable by comparing the slopes for the hydrogenolysis rates as a function of kink concentration in Figure 9.7a and b. The hydrogenolysis rate increases by an order of magnitude by changing the kink density by a factor of 3. In fact, we can conclude that the residual hydrogenolysis activity of the stepped surfaces is probably due to thermally generated kinks in the steps. It appears that while steps are active in breaking H—H and C—H bonds, the kinks are active in breaking C—C bonds in

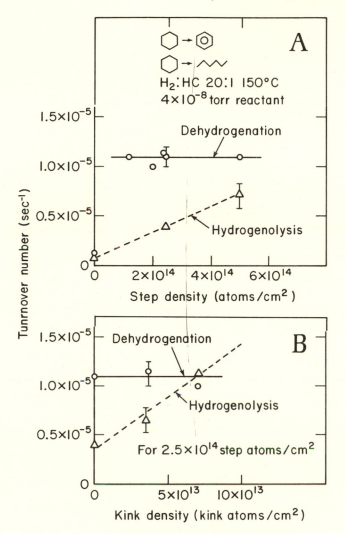

$H_2:HC$ 20:1 150°C
4×10^{-8} torr reactant

Figure 9.7. Cyclohexane dehydrogenation to benzene and hydrogenolysis to *n*-hexane (A) as a function of step density, and (B) as a function of kink density at a constant step density of $2.0 \times 10^{14}/$ cm^2.

addition to C—H and H—H bonds. Thus there are two structurally different active sites on platinum that are distinguishable.[33]

The unique ability of kink sites on platinum to break strong chemical bonds is well demonstrated by the photoelectron spectroscopy studies of CO chemisorption on kinked platinum surfaces.[34] At low surface coverages, where the CO surface concentration is below the kink concentration, only one photoelectron peak corresponding to the carbon 1s binding states is detectable. As the CO coverage is increased above the kink concentration, a second photoelectron peak, corresponding to photoemission from carbon in the CO molecules, is also detectable. The photoelectron spectra of CO on platinum obtained at different exposures is shown in Figure 9.8. It appears that CO dissociates as long as free kink sites are available. As soon as all the kink sites are covered with carbon, which appears to block these sites, molecular CO appears. In fact, platinum is widely known for not being able to dissociate carbon monoxide. The small concentration of kink sites can break this very strong bond (dissociation energy ~ 256 kcal) as compared to the H—H, C—H, and C—C bonds, but they are blocked and rendered inactive.

It is frequently found that the addition of a small amount of a second, noncatalytic metal component to a metal catalyst, such as platinum, iridium, or ruthenium, rapidly reduces the hydrogenolysis activity (C—C bond breaking) while influencing only slightly the dehydrogenation or isomerization activity.[26,35] The selectivity that can be introduced by alloying has found significant applications in many industrial catalyst systems. For example, the effect of adding copper to ruthenium on the

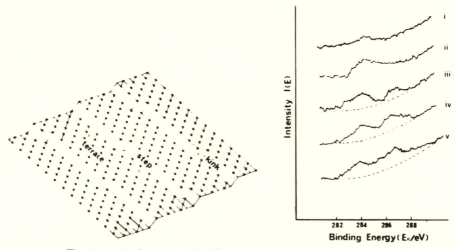

The average atomic arrangement at the Pt(S)–(6(111)×(710)) surface.

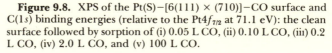

Figure 9.8. XPS of the Pt(S)−[6(111) × (710)]−CO surface and C(1s) binding energies (relative to the Pt4$f_{7/2}$ at 71.1 eV): the clean surface followed by sorption of (i) 0.05 L CO, (ii) 0.10 L CO, (iii) 0.2 L CO, (iv) 2.0 L CO, and (v) 100 L CO.

dehydrogenation and hydrogenolysis activity of this metal[36] is shown in Figure 8.7. While the dehydrogenation of cyclohexane is unaffected, the hydrogenolysis rate of ethane is reduced substantially by the addition of a small amount (20 atom percent) of copper. Similar results were obtained by the addition of small amounts of gold to iridium and platinum surfaces.[35] We can associate the dramatic decrease in hydrogenolysis activity with the addition of a second metal constituent—gold or copper—to the surface of a catalyst, thereby blocking the kink sites on the catalyst surface where the C—C bond breaking occurs with great efficiency. The surface concentration of kink sites is fairly small, no more than 5 to 10 percent per monolayer. When these sites are blocked, the C—C bond-breaking reactions are largely eliminated, while the other reactions that do not require C—C bond scission (such as isomerization and dehydrogenation) are not affected much by the small decrease in the available active surface area.

The atomic structure of steps can be varied depending on the angle and orientation of the crystal face. In Figure 4.15, stepped surfaces that have different step structures are displayed. One is composed of (111) orientation microfacets in the step, the other of (100) orientation facets. These different step structures may exhibit different chemical activities for bond breaking, binding, or rearrangements, which are yet to be explored and identified. Similarly, kinked surfaces may be prepared with a variety of atomic structures that may have different chemical activities, which also remain to be explored.

Identical Reaction Rates for Single-Crystal and Dispersed Platinum Catalysts for Structure-Insensitive Reactions at High Pressures: The Ring Opening of Cyclopropane

The ring opening of cyclopropane was studied at 1 atm total pressure on the stepped single-crystal surfaces of platinum as well as the platinum (111) crystal face.[22] Hydrogenolysis of cyclopropane was chosen as a test reaction because of the considerable amount of data and experience that have been collected by various laboratories in studies of this reaction. The rate is relatively high, even at room temperature, on supported platinum catalysts, and only one product, propane, is formed below 150°C, thereby simplifying the analysis of the results. Table 8.4 summarizes the results that were obtained and compares the results on stepped single-crystal surfaces at atmospheric pressure with those of others obtained using supported platinum catalysts. It appears that at 1 atm pressure, the platinum stepped single crystal behaves very much like the highly dispersed supported platinum catalyst for the cyclopropane hydrogenolysis. In addition, the same studies that were carried out on the Pt(111) crystal face result in reaction rates identical to those found on stepped crystal surfaces of platinum. These observations support the contention that well-defined crystal surfaces can be excellent models for polycrystalline-supported metal catalysts. They also tend to verify Boudart's classification of cyclopropane hydrogenolysis as an example of a structure-insensitive reaction. The initial specific reaction rates on the crystal surfaces, which were reproducible within 10 percent, are, within a factor of 2, identical to earlier published values for this reaction on dispersed platinum catalysts. The activation energies that were observed for this reaction, in addition to the turnover number, are similar enough on the various platinum surfaces that we may call the agreement excellent.

Effects of Additives That Modify the Catalytic Behavior of Platinum

Two types of additives appear to play important roles in the heterogeneous catalysis by platinum and by other transition metals.[27,28] The first type is deposited on the surface by the reaction mixture under the conditions of the reaction (given pressure and temperature) during hydrocarbon reactions. This is a carbonaceous deposit that is adsorbed reversibly or irreversibly on the metal surface. Under oxidizing conditions the deposit may be an oxide layer. Since the catalytic surface reaction takes place on top of or in exchange with atoms and molecules in this deposit, its chemical behavior is an integral part of the catalytic process.

The second type of additive that exerts a profound influence on the catalytic activity of platinum and of other transition metals is that which is introduced during catalyst preparation before the reaction. Pretreatment using oxygen, chlorine, or the addition of potassium, cesium, and other alkali metals is carried out most frequently. These additives can then markedly influence the lifetime of a catalyst, its activity, and its selectivity. In these circumstances the additives are called *promoters*. Undesirable additives that deactivate the surface are called *catalyst poisons*.

The Carbonaceous Deposit

In Figure 9.5 the deposition of carbon on the platinum surface is displayed as a function of time. Using AES, the carbon buildup on the metal surface can be readily monitored during reaction studies at low pressures ($<10^{-4}$ torr) and at any reaction temperature. At high reactant pressures (≥ 10 torr) the near-monolayer carbonaceous deposit forms in seconds in the usual hydrogen–hydrocarbon mixtures used during catalyzed hydrocarbon reactions. Its removal by desorption or by rehydrogenation can be slow compared with the turnover frequency of the catalytic reactions that were responsible for its formation. Thus the hydrocarbon reactions may take place on top of or in exchange with this carbon-containing deposit. The deposit on the metal is schematically represented in Figure 9.9. This overlayer, together with its interaction

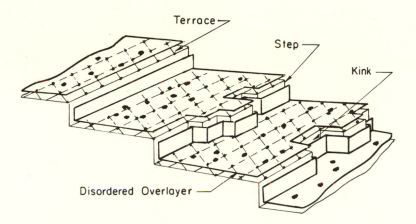

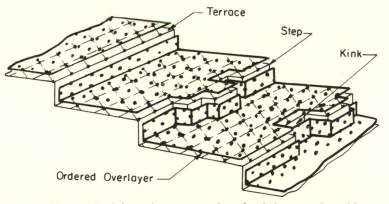

Figure 9.9. Schematic representation of a platinum catalyst with a monolayer of carbonaceous overlayer, showing the exposed platinum clusters.

with the adsorbates and its chemical bonding characteristics, is then an essential part of hydrocarbon conversion reactions. The platinum catalyst that operates selectively, in steady state, at high pressures for thousands of hours without any sign of deactivation, is covered with this carbonaceous deposit that forms instantly as the reaction commences.[29]

Let us concentrate on the nature of the carbonaceous deposit that forms on platinum surfaces during hydrocarbon reactions. It was found that the deposition of this carbon layer always precedes the desorption of benzene or olefinic or paraffinic products of the hydrocarbon reaction. There is always an induction period before the steady-state rate of production of the hydrocarbon product is established.[37] Mass-balance calculations and Auger electron spectroscopy indicate the deposition of 0.1 to 1.0 monolayer of carbon on the surface during this period. The chemical nature of the deposit depends on the temperature. The changing chemical nature of the deposit is revealed by thermal desorption studies, for example. The carbonaceous deposit that forms during the dehydrogenation of cyclohexane can be removed reversibly from the surface by thermal desorption or by displacement by other, more strongly bound molecules at temperatures below 450 K.[38] Above this temperature the carbonaceous layer is irreversibly adsorbed and can no longer be removed by heating the platinum crystal. The thermal desorption of benzene that forms during cyclohexane dehydrogenation is shown in Figure 9.10. The nature of bonding of the adsorbed aromatic molecule changes markedly in the presence of the surface carbon. A new, stronger binding state for benzene is created on the carbon-covered platinum. The platinum surfaces covered with carbonaceous deposits that form at the reaction temperatures exhibit high steady-state catalytic activity. This indicates that while some of the carbon may deactivate part of the platinum surface, much of the surface carbon does not impede the catalytic activity of the crystal. In fact, it appears that the chemical activity of this metal–carbon surface layer, and not the metal surface alone, controls the activity and selectivity of the hydrocarbon conversion.

One of the important properties of this carbonaceous deposit is hydrogen storage.[33] Thermal desorption studies indicate that a monolayer of deposit desorbs at least as much hydrogen upon heating as does the clean platinum surface. This hydrogen is also bound more strongly, as indicated by the higher-temperature thermal desorption peak, as compared with hydrogen thermal desorption from the clean platinum.[33] Once desorbed, the hydrogen (or deuterium) can be reintroduced by heating the deposit-covered platinum crystal in hydrogen in the temperature range 400 to 600 K. If the carbon deposit is saturated with

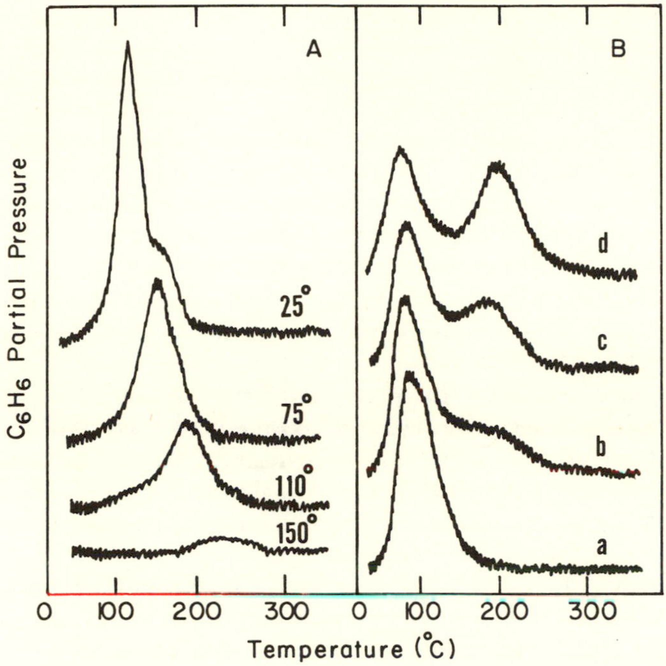

Figure 9.10. Benzene thermal desorption (10 K/sec): (A) after C_6H_{10} reactions at indicated T (°C); (B) C_6H_6 chemisorbed on (a) clean and (b–d) progressively carbon covered Pt(654).

deuterium, the adsorption of hydrocarbons on top of it leads to rapid exchange of H and D between the two layers of organic deposits.[38] Thus the hydrogen in the deposit is readily available to the adsorbed molecules on top of it.

The nature of this carbon-containing deposit has been subjected to experimental scrutiny by surface crystallography and by high-resolution electron energy loss spectroscopy as well as by recent reaction studies on transition-metal surfaces.

If the temperature is low enough, any hydrocarbon molecule will adsorb intact on any platinum surface.[39] If the platinum catalyst is heated to a high-enough temperature in the vapor of any hydrocarbon ($\geqslant 800$ K), a layer of graphitic carbon forms that poisons all catalytic activity. In the intermediate temperature range the adsorbed molecules undergo rearrangements—sequential bond breaking—that result in the formation of chemically active hydrocarbon fragments that constitute the catalytically active surface. The chemical bonding and the carbon-to-hydrogen ratio in this carbonaceous deposit change with temperature and reactant pressure.

A carbenelike metal–carbon bonding was uncovered by low-energy electron diffraction studies of C_2H_4 and C_2H_2 adsorbed on platinum surfaces.[40] These molecules form ordered surface structures on the Pt(111) surfaces. Intensity analysis of the diffraction beams and the electron energy loss spectrum at various surface temperatures revealed that the nature of the bonding changed markedly with increasing temperature. Above 375 K the molecules assume a stable surface structure that is depicted in Figure 5.17. This is an ethylidyne molecule with a C—C internuclear axis perpendicular to the platinum surface. The carbon atom closer to the metal is bonded to three platinum atoms at a Pt—C bond distance of 2.0 Å. This short metal–carbon bond is characteristic of carbenelike molecules (the Pt—C covalent bond distance is 2.2 Å), which exhibit unique chemical activity in many displacement reactions. The C—C bond is single-bond (1.5 Å) in character and is perpendicular to the surface plane. This structure resembles the structure of several trinuclear metal acetylene complexes.

In order to place the two carbon atoms of such symmetric molecules as C_2H_2 or C_2H_4 into asymmetric positions on the surface, shown in Figure 5.17, very strong metal–carbon bonds are needed. The carbenelike Pt_3—C bond can be estimated to be stronger than 68 kcal/mol from thermodynamic arguments. However, on platinum the Pt—C bond is not strong enough to cause breaking of the C—C bond under 400 K. This is likely to happen at or above 450 K, where the formation of stable CH or C_2H or other types of fragments occur. There is evidence from UV photoelectron spectroscopy studies that C_2H_4 breaks its C—C bond on iron surfaces at 300 K and CH_2 units remain on the surface.[41] High-resolution electron energy loss spectroscopy indicates the presence of C_2 and CH units on nickel surfaces as well.[42] The presence of a chemically active form of carbon is also reported on nickel surfaces that form as a result of the disproportionation of carbon monoxide in the temperature range 200 to 400°C. This active form of carbon yields methane by direct hydrogenation.[43] This appears to be the predominant mechanism of methane formation over nickel during the hydrogenation of CO.

Since the hydrogenation of the active carbon deposit on platinum occurs at a much slower rate compared with the hydrogenation of the active carbon on a nickel surface, the nature of the carbon fragments on the two metal surfaces must be different. On Pt surfaces there may be many active organic fragments with several carbon atoms linked together, while on Ni only isolated active carbon atoms may be present.

The active surface carbon with its carbenelike bonding appears to play an important role during hydrocarbon catalysis on transition-metal surfaces. It is not clear whether the surface reaction takes place over the active carbon-covered surfaces that play only the role of the catalyst, or whether the active carbon builds into the reaction product and is being constantly regenerated from the reactant. Isotope-exchange studies will be of help in the near future in elucidating the nature of participation of the active carbon deposit in heterogeneous catalysis.

The changes of bonding of acetylene with small variations of temperature and the sequential bond breaking of molecules with increasing temperature demonstrate one of the most important properties of the surface chemical bond, which is essential in determining the catalytic behavior of transition metals—its temperature-dependent character. On platinum, there are three temperature regimes distinguishable, with widely differing bonding of hydrocarbon molecules and chemical activities in hydrocarbon conversion reactions:

1. Low temperature (<340 K), high hydrogen pressures; clean metal catalysis; reversible adsorption of hydrocarbons.
2. Medium temperature (340 to 700 K), hydrogen pressures; catalysis by active C_xH_y fragments.
3. High temperature (>700 K), hydrogen pressures; multilayer carbon buildup; poisoning by graphite.

Investigation of the surface chemistry of the active carbon and carbonaceous species as a function of temperature on transition-metal surfaces appears to be a rich new field of catalytic chemistry. It is likely that the exploration of the surface chemistry of the deposit will help to build a bridge between metallorganic and heterogeneous catalytic chemistry.

Oxygen Additive to Platinum Surfaces

When platinum crystal surfaces of various atomic surface structures were heated in oxygen prior to the catalytic studies of hydrocarbon conversion reactions, marked changes of catalytic activity and selectivity occurred as a function of oxygen surface concentration on some of the crystal faces.[44] The rates of dehydrogenation of cyclohexene and cyclohexane and the rate of hydrogenation of cyclohexene were greatly enhanced by changes of the surface oxygen concentration during low-pressure studies. At high hydrocarbon and hydrogen pressures the rates of selective hydrogenolysis increased markedly in the presence of strongly bound surface oxygen.[45] Interestingly, only kinked surfaces of platinum showed substantial changes of reaction rates with changing oxygen surface coverage. It appears that the combination of the surface defect (kink site) and an additive at that site (oxygen) are needed to produce the observed changes in catalytic behavior.

Surprisingly, the oxygen surface concentration of the oxygen-pretreated catalysts remain unchanged during the low- or high-pressure reaction studies, which are always carried out in reducing atmospheres, in excess hydrogen. AES and thermal desorption studies have revealed the presence of a strongly bound surface oxygen on platinum that is not removed when heated in hydrogen. The thermal desorption spectrum of oxygen from platinum is shown in Figure 9.11.[33,44] This strongly bound oxygen desorbs at temperatures in excess of 1000°C, indicating a heat of desorption of 65 kcal/mol or larger. Thus oxygen pretreatment of platinum under appropriate conditions affixes a permanent surface concentration of oxygen during hydrocarbon catalysis. Analysis of these results indicates that a change of the electronic structure of the platinum surface through oxidation provides the best general model for explaining the oxygen effects, although surface reconstruction during oxidation or complex formation involving the adsorbed oxygen may also be important.[44] These results also demonstrate that additives such as oxygen, in

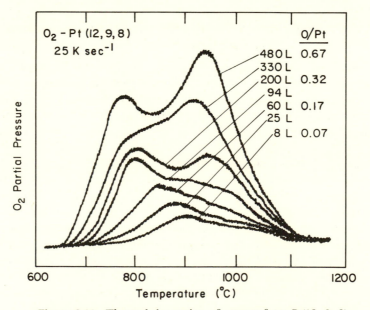

Figure 9.11. Thermal desorption of oxygen from Pt(12, 9, 8) crystal surface after exposure to different amounts of oxygen. (1 L = 1 Langmuir = 10^{-6} torr-sec.)

addition to surface irregularities, can play important roles in optimizing the rates and selectivity of hydrocarbon reactions over platinum catalysts. For supported catalysts strongly bound oxygen could be introduced by oxygen pretreatments or might arise naturally through interactions of platinum with the oxide support.

In Figure 9.12, the maximum turnover for benzene production from cyclohexene as a function of oxygen coverage is shown at low reactant pressures. The largest effect on reactivity of adsorbed oxygen was observed on the kinked Pt(S)-[8(111) × (310)] surface. In this surface the benzene production exhibits a maximum at an oxygen coverage of about half a monolayer (O/Pt Auger peak ratio, 0.25), which represented an increase in reactivity of about six times over the clean surface. At oxygen coverages greater than one monolayer, the benzene production fell off to a relatively constant value that was less than that observed for the clean kinked surface.

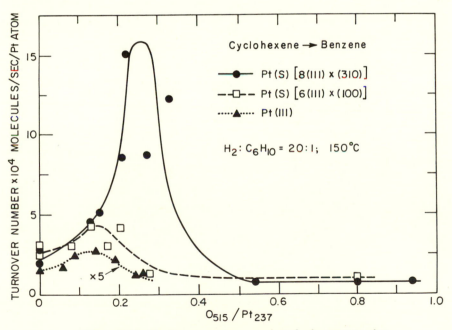

Figure 9.12. Maximum turnover numbers for benzene production from cyclohexene as a function of oxygen coverage ($O_{510}/Pt_{237} = 0.5$ corresponds to about 5×10^{14} oxygen atoms/cm^2) on Pt(S)-[8(111) × (310)], Pt(S)-[6(111) × (100)], and Pt(111). Reaction conditions: catalyst temperature 150°C, cyclohexene pressure 6×10^{-8} torr, hydrogen pressure 1×10^{-6} torr.

The maximum turnover numbers for cyclohexane production from cyclohexene are plotted as a function of oxygen coverage in Figure 9.13 for the three platinum samples. At the low-pressure conditions under which these reactions were carried out, the dehydrogenation product is thermodynamically favored over cyclohexane production. Cyclohexane formation was below the level of detectability in all the experiments over the clean platinum surfaces. With the addition of adsorbed oxygen the product, cyclohexane, was observed with the kinked Pt(S)-[7(111) × (310)] and the stepped Pt(S)-[6(111) × (100)] but remained undetectable with the Pt(111). On both the kinked and stepped surfaces, the maximum hydrogenation activity occurred at about one-third monolayer oxygen coverage (O/Pt Auger peak ratio, 0.15) and was greater for the kinked surface. Hydrogenation activity on the kinked surface decreased to undetectable levels at oxygen coverages greater than one monolayer.

The detection of two products, benzene and cyclohexane, from cyclohexene on oxygen-covered platinum made it possible to consider the

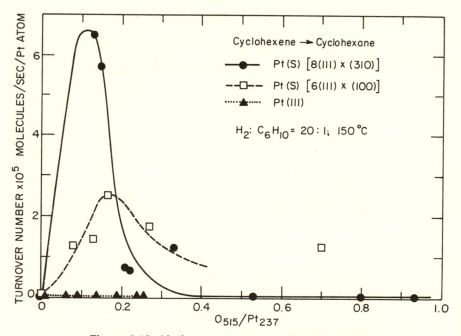

Figure 9.13. Maximum turnover numbers for cyclohexane production from cyclohexene as a function of oxygen coverage on Pt(S)−[8(111) × (310)], Pt(S)−[6(111) × (100)], and Pt(111). Reaction conditions: catalyst temperature 150°C, cyclohexene pressure 6×10^{-8} torr, hydrogen pressure 1×10^{-6} torr.

selectivity of dehydrogenation over hydrogenation as a function of oxygen coverage. The turnover numbers for benzene and cyclohexane production on the kinked surface have been replotted together as a function of oxygen coverage in Figure 9.14. It can be seen that the ratio of benzene to cyclohexane varies from greater than 40:1 on the clean surface, to about 10:1 at an oxygen/platinum Auger peak ratio of 0.13, to about 100:1 at an oxygen/platinum ratio of 0.22. The selectivity of the kinked catalyst changes markedly with small changes in oxygen coverage during the buildup of half a monolayer of oxygen on platinum.

The reaction of cyclohexane in the presence of excess hydrogen was also studied on the three platinum samples with clean surfaces and for a series of oxygen coverages. The maximum turnover numbers for benzene production from cyclohexane are plotted as a function of oxygen coverage in Figure 9.15 for the three platinum samples. The effect of

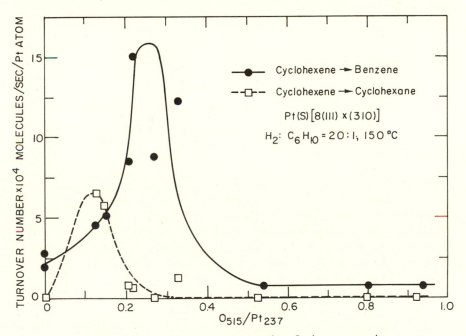

Figure 9.14. Maximum turnover numbers for benzene production from cyclohexene and for cyclohexane production from cyclohexene as a function of oxygen coverage on Pt(S)−8(111) × (310), showing the change in selectivity of cyclohexene dehydrogenation over hydrogenation. Reaction conditions: catalyst temperature 150°C, cyclohexene pressure 6 × 10⁻⁸ torr, hydrogen pressure 1 × 10⁻⁶ torr.

Cyclohexane → Benzene

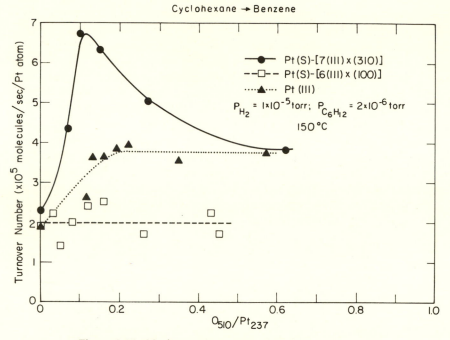

Figure 9.15. Maximum turnover numbers for benzene production from cyclohexane as a function of oxygen coverage on Pt(S)−[8(111) × (310)], Pt(S)−[6(111) × (100)], and Pt(111). Reaction conditions: catalyst temperature 150°C, cyclohexane pressure 2×10^{-6} torr, hydrogen pressure 1×10^{-5} torr.

adsorbed oxygen on the cyclohexane dehydrogenation reactivity of the kinked surface is similar to the effects previously observed for cyclohexene dehydrogenation and hydrogenation. The reactivity goes through a maximum at an oxygen/platinum Auger peak ratio of about 0.12 (approximately one-fourth of a monolayer), then decreases again.

At high hydrogen and hydrocarbon pressures the presence of strongly bound surface oxygen did not alter the dehydrogenation or hydrogenation activity markedly under steady-state reaction conditions. However, the hydrogenolysis activity was enhanced greatly over oxygen-treated kinked platinum surfaces.[45] Figure 9.16 shows the ratio of turnover numbers for dehydrocyclization to toluene and for hydrogenolysis to light alkanes (methane, ethane) as a function of oxygen surface concentration. Again, the kinked platinum surfaces were most active for hydrogenolysis when pretreated with oxygen. As a result, the selectivity for hydrogenolysis of *n*-heptane and cyclohexane was greatly enhanced over the competing dehydrocyclization and dehydrogenation reactions.

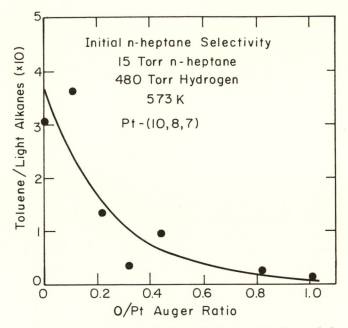

Figure 9.16. Ratio of dehydrocyclization and hydrogenolysis rates of *n*-heptane to toluene and like alkane products, respectively, as a function of oxygen/platinum Auger peak ratios.

It should be remembered that low-pressure catalytic studies explore the activity and selectivity of the clean platinum surface. High-pressure studies investigate the catalytic behavior of the metal in the presence of a carbonaceous overlayer. It is thus not surprising that the catalytic behaviors of the same surface may be different under low-pressure and high-pressure conditions.

The experimental results lead to two important conclusions:

1. Low coverages of strongly bound nonreactive oxygen on the otherwise clean platinum catalysts enhance the dehydrogenation rates of both cyclohexene and cyclohexane to benzene and change the selectivity of dehydrogenation over hydrogenation for the cyclohexene reaction at low pressures. At high reactant pressures the rates of hydrogenolysis of hydrocarbons were enhanced by the presence of the strongly bound surface oxygen.

2. The effect of preoxidation on the reactivity of the platinum surface is structure-sensitive, the activity being most enhanced on the kinked surface.

The role that oxygen plays in changing the reactivity and selectivity of

505

platinum catalysts is as yet poorly understood. There are three models that explain the observed effects. These models have been discussed by McCabe and Schmidt[46] with respect to the enhanced bonding of H_2 and CO that they observed on oxidized platinum surfaces. The first model postulates that the formation of the surface layer of oxide will result in a change in the electronic structure of the surface platinum atoms. In the oxide, the adsorbed oxygen atoms would tend to remove platinum valence electrons, and the surface platinum atoms would become more positively charged. This change in electronic structure could affect the binding of hydrogen and hydrocarbon reactants, intermediates, and products to the catalyst surface, which in turn could change the rates and the selectivity of the observed reactions. Since the presence of high concentrations of step and kink defect sites has a significant effect on the electronic structure of the clean platinum surfaces, the formation of a surface oxide might change the electronic structure of each surface site differently, giving rise to the observed surface structure sensitivity of the preoxidation on the reactions. McCabe and Schmidt[46] have determined that oxide-covered platinum surfaces have new binding sites for both hydrogen and CO, with significantly higher binding energies than on the clean surfaces. In addition, they observed that the initial sticking coefficient for hydrogen was higher by almost a factor of 2 on the oxidized surface, falling to a low value as soon as the higher binding-energy state was populated. This enhanced bonding of hydrogen could explain the large increase in hydrogenation activity of cyclohexene to cyclohexane observed on two of the platinum catalysts studied. Upon low-temperature adsorption of oxygen on platinum, the work function increases by about 1 eV; with high temperature oxidation, the work function has been shown to decrease by about 1 eV. This decrease would seem to indicate that oxidation leads to adsorbed oxygen atoms beneath the surface platinum atoms. The positively charged surface metal atoms would thus be readily available for bonding with hydrogen and hydrocarbons. Recent ion-scattering studies identified the strongly bound oxygen atoms as being located beneath the Pt surface atoms.[47]

The second model proposes that the strongly adsorbed oxygen atoms are active in compound formation with other adsorbates, such as hydrogen and hydrocarbons; such oxygen-containing compounds could provide alternative pathways for dehydrogenation or hydrogenation; thus changing observed reaction rates and selectivity. For example, the formation of hydroxyl groups on the surface might enhance the hydrogenation activity of cyclohexene to cyclohexane. An oxygen atom strongly adsorbed at a kink or stepped platinum site might show different activity for compound formation than would an oxygen atom at a (111) terrace site; this could explain the observed structure sensitivity of the reactivity and selectivity on oxidized platinum catalysts.

The third model proposes that the oxidation of the platinum surface results in a reconstruction or rearrangement of the surface atoms. The enhancement of dehydrogenation activity and the change in selectivity could then be postulated to result from the creation of new active sites during this rearrangement. The structure sensitivity could arise from a variation, in the case of reconstruction, from surface to surface. Platinum dispersed as small particles on oxide supports has been observed to exhibit increased mobility under oxidizing atmospheres, and LEED observations indicate that the Pt(110) surface reconstructs after extensive heating in oxygen. A similar surface rearrangement has been observed by Pareja et al.[48] to occur on platinum and nickel during the synthesis of water from oxygen and hydrogen, and they postulate that sites formed by this reaction are active for ethylene hydrogenation.

The enhancement of reactivity for all these hydrocarbon reactions by low oxygen coverages seems too large, particularly on the kinked surfaces, to be explained by surface reconstruction with the creation of new active sites, especially in view of the observation that the LEED pattern from the platinum substrate did not change significantly with oxygen coverage for any of the surfaces studied. It seems that change in the electronic structure of platinum surface atoms, due to the preoxidation and/or the formation of the surface compounds involving oxygen, must be largely responsible for the observed enhancement and change in selectivity. Apparently, the presence of platinum kink sites promotes a particularly favorable surface electronic change and enhances compound formation. Although the formation of hydroxyl groups might be important in the observed hydrogenation activity, the decrease in the work function observed with the strongly bound oxygen leads to the interpretation that a significant proportion of the adsorbed oxygen is below the platinum surface atoms, and prompts one to believe that a compound formation is not the primary mechanism for enhancement of dehydrogenation activity.

Reconstruction of the Platinum Catalyst Surface Structure under the Reaction Conditions

The atomic surface structure of the freshly prepared catalyst is drastically changed under the reaction conditions and may be changed again in the presence of hydrocarbons or upon cleaning in oxygen or heating in hydrogen. This was revealed during studies of 22 single-crystal surfaces of platinum that were studied in ultrahigh vacuum when clean, and in the presence of a monolayer of chemisorbed oxygen or carbon, by low-energy electron diffraction.[49] The results of these investigations have been described in detail in Chapter 3.

Most crystal surfaces restructure as the surface composition is changed. Some of the surface structures are stable in ultrahigh vacuum and in the presence of oxygen but reconstruct in the presence of carbon, whereas others are stable when clean and when carbon-covered but restructure when covered with oxygen. The types of changes of surface structure that occur most commonly are depicted schematically in Figure 4.16. A surface that exhibits the one-atom-height stepped terrace configuration reconstructs into a multiple-height stepped structure with wider terraces as the surface composition is changed. Other surfaces develop a "hill-and-valley" configuration, consisting of large facet planes detectable by low-energy electron diffraction. Interestingly, most of these restructuring processes are reversible. Once the adsorbate, oxygen or carbon, is removed, the surface returns to its original clean surface structure. There are a few stable surfaces that do not restructure at all under any conditions of the experiments. Besides the low-Miller-index

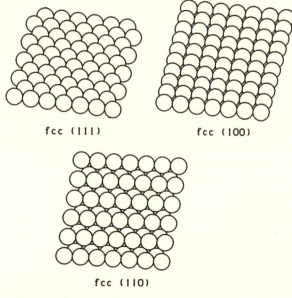

fcc (111)   fcc (100)

fcc (110)

Figure 9.17a. Low-Miller-index platinum crystal faces.

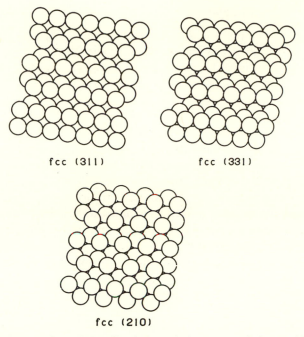

fcc (311)   fcc (331)

fcc (210)

Figure 9.17b. High-Miller-index platinum crystal faces which remain stable under all conditions of experiments in the presence of carbon as well as in oxygen.

surfaces (111), (100), and (110), these are the (210), (331), (311), and (211) orientation crystal faces. They are shown in Figure 9.17. These crystal planes are characterized by a very high density of periodic steps, one atom in height, or a complete lack of steps. Because of their structural stability, it is expected that they play important roles in the catalytic chemistry of transition-metal surfaces. The stepped and kinked platinum crystal surfaces that were utilized during the dehydrogenation and hydrogenolysis studies reported also maintained their structural integrity.

Restructuring in the presence of adsorbates serves as a mechanism for redispersion. Chlorine is a known adsorbate that aids redispersion of transition metal catalyst particles. The structural changes that must accompany redispersion are desribed in detail in Chapter 3. Poisoning of catalytic activity is often due to the reconstruction of the active catalyst structure upon adsorption of impurities. Sulfur is known to cause gross restructuring on metal surfaces.[50]

Building of Platinum Catalysts

The atomic-scale investigation of the reactivities of well-characterized platinum crystal surfaces revealed the importance of irregularities (steps and kinks) and additives (carbonaceous deposits or oxygen, etc.) in tailoring the activity and selectivity of the catalyst. It appears that the catalyzed hydrocarbon conversion does not occur on platinum alone but takes place on a complex platinum-additive system in which both the surface structure and composition are finely adjusted to optimize activity and selectivity. In dispersed catalyst systems the support can also play a significant role in adjusting the catalytic behavior by aiding or eliminating side reactions or by stabilizing the structure and/or the oxidation state of the catalyst surface atoms by the metal support interaction.

Once these ingredients of hydrocarbon catalysis by platinum are uncovered, the question arises as to how we can put this information to best use. It would be desirable to be able to deposit platinum particles with controlled surface structure and additive composition. One might be able to minimize the amount of catalyst needed by depositing particles with surface structures with optimum activity, thereby saving much of this precious metal for other applications. Alternately, it might be possible to explore ways to substitute for platinum by systematically probing the effects of other metals and other additives on chemical reactions catalyzed by platinum. Studies to build new platinum and other active catalysts to perform desired selective chemistry or to find substitutes for platinum or other precious catalysts (rhodium, palladium, and iridium, for example) are in their infancy. Nevertheless, several approaches are being attempted, and some of these are described next.

Ultimately, one would like to maintain control over both the atomic surface structure and the surface composition of the synthetic catalyst system. One way to achieve this is by epitaxial deposition of the metal layer by layer on ordered substrates (supports). This is shown schematically in Figure 9.18. Under proper circumstances, the atomic surface structures of the substrate would control the surface structure of the deposited catalyst monolayer and multilayers. The catalytic activity of the condensed metal film could then be tested by varying the concentration from 10 percent of a monolayer to several layers and by varying its surface structure using different atomic-surface-structure substrates.

Studies of this type have been carried out using gold single crystal surfaces as substrates for the deposition of single crystal platinum films and platinum (100) crystal faces as substrates for the deposition of gold films.[51] The results are shown in Figure 9.19 and 9.20. Maximum rates

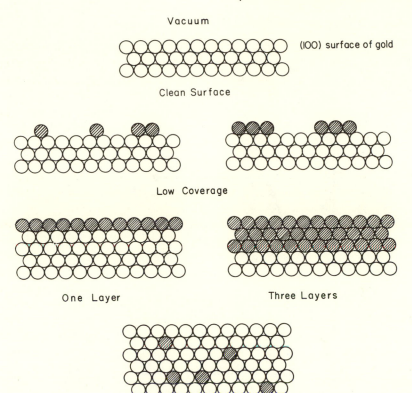

Vacuum

(100) surface of gold

Clean Surface

Low Coverage

One Layer   Three Layers

After Heating at High Temperature

Figure 9.18. Scheme of building catalyst monolayers of well-characterized structure and composition. Metal atoms are condensed from the vapor phase on single-crystal metal surfaces until desired amounts and atomic structures are obtained.

have been obtained for the dehydrogenation of cyclohexene to benzene at low pressures only after the deposition of two or more layers of platinum (Figure 9.19). However, the activity of the deposited Pt film is then at least fivefold higher than that of the Pt(100) crystal face under the same conditions. Figure 9.20 shows that the dehydrogenation activity of the Pt(100) face that is covered with one monolayer of gold is about five times higher than that of the clean Pt(100) surface. It should be noted that gold alone is inactive for this reaction.

Deposition of platinum thin films on oxide surfaces has also been carried out. Because of the difference in surface free energies of the oxide and the condensing metal, layer-by-layer growth does not readily

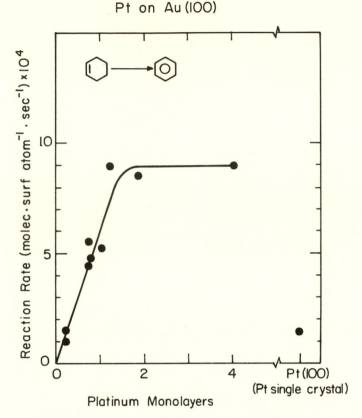

Figure 9.19. Rate of dehydrogenation of cyclohexene to benzene on ordered platinum (100) orientation monolayers deposited on the (100) crystal face of gold as a function of the number of monolayers of deposited platinum at low reactant pressures.

occur. Nevertheless, changes in chemical behavior have been identified as the grain size of the platinum particles in the evaporated films has been altered.[10]

The effects of adding a second metal to the active transition-metal catalyst have been investigated in several laboratories. The two metals may be codeposited from the vapor phase or from solution, or deposited separately, one after the other. Perhaps the most successful approach in this direction of research on new catalyst systems was invented by Sinfelt,[52] who has discovered drastically different properties of small alloy particles as compared to the behavior of macroscopic alloy systems com-

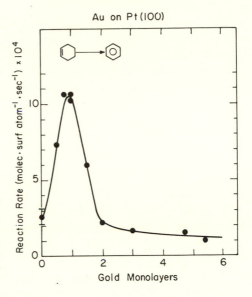

Figure 9.20. Rate of dehydrogenation of cyclohexene to benzene on ordered gold (100) orientation monolayers deposited on the (100) crystal face of platinum as a function of the number of monolayers of deposited gold at low reactant pressures.

posed of the same metals. Gold and iridium, for example, when deposited as small particles, exhibit complete miscibility as proven by chemisorption studies, while they are almost immiscible according to their bulk-phase diagram. Although the structure of these bimetallic clusters, as they have become known, cannot readily be controlled, interesting systematic changes in their catalytic chemistry have been detected as the composition and their conditions of preparation are varied.

Another interesting approach to the synthesis of new catalysts has been pursued by Leclercq et al.[53] Noting the similarities in the electronic structure and chemical activities of platinum and tungsten carbide, they explored the catalytic chemistry of carbides in atomic detail in a systematic manner.

Another approach to building new catalysts was reported by Touster et al.[54] The deposition of transition metal particles on basic oxide supports (La_2O_3, IrO_2, V_2O_5) and their subsequent heat treatment in hydrogen yielded an entirely different catalytic behavior than when the metal particles were deposited on alumina or silica supports. A strong metal support interaction is responsible for this effect as shown by XPS studies of Kao et al.[55] which indicate the transfer of negative charge to the metal from the basic oxides. Thus the oxide support can be used to control the oxidation states of atoms of the surface of the transition metal catalyst particles.

As more information becomes available on the role of surface structure, composition, oxidation state, and additives in catalyst systems, new avenues to synthesize catalysts of desired selectivity and activity will be explored. There are many exciting discoveries in store in the near future for those pursuing these paths of catalytic investigation.

References

1. F.G. Ciapetta and D.N. Wallace, Catal. Rev. **5,** 67 (1972); J.H. Sinfelt, Progr. Solid State Chem. **10,** 55 (1975); G.A. Mills, H. Heineman, T.H. Milliken, and K.G. Oblad, Ind. Eng. Chem. **45,** 135 (1953).
2. J.K. Dixon and J.E. Longfield, in *Catalysis,* ed. P.H. Emmett, vol. 7, Reinhold, New York, 1960, p. 281; N.I. Ilchenko and G.I. Golodets, J. Catal. **39,** 57 (1975).
3. M. Shelef, Catal. Rev. **11,** 1 (1975); C.C. Chang and L.L. Hegedus, J. Catal. **57,** 361 (1976).
4. A.J. Appleby, Catal. Rev. **4,** 221 (1970); P. Stonehart and P.M. Ross, Catal. Rev. **12,** 1 (1975); E. Segal. R.J. Madon, and M. Boudart, J. Catal. **52,** 45 (1978); J.C. Schlatter and M. Boudart, J. Catal. **24,** 482 (1972).
5. M. Kraft and H. Spindler, Proc. 4th Int. Congr. Catal., Moscow (1968); J. Haro, R. Gomez, and J.M. Ferreira, J. Catal. **45,** 326 (1976).
6. J.R. Anderson and N.R. Avery, J. Catal. **5,** 446 (1966); C. Leclercq. L. Leclercq, and R. Maurel, J. Catal. **50,** 87 (1977).
7. M. Boudart, A.W. Aldag, L.D. Ptak, and J.E. Benson, J. Catal. **11,** 35 (1968).
8. W.C. Gillespie, S.M. Davis, R.K. Herz, and G.A. Somorjai, J. Catal., to be published 1980.
9. J.K.A. Clark and J.J. Rooney, Adv. Catal. **25,** 125 (1976).
10. J.R. Anderson and N.R. Avery, J. Catal. **8,** 48 (1967); J. Catal. **5,** 446 (1966); J. Catal. **7,** 315 (1967).
11. C. Corolleur, S. Corolleur, and F.G. Gault, J. Catal. **24,** 385 (1972).
12. S.M. Csicsery and R.L. Burnett, J. Catal. **8,** 75 (1967).
13. J.R. Anderson, R.J. MacDonald, and Y. Shinoyama, J. Catal. **20,** 147 (1971).
14. J.R. Anderson, Adv. Catal. **23,** 1 (1973).
15. C. Corolleur, D. Tomanova, and F.G. Gault, J. Catal. **24,** 401 (1972).
16. M. Boudart, Adv. Catal. **20,** 153 (1969).
17. W.H. Monogue and J.R. Katzer, J. Catal. **32,** 166 (1974).
18. D.W. Blakely and G.A. Somorjai, J. Catal. **42,** 181 (1976).
19. J.A. Cusumano, G.W. Dembrinski, and J.H. Singelt, J. Catal. **5,** 471 (1968).
20. T.A. Dorling and R.L. Moos, J. Catal. **5,** 111 (1966).
21. M. Boudart, A. Aldag, J.E. Benson, N.A. Dougharty, and C. Girvin-Harkin, J. Catal. **6,** 92 (1966).
22. D.R. Kahn, E.E. Petersen, and G.A. Somorjai, J. Catal. **34,** 294 (1974).
23. S.L. Bernasek and G.A. Somorjai, J. Chem. Phys. **62,** 3149 (1975).
24. O.M. Poltorak and V.S. Boronin, Russ. J. Phys. Chem. **40,** 2671 (1966).
25. G.A. Somorjai, Progr. Anal. Chem. **1,** 191 (1968).
26. G.A. Somorjai, Adv. Catal. **26,** 1 (1977).
27. G.A. Somorjai, Catal. Rev. **18,** 173 (1978).
28. G.A. Somorjai, Surf. Sci. **89,** 496 (1979).
29. S.M. Davis and G.A. Somorjai, J. Catal. **65,** 78 (1980).

30. R.J. Gale, M. Salmerón, and G.A. Somorjai, Phys. Rev. Lett. **38**, 1027 (1977); J. Chem. Phys. **67**, 5324 (1977);
31. M. Salmerón, R.J. Gale, and G.A. Somorjai, J. Chem. Phys. **70**, 2807 (1979).
32. D.W. Blakely and G.A. Somorjai, Nature **258**, 580 (1975).
33. S.M. Davis and G.A. Somorjai, Surf. Sci. **91**, 73 (1980).
34. R. Mason and G.A. Somorjai, Chem. Phys. Lett. **44**, 468 (1976).
35. D.I. Hagen and G.A. Somorjai, J. Catal. **41**, 466 (1976).
36. J.H. Sinfelt, J.L. Carter, and D.J.C. Yates, J. Catal. **24**, 283 (1972).
37. R.K. Herz, Ph.D. thesis, University of California, Berkeley, 1977.
38. S.M. Davis and G.A. Somorjai, J. Catal., to be published 1981.
39. G.A. Somorjai, Angew. Chem. **16**, 92 (1977).
40. L.L. Kesmodel, L.H. Dubois, and G.A. Somorjai, J. Chem. Phys. **70**, 2180 (1979).
41. T.N. Rhodin and J.W. Gadzuk, in *The Nature of the Surface Chemical Bond*, eds. T.N. Rhodin and G. Ertl, North-Holland, Amsterdam, 1979.
42. J.E. Demuth and H. Ibach, Surf. Sci. **78**, L238 (1978).
43. J.A. Rabo, A.P. Risch, and M.L. Boutsma, J. Catal. **53**,295 (1978).
44. C.E. Smith, J.P. Biberian, and G.A. Somorjai, J. Catal. **57**, 426 (1979).
45. W.D. Gillespie and G.A. Somorjai, J. Catal., to be published 1980.
46. R.W. McCabe and L.D. Schmidt, Surf. Sci. **60**, 85 (1976); Surf. Sci. **65**, 189 (1977).
47. H. Niehus and G. Comsa, Surf. Sci. **93**, L147 (1980).
48. P. Pareja, A. Amariglio, G. Piquard, and H. Amariglio, J. Catal. **46**, 225 (1977).
49. D.W. Blakely and G.A. Somorjai, Surf. Sci. **65**, 419 (1977).
50. G.A. Somorjai, J. Catal. **27**, 453 (1972).
51. J.W.A. Sachtler and G.A. Somorjai, Phys. Rev. Lett. **45**, 1601 (1980).
52. J. Sinfelt, Catal. Rev. **9**, 147 (1974).
53. L. Leclercq, K. Imura, S. Yoshida, T. Barbee, and M. Boudart, in *Preparation of Catalysts*, vol. 2, Elsevier, Amsterdam, 1979.
54. (a) S.J. Tauster, S.C. Fung, and R.L. Garten, J. Amer. Chem. Soc. **100**, 170 (1980); (b) M.A. Vannice and R.L. Garten, J. Catal. **63**, 255 (1980).
55. C.C. Kao, S.C. Trei, M.K. Bahl, Y.W. Chung, and W.J. Lo, Surf. Sci. **95**, 1 (1980).

10. Catalytic Hydrogenation of Carbon Monoxide

Brief History

Few reactions have a more distinguished history than the catalytic hydrogenation of carbon monoxide. It was first reported in 1902 by Sabatier and Senderens,[1] who produced methane from CO and H_2 over nickel. As early as 1913, Badische Anilin und Soda Fabrik (BASF) patented the production of longer-chain and oxygenated hydrocarbons at over 100 atm pressure at about 350°C over alkali-metal-activated transition-metal catalysts of cobalt and osmium oxides. Then, in 1923, BASF announced the successful synthesis of methanol from CO and H_2 with the exclusion of other products.[2] In 1926, Fischer and Tropsch[3] published their classical work reporting on the synthesis, near atmospheric pressures and at 200°C, of various higher-molecular-weight hydrocarbons from CO and H_2. The reaction carried out in this manner is called the Fischer–Tropsch synthesis. The original catalysts were both iron and cobalt, with added K_2CO_3 and copper as promoters. The catalyst underwent many improvements in the years that followed, including the addition of thoria, the use of pressures in the range 10 to 50 atm, and better means of dispersion and stabilization. The first catalyst to become standard in German commercial plants contained cobalt, thoria, magnesium oxide, and kieselguhr.[4,5] Germany then embarked on a program to rapidly expand her synthetic fuel production. In 1943, at the height of the war, production was over 100,000 barrels per day. The typical product distribution obtained at that time is shown in Table 10.1.

In the United States during the 1940s and 1950s a great deal of research was carried out by the Bureau of Mines to develop new catalysts and to explore the mechanism of this reaction, using a variety of catalysts ranging from iron to molybdenum disulfide (MoS_2). Review papers at-

Table 10.1. Typical product distribution obtained during commercial-scale Fischer–Tropsch synthesis over cobalt catalysts

Constituent	Total products listed (wt%)[a]	Olefins (vol %)	Number of carbon atoms	Octane number, research method
Normal-pressure synthesis[b]				
Gasol ($C_3 + C_4$)	12	50	$C_3 + C_4$	
Gasoline, to 185°C	49	37	C_4-C_{10}	52
Gasoline, to 200°C	54	34	C_4-C_{11}	49
Diesel oil, 185–320°C	29	15	C_{11}-C_{18}	
Diesel oil, 200–320°C	24	13	C_{12}-C_{19}	
Soft paraffins, 320–450°C	7	iodine value, 2	$>C_{19}$	
Hard paraffins, >450°C				
Medium-pressure synthesis[c]				
Gasol ($C_3 + C_4$)		30	66% C_4 33% C_3	
Gasoline, to 185°C	35	20	C_4-C_{10}	28
Gasoline, to 200°C	40	18	C_4-C_{11}	25
Diesel oil, 185–320°C	35	10	C_{11}-C_{18}	
Diesel oil, 200–320°C	35	8	C_{12}-C_{19}	
Soft paraffins, 320°C	30	iodine value, 2	C_{18}	
Soft paraffins, 330°C	25	2	C_{19}	

[a] Total yield per cubic meter of synthesis gas: normal-pressure synthesis, 148 g; medium-pressure synthesis, 145 g of liquid products and 10 g of gasoline.
[b] At 1 atm; 180–195°C; catalyst, 100 Co: 5 ThO_2:7.5 MgO:200 kieselguhr; 1 CO:2 H_2 (18–20% inert component); throughput 1 m^3 synthesis gas/hr) (kg Co); two-stage; no recycle.
[c] At 7 atm abs.; 175–195°C; catalyst, 100 Co:5 ThO_2:7.5 MgO:200 kieselguhr; 1 CO: 2 H_2 (18–20% inert components); throughput 1 m^3 synthesis gas/(hr) (kg Co); two-stage; no recycle.

test to the detailed studies that were carried out at the time.[6,7] In Emmett's laboratory at Johns Hopkins University clever isotope-substitution studies revealed the existence of plausible enol reaction intermediates on iron surfaces.[8,9] A commercial-scale pilot plant was built in Texas with a capacity of about 6500 barrels of synthetic fuel per day. The pilot plant was closed in 1957. Development of synthetic-fuel technology was pursued only in South Africa, where, with the use of promoted iron catalyst, large-scale production of liquid and gaseous hydrocarbons from CO and H_2 was established. In 1974 about 9000 barrels of fuel per day were produced, and by 1985 the production will be expanded to 100,000 barrels per day. Present daily oil consumption in the United States is about 16 million barrels. To be self-sufficient in fuel we need to build a synthetic-fuel industry with over 160 times the production capacity of Germany during World War II. This is a formidable challenge—a difficult but worthwhile undertaking.

With the renewed interest in coal as an energy source, research and development of the catalytic hydrogenation of carbon monoxide have proceeded rapidly in recent years. With the use of modern surface diagnostic techniques, the atomic-level mechanisms of this reaction are being unraveled. As a result, it will become possible to produce single-component hydrocarbons from methane to ethylene glycol or benzene instead of obtaining an undesirable mixture of hydrocarbons as was obtained 40 years ago. This reaction is attractive to researchers in the field of catalysis science for many reasons:

1. With the utilization of small, easily available molecules (CO, H_2O, CO_2, and H_2), smaller or more complex hydrocarbon molecules may be synthesized. The formation of almost every hydrocarbon molecule is thermodynamically feasible using these reactants. For example, the standard free energies of three of the four reactions that produce methane, CH_4, from these molecules are negative:

$$CO + 3H_2 = CH_4 + H_2O \qquad \Delta G° = -33.4 \text{ kcal/mol}$$

$$4CO + 2H_2O = CH_4 + 3CO_2 \qquad \Delta G° = -54.1 \text{ kcal/mol}$$

$$CO_2 + 4H_2 = CH_4 + 2H_2O \qquad \Delta G° = -26.2 \text{ kcal/mol}$$

$$CO_2 + 2H_2O = CH_4 + 2O_2 \qquad \Delta G° = +19.1 \text{ kcal/mol}$$

The only reaction that is thermodynamically uphill is the one between CO_2 and H_2O, which produces oxygen as well. This type of reaction is the basis of photosynthesis, which is discussed in more detail in Chapter 11.

2. Transition metals, some of which are readily available at low cost, act as catalysts for these reactions. Among those most frequently used are nickel, iron, cobalt, rhodium, and ruthenium. Readily available oxides (ThO_2, ZrO_2), sulfides (MoS_2), and nitrides (FeN_3) have also been used as catalysts to produce hydrocarbons.

Next, we review the thermodynamics of the hydrogenation of carbon monoxide and carbon dioxide, which also dictate the conditions necessary to carry out these reactions. Then, we describe atomic-scale studies over iron, rhodium, and nickel surfaces, which reveal the various mechanisms that operate.

Thermodynamic Considerations

The gasification of coal by steam at high temperatures (~ 1100 K) produces predominantly carbon monoxide and hydrogen, a gas mixture that is appropriately called "water gas":

$$\text{coal} + H_2O \rightarrow CO + H_2$$

Using the water-gas-shift reaction,

$$CO + H_2O \rightleftharpoons CO_2 + H_2$$

the $CO-H_2$ mixture can be enriched with hydrogen, which is desirable in many of the chemical reactions of these two molecules. Table 10.2 lists the thermodynamic data for the coal gasification and water-gas-shift reactions.

Using various ratios of carbon monoxide and hydrogen, the production of hydrocarbons of different types is thermodynamically feasible.

Let us consider the formation of alkanes according to the reactions

$$(n + 1)H_2 + 2nCO = C_nH_{2n+2} + nCO_2 \tag{1}$$

$$(2n + 1)H_2 + nCO = C_nH_{2n+2} + nH_2O \tag{2}$$

Both reactions are thermodynamically feasible, although (1) has a somewhat lower negative free energy of formation. Thus the by-product of alkane formation is either CO_2 or H_2O. These reactions are not independent but are related through the water-gas-shift reaction. Catalysts that carry out the water gas shift readily (iron, for example) may produce alkanes and both CO_2 and H_2O, depending on the reaction conditions. Other catalysts that are poor for catalyzing the water gas shift may produce alkanes and mostly water or mostly CO_2. Catalysts that produce

Table 10.2. Selected thermodynamic data for the coal gasification and the water-gas-shift reactions

Carbon gasification reaction: C (graphite) + H_2O = CO + H_2

$\Delta H_{500K} = 32.0$ kcal/mol
$\Delta G_{500K} = 15.2$ kcal/mol
$\Delta G_{1000K} = -1.9$ kcal/mol

Water-gas-shift reaction: CO + H_2O = CO_2 + H_2

$\Delta H_{500K} = -9.5$ kcal/mol
$\Delta G_{500K} = -4.8$ kcal/mol
$\Delta G_{1000K} = -0.6$ kcal/mol

alkanes and CO_2 are often more desirable, because less hydrogen is used up in this circumstance. Hydrogen is in general the costlier of the two reactants. Let us write only one of these reactions for the formation of alkanes, alkenes, and alcohols, which are also produced from CO and H_2, and compare their free energies of formation.

$$(2n + 1)H_2 + nCO = C_nH_{2n+2} + nH_2O \qquad (2)$$

$$2nH_2 + nCO = C_nH_{2n} + nH_2O \qquad (3)$$

$$2nH_2 + nCO = C_nH_{2n+1}OH + (n - 1)H_2O \qquad (4)$$

The standard free energies of formation of the various products, as a function of temperature, are shown in Figures 10.1 to 10.3. Since these are all exothermic reactions, low temperatures favor the formation of the products. At present, however, none of the known catalysts for the hydrogenation of CO can produce hydrocarbons at high-enough rates to approach the concentrations that are predicted from thermodynamic equilibrium consideration. In fact, the reaction rates are orders of magnitude lower than the maximum rates calculable at equilibrium. Thus these thermodynamic calculations provide only guidance and boundary condi-

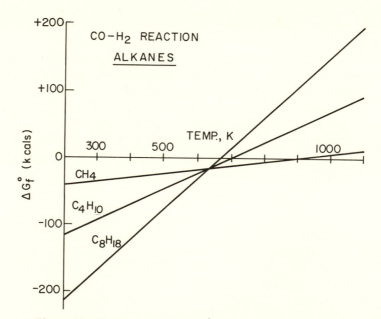

Figure 10.1 Free energies of formation of alkanes as a function of temperature.

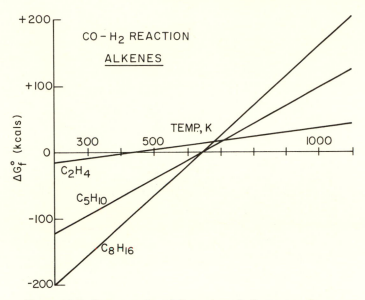

Figure 10.2. Free energies of formation of alkenes as a function of temperature.

tions of the product distribution that may be produced under various experimental conditions. Since a slow surface reaction step, or perhaps several steps that have large activation energies, control the rate of the reaction as well as the product distribution (they have low turnover frequencies, 10^{-6} to 10 molecules/surface atom/sec), higher temperatures, in the range 500 to 700 K, are usually employed to optimize the rates of formation of the products.

Another reaction that appears to play an important role in the synthesis of hydrocarbons from CO and H_2 is the disproportionation of carbon monoxide,

$$2CO \rightleftharpoons C + CO_2 \qquad \Delta G_{500K} = -32 \text{ kcal} \qquad (5)$$

There is a great deal of experimental evidence, which is presented later in the section on methanation, that the hydrogenation of the active form of carbon that deposits as a result of the reaction in Eq. (5)—frequently called the Boudouard reaction—leads to the formation of hydrocarbons.

According to the Le Châtelier principle, high pressures favor the association reactions, which are accompanied by a decrease in the number of moles in the reaction mixture as the product molecules are formed. Thus

521

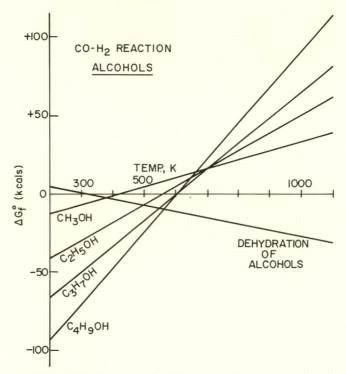

Figure 10.3. Free energies of formation of alcohols as a function of temperatures.

the formation of higher-molecular-weight products is more favorable at high pressures. To demonstrate this,[10] let us consider the reaction $aA + bB = cC + dD$. The equilibrium constant in terms of partial fugacities is

$$K_f = \frac{(f_C)^c \, (f_D)^d}{(f_A)^a \, (f_B)^b}$$

In terms of partial pressures, this becomes

$$K_p = \frac{(Px_C)^c \, (Px_D)^d}{(Px_A)^a \, (Px_B)^b}$$

where P is the total pressure and x_A, x_B, and so on, are the mole fractions. It follows that $K_f = K_p K_\gamma$, where $\gamma = f/p$ and

$$K_\gamma = \frac{(\gamma_C)^c (\gamma_D)^d}{(\gamma_A)^a (\gamma_B)^b}$$

The approximation $K_\gamma \simeq 1$ for Fischer–Tropsch reaction conditions (less than 100 atm) yields

$$\frac{(x_C)^c (x_D)^d}{(x_A)^a (x_B)^b} \simeq K_f P^{-\Delta n}$$

where $-\Delta n = a + b - c - d$. Thus, associative reactions, where $(a + b)$ is larger than $(c + d)$, are favored by a pressure increase. For all of the Fischer–Tropsch reactions, $(a + b)$ is larger than $(c + d)$, as a general rule. As an example, for the reaction $3H_2 + CO = CH_4 + H_2O$, ΔG_f at 730 K and 1 atm equals -11.42 kcal/mole and K equals 3.68×10^3. At 10^{-4} torr total pressure, $K_f \cdot P^2 = 3.68 \times 10^3 \times 1.73 \times 10^{-14} = 6.4 \times 10^{-11}$.

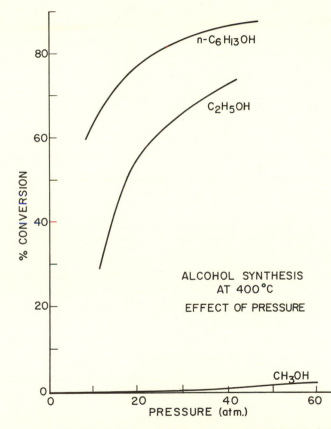

10.4. Calculated yields of higher molecular-weight alcohols from CO and H_2 as a function of total reactant pressures.

The equilibrium methane concentration under these conditions would therefore be very low.

Figures 10.4 and 10.5 show that pressures in excess of 20 atm are desirable to produce high-molecular-weight alcohols or benzene. If reactions are carried out at 1 atm, for example, the catalyst cannot exhibit its real performance because of thermodynamic limitations. We must use high-pressure batch or flow reactors capable of carrying out the reaction of CO–H_2 mixtures up to 100 atm.

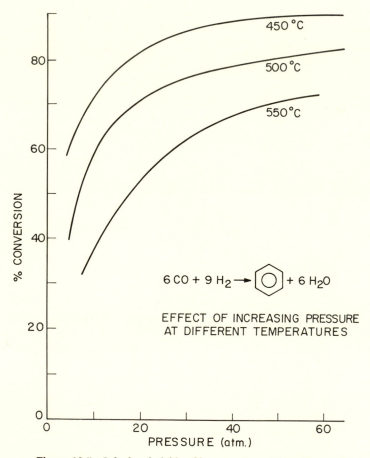

Figure 10.5. Calculated yields of benzene from CO and H_2 as a function of total reactant pressures.

Different Reaction Paths and Selectivity during the Hydrogenation and Insertion of Carbon Monoxide

The hydrogenation of CO can lead to the selective formation of C_1 hydrocarbons, methane, or methanol, or to the formation of longer-carbon-chain hydrocarbons with the exclusion of other products. Methane is produced selectively with a relatively high rate over nickel catalysts:[11]

$$2CO + 2H_2 \xrightarrow[500-600\ K]{Ni} CH_4 + CO_2 \tag{6}$$

Methanol synthesis from CO and H_2 can be carried out over zinc chromate–copper chromate catalysts,[12]

$$CO + 2H_2 \xrightarrow[450-600\ K]{Zu-Cr-Cu} CH_3OH \tag{7}$$

Recently, palladium and platinum catalysts were also found to produce methanol selectively at higher pressures (~ 12 atm).[13] Both of these catalytic reactions form the basis of important industrial technologies for methanation and methanol production.

One of the earliest reactions involving the insertion of only one carbon monoxide molecule in a C_n olefin molecule to produce an aldehyde with one greater C_{n+1} carbon number is the "hydroformylation" or "oxo" reaction. Oxo reactions are carried out over homogeneous rhodium $[HRh(CO)_2]$ or cobalt carbonyl $[HCo(CO)_4]$ and are important industrial processes.[14]

$$RCH = CH_2 + CO + H_2 = RCH_2CH_2CHO + \underset{\underset{CH_3}{|}}{R-CHCHO} \tag{8}$$

Once the aldehyde has been synthesized by carbonylation, reduction leads to the formation of a variety of important alcohols. The rhodium hydroformylation catalyst is 10^4 times more active than the cobalt catalyst.

Another important example of the selective insertion of only one CO molecule is the production of acetic acid, CH_3COOH, from methanol,

$$CO + CH_3OH \xrightarrow{[Rh(CO)_2I_2]^-} CH_3COOH \tag{9}$$

This homogenous catalytic reaction again uses rhodium or cobalt as a catalyst in the presence of iodide ions as promoters. Recently, the selective production of acetic acid and acetaldehyde[16] and glycol[17] from CO and H_2 over heterogeneous and homogeneous rhodium catalysts has been reported.

These examples clearly indicate that starting from CO and H_2, the selective synthesis of various types of organic molecules is feasible. Some of these products may be straight-chain saturated or unsaturated hydrocarbons; others are oxygenated (alcohols, aldehydes, and acids). However, to accomplish this for other molecules, we must gain an understanding of the elementary surface processes on the molecular scale that control the formation of the reaction intermediates and the reaction products. For this purpose it is appropriate to subdivide the processes that occur during the hydrogenation of carbon monoxide as follows:

1. Methanation.
2. Methanol formation.
3. Polymerization reactions that yield a mixture of high-molecular-weight hydrocarbons.
4. Formation of oxygenated organic molecules.
5. The insertion of CO molecules (carbonylation) to form C_2 or C_3 products.

We would like to know the mechanisms for forming these different products under a variety of experimental conditions. This includes a molecular-level understanding of the catalyst surface structure, the oxidation states of the active sites, the structure of the reaction intermediates, and the role of the various additives that have been found useful in extending catalyst life and improving catalyst performance. Although such a complete understanding of the reaction mechanism does not exist, many of the elementary reaction steps and the surface structures and composition of the active catalysts have been verified by careful studies of a given catalyst system. Next, we discuss what is known about the various CO hydrogenation reactions by reviewing the results of experiments investigating the active catalysts.

Methanation

One of the main products of the hydrogenation of CO is methane. It is produced almost exclusively over nickel, while it forms together with higher-molecular-weight hydrocarbons over many other transition-metal surfaces. Vannice[18] has determined the relative activity of various transition metals for methanation at 1 atm total pressure under conditions in which most other hydrocarbon molecules are not likely to form because of thermodynamic limitation. The order of decreasing activity is ruthenium > iron > nickel > cobalt > rhodium > palladium > platinum > iridium. The activation energy for methanation from CO and H_2 is in the range 23 to 25 kcal/mol for ruthenium, iron, nickel, cobalt, and rhodium metals, for which this has been determined. The nearly identical activation energies indicate that the mechanism of methanation is likely to be similar. Recent studies in several laboratories clearly show that the dominant mechanism involves the dissociation of CO followed by the hydrogenation of the surface carbon atoms to methane. The adsorbed oxygen is removed from the surface as CO_2 by reaction with another CO molecule. The net process by which the active surface carbon that is to be hydrogenated forms is often described as the disproportionation of CO, $2CO = C + CO_2$, called the Boudouard reaction. This mechanism has been confirmed in several ways. The formation of a carbonaceous overlayer has been detected on polycrystalline rhodium,[10] iron and nickel surfaces[19] in CO–H_2 mixtures in the temperature range 500 to 700 K. After pumping out the reaction mixture and introducing hydrogen, methane is produced at the same rate as in the presence of water gas. Ventrcek et al.[20] and Rabo et al.[21] have been able to titrate the amount of surface carbon by quantitative measurement of the amount of CO_2 evolved ($2CO = C + CO_2$) over nickel, ruthenium, and cobalt catalyst surfaces, respectively. Rabo et al.[21] have introduced pulses of H_2 after forming the surface carbon to produce predominantly methane. Biloen et al.[22] have deposited the active surface carbon on nickel, cobalt, and ruthenium by disassociating labeled ^{13}CO. The radioactive carbon layer is readily hydrogenated subsequently in the presence of a CO–H_2 mixture to yield labeled $^{13}CH_4$.

The active surface carbon that forms from the dissociation of CO maintains its activity to produce methane only in a rather narrow temperature range. Above 700 K the carbon layer becomes graphitized and loses its reactivity with hydrogen. At temperatures below 450 K, the dissociation rate of CO to produce the active carbon is too slow to produce the active surface carbon in high enough concentrations. The temperature dependence of the nature of the CO and carbon chemical bonds introduce a narrow range of conditions for the production of methane. The fact that the dissociation of molecules on surfaces is an activated process is well established by many studies of the formation of surface chemical bonds.

However, changes in the chemical activity of the surface carbon that forms are less well established. The unique hydrogenation activity of the carbon that forms upon the dissociation of carbon monoxide on transition-metal surfaces in the range 450 to 700 K indicates the formation of active carbon–metal bonds that deserve further experimental scrutiny. The formation of reactive multiple carbene or carbyne species is not unlikely, as these active metal–carbon bonds can yield the hydrogenation activity that was detected. Araki and Ponec[19b] have compared the catalytic activity of nickel and nickel–copper alloys for methanation. Upon the addition of less than 10 atom percent of copper, the activity drastically decreased. Their results indicate that more than one nickel atom is involved in forming the strong and active metal–carbon bond that yields methane by direct hydrogenation. Since ethylidyne molecules were detected on the Pt(111) crystal face upon adsorption of C_2H_4 and C_2H_2, where the strongly bound carbon is in a threefold site,[23] a similar location for the carbon or CH fragments on nickel surfaces,[24] which would bind them to three nickel atoms, seems likely. The active carbon is metastable with respect to the formation of graphite, however. Heating to above 700 K produces a stable graphite surface layer that is unreactive with hydrogen. Once the graphitic carbon is formed, the catalyst loses its activity for the formation of hydrocarbons of any type.

While the hydrogenation of the active surface carbon that forms from CO dissociation appears to be the predominant mechanism of CH_4 formation, it is not the only mechanism that produces methane. Poutsma et al. have detected the formation of CH_4 over palladium surfaces that do not dissociate carbon monoxide.[13] They also observed methane formation over nickel surfaces at 300 K under conditions in which only molecular carbon monoxide appears to be present on the catalyst surfaces.[21] Vannice[11] also reported the formation of methane over platinum, palladium, and iridium surfaces, and independent experiments indicate the absence of carbon monoxide dissociation over these transition metal catalysts in most cases. It appears that the direct hydrogenation of molecular carbon monoxide can also occur but that this reaction has a much lower rate than methane formation via the hydrogenation of the active carbon that is produced from the dissociation of carbon monoxide in the appropriate temperature range.

Another mechanism, proposed by Pichler[4a] and Emmett,[8,9] involves the direct hydrogenation of molecular carbon monoxide to an enol species, followed by dehydration and further hydrogenation to produce methane. It is likely that this mechanism provides an additional reaction channel that may compete over certain transition-metal catalysts with CH_4 formation via the dissociation of carbon monoxide. Although this latter process seems to predominate under the experimental conditions used for methanation (1 to 5 atm, 450 to 700 K over nickel, cobalt, and ruthenium), it is by no means the only reaction mechanism that produces methane from carbon monoxide and hydrogen, as many investigations indicate.

The chemistry of C_{Ni} formed from CO disproportionation and from other carbon sources has been further investigated by Rabo et al.[25] It was found that the C_{Ni} reacts readily with water. Upon injecting a pulse of steam at 600 K over freshly prepared C_{Ni}, this species rapidly reacted with water to form equimolar $CO_2 + CH_4$ according to the equation

$$2C_{Ni} + 2H_2O \xrightarrow{\text{600 K}} CO_2 + CH_4 \tag{10}$$

These experimental results are consistent with the thermodynamics of the reaction between carbon and water. At low temperatures they favor the formation of methane, in contrast to the same reaction occurring at high temperatures where the product is $CO + H_2$. The fresh C_{Ni} species reacts readily with both H_2 and H_2O, while this species aged at higher temperatures is rendered substantially inert to both. The reaction of C_{Ni} with H_2O, similar to the reaction of C_{Ni} with H_2, is rapid at about 600K, reaching $\geqslant$90 percent conversion of the C_{Ni} layer in a few minutes.

The formation of C_{Ni} from CO, according to the Boudouard reaction, is exothermic. The reaction of C_{Ni}^{CO} with H_2O is also exothermic. This latter observation is in contrast to the reaction of graphite and water, which is calculated to be endothermic at about 600 K by about 3 kcal/mol. The exothermic nature of the reaction between C_{Ni} and H_2O indicates a higher-energy state for C_{Ni} relative to graphite.

An interesting change in the kinetics of methanation was observed by Castner et al.[26] when the reaction rates were monitored over clean rhodium and oxidized rhodium surfaces in $CO-H_2$ and CO_2-H_2 gas mixtures. The rates obtained at 600 K, the activation energies, and the preexponential factors for methanation are listed in Table 10.3. The turnover frequencies are much greater and the activation energies are much lower over the preoxidized metal surface. It appears that the oxidized metal surface is not only a better catalyst, but the mechanism of methanation is very different, as indicated by the large change in the kinetic parameters. The activation energy for methane formation is in the 12 to 15 kcal range over the oxidized surface and also when CO_2-H_2 gas mixtures are used for the reaction instead of CO and H_2, in contrast with the 24 kcal activation energy for this reaction on clean metal surfaces. High-resolution electron spectroscopy studies revealed that

Table 10.3. Comparison of the kinetic parameters (TN = $A \exp(-E_a/RT)$ for CO and CO_2 hydrogenation over polycrystalline rhodium foils

Reaction conditions	Surface pretreatment	CH_4TN at 300°C (molecules/site/sec)	A (molecules/site/sec)	E_a (kcal/mol)
3 H_2:1 CO, 0.92 atm	clean	0.13 ± 0.03	10^8	24 ± 3
3 H_2:1 CO_2, 0.92 atm	clean	0.33 ± 0.05	10^5	16 ± 2
3 H_2:1 CO, 6 atm	preoxidized	1.7 ± 0.4	10^5	12 ± 3

CO_2 dissociates on the clean rhodium surface to CO and O, and thus the molecule may act as an oxidizing agent on the clean metal surface. It is then likely that the CO_2–H_2 reaction occurs on a partially oxidized rhodium surface, and for this reason it exhibits similar kinetics to the CO–H_2 reaction on the oxidized metal surface.

Methanol Formation

The production of CH_3OH from carbon monoxide and hydrogen is a commercial process that utilizes a mixture of zinc, copper, and chromium oxides at relatively low pressures (below 100 atm). Recent studies by Mehta et al.[12] show that interesting changes take place in the catalyst morphology under the reaction conditions. The Cu^+ ion that is present in the zinc oxide matrix appears to be the active component for methanol production. Chromium appears to increase the solubility of copper, and the increased Cu^+ concentration leads to higher catalytic activity. Copper dissolves in zinc oxide, and this solid-state reaction is assisted by the presence of Cr^{3+} ions. The production of methanol from carbon monoxide and hydrogen over paladium, platinum, and iridium at high pressures (larger than 150 atm) was reported by Poutsma et al.[13] Paladium appeared to be 10 times more active than platinum or iridium for this reaction. Since these surfaces do not dissociate carbon monoxide readily, the formation of CH_3OH appears to be the result of direct hydrogenation of molecular carbon monoxide over the transition metals. Although similar conclusions cannot be reached for the CH_3OH formation over the zinc/copper/chromium oxide catalyst system in the absence of more detailed studies, direct hydrogenation of carbon monoxide does provide a reaction channel for the formation of CH_3OH, at least at high pressures.

Polymerization Reactions

The hydrogenation of carbon monoxide over iron, cobalt, and ruthenium surfaces produces a mixture of hydrocarbons with a wide range of molecular-weight distribution. An example of this is shown in Table 10.1, where a typical product distribution obtained over promoted cobalt catalysts in the 1940s is displayed. Most of the hydrocarbons produced were normal paraffins; however, olefins and alcohols in smaller concentrations were also obtained.

The wide product distribution indicates that a polymerization mechanism may be operative. Some of the reaction intermediates serve as chain initiators; then the chain propagation proceeds rapidly until termination by hydrogen occurs before the molecule desorbs from the catalyst surface. The distribution of reaction products, which has been shown to follow a Schultz–Flory[27,28] distribution of molecular weights frequently encountered in polymerization processes, is given by

$$M(P) = (\ln^2 \alpha)P\alpha^P \tag{11}$$

where $M(P)$ is the weight fraction of hydrocarbons containing P carbon atoms. The chain-growth probability factor is defined as

$$\alpha = \frac{r_p}{r_p + r_t} \tag{12}$$

where r_p and r_t are the rate of propagation and termination, respectively. Eq. (11) can be expressed in logarithmic form as

$$\log \frac{M(P)}{P} = \log (\ln^2 \alpha) + P \log \alpha \tag{13}$$

A plot of $\log [M(P)/P]$ versus P yields a value of α from either the slope or the ordinate intercept. Agreement between the slope and intercept is used as a criterion of the soundness of Schultz–Flory fit.

Dwyer and Somorjai[29] have studied the Fischer–Tropsch reaction using an iron polycrystalline foil of ~1 cm^2 surface area, and the reaction was carried out with a hydrogen/carbon monoxide ratio of 3:1, at 6 atm and 600 K. At the low conversions (below 1 percent) obtained under these conditions, the products are primarily methane and ethylene, with trace amounts of other α-olefins up to C$_5$. This product distribution is compared in Figure 10.6 with that obtained from pilot-plant studies over iron catalysts under industrial conditions and at high conversions (85 percent).

Under industrial conditions, high-molecular-weight paraffins are ob-

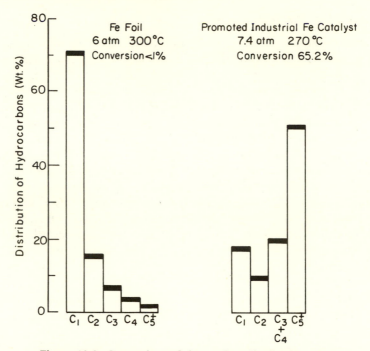

Figure 10.6. Comparison of the product distribution obtained at low conversion from ref. 29 with that obtained at high conversion, ref. 5.

tained in large concentrations. In order to simulate the experimental conditions that exist at high conversions, Dwyer and Somorjai[29] have added ethylene to the synthesis gas, since C_2H_4 was one of the products detected. The fate of the ethylene was then monitored as a function of reaction time. The majority of the ethylene was hydrogenated to ethane, as shown in Figure 10.7. However, about 10 percent of the added ethylene was converted to other hydrocarbons. The conversion of ethylene to other hydrocarbons had a significant impact on the product distribution of the $CO-H_2$ reaction shown in Figure 10.8. The relative amount of C_3 to C_5 hydrocarbons increased due to the presence of ethylene in the synthesis gas. The influence of ethylene concentration on the product distribution was investigated by varying the partial pressure of ethylene between 2 and 150 torr, while the H_2/CO ratio was held constant at 3:1 and at a total pressure of 6 atm. In Figure 10.9, the product distribution is given as a function of ethylene partial pressure. As the initial ethylene partial pressure is increased, the relative amount

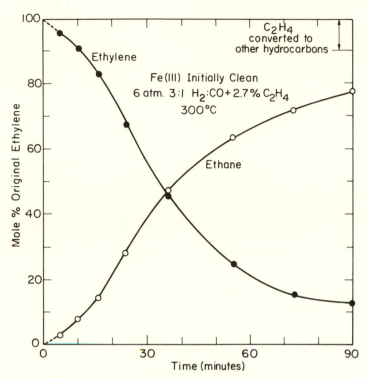

Figure 10.7. Conversion of 2.7 mole percent added ethylene to ethane as a function of time. Note that some of the ethylene is converted to other hydrocarbons.

of methane in the product distribution decreased, although the amount of methane formed remains largely unchanged. The C_5^+ fraction, however, increases with increasing ethylene in an almost linear fashion. The C_3 and C_4 fractions increase to limiting values.

Experiments in which propylene was added to synthesis gas produced results similar to those when ethylene was added. Propylene seemed to produce larger molecules than did the same amount of added ethylene. This is demonstrated in Figure 10.10. Substantial amounts of C_6 and C_7, and even trace amounts of C_8 hydrocarbons, were observed. The product distribution obtained is compared in Figure 10.11 with the distribution typical of the high-conversion experiments. The similarity of the distributions is apparent; by adding small concentrations of propylene, it is possible to obtain the product distribution found under high-conversion conditions.

The hydrocarbon distribution obtained after adding C_2H_4 to carbon

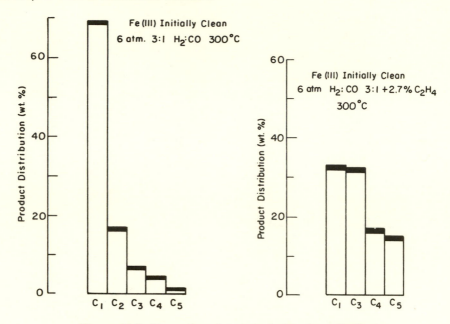

Figure 10.8. Comparison between the product distribution obtained from initially clean Fe(111) with and without added ethylene. Ethylene concentration is in mole percent.

monoxide and hydrogen is presented in Figure 10.12, assuming Schultz–Flory distribution. The chain-growth parameter varies from 0.30 at 2 torr of ethylene to 0.56 at 125 torr. Excellent agreement was obtained between the slope and intercept values, confirming that the distribution follows the Schultz–Flory equation. The increase in the chain-growth probability with increasing ethylene partial pressure indicates that ethylene participates directly in the propagation step. The Schultz–Flory plot of the hydrocarbon distribution observed upon adding propylene is shown in Figure 10.13. The slope and intercept yield a value of $\alpha = 0.70$. This value compares well with the value of 0.80 obtained under high-conversion conditions over industrial iron catalysts.

Considerable work has been published concerning the incorporation of radioactive-isotope labeled olefins in hydrocarbons during Fischer–Tropsch reactions. The pioneering work of Kummer and Emmett[8] and Hall et al.[9] suggested that ethylene acted as a chain initiator over iron catalysts. This conclusion was based on a constant molar radioactivity of the C_3 and C_5 fractions. The same results were obtained over cobalt catalyst by Eidus et al.[30]

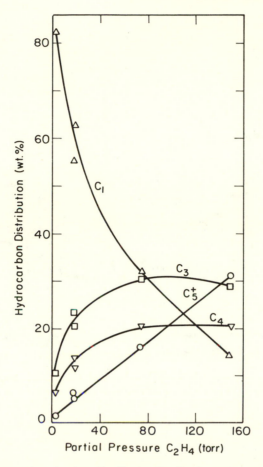

Figure 10.9. Product distribution for fixed reaction conditions (6 atm, 3:1 H_2 CO, 300°C) as a function of added ethylene.

It has long been suspected that α-olefins are the primary products of Fischer–Tropsch synthesis, although they are thermodynamically unstable under the reaction conditions. It has been shown that the concentration of α-olefins in the gas stream increases with increasing space velocity.[31] It appears that readsorption and subsequent secondary reactions of α-olefins occur readily under Fischer–Tropsch conditions. Readsorption and secondary reaction of these olefins may be a major reaction pathway leading to the growth of hydrocarbon molecules during Fischer–Tropsch synthesis. In standard flow reactors with large surface-area catalysts it is expected that at the leading edge of the bed,

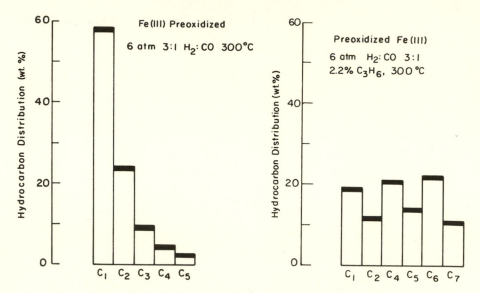

Figure 10.10. Product distribution obtained over the oxidized iron crystal surface in the presence of propylene.

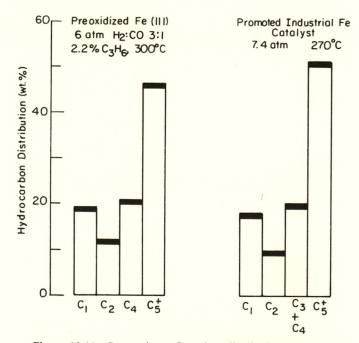

Figure 10.11. Comparison of product distribution obtained by the addition of propylene to CO and H_2 in low-conversion experiments and that from high-conversion experiments.

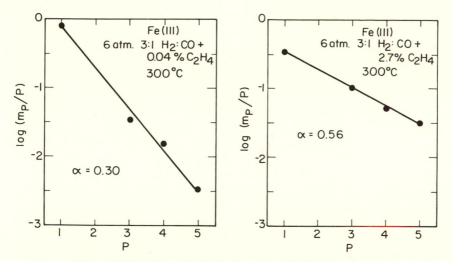

Figure 10.12. Plot of the hydrocarbon distributions in the CO–H_2 reaction over iron at two different partial pressures of added ethylene according to the Schulz–Flory distribution defined in Eq. (13). Note the increase in the chain-growth parameter α at higher ethylene pressure.

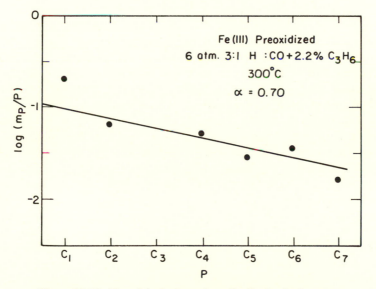

Figure 10.13. Plot of the hydrocarbon distribution in the CO–H_2 reaction over iron obtained after the addition of propylene according to the Schulz–Flory distribution defined in Eq. (13).

537

the product distribution will be similar to that obtained at low conversions. As these initial reaction products proceed along the bed, they will be readsorbed and undergo secondary reactions, leading to higher-molecular-weight products. As a result of the changing product distribution along the catalyst bed, the surface composition of the catalyst is also likely to change. The presence of readsorption as an important reaction step should permit one to devise ways of controlling the product distribution. Various additives to the reactant mixture, changing the size and geometry of the catalyst bed, and mixing of catalysts are among the experimental variables that may be used to tailor product distribution in the Fischer–Tropsch reaction.

We may then write the formation of high-molecular-weight hydrocarbons over iron or ruthenium surfaces as a two-step process, starting with olefin production:

$$2CO + 4H_2 \xrightarrow[\text{250-350°C}]{\text{Fe;6 atm}} C_2H_4(C_3H_6) + 2H_2O$$

This is followed by the readsorption of olefins that induces polymerization,

$$CO + H_2 \xrightarrow[\text{250-350°C}]{\text{Fe,C}_2\text{H}_4(\text{C}_3\text{H}_6)} C_{5-9}H_{12-20} + H_2O$$

Recent studies of the CO–H$_2$ reaction on ruthenium surfaces have also shown the importance of readsorption on the metal catalyst surface. The presence of a multiple-step reaction that proceeds via the readsorption of the initial products provides opportunities for altering the product distribution by using several different catalysts simultaneously in the reaction mixture. By physical mixing of two catalysts, for example, experimental conditions can be realized where the olefins readsorb and further react on the other catalyst instead of on the iron catalyst surface. This way the product distribution can be changed to obtain molecules that are more desirable than the saturated straight-chain hydrocarbons. Recently, Lachthaler et al.[32,33] have reported on a process that converts CO and H$_2$ to aromatic molecules or to high-octane-number gasoline. First, methanol and olefins are produced by the catalytic reactions of CO and H$_2$, as discussed above. Then, using a zeolite shape-selective catalyst that is introduced along with the ruthenium or other metal catalyst in the same reaction chamber, methanol and the olefins are converted to aromatic molecules, cycloparaffins, and paraffins. The mechanism involves the dehydration of methanol to dimethyl ether. The light olefins that also form are alkylated by methanol and by the dimethyl ether[34] to produce higher-molecular-weight olefins and then the final cyclic and aromatic products.

The formation of aromatic molecules from CO and H_2 over ThO_2 surfaces has been reported at higher temperatures[4a] while C_4 isomers were produced at lower temperatures and high pressures over the same catalyst (isosynthesis). However, the mechanism of this reaction has not yet been subjected to detailed scientific scrutiny.

Formation of Oxygenated Organic Molecules

The production of alcohols, aldehydes, and acids by carbon monoxide hydrogenation is an important aim of researchers. The insertion of CO and H_2 into an olefin with a C_n carbon number to produce aldehydes with C_{n+1} carbon atoms, called "hydroformylation" or the "oxo" process, finds major industrial applications.[14] This carbonyl insertion process is catalyzed by cobalt hydrocarbonyl or by rhodium carbonyl hydride with triphenyl phosphine as ligand modifier. Both catalysts are active in solution, and the rhodium compound operates at lower pressures (~ 10 atm) and temperature (~ 420 K). Pruett[17] has reported the synthesis of ethylene glycol from CO and H_2 at temperatures of 510 to 550 K, using rhodium carbonyl as a homogeneous catalyst. High pressures, on the order of 1200 atm, are used to carry out the reaction, since 5 mol of gas is required to produce 1 mol of product.

$$2CO + 3H_2 = HOCH_2CH_2OH$$

Infrared spectroscopy studies identify the large rhodium complex $[Rh_{12}(CO)_{34}]^{2-}$ anion as the likely catalyst in this reaction.

Selective rhodium catalysts have been prepared by anchoring metalloorganic rhodium complexes to oxide or polymer surfaces. The production of propionaldehyde from ethylene can be carried out by a $Rh(PPh_3)(CO)_2Cl$ catalyst anchored to silica or to styrene polymer.[35] Recently, Perkins and Vollhardt[36] have reported methanation and Fischer–Tropsch activity by polymer-supported cyclopentadienyl cobalt. Ichikawa[37] has reported the synthesis of methanol from H_2 and CO over partially decomposed rhodium carbonyl clusters that were deposited on MgO and ZnO supports. Interestingly, when the same rhodium clusters were placed on silica support only methanation could be detected.

Heterogeneous rhodium catalysts have been utilized to produce acetic acide and acetaldehyde[16] from CO and H_2. Rhodium deposited on a silica support produces these oxygenated C_2 products at 600 K and 70 atm. It should be noted that the direct carbonylation of methanol to acetic acid takes place in the presence of a homogeneous rhodium catalyst, $[Rh(CO)_2I_2]^-$, at 480 K and 30 to 40 atm.

Rhodium is an excellent catalyst to produce small hydrocarbon molecules (C_2 or C_3) with high oxygen content and without exhibiting chain growth or polymerization activity. Recent studies by Castner et al.[26] indicate that the addition of alcohols or olefins to the $CO-H_2$ mixture during low-conversion experiments does not yield chain growth, unlike on iron or ruthenium.

Low-conversion studies by Sexton and Somorjai[10] also indicate that the clean, unsupported rhodium metal surface does not produce oxygenated products. Preoxidation of the surface was necessary to obtain alcohols, aldehydes, or acids. It appears the zero-valent rhodium is not the catalyst for oxygen insertion, but that other oxidation states of the metal are implicated. The use of basic oxide supports that are electron donors for the rhodium catalyst results in the formation of alcohols, while more acidic supports produce olefins and methane. The importance of promoters, such as alkali metals or manganese, to enhance the formation of oxygenated products, could also imply the necessity to stabilize a unique rhodium oxidation state to obtain the oxygenated molecules.

A striking confirmation of the need to stabilize an oxidation state that is different from that of the metal to form oxygenated products is the study by Anderson[38] of the catalytic behavior of iron nitrides. These compounds are prepared by heating iron in ammonia, which produces predominantly alcohols. Although the nitride slowly decomposes in the $CO-H_2$ mixture, it can readily be regenerated. The product distribution obtained over the iron nitrides is so different from that obtained over iron or iron oxides that the oxidation state of the iron must play a key role in controlling the reaction paths.

The addition of electron donor alkali metals to the active catalyst can markedly influence the product distribution. Recent studies of potassium pretreatment of iron oxides[39] indicate that the metal stabilizes the oxide, which would otherwise be unstable in the $CO-H_2$ reaction mixture. Auger spectroscopy studies indicate that the iron or iron oxide catalyst surface becomes covered with a multilayer of carbon after short reaction times in the $CO-H_2$ mixture. At that point the catalyst surface loses its Fischer–Tropsch reaction activity and becomes a slow methanation catalyst. The alkali-pretreated iron oxide surface retains its surface composition, and there is no evidence for carbon buildup on the surface, even after extensive reaction times. Veraa and Bell[40] have suggested that potassium catalyzes the removal of surface carbon by the following sequence of reactions.

$$
\begin{array}{ll}
(1) & K_2CO_3 + 2C = 2K + 3CO \\
(2) & 2K + 2H_2O = 2KOH + H_2 \\
(3) & CO + H_2O = CO_2 + H_2 \\
(4) & \underline{2KOH + CO_2 = K_2CO_3 + H_2O} \\
& \quad 2C + 2H_2O = 2CO + 2H_2
\end{array}
$$

Since carbon is removed from the surface, the direct rehydrogenation of surface carbon to methane becomes less likely. Thus the methanation rate is much diminished. There is also evidence that the presence of alkali metal on heterogeneous and homogeneous rhodium catalysts[16,17] facilitates the production of oxygenated products, perhaps by the stabilization of an active oxidation state.

Insertion of CO molecules (Carbonylation) to Form C_2 or C_3 Products

With the exception of the methanation and methanol formation reactions, the formation of carbon–carbon bonds is one of the most important steps during the hydrogenation of carbon monoxide. One mechanism for this process is the CO insertion, which is a key step during hydroformylation. This process generates an acyl species[41] and increases the carbon chain length by one:

$$
CO + R\!-\!M \rightleftharpoons R\!-\!M\!-\!CO \rightleftharpoons R\!-\!CO\!-\!M
$$

This process is followed by hydrogen reduction of the inserted carbonyl to $-CH_2-$. There is also a possibility of CO insertion to multiply bonded methylene or carbene species that exist as active carbon species on the metal surfaces during the Fischer–Tropsch reaction,[42]

$$
M\!-\!CH_x + CO \rightarrow M\!-\!CO
$$
$$
\qquad\qquad\qquad\; |
$$
$$
\qquad\qquad\quad CH_x
$$

The reduction of CO and the incorporation of the oxygen-free carbon in the growing chains could then occur at a rate consistent with the known kinetics of the Fischer–Tropsch process.

High-resolution electron loss spectroscopy provided evidence for the presence of stable CH and CH_2 groups on nickel, iron, and rhodium[43] surfaces. It is possible that these groups have a reasonable mobility on the transition-metal surface that could lead to C—C bond formation.

At present, there is no suitable experimental evidence that reveals the predominant mechanism of C—C bond formation on the molecular level. Future studies must focus on this important reaction step that could be the key to controlling chain growth and chain termination to alter the product distribution during catalytic hydrogenation of carbon monoxide.

Research Directions in the Near Future

Several important questions emerge from atomic-scale studies of the hydrogenation of carbon monoxide over transition-metal surfaces. These involve the chemical bonding and reactions of active carbon fragments, identification of the metal surface oxidation states, and the formation of C—C bonds.

1. There is a great deal of experimental evidence that highly reactive carbon fragments play important roles in the $CO-H_2$ reaction.[44] HREELS studies have identified the presence of stable C, CH, and CH_2 species on nickel, iron, and rhodium surfaces. The reactions of these active species with hydrogen or water lead to the production of methane and other hydrocarbons. These fragments are active in a finite temperature range; from the temperature at which CO decomposes (300 to 500 K) to that where the active fragments are converted to graphite (>700 K).[44] The chemical bonding, mobility, and reactivity of these fragments seem to determine many of the reaction paths that are operative during the catalyzed $CO-H_2$ reactions.

2. There are many experimental indications that implicate a unique surface oxidation state in the production of molecules selectively from CO and H_2. Preoxidation of rhodium and iron leads to greatly increased activity, different mechanisms for methanation, and a different product distribution. Iron nitrides yield alcohols during the Fischer–Tropsch reaction, while other iron compounds (carbides, oxides) produce straight-chain saturated hydrocarbons. The deposition of rhodium onto basic supports aids the formation of methanol; when rhodium is deposited on more acidic supports, the formation of methane predominates. Additives that are electron donors (alkali metals) or electron acceptors (oxygen) also markedly influence the product distribution. These results provide strong, although indirect, evidence that changes in the oxidation state of the metallic catalyst surfaces strongly influence the reaction paths and thus the product distribution. In the near future the catalyst surface will undergo close scrutiny by electron spectroscopy and by other techniques to establish a better understanding of relationships between the surface atom oxidation states and the product distribution during the Fischer–Tropsch reaction.

3. Perhaps the most important reaction step in producing hydrocar-

bon molecules from CO and H_2 selectively is the formation of the carbon–carbon bond. Although secondary reactions involving the readsorption of olefins that are produced in the primary step and their polymerization activity have been identified, the reaction steps leading to the production of the olefins have not yet been verified. Electron spectroscopic as well as isotope labeling studies should be of help in gaining a better understanding of C—C bond-forming reactions.

References

1. P. Sabatier and J.B. Senderens, C.R. Acad. Sci. Paris **134,** 514 (1902).
2. (a) M.E. Dry, Ind. Eng. Chem. Prod. Res. Dev. **15,** 282 (1976); (b) G.A. Mills and F.W. Sheffgen, Catal. Rev. **8,** 159 (1973).
3. F. Fischer and H. Tropsch, Brennst.-Chim. **7,** 97 (1926).
4. (a) H. Pichler, Adv. Catal. **4,** 271 (1952); (b) H. Kolbel, Chem.-Ing-Tech. **29,** 505 (1957).
5. J.H. Storch, N. Golumbic, and R.B. Anderson, *The Fischer–Tropsch and Related Syntheses,* Wiley, New York, 1951.
6. R.B. Anderson in *Catalysis,* vol. 4, ed. P.H. Emmett, Reinhold, New York, 1956.
7. G. Natta, U. Columbo, and I. Pasquon in *Catalysis,* vol. 5, ed. P.H. Emmett, Reinhold, New York, 1957.
8. J.T. Kummer and P.H. Emmett, J. Am. Chem. Soc. **75,** 5177 (1953).
9. W.K. Hall, R.J. Kokes, and P.H. Emmett, J. Am. Chem. Soc. **82,** 1027 (1960).
10. B.A. Sexton and G.A. Somorjai, J. Catal. **46,** 167 (1977).
11. M.A. Vannice, Catal. Rev. **14,** 153 (1976).
12. (a) R.G. Herman, K. Klier, G.W. Simmons, B.P. Finn, and S.B. Bulko, J. Catal. **56,** 407 (1978); (b) S. Mehta, G.W. Simmons, K. Klier, and R.G. Herman, J. Catal. **57,** 339 (1979).
13. M.L. Poutsma, L.F. Elek, P. Ibarbia, H. Risch, and J.A. Rabo, J. Catal. **52,** 157 (1978).
14. R.L. Pruett, Adv. Organomet. Chem. **17,** 1 (1979).
15. G.W. Parshall, J. Mol. Catal. **4,** 243 (1978).
16. M.M. Bhasin, W.J. Bartley, P.C. Ellgen, and T.P. Wilson, J. Catal. **54,** 120 (1978).
17. R. Pruett, Ann. N.Y. Acad. Sci. **295,** 239 (1977).
18. M.A. Vannice, J. Catal. **37,** 449 (1975).
19. (a) D.J. Dwyer and G.A. Somorjai, J. Catal. **52,** 291 (1978); (b) M. Araki and V. Ponec, J. Catal. **44,** 439 (1976).
20. P.R. Ventrcek, B.J. Wood, and H. Wise, J. Catal. **43,** 363 (1976).
21. J.A. Rabo, A.P. Risch, and M.L. Poutsma, J. Catal. **53,** 295 (1978).
22. P. Biloen, J.N. Helle, and W.M.H. Sachtler, J. Catal. **58,** 95 (1979).
23. L.L. Kesmodel, L.H. Dubois, and G.A. Somorjai, J. Chem. Phys. **70,** 2180 (1979).
24. J.E. Demuth and H. Ibach, Surf. Sci. **78,** L238 (1978).

25. J.A. Rabo, J.N. Francis, and L.F. Elek, Proc. Catal. Congr., Tokyo (1980).
26. D.G. Castner, R. Blackadar, and G.A. Somorjai, J. Catal., to be published 1980.
27. G.V. Schulz, Z. Phys. Chem. **B43,** 25 (1939).
28. P.J. Flory, *Principles of Polymer Chemistry,* Cornell University Press, Ithaca, N.Y., 1953.
29. D.J. Dwyer and G.A. Somorjai, J. Catal. **56,** 249 (1979).
30. Y.T. Eidus, N.D. Zelinski, and N.I. Ershov, Dokl. Skad. Nank, USSR **60,** 599 (1948).
31. (a) H. Richler, H. Schulz, and M. Elstner, Brennst.-Chim. **48,** 78 (1967); (b) J.G. Ekerdt and A.T. Bell, J. Catal. **62,** 19 (1980).
32. S.L. Meisel, J.P. McCullough, C.H. Lechshaler, and P.B. Weisz, Chemtech **6,** 86 (1976).
33. C.D. Chang and H.J. Silvestri, J. Catal. **47,** 249 (1977).
34. W.W. Keading and S.A. Butter, J. Catal., to be published 1980.
35. H. Arai, T. Keneko, and T. Kumigi, Chem. Lett. 265 (1975).
36. P.K. Perkins and P.C. Vollhardt, J. Am. Chem. Soc. **101,** 3985 (1979).
37. M. Ichikawa, J. Chem. Soc. Chem. Commun. 566 (1978); Ball. Chem. Soc. Jap. **51,** 2273 (1978).
38. R.B. Anderson, Adv. Catal. **5,** 355 (1953).
39. G.A. Somorjai, Catal. Rev. **18,** 173 (1978).
40. M.J. Veraa and A.T. Bell, Fuel **57,** 194 (1975).
41. R.F. Heck, *Organotransition Metal Chemistry,* Academic Press, New York, 1974.
42. E.L. Muetterties and J. Stein, Chem. Rev. **79,** 479 (1979).
43. L.H. Dubois and G.A. Somorjai, J. Chem. Phys. **72,** 5234 (1980).
44. H.P. Bonzel and H.J. Krebs, Surf. Sci. **91,** 499 (1980).

11. Photochemical Surface Reactions

Introduction, Photosynthesis, and the Photoelectrochemical Dissociation of Water

According to our definition of catalysis, a chemical reaction must be thermodynamically feasible before we should consider catalyzing it to achieve or approach equilibrium. However, a thermodynamically uphill reaction may be carried out with the aid of an external source of energy. In fact, one of the most important chemical reactions of our planet, photosynthesis, requires the input of a great deal of energy to convert water and carbon dioxide to hydrocarbons and oxygen:

$$CO_2 + H_2O \xrightarrow[\text{chlorophyll}]{\text{light}} \tfrac{1}{6}C_6H_{12}O_6 + O_2 \qquad (1)$$

The heat of this reaction is 120 kcal/mol of oxygen. It can also be rewritten to produce 1 mol of sugar ($C_6H_{12}O_6$) with a heat of reaction of 720 kcal/mol. It is interesting to note that of the four reactions of water, hydrogen, carbon monoxide, and carbon dioxide that produce hydrocarbons,

$$2CO + H_2 = -CH_2- + CO_2 \qquad (2)$$

$$CO_2 + 3H_2 = -CH_2- + 2H_2O \qquad (3)$$

$$CO + H_2O = -CH_2- + CO_2 \qquad (4)$$

$$CO_2 + H_2O = -CH_2- + \tfrac{3}{2}O_2 \qquad (5)$$

the first three are exothermic and thermodynamically feasible. Yet nature chose the last reaction to produce most of the hydrocarbons on our planet. The photosynthetic reaction is the only one that produces oxygen. It appears that the oxidizing atmosphere that was created may have eliminated the other reactants, CO and H_2, from the atmosphere.

It is useful to consider light as one of the reactants in photosynthesis. By adding the light energy to Eq. (1), the reaction becomes athermic or even exothermic if excess light energy is utilized, $h\nu + H_2O + CO_2 =$ $-CH_2- + \frac{3}{2}O_2$. We may consider photon-assisted or photochemical reactions of many types that lead to the formation of lower-molecular-weight hydrocarbons and of other products. One of the simplest of these important new classes of reactions leads to the dissociation of water:

$$hv + H_2O = H_2 + \tfrac{1}{2}O_2 \tag{6}$$

Another leads to the formation of methane:

$$hv + CO_2 + 2H_2O = CH_4 + \tfrac{3}{2}O_2 \tag{7}$$

or to the fixation of nitrogen:

$$hv + 3H_2O + N_2 = 2NH_3 + \tfrac{3}{2}O_2 \tag{8}$$

Light as a reactant may be employed in two ways. The adsorbed molecules can be excited directly by photons of suitable energy to their higher vibrational or electronic states. The excited species then may undergo chemical rearrangements or interactions that are different from those in the ground vibrational or electronic states. Alternatively, the solid can be excited by light in the near-surface region. Photons of band-gap or greater energy may excite electron–hole pairs at the surface. As long as these charge carriers have a relatively long lifetime (i.e., they are trapped at the surface, so that their recombination is not an efficient process), there is a high probability of their capture by the adsorbed reactants. These, in turn, can undergo reduction or oxidation processes using the photogenerated electrons and holes, respectively. The photographic process is one example of this type of surface photochemical reaction. However, one would like the photogenerated electrons and holes to be captured by the adsorbed molecules in order to carry out photochemical surface reactions of the adsorbates instead of the photodecomposition of the solid at the surface. The cross sections for adsorption of band-gap or higher-than-band-gap energy photons are so large that the photogeneration of electron–hole pairs is a most efficient process. At present, this cannot be readily matched by the efficiency of direct photoexcitation of vibrational or electronic energy states of the adsorbed molecules.

Many solid surfaces efficiently convert light to long-lived electron–hole pairs that can induce the chemical changes leading to the reactions in Eqs. (6), (7), and (8). In fact, inorganic photoreaction is one of the exciting new fields of surface science and heterogeneous catalysis.

It is important to distinguish between thermodynamically uphill photochemical reactions and thermodynamically allowed photon-assisted reactions. The latter reactions are thermodynamically feasible without any external energy input, but light is used to obtain certain product selectively. Excitation of selected vibrations, rotations, or electronic states of the incident or adsorbed molecules by light permits one to change the reaction path or increase the reaction rate. For example, the hydrogenation of acetylene[1] or the oxidation of ammonia[2] can be photon-assisted, leading to different reaction rates than in the absence of light. We consider here only those photochemical processes where the reaction would not occur at all without light because it is thermodynamically prohibited.

Let us briefly review the photochemical reactions that are most frequently utilized and that are the focal point of most research efforts in this area at present. These are (1) photosynthesis and (2) the photoelectrochemical dissociation of water.

1. The photosynthetic path whereby sugars and oxygen are produced from water and carbon dioxide with the assistance of chlorophylls, which convert light to other forms of energy that can be utilized by the plant, has been explored by many researchers over several decades.[3] This efficient process uses eight quanta of light for every oxygen molecule produced or for every carbon dioxide molecule that is reduced and converted to sugar. The chemical reaction for oxygen production in the chlorella algae, for example, takes place in about 40 msec, and one oxygen molecule is produced by about every 2500 chlorophyll molecules. Chlorophyll a and chlorophyll b molecules have absorption maxima at 6700 Å and 6500 Å, respectively, and with the help of other pigments (such as caratenoids, with a maximum absorption of 4000 to 5500 Å) they can convert much of the available visible light to excitation.[4] Higher-energy (ultraviolet) radiation is efficiently screened from the planet surface by the ozone layer in the stratosphere, and the lower-energy (infrared) end of the spectrum is absorbed by water. Hill and others have established that the oxidation and reduction step leading to the formation of O_2 and reduced CO_2 are separable.[3] Oxygen evolution is detectable only if there are electron acceptors, such as benzoquinone, present. Light is utilized in the photosynthetic system to convert adenosine diphosphate, ADP, to adenosine triphosphate, ATP, and to produce reduced nicotinamide adenine dinucleotide phosphate, NADPH. These molecules, ATP and NADPH, can catalyze the se-

547

quences of reaction steps that bind and convert CO_2 into sugar and produce O_2 while reconverting to ADP and NADP, respectively.[5] Then the whole process is repeated.

The overall photosynthetic scheme involves many steps. Ribulose-1,5-biphosphate reacts with carbon dioxide in the leaf chloroplasts to produce two molecules of 3-phospho-D-glycerate. This process is catalyzed by an enzyme, ribulose bisphosphate carboxylase.[6] Then the phosphoglycerate is reduced to triose phosphate in the presence of both ATP and NADPH, the high-energy phosphates produced by the photosynthetic electron transport. Finally, the ribulose bisphosphate is regenerated through several enzyme-catalyzed reaction steps to be available again to accept CO_2 together with the ADP, NADP, and an inorganic phosphate, P. The net reaction for the cycle can be expressed as[6]

$$3CO_2 + 9ATP + 6NADPH + 6H^+ + 5H_2O$$
$$\rightarrow \text{triose-P} + 9ADP + 8P + 6NADP$$

Thus for each CO_2 molecule fixed, 3ATP and 2NADPH are needed. Much of the reduction cycle to convert CO_2 to sugar has been verified by Calvin and his associates. This complex and efficient reaction is responsible for most of the plant chemistry on this planet.

A key feature of photosynthesis is the efficient storage of photon energy in the long-lived, excited states of organic macromolecules. Within the long trapping times of the excited electrons, the chemical reaction occurs without much competition from recombination or back-electron-transfer processes.[7] The excited molecules are also powerful oxidants and reducing agents.

2. Another important area where studies of thermodynamically uphill reactions have been carried out is that of the photoelectrochemical dissociation of water. It was reported by Fujishima and Honda[8] in 1972 that upon illumination of reduced titanium oxide (TiO_2), which served as the anode in basic electrolyte solution, oxygen evolution was detectable at the anode, while hydrogen evolved at a metal (platinum) cathode. This reaction requires an energy of 1.23 V/electron (a two-electron process per dissociated water molecule). In the presence of light of energy equal to or greater than the band-gap energy of titanium oxide (3.1 eV), an external voltage as low as 0.2 V was sufficient to dissociate water. The process stopped as soon as the light was turned off, and started again upon re-illumination. Shortly after, several other systems showed the ability to carry out photon-assisted dissociation of water.[9–12] When p-type gallium phosphide, GaP, was used as a cathode instead of platinum upon illumination of the TiO_2 anode, O_2 and H_2 could be

generated at the semiconductor anode and cathode, respectively, without the need of applying any external potential.[12] When strontium titanate, $SrTiO_3$, was substituted for TiO_2 as the anode, H_2O photodissociation was found to take place without external potential even when a platinum cathode was employed.

Figure 11.1 shows a schematic energy diagram to indicate the conditions necessary to carry out photoelectrochemical reactions efficiently. If the band-gap energy is greater than the free energies for the reduction and oxidation reactions, the photoelectron that is excited into the conduction band by light could reduce B^+ to B by electron transfer from the surface to the molecule. The photogenerated electron vacancies (holes) could also oxidize the A^- anions to A by capturing the electron. For the photodissociation of water, the conduction band must be above the H^+/H_2 potential and the valence band below the O_2/OH^- potential to be able to carry out the photoreaction without an external potential. The band gap has to be greater than 1.23 V and the "flat band" potential of the conduction and valence bands energetically well placed with respect to the (H^+/H_2) and O_2/OH^- couples.[13] The flat-band potentials can be obtained by capacitance measurements as a function of external potential;[14] these potentials are shown in Figure 11.2 for several solids and are compared with the H^+/H_2 and O_2/OH^- couples, using the electrochemical scale. The flat-band potentials become more negative with increasing pH at a rate of about (-0.59 eV/pH unit), according to the Nernst equation.

There is, of course, considerable band bending of the conduction and valence bands of any semiconductor at the surface.[13,14] This is due to the

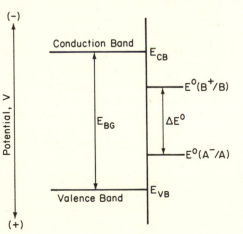

Figure 11.1. Energy conditions needed to reduce B^+ to B and oxidize A^- to A at a semiconductor surface. Electrons that are excited by photons into the conduction bond E_{CB} must be able to reduce B^+, and electron vacancies (holes) in the valence band E_{VB} must be able to oxidize A^-.

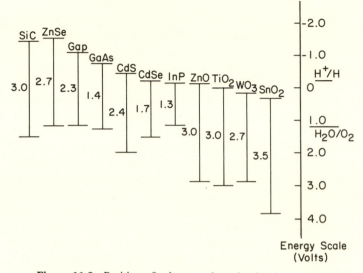

Figure 11.2. Position of valence and conduction band edges for several semiconductors in contact with aqueous electrolyte at pH 1.0. The position of the H^+/H_2 and H_2O/O_2 redox couples are indicated at the right. (After A. J. Nozik, ref. 12.)

presence of localized electronic surface states and to charge transfer between the adsorbates and semiconductor. Potential-energy diagrams that show the band positions schematically at an n-type or p-type semiconductor liquid interface are shown in Figure 11.3. The band bending provides an efficient means of separating electron–hole pairs, since the potential gradient as shown for the n-type semiconductor drives the electrons away from the semiconductor surface while it attracts the holes in the valence band toward the semiconductor electrolyte interface. As a result, the oxidation reaction takes place at the oxide anode while the reduction reaction takes place at the cathode to which the photoelectron migrates along the external circuit. The magnitude of the band bending at the surface depends primarily on the carrier concentration in the semiconductor and on the electron-donating or -accepting abilities of the adsorbates at the surface. Semiconductors that are not likely to carry out the photodissociation of water, according to the location of their flat-band potential, may become photochemically active as a result of strong band bending at the surface.

When a $SrTiO_3$ single crystal, that was partially reduced in hydrogen, was platinized on one side and then illuminated by band-gap radiation from the other side after immersion in a basic electrolyte solution, hy-

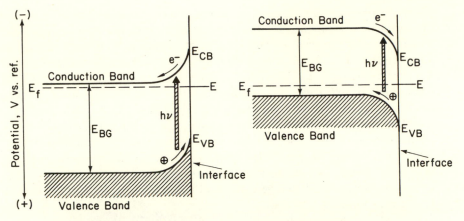

Figure 11.3. Band bending at the n-type and p-type semiconductor interfaces. The surface space charge induces a potential gradient that drives the photogenerated electrons away (n-type) or toward (p-type) the interface in the conduction band. The photogenerated hole is driven in the opposite direction in the valence band, resulting in efficient charge separation. E_f is the Fermi level.

drogen and oxygen evolution was detectable.[13] The photochemical reaction seems to take place even if the anode and cathode are short-circuited (touching each other). Suspensions of partially platinum-coated $SrTiO_3$ particles in aqueous solution have also produced hydrogen and oxygen.[12,14]

Often the oxidation or reduction photoreactions lead to the decomposition of the semiconductor electrode material. Thus instead of the photoreactions of adsorbate ions or molecules, a solid-state photoreaction occurs. This is particularly noticeable at the surfaces of illuminated CdS, Si, and GaP. Much of the research is therefore directed toward stabilizing these photoelectrode materials by suitable adsorbates[15] that could prevent the occurrence of photodecomposition by providing an alternative chemical route for the photoreduction or photooxidation.

Photocatalytic Dissociation of Water Using SrTiO₃ Crystals

We have embarked on a program to explore the kinetics and mechanisms of photochemical reactions that take place at the solid–vapor and solid–liquid interfaces. There are several advantages in performing photochemical reactions using water vapor and other gaseous molecules. The surface composition and its changes can be analyzed readily by the modern surface diagnostic techniques of surface science that require high vacuum; thus the atomic-scale reaction mechanism can be studied more easily. The photochemical reactions may be carried out at higher temperatures, for example, in the range 140 to 400°C, where most catalyzed hydrocarbon reactions take place. As a result, one may combine the photochemical reaction with catalyzed hydrocarbon conversion processes. The adsorption of light and the chemical dissolution of the active surface by the electrolyte are absent in photoreactions at the solid–vapor interface. In addition, diffusion of reactants and products to and from the surface is more rapid and is not likely to control the rate of the photochemical reaction.

The photochemical dissociation of water to hydrogen and oxygen over $SrTiO_3$ surfaces was studied in two ways:[16] (1) by reaction-rate measurements over single-crystal surfaces in a variety of experimental circumstances, and (2) by electron spectroscopy techniques to identify the elementary reaction steps and the active surface sites for photoreduction and photooxidation. These two types of studies will be discussed in some detail below.

The use of well-characterized single-crystal surfaces instead of large surface-area powders was deemed necessary to be able to correlate the electronic and atomic surface structures to the reaction rates. In addition, the high-surface-area powders are generally contaminated with uncontrolled impurities, principally carbon. The formation of H_2 by the side reaction $C+2H_2O \xrightarrow{h\nu} 2H_2+CO_2$ would be facile. The small-area, high-purity single crystal can be readily cleaned, and the small amount of carbon that may be present on its surface will not influence the photoreaction studies to any appreciable extent.

We have found that photodissociation of water introduced through the vapor phase occurs over an $SrTiO_3$–Pt sandwich as well as over a metal-free $SrTiO_3$ single-crystal sample in the presence of alkali hydroxides (NaOH, KOH, etc.).[17] At least two reaction mechanisms are operating: one is a photoelectrochemical type which leads to oxidation at the oxide (O_2 evolution) and reduction at the metal surface (H_2 evolution); the other is a photocatalytic process which leads to both oxidation and reduction at the same illuminated oxide surface. The presence of hydroxide ions is essential to obtain water photodissociation. At high (OH^-) concentration (5 N to 20 N), the photoreaction rates are greatly accelerated. Electron spectroscopy studies (AES, UPS, and XPS) revealed that the Ti^{3+} ions, which are present in abundance at the reduced $SrTiO_3$ surface, interact strongly with oxygen molecules and are oxidized by them.[18] Illumination by band-gap energy light causes photodesorption of oxygen and partial regeneration of the Ti^{3+} surface sites. The formation of surface hydroxyls upon the dissociative adsorption of water or hydrogen on the reduced $SrTiO_3$ surface could be identified and monitored.[18]

All hydrogen production experiments were performed in reaction chambers capable of being evacuated and then backfilled with water vapor at room temperature. Products were detected by gas-chromatographic sampling of a gas phase consisting of water vapor, argon, and products that circulated in a closed loop by the $SrTiO_3$ crystal or, in some experiments, over the electrolyte in which the crystal was immersed. Hydrogen production rates as low as 12×10^{15} molecules/hr were readily measurable. Technical difficulties placed the smallest detectable oxygen evolution rate above 10^{17} molecules/hr.

The photochemical reaction-rate measurements were carried out in two different reaction cells. Rate studies utilizing water vapor were carried out using the low-pressure/high-pressure reaction cell[19] described in Chapter 2. Solid–liquid interface studies were carried out in a Pyrex flask that was inserted into the vacuum line. Crystal wafers about 1 mm thick were cut from single-crystal boules. Wafers were oriented to within 1° of the (111) plane by Laué back-reflection X-ray diffraction. Prereduced crystals were treated in a hydrogen furnace at 1000°C for 4 hr and appeared blue-black and almost opaque. Stoichiometric crystals received no heat or hydrogen treatment and were clear and colorless when polished. The backs of "platinized" crystals were coated with platinum metal via the thermal decomposition of aqueous chloroplatinic acid. A 500-W high-pressure mercury lamp filtered through water to remove infrared radiation was used. Corning glass filters allowed only band-gap or sub-band-gap radiation to impinge upon the crystals. The flux of photons with greater-than-band-gap energy (>3.2 eV) was in the range 10^{14} to 10^{16} sec^{-1}/cm^2. The photoelectron spectroscopy studies were carried out in a commercial ultrahigh-vacuum chamber with base pressure in the low-10^{-10}-torr range.[20,21] Electron energy analysis was performed using a double-pass cylindrical mirror analyzer. Ultraviolet photoelectron spectroscopy was carried out using the He I emission (21.2 eV). A Physical Electronics X-ray source with a Mg anode provided photons for XPS measurements. Auger spectra served to monitor surface cleanliness.

Studies of Hydrogen Photoproduction on SrTiO$_3$ Crystals

We have observed sustained hydrogen photoproduction on metal-free as well as on platinized SrTiO$_3$ crystals when covered with a water-vapor-saturated film of NaOH, KOH, CsOH, or Cs$_2$CO$_3$, or when immersed in an alkaline aqueous electrolyte.[16] Figure 11.4 compares hydrogen photoproduction from platinized and metal-free prereduced

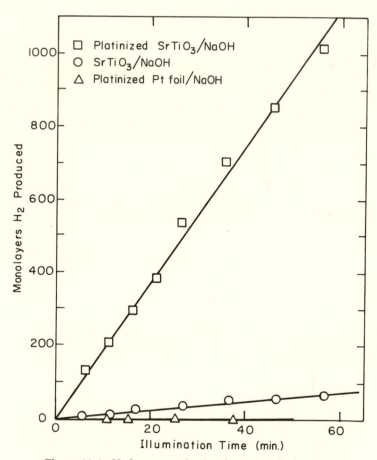

Figure 11.4. Hydrogen production from platinized and metal-free n-type SrTiO$_3$(111) crystals. Both crystals were coated with >30 μm of NaOH film and were illuminated in a saturation pressure (~20 torr) of water vapor (1 monolayer = 10^{15} molecules H$_2$/cm^2 of illuminated surface).

SrTiO$_3$ crystals. These crystals and their Vycor glass mounting brackets were coated with NaOH crusts of thickness greater than 30 μm. The crystals were inserted into the reaction cell, saturated with water vapor, and illuminated. Under these conditions hydrogen production rates of up to 1600 monolayers/hr could be obtained on platinized crystals. Metal-free crystals yielded hydrogen at rates up to 100 layers/hr. Prereduced and stoichiometric metal-free crystals yielded hydrogen at similar rates. Similar results were obtained in bulk concentrated NaOH solutions. Table 11.1 shows maximum hydrogen production rates observed under a variety of conditions. In solution, these hydrogen evolution rates could be maintained for tens of hours. Experiments carried out in NaOH films exhibited slowly decreasing hydrogen evolution rates, probably due to loss of NaOH from the crystal by gravity flow. No hydrogen production was observed in the dark, under illumination with photons of less-than-band-gap (3,2 eV) energy, in the absence of water vapor or electrolyte solution, or in the presence of water vapor when there was no film of basic deliquescent compound on the crystal surface. Figure 11.4 also shows that no hydrogen was produced when a piece of platinized platinum foil was mounted in place of the SrTiO$_3$ crystal, coated with NaOH, saturated with water vapor, and illuminated.

On platinized crystals photoproduction of oxygen was observed at rates relative to hydrogen evolution somewhat lower than those expected from the stoichiometric ratio in water, but within the calibration

Table 11.1. Catalytic hydrogen production by the photodissociation of water over SrTiO$_3$ crystal surfaces under a variety of experimental conditions

Crystal	Monolayers[a] H$_2$/hr
Hydrogen production from SrTiO$_3$ crystals covered by thick (>30 μm) NaOH films saturated with water vapor	
Prereduced, platinized	1580
Prereduced, metal-free	100
Stoichiometric, metal-free	30
Hydrogen production from SrTiO$_3$ crystals in 20 M NaOH	
Prereduced, platinized	4500
Stoichiometric, platinized	120
Prereduced, metal-free	30
Stoichiometric, metal-free	50

[a] 1 monolayer $\equiv$ 1 $\times$ 10^{15} molecules/cm^2 of illuminated surface.

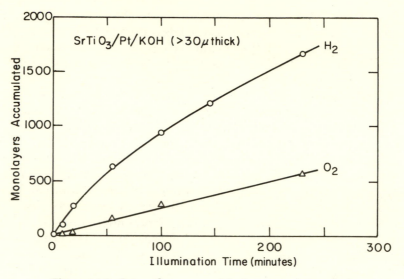

Figure 11.5. Rates of oxygen and hydrogen evolution during the photodissociation of water over KOH covered and platinized SrTiO₃ crystal surfaces.

uncertainty. This is shown in Figure 11.5. If water stoichiometry were followed, oxygen photoproduction rates from metal-free crystals would be too slow to be readily detectable with the apparatus that was used. We display the experimental data by plotting only the detected hydrogen concentration as a function of different experimental variables.

Table 11.2 summarizes results obtained with films of various compounds on platinized SrTiO₃. Hydrogen production was observed only with films of deliquescent basic compounds. A deliquescent compound,

Table 11.2. Hydrogen production from platinized SrTiO₃ crystal structures in various bulk liquid electrolytes

Electrolyte		H₂ produced	H₂ not produced
0.001–20 N	NaOH	×	
1–10 N	HClO₄		×
10 N	H₂SO₄		×
1 N	NaClO₄		×
1 N	NaF		×
10 N	LiCl		×
18 N	NH₃	×	

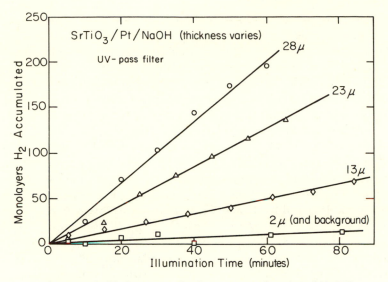

Figure 11.6. Hydrogen production rates as a function of NaOH film thickness over platinized SrTiO₃ crystal surfaces during the photodissociation of water.

when left in contact with moderately humid air, turns into an aqueous solution. Hygroscopic compounds, which also pick up water from the air but in insufficient quantity to actually dissolve, proved inactive for hydrogen production, even in the case of the base Na_2CO_3. Nonbasic deliquescent compounds, such as $CaCl_2$, also proved ineffective.

Figure 11.6 shows the dependence of hydrogen production on the average dry thickness of the NaOH film as calculated from the known number of micromoles of NaOH spread evenly over the surface. Hydrogen production decreases as the thickness of the NaOH film becomes smaller. The hydrogen production for the 2-μm film is indistinguishable from background hydrogen.

Studies of the Location of Hydrogen Production on Platinized and Metal-Free SrTiO₃ Crystals

When a platinized prereduced SrTiO₃ crystal was placed in a concentrated NaOH solution and illuminated, visible gas evolution occurred from both the illuminated SrTiO₃ surface and the nonilluminated Pt surface. When a metal-free crystal was illuminated, gas bubbles formed only at the illuminated surface. Sealing of the nonilluminated surfaces of metal-free crystals with epoxy caused no diminution of hydrogen production. Blank experiments show that the epoxy was not a source of hydrogen. Thus hydrogen production on metal-free crystals occurs on the illuminated $SrTiO_3$ surface. Since stoichiometric metal-free crystals evolve hydrogen at rates similar to those observed from prereduced crystals, good electrical conductivity is not a precondition for hydrogen production in the absence of platinum.

Sealing off the metal-coated surfaces of platinized crystals attenuated hydrogen production to the rates seen for metal-free crystals. Platinization of the backs of stoichiometric crystals caused, at most, a twofold increase in hydrogen evolution rates. Platinum foil in pressure contact with a metal-free reduced crystal did not increase hydrogen production rates above those observed in experiments where no platinum was present. A good electrical contact between the $SrTiO_3$ and platinum is thus necessary for rate enhancement by the platinum. Platinized crystals, therefore, do appear to behave as short-circuited photoelectrochemical cells,[13] wherein oxygen is produced at the illuminated $SrTiO_3$ surface and hydrogen is produced primarily at the platinum surface. On the other hand, hydrogen production from metal-free $SrTiO_3$, which takes place at the illuminated surface, appears to proceed via a mechanism distinct from the major reaction path operating in photoelectrochemical cells.

Hydroxide Concentration Dependence of Hydrogen Photoproduction

Films of many water-soluble ionic compounds were tested on platinized and metal-free $SrTiO_3$ crystals. Upon saturation with water vapor and illumination of the crystal, only films of compounds that were both basic and deliquescent gave detectable hydrogen yields. The requirement of deliquescent materials plus the similarity of results obtained with water-saturated films and in bulk aqueous electrolytes indicate that the films turn into thin layers of aqueous electrolyte upon saturation.

The requirement of the presence of basic compounds for the hydrogen photoproduction led to the experiments summarized in Figure 11.7. Here the rate of hydrogen production from a platinized, prereduced $SrTiO_3$ crystal is plotted against the (OH^-) concentration in a NaOH electrolyte. The hydroxide concentration dependence is rather weak below 5 N but becomes remarkably strong above this concentration.

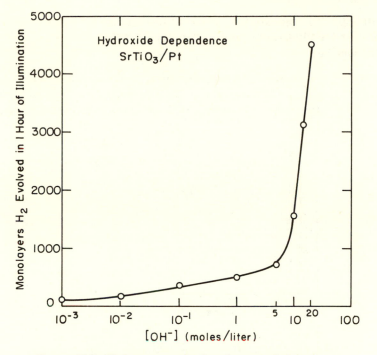

Figure 11.7. Hydrogen production from a platinized, pre-reduced $SrTiO_3(111)$ crystal during 1 hr of illumination as a function of NaOH electrolyte concentration (1 monolayer $\equiv 10^{15}$ molecules H_2/cm^2 of illuminated surface).

Hydrogen production from metal-free crystals shows similar (OH^-) concentration-dependent rate behavior. No hydrogen production from platinized crystals was seen in 1 to 10 N $HClO_4$, 10 N H_2SO_4, Mallinckrodt pH 4.01 buffer, 1 N NaF, or 10 N LiCl. The observed change in hydrogen production rate is therefore not simply a matter of ionic strength or anion size, and seems quite specific to hydroxide or at least to a base.

Thus we have evidence for a hydrogen-producing photocatalytic process in which all chemistry occurs on the same illuminated metal-free $SrTiO_3$ surface. We have also shown that the reaction rate is dependent on the hydroxide concentration of the electrolyte. We have then undertaken photoemission studies of the chemisorption and photochemistry of H_2, O_2, and H_2O on $SrTiO_3(111)$ surfaces in an attempt to better understand the active surface species and the elementary steps involved in the $SrTiO_3$ photochemistry.

Electron Spectroscopy Studies of the Active Surface Species for the Photodissociation of Water on $SrTiO_3$ Crystal Surfaces

Figure 11.8 shows the Auger spectrum, the UV photelectron spectrum, and the X-ray photoelectron spectrum from a reduced $SrTiO_3$ crystal. The large Sr/O ratio in the AES spectrum indicates the segrega-

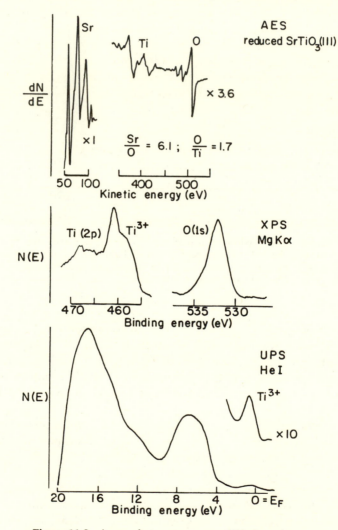

Figure 11.8. Auger electron spectrum, UV photoelectron spectrum, and X-ray photoelectron spectrum from a clean, reduced $SrTiO_3$ crystal surface.

tion of strontium to the surface when compared with the Auger spectrum from the stoichiometric $SrTiO_3(111)$ crystal.[21] The UPS spectrum exhibits a peak at 0.8 eV below the Fermi level associated with the presence of the reduced Ti^{3+} oxidation state ions at the surface.[21] The titanium $3d$ peak in the XPS spectrum also exhibits a shoulder that is also due to the presence of Ti^{3+}. These peaks almost completely disappear for the stoichiometric crystal surfaces.[22] In previous studies of TiO_2 the nature of the Ti^{3+} sites was thoroughly studied by electron spectroscopies.[20,23] By oxidation of titanium metal and from studies of TiO and Ti_2O_3, the low-energy 0.8-eV UPS peak and a corresponding 1.6-eV ELS peak were identified as due to the presence of Ti^{3+}. For $SrTiO_3$ the same conclusions may be drawn from this and from previous studies. From the Ti^{4+}/Ti^{3+} peak ratios in the XPS spectrum, it could be concluded that most of the surface titanium ions are in the Ti^{3+} oxidation state.[22]

Interaction of Ti^{3+} Surface Sites with Oxygen, Hydrogen, and Water in Dark and in Light

The Ti^{3+} signal is drastically reduced upon adsorption of oxygen in the dark. This is shown in Figure 11.9. There is a rapid drop in the UPS Ti^{3+} signal intensity with increasing exposure to oxygen; then a low steady-state value is reached. When the surface is illuminated with band-gap- or higher-energy light, the Ti^{3+} signal is partially restored.[22] Hydrogen and water have weaker interactions with the Ti^{3+} surface sites.

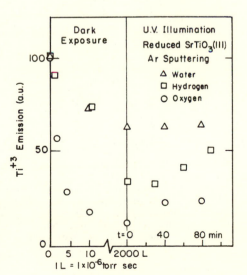

Figure 11.9. Intensity of the UPS Ti^{+3} emission as a function of H_2O, O_2, and H_2 exposures up to 2000 L in dark, and as a function of the irradiation on the H_2O, O_2, and H_2 surfaces after having pumped away these gases.

Although the adsorption of both reduce the Ti^{3+} signal intensity, the effect is much smaller than that for oxygen adsorption, as shown in Figure 11.9. There is also no detectable photodesorption of water and only minor photodesorption of hydrogen.

It appears that one of the important reaction steps involves the interaction of the reduced Ti^{3+} surface site with oxygen. Light of band-gap energy is effective in photodesorbing the oxygen and in regenerating the Ti^{3+} sites.[22,24]

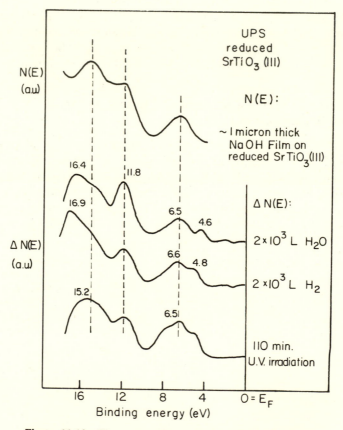

Figure 11.10. (Top curve) UPS, *N(E)* spectrum of a 1-μm-thick NaOH film on the $SrTiO_3$ reduced crystal. (Second curve) Difference spectrum due to 2000 L of water exposure. (Third curve) Same as second curve, but with H_2 rather than H_2O. (Fourth curve) Difference spectrum due to the band-gap irradiation of the clean surface.

Identification of Hydroxyl Groups on the SrTiO$_3$ Surface

A series of UPS studies were carried out to identify the presence of hydroxyl groups on the SrTiO$_3$ surfaces.[22] The UPS spectra of solid NaOH and that of the NaOH-monolayer-covered reduced SrTiO$_3$ surface are shown in Figure 11.10. The peaks at 6.5 eV, 11.8 eV, and 15.1 eV are due to the hydroxyl groups. These peaks can be used to identify the presence of OH$^-$ groups on the oxide surface if they are produced during the adsorption of water or hydrogen. Figure 11.10 shows the UPS difference spectra that were obtained upon the adsorption of water and hydrogen, respectively, on the reduced SrTiO$_3$ surfaces. The presence of hydroxyl groups is clearly discernible by comparison with the reference sodium hydroxide spectra. OH$^-$ surface species formed when H$_2$O or H$_2$ was adsorbed on the reduced SrTiO$_3$ surface. When the stoichiometric SrTiO$_3$ was exposed to H$_2$O or H$_2$ the surface hydroxyl groups either did not form or their rate of production was very slow indeed.[22]

Annealing the reduced SrTiO$_3$ crystals by heating to 500°C in vacuum also produced OH$^-$ groups at the surface, as indicated by changes in the UPS spectrum (Figure 11.10). It appears that hydrogen diffuses to the surface from the bulk in these circumstances and reacts with the surface oxygen sites to form OH$^-$. This phenomenon has also been reported by others.[25]

There are several different hydroxyls with differing binding energies reported on TiO$_2$.[25] Although more detailed studies of the chemical nature of surface hydroxyls on reduced SrTiO$_3$ are necessary, it is clear that these hydroxyls form readily in the presence of water vapor and hydrogen. Thus they are likely reaction intermediates or reaction centers during the photodissociation of water over these surfaces.[26]

Mechanistic Considerations

Sustained hydrogen production has been obtained from platinized and metal-free SrTiO$_3$ crystals illuminated in water-vapor-saturated NaOH films or in concentrated aqueous NaOH. Whereas hydrogen photoproduction from platinized crystals appears largely analogous to the operation of a photoelectrochemical cell with distinct electrodes, hydrogen production from metal-free crystals appears to proceed by a somewhat different mechanism. This distinction bears further consideration.

We observe hydrogen production and infer the production of

oxidized products at the illuminated surface of metal-free $SrTiO_3$ crystals. Active surface sites for oxidation and reduction may be interdispersed on an atomic scale on the same oxide surface. This could prove advantageous to gas-phase reactions, since it would obviate the necessity of ion transport over long distances. However, we have observed no hydrogen production upon illumination of $SrTiO_3$ in room-temperature water vapor in the absence of an ionic compound film. Our hydroxide dependence of the reaction rate may give a clue to a possible kinetic barrier to such gas–surface photocatalytic reactions.

Schrauzer and Guth[27] have reported photodissociation of water vapor on TiO_2 powders. However, the total hydrogen production they observed was on the order of a monolayer. Boonstra and Mutsaers[28] have observed hydrogenation of acetylene on hydroxylated TiO_2. Only a monolayer of hydrogenated products was observed, and dehydroxylated powders showed no hydrogenation activity. Van Damme and Hall[29] have proposed that the gas–solid photo- "catalytic" reactions reported to date have actually been stoichiometric reactions of surface hydroxyl groups. They believe that immeasurably slow rates of surface rehydroxylation by water vapor have prevented true photocatalytic activity. Munuera[26] has implicated surface hydroxyl groups in the oxygen adsorption and photodesorption behavior of TiO_2 powders. Using infrared spectroscopy, he has shown that certain hydroxyl groups are more photoactive than others. After thermal dehydroxylation of the surface, the most active hydroxyls cannot be restored by immersion in water vapor or liquid water at room temperature.[30] Our photoemission studies have shown that adsorption of either water vapor or hydrogen on clean reduced $SrTiO_3(111)$ gives rise to a hydroxylated surface. Illumination of the surface exposed to hydrogen increased the Ti^{3+} signal, whereas no photoeffects were seen on the surface exposed to water. This agrees with Munuera's observations that different treatments lead to surface hydroxyl groups of varying photoactivity.

We have shown that the rates of hydrogen production from metalfree, as well as from platinized, crystals increase with increased concentration of a NaOH electrolyte. This increase may be due to an increased rate of surface hydroxylation. A number of other explanations are possible, such as increased band bending or stabilization of new oxidized intermediates or products in highly alkaline media. Here we can only note that the equivalent photochemical behavior of stoichiometric and prereduced metal-free crystals with increasing OH^- concentration argues against an explanation of the hydroxide concentration dependence based entirely on changes in band bending.

We thus have evidence for a mechanism of hydrogen photoproduction which could prove to be more useful for the development of heterogeneous solid–gas photochemical reactions than is the major reaction pathway operating in electrochemical cells. Our results and those of others show that surface hydroxylation may be crucial to the kinetics of gas-phase heterogeneous photocatalysis.

Photoelectron spectroscopic techniques can identify and monitor Ti^{3+} and hydroxyl surface species on $SrTiO_3(111)$. We have shown that Ti^{3+} is involved in oxygen adsorption–photodesorption chemistry on reduced $SrTiO_3(111)$ surfaces.[22] Photodesorption of the product oxygen may be important to sustain water photolysis on semiconductor surfaces. Ti^{3+} may be directly involved in the reductive chemistry observed in aqueous electrolyte at the illuminated $SrTiO_3$ surface.

Local active sites do appear to be important to large-band-gap semiconductor photochemistry. Continued parallel reaction and surface studies may lead to a better understanding of the mechanisms of photocatalysis on these and on other materials.

Titanium-containing minerals make up over 2 percent of the earth's mantle. It is tempting to speculate on the importance of the photochemical activity of these compounds in the planet's evolution. Hydrogen may further react with CO_2 to produce hydrocarbons, while the accumulation of oxygen would make the primordial atmosphere more oxidizing. The simple photochemical reactions of these oxides could have played a significant evolutionary role before the evolution of the chlorophylls and other complex chromophore systems. This hypothesis would certainly receive more support if other minerals that are more concentrated in the earth's mantle and whose band gaps better overlap the solar spectrum also exhibit similar photochemical activities. Future studies will certainly help to verify the role of photocatalytic processes leading to the production of hydrogen and other small molecules in the evolution of our planet, and of others as well.

References

1. A.H. Boonstra and C.A.H.A. Mutsaers, J. Phys. Chem. **79**, 2025 (1975).
2. I.E. Den Besten and M. Qasim, J. Catal. **3**, 387 (1964).
3. R.K. Clayton, *Light and Living Matter*, vol. 1 and 2, McGraw-Hill, New York, 1971.
4. K. Sauer, in *Bioenergetics of Photosynthesis*, Academic Press, New York, 1975, p. 116.
5. M. Calvin and J.A. Bassham, *The Photosynthesis of Carbon Compounds*, W.A. Benjamin, New York, 1962.

6. E. Rabinowitch and Govindjee, *Photosynthesis,* Wiley, New York, 1969.
7. K. Sauer, Acc. Chem. Res. **11,** 257 (1978).
8. A. Fujishima and K. Honda, Nature **238,** 37 (1972).
9. J.G. Mavroides, J.A. Kafalas, and D.F. Koesar, Appl. Phys. Lett. **28,** 241 (1976).
10. M.S. Wrighton, A.B. Ellis, P.T. Wolczanski, D.L. Morse, H.B. Abrahamson, and D.S. Ginley, J. Am. Chem. Soc. **98,** 2774 (1976).
11. T. Watanabe, A. Fujishima, and K. Honda, Bull. Chem. Soc. Jap. **49,** 355 (1976).
12. A.J. Nozik, Ann. Rev. Phys. Chem. **29,** 189 (1978).
13. M.S. Wrighton, P.T. Wolczanski, and A.B. Ellis, J. Solid State Chem. **22,** 19 (1977).
14. L.A. Harris and R.H. Wilson, Am. Rev. Mater. Sci. **8,** 99, 1978).
15. J.M. Bolts, A.B. Bocarsly, M.C. Palazzotto, E.G. Walton, N.S. Lewis, and M.S. Wrighton, J. Am. Chem. Soc. **101,** 1378 (1979).
16. F.T. Wagner and G.A. Somorjai, Nature **285,** 559 (1980).
17. F.T. Wagner and G.A. Somorjai, J. Am. Chem. Soc. **102,** 5494 (1980).
18. S. Ferrer and G.A. Somorjai, Surf. Sci. **94,** 41 (1980).
19. J.C. Hemminger, R. Carr, and G.A. Somorjai in *Interfacial Photoprocesses: Energy Conversion and Synthesis,* Adv. Chem. Ser. No. 184, American Chemical Society, Washington, D.C., 1979, 233.
20. W.J. Lo, Y.W. Chung, and G.A. Somorjai, Surf. Sci. **71,** 199 (1978).
21. W.J. Lo and G.A. Somorjai, Phys. Rev. **B17,** 4942 (1978).
22. S. Ferrer and G.A. Somorjai, Surf. Sci. **97,** L304 (1980).
23. V.E. Henrich, G. Dresselhaus, and H.J. Zeiger, Phys. Ref. **B17,** 4908 (1978).
24. S.M. Cox and D. Lightman, Surf. Sci. **54,** 675 (1976); M. Formenti, H. Courbon, F. Juillet, A. Lissatchenko, J.R. Martin, P. Meriaudeau, and S.J. Teichner, J. Vac. Sci. Technol. **9,** 947 (1972).
25. M.L. Knotek and P.J. Feibelman, Phys. Rev. Lett. **40,** 964 (1978).
26. G. Munuera, V. Rives-Arnau, and A. Saucedo, J. Chem. Soc. Farad. Trans. I **75,** 736 (1979).
27. G.N. Schrauzer and T.D. Guth, J. Am. Chem. Soc. **99,** 7189 (1977).
28. A.H. Boonstra and C.A.H.A. Mutsaers, J. Phys. Chem. **79,** 2025 (1975).
29. H. Van Damme and W.K. Hall, J. Am. Chem. Soc. **101,** 4373 (1979).
30. G. Munuera et al., op. cit., and personal communication.

566

Index

Absolute rate theory, 357
Acetylene, 245, 277
Activation energies, 78, 332
  for adsorption, 357
  for bulk diffusion, 29, 181
  for desorption, 285
  for exchange, 311
  of hydrogen dissociation, 357
  for surface diffusion, 29, 168, 181, 187, 359
  of surface and bulk diffusion, 29, 194
  for surface reactions, 8
Active sites, 392, 484
Active surface carbon, 499, 527, 528
Adatoms, 26
Additives, 28, 398, 484, 494
Adhesion, 7, 31
Adsorbate–adsorbate interactions, 182, 249
Adsorbates
  diatomic, effect on the $T-V_s$ energy transfer, 338
  effect of, on bonding of adsorbed gases, 305
  organic, 242, 249
  primitive unit cell of, 133
Adsorbate–substrate forces, 194
Adsorbate–substrate interactions, 183, 249, 251
Adsorbate surface structures, 209
Adsorbed monolayers, 10, 187
  principles of ordering, 181
  structure of, 179
  unit cells of, 192
Adsorption, 8, 27, 29, 244, 275, 336, 346, 353, 364
  multilayer, 187
  physical, 178, 299

Adsorption bond length, 267, 272
Adsorption coefficient, 177
Adsorption geometry, 85, 274
  of atoms, 267, 275
  of molecules, 276
Adsorption isotherm, 179
Adsorption layer spacing, 267
Adsorption probability, 192, 305, 350, 354, 357
Adsorption rate, 177
Adsorption registry, 271
Adsorption site, 267, 268, 270, 272, 287
Alanine, 155, 251
Alloys, 100, 116
Alumina, 153
Aluminum, 264
Amino acid monolayers, 251
Amino acids, 155, 246
Ammonia, synthesis of, 382
Ammonia oxidation, 479
Ammoxidation, 404
Angle-resolved photoemission, 44
Angular distribution, 333, 335
Angularly resolved photoelectron spectroscopy, 67
Aniline, 249
Antiferromagnetic phase transformation, 168
Antiferromagnetic surface structure, 168
Argon, 156, 191
Aromatic hydrocarbons, 247
Aromatic molecules, formed from CO and H_2, 539
Atomic diffraction, 61
Atomic scattering, 44, 61
Atomic steps and kinks, 303
Atomic structure of clean solid surfaces, 126

Index

Atom recombination reactions, 368, 376
Auger analyzer, 73
Auger depth profile, 75
Auger electron emission, 73
Auger electrons, 65
Auger Electron Spectroscopy (AES), 44, 72, 74, 82, 112, 145

Back reflection Laué technique, 37
Backscattering intensity, 48
Band-gap energy, 553
Barium titanate, 154
Benzene, 155, 248, 249
Berzelius, 381
Bimetallic clusters, 123, 307, 308
Bimetallic systems, 123
Bimolecular reactions, 361
Binary alloy systems, 116
Binding energies, 192
Binding sites, 177, 284
  sequential filling of, 284, 290
Biphenyl, 249
Block-wave methods, 51
Bond energy, 106
Bonding, 286, 303, 305
  metal–carbon, 279
Bond-length contractions, 143, 147, 264, 266
Bond-shift mechanism, 481
Boudouard reaction, 527, 529
Bound-state energies, 344
Bound-state resonance, 343, 344
Bound states of the incident helium atom, 62
Bragg equation, 161
Bragg peaks, 48
Brain, human, 21
Brønsted acid, 395
Brønsted base, 395
n-Butane, 246
n-Butylbenzene, 249
t-Butylbenzene, 249

Cadmium, 271
Cadmium sulfide, 151
Carbene, 279
Carbonaceous deposits, 39, 494, 496
Carbonaceous overlayer, 399, 505
Carbon monoxide, 8, 183, 192, 242, 276, 280
  catalytic hydrogenation of, 516, 518
Carbonyl, 279
  metal, 309

Carbonylation, 404, 525, 526, 541
Carbonyl insertion, 539
Catalysis, 21, 28, 382, 400, 402
  heterogeneous, 7, 9, 10
  homogeneous, 404
Catalysts, 21
  acid or base character of, 395
  additives, 396
  deactivation, 483
  multifunctional, 398
  platinum, 497, 510
  poisons, 494
Catalyst surface areas, 389
Catalyst systems
  multicomponent, 398
  synthetic approach to, 405
Catalytic control of car emissions, 403
Catalytic hydrogenation, 404
  of carbon monoxide, 516, 518
Catalytic reactions, 92, 367, 381
  kinetic parameters for, 386
Centered unit cell, 132
Charge density, 26, 146
Charge transfer, 21, 71, 272
Chemical selectivity, 21
Chemical shifts, 76
Chemisorbed systems, 178
Chemisorption, 183, 285, 299
  selective, 389
Chlorophyl, 547
Chopping frequency, 335, 347
Clausius-Clapeyron equation, 181
Cluster model of the surface chemical bond, 308
Clusters, 90
  metal, 263, 308
  multinuclear, 284
Coadsorption, 176
  of atoms, 275
Cobalt oxide, 264
Coherence distance, 194
Coherence length, 185, 190
Coherence width, 45
Coincidence structures, 143
Coincidence unit cell, 192
Colloid systems, 7, 21, 35
Compensation effect, 401
Compound semiconductors, 151
Condensation, rate of, 187
Contraction, 150, 263
  of topmost atomic layer, 152
Coordination number, 270
Copper, 157, 191, 209, 243, 249, 251

Copper–nickel alloys, 110
Copper phthalocyanine, 251
Corrosion, 28
Corrosion reactions, 366
Cosine scattering distribution, 334
Coverage, 192, 194
Cracking, 386, 387
Crystal-field splitting, 26
Crystal growth, 29, 33
Cyanobenzene, 249
Cyclic intermediates, 481
Cyclic ring-opening mechanism, 481
Cyclohexadiene, 248
Cyclohexane, 155, 248
Cyclohexene, 248
Cyclopentane, 249

Dangling bonds, 148, 267, 268
Deactivation, 399
de Broglie wavelength, 43, 53, 343
Debye temperature, 54, 64, 169, 170
Debye-Waller factor, 53, 169, 171
Dehydrocyclization, 382, 386, 396, 479,
    481, 483, 504
Dehydrogenation, 383, 385, 386, 387, 398,
    480, 481, 482, 483, 489, 499, 502, 507
Dehydroxylation, 564
Density, 8
Depth composition profile, 124
Desorption, 336, 346, 360
    field, 59
    first- and second-order, 352
    flash, 77
    thermal, 95, 297, 348
Desorption rate, 78, 177, 187
Desorption spectrum, 77
Desorption steps, 364
Detailed balancing, 335
Differential mode of detection, 349
Differential reaction probability, 348
Diffracted electrons, 45
Diffraction from metal surfaces, 344
Diffraction patterns, 45
Dilute binary system, 102
Dipolar charge density, 71
Dipole length, 71
Dipole–dipole interactions, 182
Dipole strengths, 274
Dispersion, 22, 32, 400
Disproportionation of carbon monoxide,
    521, 527
Dissociation reactions, 368, 374

Dissociative adsorption of carbon monox-
    ide, 276

Electron acceptors, 306, 396, 542
Electron back-scattering, 112
Electron beam damage, 156, 247, 252
Electron-beam-induced decomposition, 77
Electron beam penetration, 170
Electron beam source, 45
Electron donors, 306, 396, 542
Electron emission, 40
Electron gas, 26
Electronic structure, 42
Electron mean free path, 40
Electron microscopy, 56
Electron penetration depth, 42
Electrons, multiple scattering of, 48, 49,
    170
Electron-stimulated desorption, 76
Electron tunneling spectroscopy, 87
Elementary reaction steps, 377
Elementary surface processes, 346, 347
Ellipsometry, 79
Energy accommodation, 335, 336
Energy conversion, 9, 21
Energy spread, 86
Energy transfer, 43, 336, 342
    in gas–surface interactions, 331
    processes at the surface, 333
Epitaxial deposition, 510
Epitaxial growth, 187, 251
Equilibrium surface composition, 103
Equilibrium surface concentration, 103
Etching, 38
Ethylene, 245, 277, 278, 397
Ethylidyne, 278, 279
Evaporation, 29
    field evaporation, 59
Exchange-correlation effects, 53
Exchange correlation hole, 34
Exchange reactions, 368
Exothermic small-molecule reactions, 368

Faceted surfaces, 165
Faceting, 246
Field desorption, 59
Field emission, 35
Field evaporation, 59
Field ionization, 35
Field-ion microscopy (FIM), 58
Fischer-Tropsch synthesis, 535, 540, 542
Flash desorption, 77
Flat-band potential, 549

Fluorocarbon, 30
Formic acid, decomposition of, 361, 375
Fractional order, 143

Gallium arsenide, 151, 157, 266, 268
Gallium phosphide, 151
Gas–surface interactions, 8, 43, 331, 336, 342, 344
Gas–surface scattering process, 333
Gem-dicarbonyl species, 281
Germanium, 151, 157, 167
Gibbs adsorption equation, 102
Glycine, 155, 251
Gold, 183, 194, 209, 264, 265
Gold-copper system, 120
Gold-nickel alloys, 110
Gold-silver alloys, 107, 120
Gold-tin system, 116, 120
Growth morphology, 156

Heat
    of adsorption, 8, 26, 27, 177, 178, 179, 187, 194, 249, 280, 284, 285, 287, 290
    of chemisorption, 34, 77, 79, 177, 308, 390
    of desorption, 29, 101
    of formation, 122
    of mixing, 106, 107, 119, 121
    of segregation, 102, 103
    of vaporization, 101
Helium diffraction, 345
Heterogeneous catalysis, 7, 9, 10
Heterogeneous nucleation, 33
High-Miller-index surfaces, 126, 157, 162, 195
High-pressure reactors, 408
High resolution electron loss spectroscopy (HREELS), 43, 86, 171, 245, 274, 280
Hollow sites, 268
Homogeneous catalysts, 404
Homogeneous nucleation, 33
Hybridization, 150
Hydrocarbon conversion, 10, 479
    on platinum, 479
Hydroformylation, 525, 539
Hydrogen, 242, 271
    chemisorption of, 147
    diffraction, 345
    hydrogen-deuterium exchange, 368, 372
    photoproduction of, 554, 558
Hydrogenation, 386, 480, 482, 483, 498, 499, 502, 506, 530
    of acetylene, 547

Hydrogenation (*cont.*)
    of carbon monoxide, 7, 10, 516, 518, 530
    catalytic, 404, 516, 518
    selective, 404, 499
    selectivity for, 505
Hydrogenolysis, 382, 386, 387, 390, 394, 398, 481, 482, 483, 489, 505
    selectivity for, 504
Hydroxylated surface, 564

Ice, 155
Incommensurate lattices, 183, 272
Incommensurate overlayer, 194
Infrared spectroscopy (IR), 8, 44, 84
Inner potential, 50
Integral mode measurement, 349
Internal modes of molecules, 43
Ion bombardment, 39, 190
Ionic character, 276
Ionization potential, 72
Ion neutralization spectroscopy (INS), 43, 70
Ion scattering, 44, 60, 506
Ion scattering spectroscopy (ISS), 43, 82, 112
Ion sputtering, 75
Iridium, 183, 191, 243, 264, 265, 270
Iron, 8, 264, 276
Iron oxide, 264
Iron oxide bond length, 276
Iron-phthalocyanines, 250
Island formation, 185, 186, 242
Isomerization, 383, 387, 396, 398, 481, 482
I–V curves, 48, 51, 248

Kelvin method, 71
Kinematical theory of diffraction, 50
Kinetics of ordering, 191, 192
Kinked surfaces, 243
Kinks, 157, 159, 162
Kink sites, 26
Krypton, 156, 191

Langmuir, 27
Langmuir adsorption isotherm, 179
Langmuir-Hinshelwood mechanism, 400
Laser simulation, 144
Lattice energy, 147
Lattices, incommensurate, 183, 272
Lattice strain, 119, 121
Layer doubling, 50, 52
Layered compounds, 152

Lead, 194
Lead-indium system, 107
LEED, 42, 43, 45, 127, 143, 155, 156, 157, 185, 186, 189, 195, 243, 245, 249, 251, 252, 265, 271, 272, 274, 283
    polarized, 168
    three-dimensional, 47
    two-dimensional, 46
LEED intensities
    analysis, 48
    studies, 144
LEED structure analysis, 151
LEED theory for disordered surfaces, 50
Lennard-Jones potential, 342
Lewis acid, 395
Liquid-gas interface, 31
Lithium fluoride, 153
Lithium hydroxide, 153
Long-bridge sites, 268
Long-range order, 119, 148
Low-energy atomic and ion scattering, 44
Low energy electron diffraction. See LEED
Low-, medium-, and high-energy ion scattering (LEIS, MEIS, HEIS), 60
Low-Miller-index surfaces, 126, 134, 147, 157, 195, 252
Low-pressure reactors, 409

Magnesium oxide, 154, 264
Matrix inversion, 50, 53
Matrix notation, 131
Maxwellian distribution, 335
Mean kinetic energy, 336
Mean square amplitude of vibration, 169
Medium energy electron diffraction (MEED), 54
Mesitylene, 249
Metal carbonyls, 309
Metal clusters, 263, 308
Metals on metals, 195
Metal surfaces, 268
Metastable surface, 144, 149
Methanation, 526
Methanol, 155
    formation of, 526, 530
2-methylnaphthalene, 249
Microfacet notation, 162
Miscibility gap, 110
Missing row model, 145, 265
Model calculations, 89
Molecular beam scattering, 8
    reactive, 347
Molecular beam–surface scattering, 8, 91, 332, 334, 335, 347, 348, 349, 357

Molybdenum, 209, 266
Molybdenum sulfide, 152
Monatomic height steps, 165
Monoenergetic beam scattering, 343
Monolayer adsorption, principles, 177
Monolayer coverage, 179
Monolayer of adsorbate, 27
Monolayers, 26, 27, 248
Monolayer surface structures, 249, 250
Muffin-tin potential, 52
Multicomponent catalyst systems, 398
Multifunctional catalysts, 398
Multilayer adsorption, 187
Multinuclear cluster, 284
Multiple-height steps, 165
Multiple scattering of electrons, 48, 49, 170

Naphthalene, 155, 187, 248, 249
Neel temperature, 168
Nernst equation, 549
Nickel, 8, 168, 192, 209, 243, 264, 266, 268, 271, 272, 274, 275, 279, 280
Nickel oxide, 168, 264
Niobium, 156, 157, 243
Niobium selenide, 152
Nitric oxide, 242
Nitrobenzene, 249
Nitrogen, 179, 242
Nonstoichiometry, 122, 153
Normal paraffin monolayers, 246
Notation of surface structures, 127
Nozzle-beam sources, 42
Nucleation, 32, 190
    heterogeneous, 33
    homogeneous, 33
Nucleation sites, 190

n-Octane, 155, 247
Order–disorder transformations, 47, 147
Ordered adsorbate structures, 194
Ordered carbonaceous layers, 191
Ordered domains, 190
Ordered monolayers, 209, 246, 249
Ordered multilayer deposits, 251
Ordered organic monolayers, 245
Ordering, 184, 187, 189, 194
    of the adsorbate layer, 176
    of small molecular adsorbates, 191
    of vacancies, 154
Organic adsorbates, 242, 249
Organic deposits, 497
Organic multilayers, 246
Overlayer, incommensurate, 194
Overlayer unit cell, 192

Index

Oxidation, 67, 404
  of ammonia, 404, 547
  of carbon monoxide, 479
  of sulfur dioxide, 404
Oxychlorination, 404
Oxygen, 242, 270, 271, 272, 274
Oxygen additive, 499
Oxygenated organic molecules, 526, 539

Palladium, 192, 243, 279, 280, 403
Paraffins, 246
Petroleum refining, 403
Phase separation, 110
Phase shifts, 53
Photocatalysis, 565
Photocatalytic dissociation of water, 551
Photocatalytic process, 554, 559
Photochemical reactions, 10, 551, 565
Photochemical surface reactions, 545
Photodecomposition, 551
Photodesorption, 565
Photodissociation of water, 552, 560, 564
Photoelectrochemical cells, 558
Photoelectrochemical dissociation of water, 547
Photoelectrons, 67
Photoelectron spectroscopy, 66, 74
Photoemission, 44
Photoexcitation, 546
Photogenerated electrons and holes, 546
Photogenerated electron vacancies, 549
Photographic process, 546
Photooxidation, 551
Photoreduction, 551
Photosynthesis, 21, 546, 547
Phthalocyanine monolayers, 249
Phthalocyanines, 246
  metal-free, 155
Physical adsorption, 178, 299
Physisorption, 182, 183, 243, 264
Platinum, 7, 155, 162, 164, 167, 183, 187, 191, 192, 209, 243, 246, 248, 249, 251, 264, 277, 280, 403, 479
Platinum catalysts, 497, 510
Platinum high-Miller-index planes, 157
Point defects, 26
Poisoning, 399
Polar faces, of compound semiconductors, 151
Polarizability, 72
Polarized LEED, 168
Polymerization, 404, 526, 530
Pore structure of the support, 406
Potassium, 209
Potential-energy barriers, 29, 181

Precursor model, 356
Preexponential factor, 78, 332, 357, 360, 361, 364, 365, 386
Primitive unit cell, 133
Promoters, 494
  textural, 397
Promotion energy, 147
Propylene, 249
Pseudomorphism, 156, 247
Pyridine, 249

Quasi-dynamical theory, 50

Radiation damage, 76, 113
Rainbow scattering, 61, 345
Random-walk type of analysis, 359
Rate constants, 332
Reaction probability, 347, 366, 367, 375, 384, 385, 386
Reaction rate, 93
Reactions
  corrosion, 366
  low- and high-pressure, 93
  exchange, 368
  photochemical, 10, 551, 565
  at solid-liquid interfaces, 93
  structure-sensitive and -insensitive, 400, 483, 493
  surface, 29, 346, 361, 365, 367, 545
  unimolecular, 361, 366
Reaction selectivities, 484
Reactive carbon fragments, 542
Reactive molecular beam scattering, 347
Reactive scattering, 92
Reactors, 408, 409
Real-space vectors, 128
Reciprocal space vectors, 127, 128
Reconstruction, 127, 143, 148, 151, 154, 265, 266, 507
  on metal surfaces, 145
Reconstruction geometries, 264
Rectilinear plot, 333
Redispersion, 397, 483
Reflection high energy electron diffraction (RHEED), 54
Re-forming reactions, 396, 403, 483
Refractive index, 79
Regeneration, 399
Regular solution parameter, 106, 109, 114, 121
Regular solution approximation, 109
Relaxation, 127, 135, 144
  of interlayer spacing at step edges, 167
Renormalized forward scattering (RFS), 50, 52

Residence times, 29, 332, 367
Restructured (100) surfaces, 143
Restructuring, 165, 509
   of the topmost atomic layer, 264
Rhenium, 243
Rhodium, 191, 243, 280, 403
Rideal-Eley mechanism, 401
Ring opening, 386, 387
   of cyclopropane, 493
Rotational symmetry, 195, 209, 242
Rotation–vibrational energy transfer, 331,
   336, 338, 341
Rubidium, 209
Ruthenium, 8, 243
Rutherford scattering, 60

Sample preparation, 37
Scanning electron microscope, 57
Scanning transmission electron microscope
   (STEM), 57
Schulz-Flory distribution, 531, 534
Secondary electrons, 42
Secondary-ion mass spectroscopy (SIMS),
   43, 81
Segregation isostere, 102
Selective chemisorption, 389
Selective hydrogenation, 404, 499
Selectivity, 382, 484, 492
   for hydrogenolysis, 504
   of hydrogenation, 505
Selenium, 270, 272, 274
Short-bridge sites, 268
Silicon, 151, 167, 267, 268
Silver, 191, 243, 246, 248, 270, 271, 272
Single-crystal platinum surfaces, 484
Single-crystal sample, 38
Single-scattering theory, 49
Single-step-height configuration, 243
Sodium, 275
Sodium–sulfur bond length, 276
Soft-cube model, 338
Solid–gas interface, 22, 25, 31
Solid–liquid interface, 22, 25, 31
Solid–solid interface, 22, 25
Solid–vacuum interface, 25
Space-charge, 34
Space velocity, 399
Spark erosion cutting technique, 37
Specular and cosine scattering, 334
Specular scattering, 63, 334
Spin-polarized electrons, 168
Spin-polarized LEED theory, 50
Step height, 161
Stepped crystal surfaces, 191

Stepped high-Miller-index surfaces, 157
Stepped surfaces, 158, 159, 209, 243, 245
Stepped surface stability, 167, 209, 243
Steps, 26, 157, 162
   multiple-height, 165
Step sites, different chemistry of, 191
Step-terrace structure, 159
Stereographic projections, 165
Sticking coefficient, 354
Sticking probability, 335, 342, 355
Strain energy, 111
Strongly bound surface oxide, 307
Strontium titanate, 154, 155, 550
Structure of adsorbed monolayers, 176
Structure analysis, 248, 264
Structure-insensitive reactions, 483, 493
Structure
   of adsorbed monolayers, 176
   of the binding sites, 284
   of clean surfaces, 126
   of ionic crystal surfaces, 152
   of low-Miller-index semiconductor sur-
     faces, 148
   of molecular crystal surfaces, 155
   of oxide surfaces, 153
Structure-sensitive deactivation, 486
Structure-sensitive reactions, 400, 483
Structure sensitivity, 291, 400, 483, 488,
   493, 506, 507
Sulfur, 270, 272, 274, 275
Superlattice, 143, 185, 192, 263, 264, 267
Superlattice periodicity, 148
Supported metal catalysts, 387
Surface alloy systems, 123
Surface analysis, 40
Surface annealing, 39
Surface areas, 30, 387
Surface atom fraction ratio, 103
Surface-atom vibrations, 29, 43, 169, 171
Surface bond lengths, 252, 273
Surface buckling model, 149
Surface chemical bond, 101, 284, 285
Surface composition, 100, 102, 112, 116,
   121
Surface compounds, 100, 284, 306, 307
Surface concentration, 8, 102, 177
Surface crystallography, 51, 135, 148, 245,
   246, 252, 277
   of ordered monolayers, 252, 274
Surface Debye temperature, 64, 170
Surface Debye-Waller factor, 171
Surface defects, 191
Surface diffusion, 59, 167, 187, 336, 346,
   359, 360, 364, 375
Surface dipoles, 26, 34

Index

Surface disorder, 65
Surface energy, 30, 107, 118
Surface enrichment, 105, 107, 116
Surface-free energy, 30, 101, 119, 121,
    135, 209, 265
Surface hydroxyls, 553
Surface intermediates, 391, 395
Surface irregularities, 109, 126, 167, 191,
    303, 305, 484
    effect on ordering, 189
Surface magnetization, 168
Surface mobility, 189
Surface oxidation state, 542
Surface penning ionization (SPI), 44, 70
Surface-phase diagram, 120
Surface phases, 120, 121, 284, 308
Surface processes, 91
    See also Elementary surface processes
Surface reactions, 29, 346, 361, 365
    catalyzed, 367
    photochemical, 545
Surface reconstruction, 147, 148, 268, 507
Surface relaxation, 151
Surface residence time, 178, 332, 353, 391
Surfaces
    faceted, 165
    high-Miller-index, structure of, 162
    hydroxylated, 564
    low-Miller-index, 126, 134, 147, 157,
        195, 252
    metal, 268
    physical-chemical properties of, 25
    stepped, 158, 159, 209, 243, 245
Surface segregation, 39, 101, 102, 104,
    106, 109, 114, 121, 122, 148, 306
Surface-sensitive extended X-ray absorp-
    tion fine structure (SEXAFS), 65
Surface sensitivity, 48
Surface sites, threefold or fourfold, 297
Surface space charge, 8, 35
Surface structural analysis, 48
Surface structures, 45, 126, 145, 151, 155,
    183, 194, 195, 245
    of alloys, 147
    impurity-induced changes of, 165
    monolayer, 249, 250
    notation of, 127
    of ordered metal monolayers, 195
    of small molecules, 209
    two-dimensional, 187, 246
Surface tensions, 102, 105, 107, 119, 121
    of liquids, 7
Surface-to-volume ratio, 21
Surface transformation, 148

Surface unit cell, 143, 194, 246
Surface vacancies, 150
Surface vibrations, 54, 170
Synchrotron, 9
Synchrotron radiation, 65, 66
Synthesis of ammonia, 382
Synthesis of water, 507

Tellurium, 270, 272, 274
Temperature dependence of the helium
    beam intensities, 64
Temperature-dependent changes in the
    surface chemical bonds, 299, 301
Terraces, 162
Thermal accommodation, 334, 335, 336
Thermal annealing, 190
Thermal desorption, 95, 297, 348
Thermal desorption spectra, 79, 288
Thermal desorption spectroscopy (TDS),
    44, 77
Thorium oxide, 8
Three-dimensional LEED, 47
Time of flight, 332
Titanium, 264, 271, 276
Titanium nitride, 264
Titanium oxide, 154
Toluene, 249
Total partition function, 364
Trajectory calculations, 338
Translational–vibrational energy transfer,
    331, 334, 336, 355
Transition-state theory, 364
Transmission electron microscope (TEM),
    56
Tryptophan, 155
D-Tryptophan, 251
DL-Tryptophan, 251
L-Tryptophan, 251
Tungsten, 191, 265, 266, 270
Turnover frequency, 384
Turnover number, 384
Two-dimensional LEED, 46
Two-dimensional ordering, 183
Two-dimensional phase, 29, 346
Two-dimensional surface phases, 307
Two-dimensional surface structures, 187,
    246

Ultrahigh vacuum (UHV) technology, 9
Ultrahigh vacuum chamber, 38, 39
Ultraviolet photoelectron spectroscopy
    (UPS), 245
Unimolecular reactions, 361, 366
Unit cells of adsorbed monolayers, 192

Universal curve for electron penetration in solids, 40
Uranium oxide, 157

Vacancies, 26
Vacuum technology, 40
Vanadium oxide, 153
Vapor deposition, 155
Velocity distribution, 334, 335
  at given angles of scattering, 333
  .of incident and scattered molecules, 333
Vibrational spectra, 295
Vibrational structure, 42
Vibrational–vibrational energy transfer, 331, 336, 342
Vinylidene, 278

Water, formation of, 381
  photodissociation of, 552, 560, 564
  photoelectrochemical dissociation of, 547
  synthesis of, 507

Water-gas shift reaction, 404, 519
Wetting, 31
Wood's notation for surface structures, 131
Work function, 71
  changes, 34, 248, 249, 273, 276
  of the energy analyzer, 67
  measurements, 44, 71
Wurtzite, 268

Xenon, 156, 243, 264, 272
X-ray crystallography, 46
X-ray diffraction, 8
m-Xylene, 249

Zeolites, 387, 396, 403
Zinc, 209
Zincblende, 268
Zinc oxide, 151, 167
Zinc-selenide, 151, 266

Library of Congress Cataloging in Publication Data

Somorjai, Gabor A
    Chemistry in two dimensions.

    (The George Fisher Baker non-resident lectureship
in chemistry at Cornell University)
    Includes bibliographical references and index.
    1. Surface chemistry. I. Title. II. Series:
George Fisher Baker non-resident lectureship in
chemistry at Cornell University.
QD506.S588   541.3′453   80-21443
ISBN 0-8014-1179-3